AUTOCLAVED AERATED CONCRETE

INNOVATION AND DEVELOPMENT

Proceedings of the 4th International Conference on Autoclaved Aerated Concrete, organised by the Concrete and Masonry Research Group, and held at Kingston University- London, on 8–9 September, 2005

Autoclaved Aerated Concrete

Innovation and Development

Edited by

Mukesh C. Limbachiya
Sustainable Technology Research Centre, Kingston University

John J. Roberts
Sustainable Technology Research Centre, Kingston University

Taylor & Francis
Taylor & Francis Group

LONDON/LEIDEN/NEW YORK/PHILADELPHIA/SINGAPORE

Published by: A.A. Balkema Publishers, Leiden, The Netherlands, a member of Taylor & Francis Group plc
www.balkema.nl, www.tandf.co.uk and www.crcpress.com

ISBN Set: 0 415 38356 0
ISBN Book: 0 415 38357 9
ISBN CD-rom: 0 415 38367 6

Printed in Great Britain

Autoclaved Aerated Concrete – Limbachiya and Roberts (eds)
© 2005 Taylor & Francis Group, London, ISBN 0 415 38356 0

Table of Contents

THEME 3: CODES, STANDARDS AND REGULATIONS

THEME 4: DESIGN AND PERFORMANCE

THEME 5: DEVELOPMENTS IN CONSTRUCTION TECHNIQUES

THEME 6: ENVIRONMENT AND BUILDING PERFORMANCE ASPECTS

Autoclaved Aerated Concrete – Limbachiya and Roberts (eds)
© 2005 Taylor & Francis Group, London, ISBN 0 415 38356 0

Preface

Autoclaved Aerated Concrete (AAC) has gained world-wide recognition as a high quality, innovative construction material which has been extensively used in a wide range of residential, commercial and industrial buildings.

Recently there have been a number of innovative developments in AAC manufacturing processes, product properties and construction methods, resulting in enhanced performance to meet increasingly stringent building design requirements whilst ensuring environmental compliance. Indeed, AAC products have shown potential for being a universal material that can underwrite commercial, social and sustainable development.

The Concrete and Masonry Research Group (CMRG), part of the Sustainability Technology Research Centre within the Faculty of Engineering at Kingston University, organised this 4th International Conference on AAC to share and explore existing expertise and review, and discuss the innovations and developments in AAC production and usage, offering environmentally friendly construction solutions. It considered raw materials, manufacturing techniques and product characteristics that satisfy ever more demanding energy, design, structural and environmental requirements for sustainable development. This Conference dealt with key technical and practical issues under concurrently proceeded six themes; *(i) Raw Materials and Production, (ii) Product Properties and Performance, (iii) Codes, Standards and Regulations, (iv) Design and Performance, (v) Developments in Construction Techniques, and (vi) Environment and Building Performance Aspects.* Over 60 papers were presented by authors from 15 different countries during this Conference and these are compiled in the CD and a hard bound single volume proceedings.

The event was organised in collaboration with the UK Autoclaved Aerated Concrete Products Association and the European Autoclaved Aerated Concrete Association. The Conference was sponsored by UK Quality Ash Association, NHBC- National House-Building Council and Masa-Henke Maschinenfabrik GmbH, Germany. All sponsors are gratefully acknowledged for their invaluable support. The work of Conference was an immense undertaking and help from all those involved are gratefully acknowledged, in particular, members of the Scientific and Technical Committee for their assistance from start to finish; the Authors and the Chair of Technical Sessions for their invaluable contribution. Special thanks to Lhoist Group- UK, Stork Bouwtechniek- The Netherlands, Wehrhahn GmbH- Germany, Singleton Birch- UK, Carlfors Bruk AB and on behalf of CBC Engineering AB- Sweden, Philco Technologies GmbH- Germany and Terruzzi Fercalx Spa- Italy for their support.

The Proceedings have been prepared directly from the camera-ready electronic manuscripts submitted by the authors and editing has been restricted to minor changes where it was considered absolutely necessary.

Kingston upon Thames
September 2005

Mukesh C Limbachiya
John J Roberts

Autoclaved Aerated Concrete – Limbachiya and Roberts (eds)
© 2005 Taylor & Francis Group, London, ISBN 0 415 38356 0

IN COLLABORATION WITH

SPONSORING ORGANISATIONS

UK Quality Ash Association

Masa-Henke Maschinenfabrik GmbH, Germany

National House-Building Council, UK

Autoclaved Aerated Concrete – Limbachiya and Roberts (eds)
© 2005 Taylor & Francis Group, London, ISBN 0 415 38356 0

Exhibiting Organisations

Lhoist Group, UK

Stork Bouwtechniek, The Netherlands

Wehrhahn GmbH, Germany

Singleton Birch, UK

Thin Joint Technology Ltd, UK

Carlfors Bruk AB & on behalf of CBC Engineering, AB Sweden

Philco Technologies GmbH, Germany

Terruzzi Fercalx Spa, Italy

Autoclaved Aerated Concrete – Limbachiya and Roberts (eds)
© 2005 Taylor & Francis Group, London, ISBN 0 415 38356 0

Organising Committee

Concrete and Masonry Research Group
Kingston University – London

Dr M C Limbachiya *(Chairman)*

Professor J J Roberts *(Chairman)*

Professor G Sommerville OBE

Professor N J Bright MBE

Dr A N Fried

Dr K Etebar

Mr A Koulouris

Mr Y Ouchagour

Miss S Fotiadou

Dr A Ahmed

Mr O Reutter

Mr O Kanyeto

Faculty of Technology Research Office

KINGSTON UNIVERSITY

Autoclaved Aerated Concrete – Limbachiya and Roberts (eds)
© 2005 Taylor & Francis Group, London, ISBN 0 415 38356 0

Scientific and Technical Committee

Theme One: Raw materials and production

Autoclaved Aerated Concrete – Limbachiya and Roberts (eds)
© 2005 Taylor & Francis Group, London, ISBN 0 415 38356 0

Development of factory production control guidance for European AAC masonry units

C.A. Fudge
H+H Celcon Limited, Ightham, Sevenoaks, Kent, UK

M.D. Brehme
Gueteschutz Ziegel, Munich, Germany

ABSTRACT: This paper describes the documentation for the factory production control and the factory inspection schemes for aac masonry units meeting the new European Standard (EN) for AAC masonry units, EN 771-4. A detailed explanation of the quality control clauses of the EN, as well as the guidance for use is given. The role of outside third party bodies is described and detailed information is given on the different tasks to be carried out by the manufacturer and the third party. Two systems of attestation of conformity exist for aac masonry units and the differences between them are explained. Information in the paper contains the guidance on the general requirements to carry out factory production control.

Inspection frequencies are suggested for various aspects of factory production control.

1 INTRODUCTION

The writing of European Standards (ENs) for masonry products started in earnest in the late 1980s under the auspices of the European Standards Body (CEN), Technical Committee TC 125. The mandate for the project was to write product standards for all types of masonry units, mortars and ancillary components. After years of development, public commenting, translation, comment assessment etc, the majority of ENs was given a Formal Vote of approval as "harmonised" standards (hENs) and came into effect in April 2005.

However, even within the hENs there was a lack of information on factory production control and how third parties were to go about their certification process. As a result, CEN TC 125 decided to develop separately the necessary documentation, which would suffice until such time as the requirements could be included in the "second generation" ENs.

2 SETTING UP THE ATTESTATION LEVELS AND REQUIREMENTS

The background to the whole system of requirements lies with the Construction Products Directive (CPD) published by the European Commission. The aim of the CPD was to break down technical barriers to trade in construction products between Member States in the European Economic Area (EEA). To achieve this, four main elements were provided by the CPD.

- A system of harmonised technical specifications e.g. hENs.
- An agreed system of attestation of conformity for each product family.
- A framework of "Notified Bodies".
- The CE marking of products.

The CPD set out to harmonise test methods, the method of declaration of product performance values, and the method of conformity assessment. Choices of required values for the chosen intended use are left to the regulators in each Member State.

European product standards for masonry units address characteristics that are not regulated in some Member States, but have been included for commercial reasons. As a result, all harmonised product standards under the CPD include an Informative Annex (Annex ZA), the first part of which (ZA.1) lists the regulated requirements and the clauses in the EN in which they are addressed. Some of these clauses may refer to separate supporting standards such as test methods.

As such, Annex ZA.1 in the hEN becomes a checklist for conformity marking from which the manufacturer can see all the possible requirements of his product and how they can be met.

One of the key aspects of conformity marking is that it includes technical information in the form of declared values. Where there are minimum or maximum values set in the hENs, these are not repeated in the marking. Effectively, the marking is a harmonised data sheet and together with the standard, it gives all the information needed by specifiers and regulators to assess whether the product is suitable for an intended use. In some cases, the manufacturer may declare "no performance determined". This is where regulations do not exist in the chosen market sector, either country or use, for a particular characteristic.

3 ATTESTATION OF CONFORMITY FOR AAC MASONRY PRODUCTS

The attestation system is, by definition, the term applied to the degree of involvement of third parties in assessing the conformity of the product to the relevant hEN. For aac masonry products, in common with all other family groups, the attestation system was decided collectively by the Member States and the Commission on the basis of the implications for health and safety of the product, and on the particular nature and production process for the product itself. In total there are six systems of attestation and each has a different number of tasks for the manufacturer and the Notified Body. These are shown in Table 1 below.

"Notified Bodies" are the product conformity certification bodies, FPC certification bodies, inspection bodies (in some countries) and test laboratories who are competent to carry out the attestation tasks shown in Table 1. These bodies were first approved by their respective Member States and have come together as a "Group of Notified Bodies" (GNB) to discuss practical implication matters so that there is a consistent approach to the tasks. In latter times, the GNB have produced various documents and position papers to assist in the attestation process. Various Sector Groups (SGs) of the Notified Bodies have been formed, the relevant SG for aac masonry units being SG 10.

Whilst there are six systems of attestation under the CPD, for the range of aac products for masonry, only two were deemed necessary. The more specific attestation of conformity systems

Table 1. CPD attestation tasks.

Conformity attestation	1+	1	2+	2	3	4
Tasks for the manufacturer						
Factory Production Control (FPC)	X	X	X	X	X	X
Testing of samples to a prescribed plan	X	X	X			
Initial type testing			X	X		X
Tasks for the Notified Body						
Initial type testing	X	X			X	
Certification of FPC	X	X	X	X		
Surveillance of FPC	X	X	X			
Audit testing of samples	X					

X = Task required.

for aac masonry units are given in the aac masonry unit product standard, EN 771-4, as 2+ for "Category I" masonry units and 4 for "Category II". The manufacturer is required to declare which category of product is been placed on the market. "Category I" units are defined as units with a declared compressive strength with a probability of failure to reach it not exceeding 5%. "Category II" masonry units are units not intended to comply with the level of confidence of Category I units. In some National design codes and in European Codes, for example EN 1996-1-1, Category I units attach a more favourable safety factor, allowing greater economy in design.

4 DEVELOPMENT OF THE OPERATING PROCEDURES AND FPC

The basis of the inspection schemes and general requirements for FPC was prepared by SG 10. This had been in anticipation of the needs of the certification bodies, as well as to provide a starting point for manufacturers. It was recognised that the task of agreeing inspection frequencies for end product testing and equipment had to be a delicate balance between manufacturers looking to control costs of additional testing, both in terms of financial and human resources against testing frequencies being relevant to the importance of the product and protect the needs of the end-users and designers. In addition, it had to be recognised that aac masonry units are manufactured from a wide range of plant sizes. As a result, the guidance produced could not be too prescriptive, but act as a starting point for negotiations between the manufacturer and Notified Body.

General guidance was produced for all masonry unit products, which included information such as the following:

- Responsibility for the performance of the FPC
- What to do in the case of nonconformities
- How to carry out an initial inspection of the factory and the FPC
- Information concerning the issue of a "certificate of FPC".
- The period of validity of a certificate
- Requirements for the continuous surveillance of FPC
- Marking and traceability of products

To aid the manufacturer, a model for an application form to the Notified Body was given as an annex. To aid the Notified Body, a separate annex was included giving a certificate of FPC.

Whilst the guidance was drafted for Category I products, a note was added to the effect that it could be used as a basis for establishing suitable FPC procedures for manufacturers of Category II products. The main areas covered by the CEN TC 125 guidance were a general expansion of the basic FPC clause in the product standards, as well as Schedules containing specific checklists to each type of masonry unit, which are to be considered informative, as guidance to Notified Bodies.

5 SPECIFIC INFORMATION IN THE GUIDANCE DOCUMENTS

Specific requirements for the manufacturer is given in the guidance document and contained some of the following information. In general, the manufacturer has to establish, document and maintain a factory production control system to enable continuing conformity with the standard and the declared values of the products placed on the market. The responsibility, authority and interrelation of all personnel who manage, perform and verify work affecting the quality of masonry unit products is defined in the documentation of the FPC. The "documentation of the FPC" raises itself in many places and no guidance is actually given for the document itself.

The factory production control system has to describe the procedure of production, the regular inspections by the manufacturer and his testing. Controls and tests generally include the characteristics of raw materials and finished products, the procedure of production, the production equipment or the production machines, the test equipment or the testing instruments and the marking of the product.

The results of the tests requiring action and the test results have to be recorded. Actions to be taken when the control test values or criteria do not meet those specified should be documented by the manufacturer.

Information is given on testing and measuring equipment. All relevant weighing, measuring and testing equipment, that has an influence on the declared values, has to be calibrated and regularly inspected according to the documented procedures, frequencies and as stated in the factory production control manual. In addition, all production equipment that has an influence on the declared values has to be controlled and regularly inspected according to the documented procedures, frequencies and criteria.

As for raw materials, where it is appropriate, the manufacturer should define the acceptance criteria of raw materials and the procedures operated to ensure that these are met.

The relevant features of the production processes, as appropriate, have to be defined giving the frequencies of the manufacturer's inspection checks, together with the required criteria. Actions to be taken when the criteria are not achieved are then specified by the manufacturer. It was recognised that the requirements in the inspection scheme may be varied for factories, where a higher degree of on-line routine testing takes place to fine tune the product for maximum manufacturing efficiency, rather than for testing of the finished product.

The factory production control system has to incorporate a sampling plan and the frequency of testing of the finished product, with the results of sampling and testing being recorded. It has to be noted that the sample has to be representative of the production. The tests are normally carried out in accordance with the methods described in EN 771-4, if given, and with the agreed FPC documentation.

One important aspect is that alternative methods of test to the reference methods specified in the schedules may be adopted, except for initial type testing and in case of dispute, provided that these alternative methods satisfy the following:

(a) A relationship can be shown to exist between the results from the reference test and those from the alternative test, and
(b) The information on which the relationship is based is available.

The results of testing have to meet the specified conformity criteria and be made available.

For production evaluation the manufacturer could define the conformity criteria in the FPC documentation. One method of satisfying this conformity criterion is to use the approach given in ISO 12 491.

The marking and stock control has to be documented and batches of products have to be identifiable and traceable.

The procedure for dealing with nonconforming products also has to be documented. Products that do not conform to the requirements have to be segregated and marked accordingly. However, these may be reclassified or downgraded by the manufacturer and given different declared values. As expected, the manufacturer has to take action to avoid a reoccurrence of the nonconformity.

6 SCHEDULES FOR INSPECTION SCHEMES

6.1 *General information on inspection schemes*

One of the key aspects of the guidance is in the final section, which contains the schedules of inspection schemes. The checklists within the schedules were to be used as guidance to Notified Bodies and manufacturers. The frequency of inspections were noted to depend upon the variation in raw materials, the process and the regularity of inspection in the process compared with the one of testing of finished product. The information given in the tables within was understood to be as a starting position for discussions between the manufacturer and the Notified Body. The frequency of surveillance by the Notified Body was left to take into account the product type and composition, the manufacturing process and its complexity and the sensitivity of product features to variations in manufacturing parameters. It was agreed that the inspections should normally be carried out once per year and that the inspections should be announced in advance.

Whilst there were common areas for all products, for example, inspection of testing and measuring equipment, specific products in each EN 771 part needed specific details.

Much of the inspection frequencies relied heavily on the manufacturer's documentation of FPC, because of the wide range in manufacturing techniques. However, wherever possible frequencies have been suggested in terms of 'x' number of units per volume produced or not less than a certain period of time.

6.2 *Inspection of testing and measuring equipment*

Guidance is given for strength testing equipment, balances and dimension, temperature & moisture measuring equipment, if the equipment is relevant to testing methods. The purpose is to check the correct functioning and accuracy of the equipment, which is to be calibrated with apparatus in accordance with National calibration requirements. The suggestion is that this should be carried out at least every 2 years and in the case of strength testing equipment on installation and after major repair. It is noted that external laboratories can be used to carry out the testing. Calibration to National requirements should be to an EN or ISO standard where they exist.

6.3 *Inspection of production and controlling equipment*

For production machinery to check the correct operation and functioning, the guidance simply suggests that the intervals should be given in the documentation of the FPC. For all relevant process control devices, calibration with apparatus, which are calibrated in accordance with National calibration requirements, is suggested at time intervals given in the documentation of FPC. This may also include after major repair work, but not routine maintenance.

6.4 *Inspection of raw materials and production process*

As expected, variations on the guidance can be made depending upon the type of process and production equipment. Raw materials may be from own extraction or imported or a combination of both. The aim is to identify suitable raw materials and/or to ascertain that the consignment is from the correct supplier and is the ordered quality. This may involve a simple inspection of the delivery note and material or physical, mineralogical or chemical examinations of materials. In all cases, the frequency is as laid down in the manufacture's documentation. The following are also suggested to be given in the documentation of the FPC:

- Storage of materials, to avoid contamination by other materials
- Mix Ratio, to check conformity with the determined weight or volume of the composition of material
- Grinding, if relevant, to check conformity with the grain size
- Casting, to check conformity with the production conditions
- Cutting, to check conformity with the green strength and temperature
- Autoclaving, to check conformity with the autoclave regime determined, usually by measuring pressure and autoclaving time

6.5 *Inspection of finished products*

The following Table 2 is copied directly from the guidance document and gives an indication of the necessary testing and frequency.

The manufacturer does not necessarily have to declare a value against every property and some may be on the basis of, for example, tabulated values. Where the declared value is from a tabulated value no testing is required. In these cases certification can be based on evidence that the tables are being used correctly.

Table 2. Inspection scheme for aac masonry units.

Subject	Purpose of inspection	Method/ procedure	Frequency of inspection by the manufacturer
Dimensions	Conformity with the declared dimensions and the permissible dimensional deviations determined by EN 771-4	EN 772-16 or measurements between cutting wires	-At each change of product[2] dimension and -At least weekly of 3 units (5 different dimensions)
Flatness of bed faces[1]	Conformity with the declared value and the deviation determined by EN 771-4	EN 772-20	-At each change of product[2] and -Weekly of 3 units or -As given in the documentation of FPC
Plane parallelism of bed faces[1]	Conformity with the declared value and deviation determined by EN 771-4	EN 772-16	-At each change of product[2] and -Weekly of 3 units or -As given in the documentation of FPC
Gross dry density	Conformity with the declared gross dry density and the permissible deviations determined by EN 771-4	EN 772-13	-At each change of product[2] density and -At least weekly of 3 units
Compressive strength	Conformity with the declared compressive strength and the permissible deviations determined by EN 771-4	EN 772-1	-Daily and -At least 3 units of each 1000 m^3 produced, and -At each change of product[2] strength
Thermal conductivity[3]	Conformity with declared value	ISO 8302	Once a year
Drying shrinkage[3]	Conformity with declared value	EN 680	-Once a year or -At appropriate time intervals as given in the documentation of FPC
Water vapour permeability[3]	Conformity with declared value	EN ISO 12 572	-Once a year or -At appropriate time intervals as given in the documentation of FPC
Water absorption[3]	Conformity with declared value	EN 772-11	-Once a year or -At appropriate time intervals as given in the documentation of FPC
Bond strength[3]	Conformity with declared value	EN 1052-3	-Once a year or -At appropriate time intervals as given in the documentation of FPC
Reaction to fire[3]	Conformity with declared value	EN 13 501-1	Every 5 years
Freeze-thaw resistance[3]	Conformity with declared value, if any	Reference to the provisions valid in the intended place of use of the units	Once a year

[1] Applies only where units are classified to be used with highest level of tolerances with thin layer mortar (TLMB).
[2] A change of product is defined as a change that will have a significant impact on the relevant declared value.
[3] Only when declared by the manufacturer based on testing.

6.6 *Inspection of marking and stock control*

The guidance suggests that visual inspection is carried out to check product (or packaging etc) marking for product identification in conformity with the requirements of EN 771-4, including for traceability and to check separate storage of nonconforming or reclassified products. For product marking daily checking or after every change of product is suggested, whilst for stock control it should be as given in the documentation of FPC.

REFERENCES TO EUROPEAN STANDARDS

The references given below are published European Standards associated with aac masonry units

EN 771, Specification for masonry units – Part 4: Autoclaved aerated concrete masonry units

EN 680, Determination of the drying shrinkage of autoclaved aerated concrete

EN 772-1, Methods of test for masonry units – Part 1: Determination of compressive strength

EN 772-10, Methods of test for masonry units – Part 10: Determination of moisture content of calcium silicate and autoclaved aerated concrete units

EN 772-11, Methods of test for masonry units – Part 11: Determination of water absorption of aggregate concrete, autoclaved aerated concrete, manufactured stone and natural stone masonry units due to capillary action and the initial rate of water absorption of clay masonry units

EN 772-13, Methods of test for masonry units – Part 13: Determination of net and gross dry density of masonry units (except for natural stone)

EN 772-16, Methods of test for masonry units – Part 16: Determination of dimensions

EN 772-20, Methods of test for masonry units – Part 20: Determination of flatness

EN 998-2:2001, Specification for mortar for masonry – Part 2: Masonry mortar

EN 1052-3, Methods of test for masonry – Part 3: Determination of initial shear strength

EN ISO 12 572, Determination of water vapour transmission properties (ISO 12 572:2001)

EN 1745, Masonry and masonry products – Methods for determining declared and design thermal values

Eurocode 6: Design of masonry structures – Part 1-1: Common rules for reinforced and unreinforced masonry structures

EN 13 501-1, Fire classification of construction products and building elements – Part 1: Classification using test data from reaction to fire tests

Autoclaved Aerated Concrete – Limbachiya and Roberts (eds)
© 2005 Taylor & Francis Group, London, ISBN 0 415 38356 0

Developments and opportunities for AAC with modern production technology

W.M van Boggelen, Ing
Aircrete Technology B.V., Oldenzaal, The Netherlands

ABSTRACT: A summary of worldwide Autoclaved Aerated Concrete (AAC) products will highlight the different global production developments, market cultures and applications of AAC building materials. The recent developments in modern AAC production technology are also reviewed, including a discussion of Tilting, Flat Cake and Combination systems. This review will highlight production capability and capacity, manufacturing cost and product versatility. A brief analysis of key trends in the AAC market as well as tips on production technology concludes the paper.

1 A WORLD REVIEW OF AAC DEVELOPMENTS

1.1 *Western Europe*

Western Europe has several different building cultures all of which rigorously compete for their own market share of the building materials industry. The AAC market struggled in the last years as the construction industry in general experienced an economic downturn that enhanced the competition between manufacturers of different building materials. Currently in Western Europe, the AAC market is showing signs of growth and is generally considered a block market. For several reasons, the AAC industry has been unsuccessful increasing the reinforced element market and production levels. In the 1960's and 70's, the reinforced element market was seriously impeded due to problems with production and quality. In addition to quality issues, price levels compelled architects and contractors to use other materials, such as hollow-core concrete panels, precast concrete elements, steel constructions and wood. Therefore, blocks have become the dominating product and have the largest AAC market share in Western Europe.

The UK for instance is a typical block market where both sand based as well as Pulverised Fly ash are used in the block production. Typical block dimensions differ from the continent with a height of 215 mm, and 440 mm–620 mm length . The UK market offers a versatile range of products for inner leafs, floor blocks and foundation blocks below DPC etc. Block product tolerances vary considerably within the production standards; however, these variations have little to no impact as long as mortar is properly used in the installation. Thin mortar layer installation for quicker and improved building remains underutilized in the UK market, as blocks used for thin joint require a higher accuracy (see EN 771-4). The need to produce higher accuracy product will demand significant investments for most of the UK factories. Additionally, tongue and grooves as well as handholds are not yet a regular practice in this market. It is likely that manufacturers and resellers will bring AAC products to the market as well as building material system suppliers offering full system concepts. In the long term, we are convinced that application versus best quality and best price, integral calculated, will be the determining factor in realizing a greater AAC market share.

In the Netherlands, there is an entirely different AAC market. Competition with other building materials, such as precast elements, hollow core concrete slabs, Sand lime bricks, Self-compacting concrete (SCC), and metal stud construction, is very stiff. This competition forced AAC manufacturers to another approach in the 1980's and 90's. The AAC industry has gained and will continue to increase market share by better utilizing the specific and unique properties of AAC as well as developing systems to build quicker and better at a lower cost compared to other building materials. Currently, the Netherlands AAC market for inner leaf panels, partition, firewalls and Casco systems is considered a significant success. The AAC market is shifting from m^3 to m^2 production, which means it is moving away from pallets of standard blocks to prefabricated wall systems ready for installation at the work site. Another innovative product that has increased AAC's competitiveness and has great worldwide potential is super smooth panels that are immediately ready for finishing without any need for stucco. The other remarkable building process that has enhanced the use of AAC is the dry stapling method without mortar or thin joint, in which the AAC blocks have a groove and strip to secure position. Builders in the Netherlands view the dismountable and reusable nature of AAC blocks as a significant advantage.

In the German AAC market, which is primarily a block market, significant market increase has already been achieved due to the use of innovative jumbo-sized blocks. These large blocks were primarily developed to increase building speed. Thin joint construction is generally accepted in Germany and recognized as a considerable advantage. It is used in most building sites for quicker and higher quality building. Another innovation is the gluing of U-blocks with hot melt that allows for more efficient and less expensive production.

1.2 *Eastern and southern Europe*

The Eastern European AAC market is primarily a block market that is showing signs of significant growth with millions of m^3 consumed annually. The Russian and Polish markets are demonstrating the strongest growth; however, these markets remain dominated by blocks produced to their own standards and quality levels. The recent construction of several new manufacturing plants in the region demonstrates the market growth. Italy, Spain, and Greece are showing much slower market growth though AAC is a good material for these warm climates due to its insulation properties.

1.3 *The Middle East*

The Middle East is currently a typical block market used for gap filling and walls that has potential for a significant capacity increase for utilization of AAC. Recently, block profiling and handholds were introduced in the AAC market. The unique properties of AAC are well known and accepted as a first class building material and market research to harmonize the products with the cultural and geographical conditions of the Middle East is nearing completion. The Middle East market for reinforced material is slowly developing. Middle Eastern AAC market growth is substantiated by the construction of several production plants.

1.4 *Asia*

The use of AAC in Asia is remarkably innovative and the Asian markets are growing rapidly within the AAC industry. Specifically, Japan is one of the Asian countries where AAC developed as a high quality, flexible building material. The applications and building systems for AAC in Japan differ greatly from those in Western Europe, especially the use of AAC systems for seismic building technology. Thin reinforced panels used as cladding for high-rise construction are earthquake resistant due to both vertical and horizontal panel's ability to shift when under seismic stress. Another Japanese innovation is the use of thin panels with a perfect smooth surface, custom pattern or even coating that provides AAC products with considerably more added value and greater options for building facade finishing.

1.5 *North America*

North America remains a developing market as AAC products have a very limited market share in the building industry. For several reasons, the marketing of AAC as an alternative building material

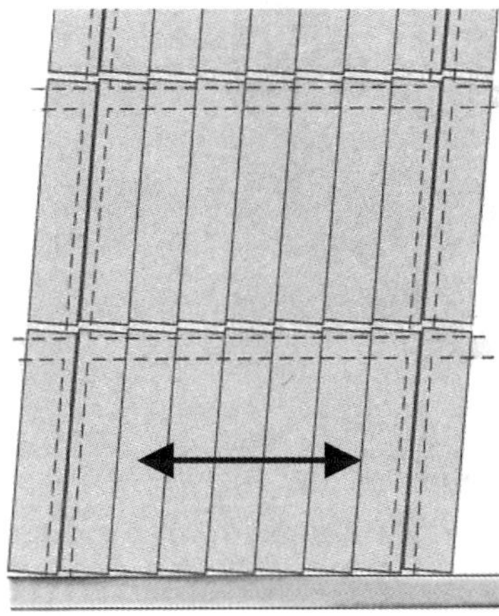

Figure 1. Shifting vertical panels for Seismic building.

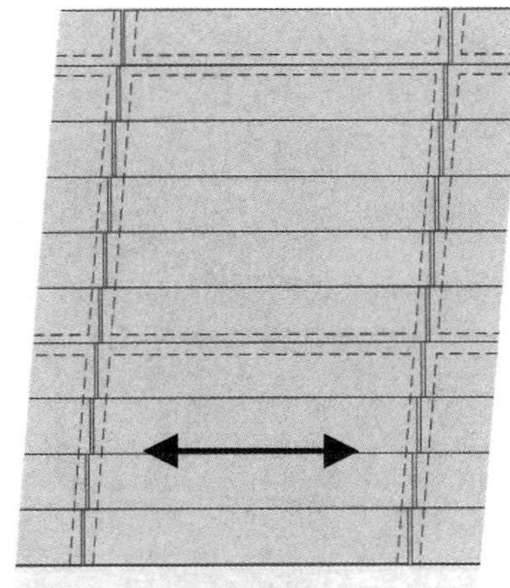

Figure 2. Shifting horizontal panels for Seismic building.

Figure 3. Labor Intensive CMU installation.

and the achievement of technological improvements is slow to succeed. In the United States block market, the standard hollow concrete CMU is the primary competitor. However, the installation of CMU in the United States continues as a labor-intensive process. In our opinion, the utilization of large size AAC blocks produced to US standards and installed with thin joint techniques would considerably increase building speed and allow for an increase in market share for AAC. With greater technology transfer and improved expertise in the US, the large sized AAC block may become a very successful building material within a short time. We envision promising market opportunities for companies offering full system concepts with top quality reinforced products and high building installation speed.

An innovative application developing in the US is the use of AAC for decorative elements. Freely formed products are creatively used for balustrades, sills, columns and wall finishings. These decorative elements are produced with special sawing and milling tools, which are reliable and quickly implemented.

Figure 4. Dry stapling of AAC blocks.

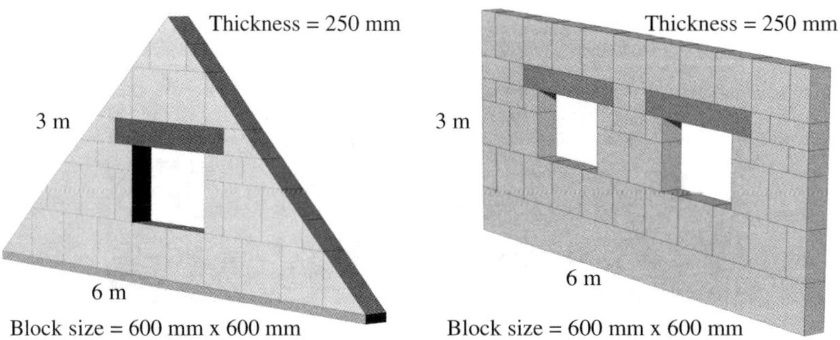

Figure 5. Prefabricated AAC wall elements.

1.6 *AAC prefabrication trend*

At the end of the 1990's significant effort was made by the Sand Lime Brick industry to develop and deliver prefabricated walls direct to the building site. The AAC industry, primarily in Western Europe, reacted to this advancement and is currently researching and testing different prefabricated options. However, developing the market is taking a longer period than is desirable. The hope is that the development of prefabricated AAC products will succeed in gaining a greater market share. At this point though, AAC as a ready-made façade has not developed enough for use in industrial buildings.

In order to gain maximum results with prefabricated AAC elements, it is necessary to have building systems that include other periphery concepts such as window glass energy. In several countries, we see a trend towards: IFD.

I = Industrial method of building (i.e. from building site to factory)
F = Flexible and individualized production
D = Dismountable and reusable product.

The increase of prefabricated elements has allowed work previously completed at the building site is now possible at the factory. Additionally, prefabrication allows for better-organized and more efficient building practices due to less stakeholder involvement in the construction process and weather not affecting building material production at a factory.

Figures 6 and 7. Competition prefabricated wall elements.

1.7 *AAC market competition*

In the Benelux countries, developments in professional high-speed building practices are occurring quickly. Building a house in a week is no longer an exceptional accomplishment. Manufacturers can deliver complete facades with other complementary building materials and components with efficient logistics directly to the building site. An example of this building system is a manufacturer in the Netherlands that is potentially able to prefabricate 95% of a building in their factory. There is considerable potential for AAC manufacturers to add complementary elements to complete façade systems. However, these prefabricated elements should not use conflicting materials, as the focus should be on AAC as much as possible.

2 AAC MODERN PRODUCTION TECHNOLOGY

2.1 *Raw materials preparation*

The technology used for raw materials preparation is essentially the same as that used 30 years ago. However, modern laboratory technology equipment has made it possible to monitor the supplied raw materials quicker and with greater accurately. This enhanced monitoring has resulted in a more consistent and precise recipes with a corresponding improvement of the entire AAC production process. Raw material preparation has also been enhanced by the increase in automation for feeding, grinding and slurry handling in many factories, so that much of the operation is now unmanned.

Good plant engineering and component selection are key factors for reliable plant operation at the lowest possible cost. New liner materials in ball mills in many plants have resulted in reduced maintenance and considerable increase in operating lifetime. The development of orbital tube-type storage slurry mixers demonstrate high performance with a significant decrease in energy consumption and enhanced reliability.

2.2 *Weighing and dosing*

New weighing and dosing technological developments have improved accuracy and flexibility in the production process. The introduction of the soft Industrial-based PC technology has simplified the mixer-control hardware systems. These systems offer nearly endless flexibility with more possibilities of raw material components, decrease installation cost and make manufacturers less dependent on ridged Programmable Logic Controller (PLC) platform systems. The improved accuracy has allowed for better consistency and leaner recipes for AAC production. Additionally, weighing and dosing improvements has resulted in increased safety of aluminum systems.

2.3 *Mixer technology*

Mixers used in the AAC production process are classified in three categories: the high, medium and low speed. Respectively, these mixers use double screw, propeller and multi-paddle types of

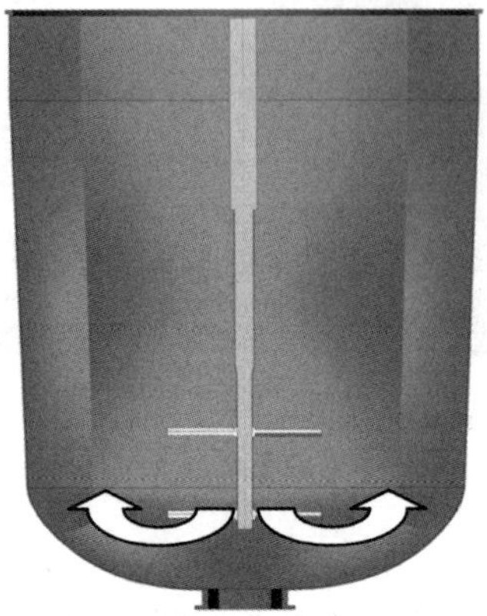

Figure 8. High speed mixer.

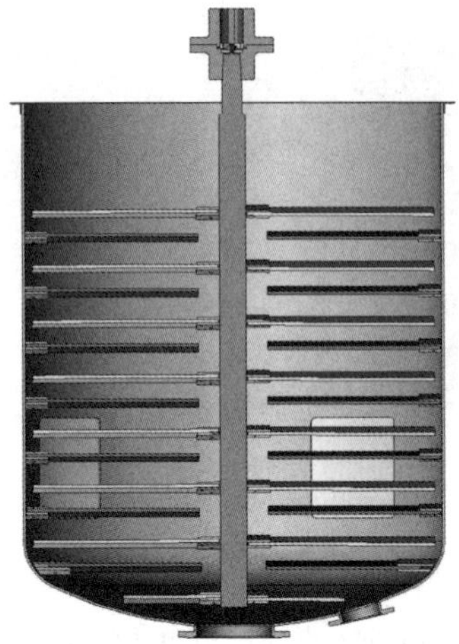

Figure 9. Low speed multi paddle mixer.

mixing mechanism. It is important for each manufacturing plant to select the proper type of mixer depending on the raw materials and specific process used for production. Special PU coatings now allow for easier cleaning and less maintenance of the mixers, which results in reduced long-term operating costs.

2.4 *Cutting technology*

'Green' cutting technology is often perceived as the heart of an AAC production plant. The cutting technology is the essential process of production as it determines the accuracy, geometry and finish of the AAC product. Many production process names created in the past, but still used to this day, are related to different cutting technology methods. In the last ten years, the performance of 'green' cutting technology has increased considerably. The highlights of the new cutting technology include, profiles, handholds, all-side cutting and hollow block cutting. Currently, the cutting systems are generally classified as:

- Flat Cake Systems (caste cakes are processed horizontally)
- Tilt Cake Systems (cakes are tilted 90° vertical for further processing)
- Combination Systems (cakes are tilted and subsequently tilted back).

Each system has now developed into a robust and reliable production process due to considerable improvement of 'green' cutting technology. However, not all products can be produced in the same quality with each of the different techniques. The resulting differences may be significant in terms of surface quality, profiling options, format size and dimensional accuracy.

2.4.1 *Tilt cake system development*

The tilt cake system was originally designed and operated as a system without the use of a bottom layer. After pre-curing the cake is turned 90° and brought to the cutting line. Tilt cake systems have had a turbulent development in Western Europe over the last eight years, but is now starting to gain greater stability. Approximately 14 factories have implemented the Durox double shaft cross cut system in their Tilt cake systems. This introduction has considerably improved block accuracy height at a critical dimension of 200 and 250 mm. Tilt cake products are now easily within the EN tolerances.

Currently, automated tilt cake systems are operating unmanned and have a high degree of flexibility in production. Cutting wire and profile knives cleaning and changing is now conducted with cell robotics within the cutting line. The cell robotics operate similarly to tool changers in the metal working industry. The use of an unmanned cutting line has resulted in less readjustment needs and maintenance. Additionally, automated cutting has significantly increased the safety level as no workers have to be in the hazard zone during production.

A disadvantage of the tilt cake system is that sticking of cured autoclaved products remains a problem. The sticking problems create a certain amount of inevitable breakage in every factory. The most common method for separating blocks is with hydraulic gripping and clamping. To resolve these sticking and fusing issues a few plants tilt the 'green' cake back after cutting and separate each layer of the 'green' shape on a grid frame. In this process, blocks are autoclaved horizontally as in the flat cake system. Most tilt cake systems currently operating are successfully used in the block production sector.

2.4.2 *'Green' waste bottom layer removal*

In order to improve AAC product quality, including accuracy, appearance and removal of concentrated oil, the bottom layer was introduced years ago. The bottom layer was used as a waste layer of approximately 40–70 mm on the bottom of curing plates. In the mid-1990's, dedicated efforts were made in the UK (Celcon) and Germany (Ytong) to remove the bottom layer in the 'green' state to save cost by avoiding unnecessary hard waste. Cakes were tilted, the bottom layer was removed and then the cakes were tilted back on a curing plate. Major manufacturers did not continue with this method because of remaining oil residues from the curing plates on the lower block layers. Appearance differences in the surface texture and color of the white sand based AAC blocks were also considered a potential marketing problem.

2.4.3 *Flat cake systems*

Flat cake production systems are available in several types. These systems are well known for their high accuracy and large sized products. One of the advantages of the flat cake systems is that the vertical cut side of the block or panel is the part of the product visible when installed. The smoothly cut surface is directly exposed whether the block or panel is used for a wall, roof or floor. Without sticking problems, long products up to 8 m with reinforcement can be produced in high quality using a flat cake system. Another remarkable advantage of the system is the possibility for the production of thin façade elements, partition panels and blocks. Products with a thickness of 28 mm are becoming more standard rather than an exception.

The Durox machines are operating in the purported One and Two section machine type. In the 2-section type, the 'green' cake moves through the wires and in the 1-section type wires move through the positioned cake. The 1-section machine was developed by Polish manufactures in the mid-1980's due to a lack of Western European spare parts. The buckle bridges below the strip bed as well as the screw type top crust remover are innovative solutions; however, the 1-section machine still has its limitations. Other than the discussed machines, there are several other flat cake systems in production, including the Siporex-Hebel and Stema systems.

2.4.4 *Super smooth surface cutting*

Cutting technology research and development over the last twenty years has led to the ability to produce super smooth surface AAC. The research first started in the mid 1980's in the Netherlands at a Durox plant under the direction of Wim Schreuders assisted by Jan Sleeuwenhoek. The research focused on the characteristics and possibilities from cutting AAC with high frequency moving wires. Prototypes were built and tested for several years and by the end of the 1980's the first systems were in production at selected Durox plants. Combined with the high-speed cutting technology it was possible to create a closed, smooth surface with narrow product tolerances and the potential for new product applications. Special steel wires typically move with a high frequency wave of 20–50 Hz that cuts the 'green' cake with reduced mechanical forces, which produces highly accurate, thin smooth surface slices. This smooth surface cutting process can be referred to as SILCRETE finishing. The product surface is closed and compacted similar to the properties of sand lime bricks, while the internal material maintains the typical AAC cell structure.

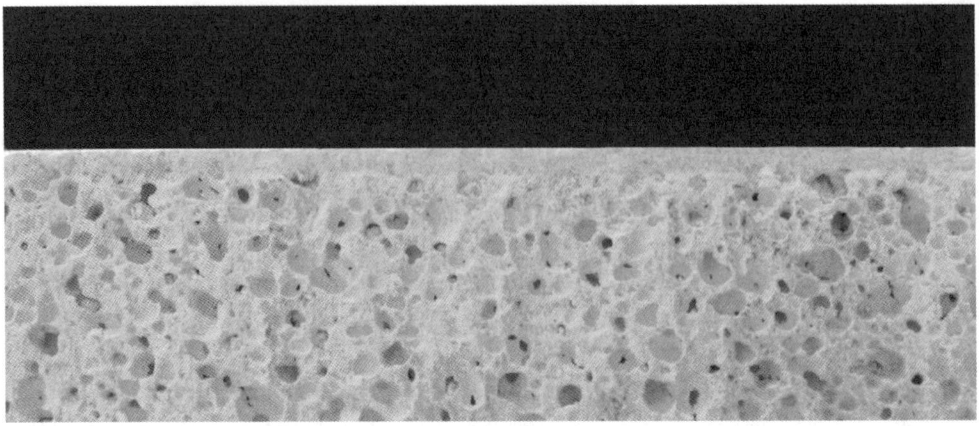

Figure 10. Cross section of smooth surface AAC product.

Figure 11. High speed cutting system for smooth surface cutting.

With smooth surface cutting, the use of stucco for finishing is generally not necessary in most cases. Only a thin film to close the panel seal is sufficient for further treatment, such as direct coating, wallpaper or sometimes absolutely no further finishing. To connect the smooth surface products the thin joint is necessary or highly recommended for satisfactory installation. In the Netherlands, Scandinavia and Japan these smooth surface products are mass produced and have taken significant market share away from other building materials. Higher quality, faster installation and lower building cost was achieved with the product, so that the AAC market in these areas is increasing slightly in a relatively flat or even decreasing building market.

In the past year, research was completed with piezo actuators on wires as well as knives. Cutting wires and knives are now actuated with several kHz, which further improved the surface result considerably so that it is possible to be smoother than paper. This advancement is a promising development, but is not yet economically feasible for mass production.

2.5 *Curing (autoclaving)*

In the last five years, the industrial PC was introduced to autoclaved silicate products market. Steam quantity, temperature, pressure and vacuum are now precisely controlled by intelligent software working on standard industrial PCs. The software allows the curing process to follow the

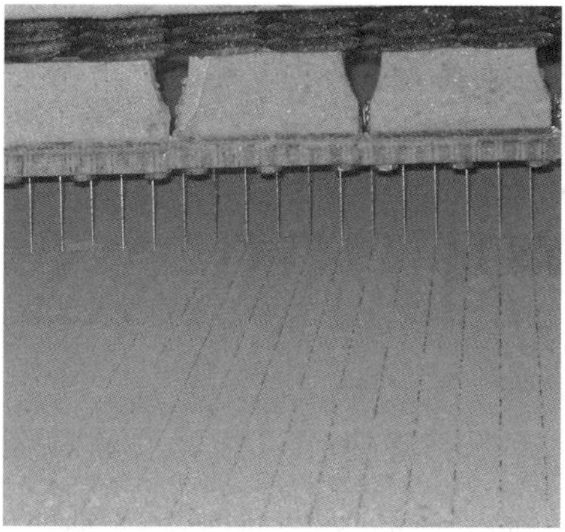

Figure 12. High speed cutting, smooth surface for board filler production.

Figure 13. Double wire smooth surface cutting.

pre-selected curves accurately for the selected products. The energy management is monitored and controlled by the system itself concerning boiler usage and steam transfer. Recent examples are available where 19% of energy was saved by controlling and managing the steam in a more efficient manner. This method of controlling autoclaves demonstrates a shift away from the usual expensive PLC platform systems with cabinets full of hardware. The capital investment for industrial PC systems is only a fraction of what it was in the past, allowing for significant improvements in this production area.

Another advancement in the curing process is the development of automatic opening and closing of the autoclave doors. The function of the autoclave operator can now shift to more of an observing operator role.

2.6 *Unloading and packaging*

Highly standardized gantry robotics up to two tons payload are the most prominent development in the unloading process. These robotics offer flexibility with smooth and fast handling and continue to gain greater acceptance in the AAC industry. It is now possible to achieve a production of 160 Euro packs per hour with the efficient handling of gantry robotics. Cell robotics are presently making a successful entrance in auxiliary functions for pallet and product handling activities. However, the payload in relation to stiffness is limited, with payloads varying from 150 to 400 kg and in exceptional cases attaining 600 kg.

2.7 *Sawing and milling*

It is generally impossible to avoid sawing and milling on finished AAC products. In the reinforced production market, it is common to mill and saw products according to customer specifications. Circle saw blades generally cut AAC products, including steel reinforced panels, to the requested dimensions. Diamond cable saws can cut special cavities and openings in reinforced produce as well as creating opportunities for free-form building walls.

An important milling development is the modern Computer Numerically Controlled (CNC) milling lines. This technology performs with the highest accuracy and flexibility against acceptable cost. In many cases where special and deep profiles are requested (flooring and roofing) a CNC milling machine performs the operation quickly and at low cost.

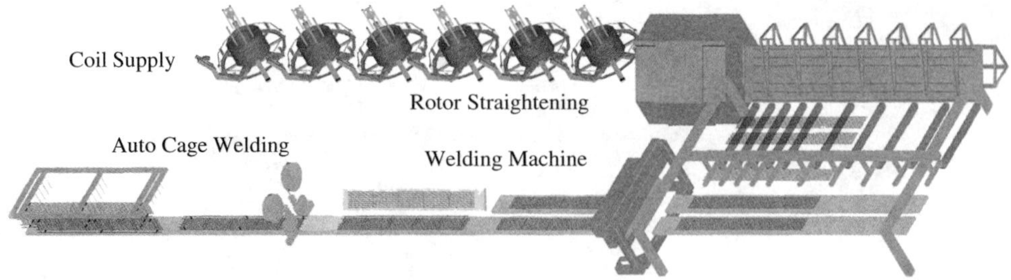

Figure 14. Advanced steel reinforcement production system.

2.8 *Steel reinforcement*

The AAC industry has been struggling already for many years with the flexibility and availability of reinforced production. There has been little economy of scale and a low degree of automation as every AAC manufacturer has attempted to develop their own solutions. This tends to result in inadequate availability and high cost. Other products, such as hollow core slabs and T beams, were often considerably less expensive to produce and purchase than reinforced AAC products.

New developments for steel reinforcement where steel from a 4–12 mm coil is straightened, cut and welded automatically for accurate cages are now possible. Currently, quantities of 2000 cages per day are achievable with this automated production process. The AAC industry should learn from the concrete industry in order to increase automation and economy of scale for reinforced products.

There is also a need for the European standardization committees to harmonize standards as soon as possible in order to make a simple cost effective and high volume production method available. Fragmented standards with regional, national or competition interests will limit the AAC reinforced market and drive demand further to other building material markets. Presently, the competitive difficulties of AAC reinforced product are due to high cost and a lack of economy of scale and specialization.

2.9 *Steel reinforcement treatment*

For technological reasons it is necessary to coat the steel reinforcement with a special treatment to protect against rust. Coating requirements caused serious issues for manufacturers as the use of bituminous coatings resulted in new specializations and the need for many health, safety and environmental precautions. New user and environmentally friendly coatings were developed in the 1990's that addressed many of these serious concerns. Currently, automatic coating lines with dipping baths and wax for needle cleaning and lubrication are in operation. The development of automated coating lines significantly enhanced capacity and made the production of steel reinforcement more cost effective. The next step to improve coating is to hang the cages automatically in the needle frames. This advancement is still in the development stage and is currently operating as a prototype.

3 CONCLUSIONS

The developments in the technology of the design, production and application of AAC offer architects, contractors and builders new possibilities for uniquely and creatively designed buildings. Using AAC blocks and, prefabricated panels and elements is very effective in reducing the overall building time. The developments that reduce manufacturing cost enhance AAC's position in the building materials market. The precision and smooth finish on AAC is an advantage over other similar building materials and is greatly valued in the Benelux, Japan and Scandinavian countries. There is a strong and growing interest for AAC in rapidly developing countries worldwide. However, the highly automated SCC precast industry creates stiff competition for AAC. Therefore, the AAC industry must continue to strive for technological developments that will result in lower cost for both the producer and consumer. It is strongly recommended that the AAC industry continue towards harmonization and standardization and to cooperate as an industry to increase the AAC market as a whole.

REFERENCES

Gassel, F., van. 2003. *Future Site*. Eindhoven: TU/e.

Kamerling, M.W. 2005. Design of precast free-form buildings. Delft: TUDelft

Michel, O. 2005. *AAC Surface Research*. Uentrop: H+H Celcon.

Autoclaved Aerated Concrete – Limbachiya and Roberts (eds)
© 2005 Taylor & Francis Group, London, ISBN 0 415 38356 0

New developments in the AAC production process

T. Dietz & P. Daschner
Wehrhahn GmbH, Delmenhorst, Germany

ABSTRACT: The global AAC market is very competitive and as a consequence requires highly efficient AAC production plants and process technologies. In the latest generation of AAC production plants, Wehrhahn has primarily focussed on the following topics:

ZERO PROCESS RELATED WASTE- The patented Wehrhahn Bed Remover for tilt cake systems recycles the 5 cm thick bottom layer before the cake is being autoclaved. With this new technology the disadvantage of the tilt-cutting technology could be eliminated. In the past the hardened (autoclaved) bed could only be recycled in the process after being milled.

With this new technology 4% of raw materials can be saved and other process related disadvantages have been overcome. All cake processing like tongue and groove cutting or handhold milling is done prior autoclaving. As a consequence there is no process related AAC waste to be expected.

AAC PRODUCTION PLANTS FOR SMALLER BUDGET AND SMALLER MARKETS- The investment for AAC production plants is relatively high in comparison to other building materials. Even if buyers prefer to start with lower capacities the initially required investment could only be reduced up to a limited extent.

For markets which require lower plant capacities Wehrhahn has developed the new SMART plant. It comprises of all advantages of the tilt cutting systems, however requires less investment in comparison to other plants. The capacity of the Wehrhahn SMART plant ranges from 150–700 m^3/day.

1 INTRODUCTION

Today two different AAC production processes can be found in the industries. The significant differences between these technologies are in the cutting area. Commonly the raw materials which are sand, the binding materials cement and quick lime, some gypsum or anhydrite and the gas generation agent aluminium are thoroughly mixed in water and finally poured into a mould.

In the mould the chemical reactions start. Cement and quick lime react with water in an exothermic reaction to calcium hydroxide and semi-crystallized calcium silicate hydrate. The aluminium reacts with water and the hydroxide ions which have been made available during the cement and quick lime reaction. Hydrogen gas H_2 is generated which produces the tiny little pores inside the AAC. Thus giving AAC its low density in comparison to other building materials.

The chemical process in the moulds is commonly described as "pre-curing of the cake". To avoid heat loss and to provide an undisturbed pre-curing moulds are parked in a climatically controlled pre-curing chamber until the cake is ready to be cut.

The question now is: How to cut the cake? And this is where the two technologies show significant differences.

2 HOW IS AAC CUT?

2.1 *Cutting of a flat cake*

A widely used size of the cake is 6,0 m length, 1,2 m width, and a height corresponding to the required block length. Typically the height is 400–625 mm.

In the flat-cake-cutting technology the cake is transferred from the mould into the cutting line. The cake remains in a flat (untilted) position. As a consequence some disadvantages during cutting are associated. Such as:

- the cake needs to be taken by a grab to be transferred from the mould into the cutting line. This requires extremely hard cakes which reduces the flexibility in the mix formulae and generally requires a longer pre-curing time
- cutting tongue and groove or handholds requires complicated machines and handling as the bottom of the cake which is difficult to access needs to be machined
- additional machines and an extra production step is required for cutting of tongue and groove and for handholds
- the cut-offs from tongue and groove and handholds are difficult to remove by suction devices
- for block height cutting long wires need to be used, longer than the cake width. Long wires are considered to be disadvantageous concerning cutting accuracies.

However, a flat cake may also offer advantages in the following parts of the process which will be discussed later in Parapraph 4. Wehrhahn SMART Plant.

2.2 *Cutting of an upright standing (tilted) cake*

Instead of cutting a flat cake, it is also very common to tilt the cake into an upright position before cutting.

Cutting of an upright standing cake offers many advantages such as:

- easy access for profiling and handhold cutting
- the cake does not need to be as hard compared to a cake which is cut in flat position. This is due to no-grab-handling
- extremely short wires can be used for block thickness and block height cutting (or panel thickness and length). This provides highest possible cutting accuracy. The tolerance for block thickness and block height in Wehrhahn AAC factories is closer than ±0,5 mm
- Most cut-offs fall down by gravity
- Cutting of tongue and groove only needs tools but no extra machines and no extra production step.

A disadvantage when cutting an upright standing cake used to be a 4–5 cm thick layer cut at the bottom of the cake (cutting bed) which in the past could not be removed prior to autoclaving. As a consequence approximately 4% of the autoclaved cake was either wasted or was after crushing returned into the mix as a filler.

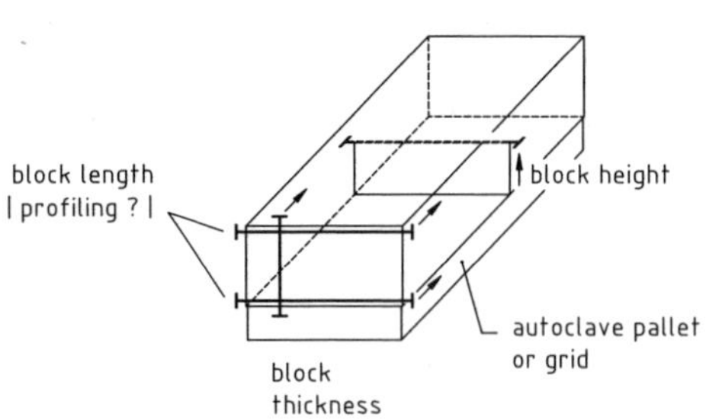

Figure 1. Cutting of a flat cake.

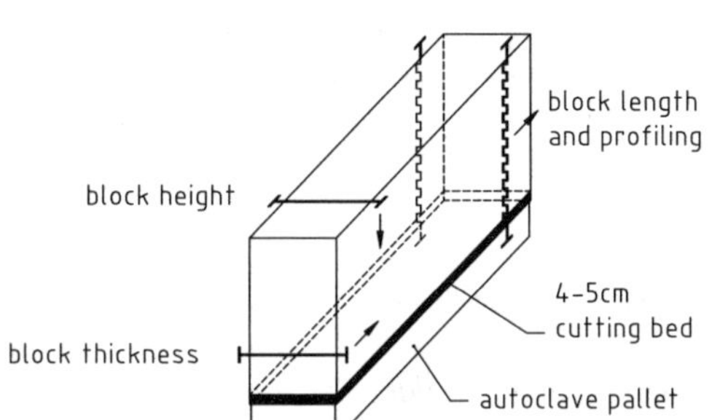

Figure 2. Cutting of an upright standing cake.

2.3 *What are the consequences?*

As a consequence, Wehrhahn exclusively offers plants where cutting is done in an upright standing cake (tilting technology) as this provides the above described advantages. Depending on the project requirements, the cut cake will either remain in upright position after cutting (PLUS plant), or can be put back into a flat position after cutting (SMART plant). These two technologies offer an utmost degree of flexibility for every customer.

2.3.1 *The new Bed Remover for the Wehrhahn PLUS plant*

Wehrhahn developed a new machine – the Bed Remover which has been patented. With this machine the 4–5 cm cutting bed can be removed prior to autoclaving. The material is recycled the same way as all other green the cut-offs.

2.3.2 *The new Wehrhahn SMART plant*

The new plant concept combines the advantages of an upright standing cake during cutting and a flat cake during the following parts of the process. The 4–5 cm cutting bed is also remarked similar to the PLUS plant. This concept will be introduced in Paragraph 4.

In the following, the Wehrhahn PLUS plant in which the cake remains in upright position is described.

3 WEHRHAHN PLUS PLANT – ELIMINATES THE DISADVANTAGE OF STANDARD TILT CUTTING SYSTEMS

3.1 *How does the PLUS plant works?*

The moulds used in a Wehrhahn plant have four hinged walls which can be opened and which are fixed to the moulds. After opening the mould, the cake is tilted into an upright *position* on a specially designed autoclave pallet. Buyers with existing AAC plants with tilting crane, interested to change to a Wehrhahn cutting line can make use of their existing box-type moulds.

The autoclave pallet where the cake rests on during cutting is not as wide as the cake to avoid that cut-offs can fall onto the pallet during cutting. With a narrow pallet the autoclave space can be better utilized as no space is wasted. The autoclave pallet with the cake rests on a high precision cutting trolley which transports the cake through the cutting line that consists of up to five individual machines.

A) Side Trimmer (block length or panel width)
   The cake moves through the side trimmer for cutting and profiling the vertical longitudinal size of the cake which either is the block length or the panel width.
B) Thickness cutter
   The cake on the cutting trolley is transported through the thickness cutter. The cutting wires are installed horizontally and in steps in order to reduce the tensions in the cake during cutting. All cutting wires are extremely short, thin and pneumatically tensioned. This is important for a very precise cut.

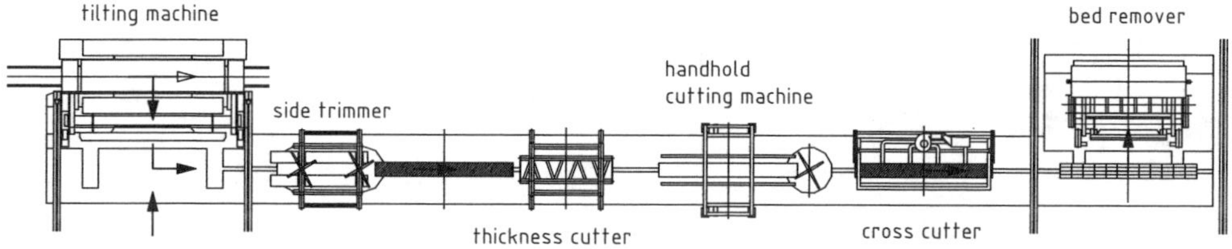

Figure 3. PLUS plant cutting line.

C) Cross cutter (block height and panel length)
The cross cutter operates with two oscillating cutting frames. The wire tension is kept constant, thus providing highest possible cutting accuracy and a longer wire life.
D) Handhold cutting machine
Handholds are of benefit for heavy and thick blocks. Cutting of handholds in the green cake (e.g. before autoclaving) avoids dust, hard waste and extra handling as well as expensive milling tools which are needed otherwise.

3.2 *The new Bed Remover*

In the latest AAC plant generation, the cutting bed (bottom cut) a 4–5 cm thin layer which remained from cutting is removed before the cake enters the autoclave.

How does it work?

The ready cut cake is horizontally moved against a platform and tilted back in a horizontal position. The autoclave pallet is carefully pulled away and the bed can now be removed. The autoclave pallet is now cleaned and moved to the cake. The cake and autoclave pallet are tilted upright.

Here the benefits of the Bed Remover:

- Facilitates "zero process related waste"
- Can reduce the sticking of the blocks after autoclaving significantly. The patented technology has already been proven in many state-of-the-art production plants
- With this technology, 4–5% of valuable raw materials can be saved.

3.2.1 *Arguments for tilting the cake back into upright position after the cutting bed has been removed*

An open question remains: Why is the cake tilted back into upright position after the cutting bed has been removed?

- Upright standing cakes require less space in the autoclave. The autoclaving efficiency is consequently increased

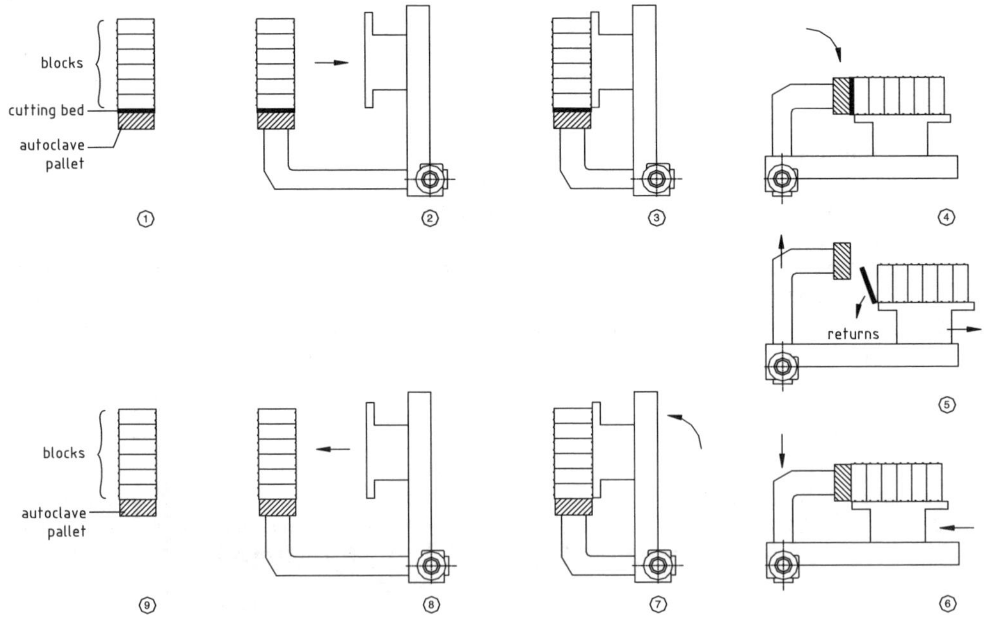

Figure 4. Function of the Bed Remover.

- Stacking of flat cakes on top of each other requires additional automation efforts. Legs which support the upper cakes section need either be inserted manually or automatically
- For many packing sizes tilting of blocks is required. Consequently additional handling of the flat cake is necessary
- Autoclave grids, the platform where the cake rests on during autoclaving, are more expensive than the much smaller autoclave pallets for upright standing cakes.

However, in plants with lower capacity and less automation, arguments can be found for the processing of a flat cake after it was cut in upright position. This concept is introduced in the following paragraph.

4 WEHRHAHN SMART PLANT

In this Paragraph a new plant design available at Wehrhahn is introduced: It features cutting in upright position, however, further processing of the cake in flat position.

It is explained why this plant is the ideal choice for smaller markets.

4.1 *How does the SMART plant work?*

The production steps up to the tilting machine are equivalent to the PLUS plant.

4.1.1 *Cutting*

In the SMART plant one "universal tilting machine" instead of two machines for tilting and bed removal is used.

The universal tilting machine tilts the cake vertically on the platform of the cutting line transport system. This system moves the cake forward with utmost precision from the universal tilting machine through the side trimmer and then backwards to the thickness cutter to the cross cutter and back to the universal tilting machine.

The following picture illustrates the cutting line of the SMART plant.

A) Side Trimmer
The cakes moves forward through the side trimmer for cutting and profiling the vertical longitudinal sides of the vertically standing cake. Cutting of block length or panel width.

B) Thickness Cutter
Once the cake is completely moved through the side trimmer, the side trimmer moves away and the thickness cutter comes on the same place.

C) The cake returns backwards and the thickness cutter cut the required thickness. The cutting wires are installed horizontally and in steps in order to reduce the tensile stress in the cake during cutting. All cutting wires are extremely short, thin and pneumatically tensioned. This is important for a very precise cut.

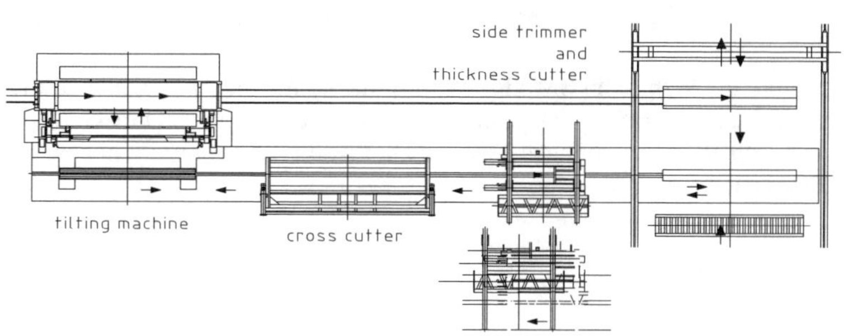

Figure 5. SMART plant cutting line.

D) Cross Cutter
   The cake moves further backwards into the cross cutter for the block height or panel length. The cross cutter operates with two adversely oscillating cutting frames. The vacuum suction head is used to remove the top cut. Tension of cutting wires is done pneumatically.
E) Handhold Cutting Machine
   Also handholds can easily be cut with this system. The handhold cutting machine can be fixed to the cross cutter for different sizes and shapes and handholds. Handholds are of benefit for heavy and thick blocks and facilitate lifting of the block on the job site.

4.1.2 *Autoclaving and Packing*

After all cutting and trimming has been done, the cake is transported back into the universal tilting machine. The cut cake is turned back into horizontal position onto an autoclave grid. The autoclave grid sits on the pick-up trolley which wheels from the universal tilting to the loading machine. The 40–50 mm bottom bed caused by the cutting process will be removed. For further recycling this is a significant economic and technological benefit, as already explained in Paragraph 3 – PLUS Plant.

The cutting ditch is underneath the cutting line. All cut-offs are collected and recycled to return slurry. As a consequence, both Wehrhahn plants, SMART and PLUS, do not produce any process related hard waste = zero process related hard waste. The reloading machine wheels on a steel substructure. The reloading machine lifts the autoclave grid with the cake on the pick-up trolley onto the curing trolley.

The autoclave trolley waits in the holding track in front of the autoclave. Several cakes are set on top of each other on an autoclave trolley. Once the autoclave trolley is full, then the autoclave trolley transport system shifts the curing trolley further into the covered and heated waiting area in front of the autoclaves. This procedure is repeated until the holding track is full.

The autoclave doors open and the autoclave is discharged via transfer car. The autoclave trolleys with the autoclave cakes are shifted onto the unloading track. From there the reloading machine discharges one autoclave trolley after the other. The hardened cake is still on the loading track. The reloading machine has a grab and takes the flat sitting cake and transports it onto the receiver table. Two cakes can be staked on top of each other on the receiver table.

A fork lift with turn grab clamps a wooden pallet together with a pack size segment of the double stack cake, turns the pack by 90° so that the pallet becomes the bottom with the pack on top of it. Instead of the receiver table and forklift with turngrab, an automatic packing system can be used in order to stack the blocks on pallets and strap or shrink wrap the packs.

4.1.3 *The SMART cutting line for existing plants with flat-cake cutting technology*

A unique feature of the Wehrhahn SMART technology is its favourable ability to improve existing flat cake system AAC plants.

In Hebel plants, Siporex plants of every type, Durox plants, and other systems with Polish and Russian technology, only the cutting line needs to be replaced at reasonable costs. Such plants almost always have a lot of equipment for precuring and autoclaving, and the existing mould systems can be kept as well as the handling system at autoclaving and the unloading und packing section.

Cakes are put into the Wehrhahn SMART cutting line by moving in the opened mould (Durox) or using the existing grab system for precured cakes (Hebel, Siporex, Durox). The cake will be tilted upright, cut with all the advantages described before and after cutting, placed on the existing grates or platforms used for autoclaving.

Result:

- more precise blocks or panels
- less maintenance cost
- easy application of tongue and groove and handholds
- re-use of green cut-offs

4.2 *Where are the savings in the SMART plant?*

- The cutting line is very compact and requires one tilting machine only
- The SMART plant is an innovative AAC plant for smaller production and low budget
- The SMART cutting line can easily be incorporated into an existing plant using flat cake cutting. The existing moulds, autoclave grids, etc. can still be used
- Many markets do not require highest possible production capacities but still highest possible AAC quality.

The cake is autoclaved in a flat position, therefore sticking of the horizontal joints well known for upright standing cakes is significantly reduced. Consequently no separating machine is needed to separate the horizontal layers after autoclaving as long as production capacity is low. The packing line requires less sophisticated machinery or packing can even be done manually depending on the plant capacity.

The total savings in comparison to a PLUS plant are approximately 20–30% based on the total budget for Wehrhahn and local supply.

5 CONCLUSION

In the following table the two plants PLUS and SMART are compared:

	PLUS	SMART
Capacity range	400–1.400 m^3/day	150–700 m^3/day
Modular concept	Yes	Yes
Investment costs	100	70–80
Cutting quality	100	100
Panel production	Yes	Yes
Process related waste	0%	0%

Depending on market conditions buyers can now choose between the Wehrhahn PLUS or SMART plant. Both plants feature *zero process related hard waste* and provide *utmost flexibility* for the production of *high quality AAC* blocks or panels.

The Wehrhahn PLUS plant offers highest possible plant capacities. With the new patented Bed Remover the known disadvantages of the tilt cutting system have been eliminated. The SMART plant opens new markets which require lower plant capacities and lower budgets, however, without compromising on quality of the finished product.

Autoclaved Aerated Concrete – Limbachiya and Roberts (eds)
© 2005 Taylor & Francis Group, London, ISBN 0 415 38356 0

Lime production for the AAC industry

F. Mücke
Rheinkalk GmbH, Wülfrath, Germany

ABSTRACT: Rheinkalk uses two types of lime kilns for the production of AAC lime. Both types of kilns are described. The produced lime qualities are compared and their application for the AAC production is described. The quality control of AAC lime is a very important part of lime production and is performed in two steps. The first step is the control of the reactivity of the lump lime before milling. The second step is the control of the milled lime according to the standard EN 459. Additional quality agreements with the AAC producers ensure a constant product for the AAC process.

1 INTRODUCTION

Rheinkalk is a young, dynamic company that evolved from the merger of two traditional companies, Rheinische Kalksteinwerke and RWK Kalk AG. It is now affiliated with the Belgian Lhoist Group, the largest producer of lime worldwide. The largest plants of Rheinkalk are located in Wülfrath-Flandersbach and Menden-Hönnetal and are well situated to supply the Rhine and Ruhr districts. The network of subsidiaries in western Germany is able to supply and tend to the needs of other important customers from locations close-by. The main markets for Rheinkalk products are presented in Figure 1. The market share of burnt products for the building industry was 10% in volume in 2004.

The main customers within the building industry are the sand-lime brick, the AAC and the mortar and plaster industries. Rheinkalk produces more than 120,000 t/y of lime for the AAC industry in Belgium, Germany and Netherlands in different qualities.

The geological conditions of the stone deposits which are used for the production of AAC lime are similar. More than 350 million years ago, gigantic limestone deposits formed from reef complexes developed by calcifying species of animals. Geologically speaking, these deposits are essentially Upper-Middle Devonian.

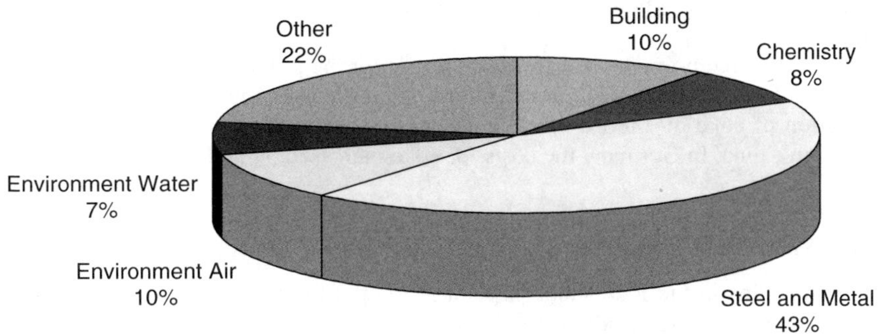

Figure 1. Rheinkalk 2004, volumes of burnt products by markets.

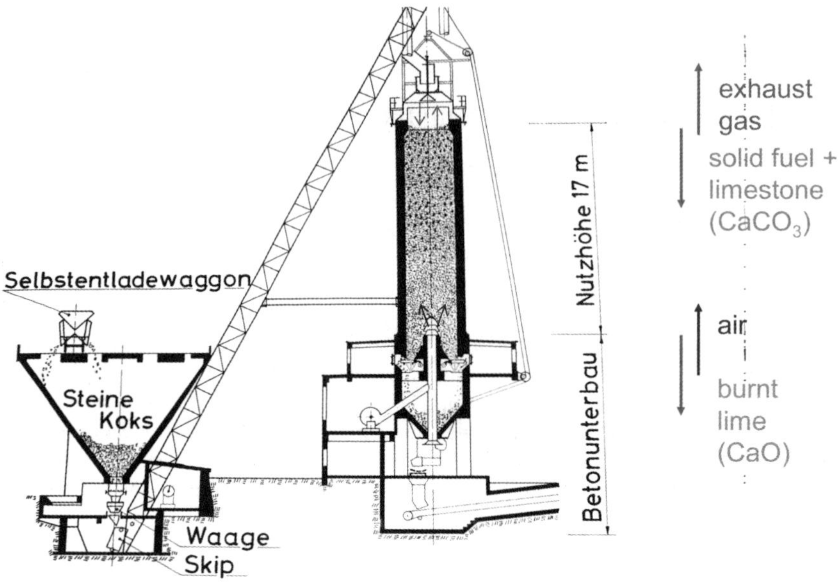

Figure 2. Schematic drawing of an NSK from plant Hönnetal.

2 LIME KILNS

The lime production for the AAC industry undergoes the same production steps up to the burning process as the other lime qualities. For the burning process Rheinkalk uses different types of kilns for the different burnt lime qualities. For the AAC lime, however, only the normal shaft kiln and the rotary kiln are used.

2.1 *Normal Shaft Kiln (NSK)*

An NSK is a vertical single-shaft kiln and works with counter-flow heating. The solid fuel coke, or anthracite, is mixed with the limestone and must be fed to the kiln as a uniform mix without segregation.

The NSK is divided into four distinct zones by imaginary horizontal planes. In sequence from top to bottom there is the stone storage zone, the preheating zone, designed to heat the stone to temperatures near dissociation, the calcining zone, where the combustion occurs, and the cooling and discharging zone at the bottom. Figure 2 presents a schematic drawing of an NSK from the Hönnetal plant.

The total thermal requirements are approx. 4.60 GJ for the production of one ton of hard-burnt lime (Table 1). The use of an NSK in the future is questioned. The reason of this development is the reduction of coke on the European coke market because of an extreme increase in coke consumption in China. In Germany the costs for coke increased by up to 400% in 2004.

2.2 *Rotary Kiln (RK)*

The four rotary kilns in the Flandersbach plant are Lepol kilns with a length of 90.00 m and a diameter of 4.40 m. They are installed at an incline of 3° on four foundation piers, on each of which they rotate on trunnions. The rotation speed is adjustable with variable-speed drives and amounts to a rate of 1.5 rpm.

Table 1. Main data of the NSK at the Hönnetal plant.

Kiln type	Normal shaft kiln
Height (m)	18.50
Shaft diameter (m)	3.85
Stone grain size (mm)	90–120
Daily output (t)	125
Fuels	Coke/anthracite
Heat consumption (GJ/t of product)	4.60
Reactivity t60 according to EN 459-2 (min)	10–14

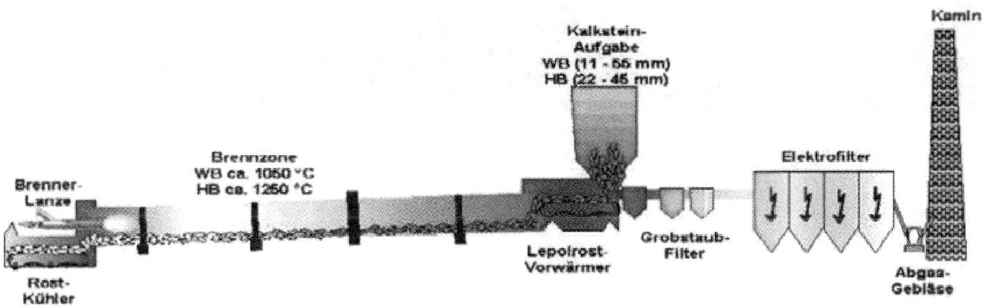

Figure 3. Schematic drawing of a Lepol kiln, Flandersbach plant.

Table 2. Main data of Lepol kiln, Flandersbach plant.

Kiln type	Rotary kiln
Length (m)	90.00
Shaft diameter (m)	4.40
Stone grain size (mm)	11–60
Daily output (t)	775–1000
Fuels	gas/pulverized lignite
Heat consumption (GJ/t of product)	5.50–6.30
Reactivity t60 according to EN 459-2 (min)	<1; 1–3; 2.5–4.0; 4.0–7.0

The limestone is charged from a storage silo into the kiln at the elevated end, and lime is discharged at the lower end, moving counter-current to the flow of combustion gases, derived from fuel injected at the lower end. These kilns are only charged with a maximum of 10% of limestone so that approx. 90% of the interior space is confined to the flame and hot gases. Figure 3 presents a schematic drawing of a Lepol kiln.

Like the NSK the kiln is subdivided into the preheating zone (Lepol grate), the combustion and the cooling zone (grate cooler). Thermal consumption is a staggering 5.50–6.30 GJ/t of lime and depends on the reactivity of the burnt lime. One of the biggest advantages of these kilns is the possibility to produce lime qualities with different reactivities from very high to medium-burnt or otherwise from soft to medium-burnt lime. Pulverized lignite and gas are used as fuel. Table 2 presents more data.

Because of the small grain size of the limestone the stone is very well calcined and the lime is presented with a very low loss of ignition. The available CaO content of lime from an RK is higher than from an NSK. Because of the use of locally sourced pulverized lignite the production costs for lime from an RK are more independent from the world energy market than the production costs from an NSK.

Figure 4. View of a Lepol kiln, Flandersbach plant.

Table 3. Chemical analysis (mass-%) of AAC lime.

Lime	CaO	MgO	SiO_2	SO_3	Fe_2O_3	Al_2O_3	CO_2	H_2O	LOI
Hönnetal	95.58	0.62	1.25	0.14	0.24	0.53	1.03	0.80	1.83
Flandersbach	96.49	0.95	1.11	0.07	0.22	0.27	0.30	0.34	0.64
Messinghausen	94.07	1.57	1.16	0.33	0.34	0.21	1.34	0.66	2.00

3 LIME FOR THE AAC INDUSTRY

Rheinkalk produces lime for the AAC industry in three plants. In the Hönnetal and Messinghausen plants the lime is produced in NSKs and in the Flandersbach plant in RKs. The lime qualities fulfill the requirements of the AAC industry with regard to the chemical analysis and fineness, but the reactivity of these limes differs very much. The lime from Hönnetal is hard-burnt, the lime qualities from the other plants are medium-burnt.

3.1 *Chemical analysis*

The chemical analysis was performed as an X-ray fluorescent analysis. Table 3 presents the results.

The available CaO content can approximately be calculated by using the results from the X-ray analysis as follows

$$CaO_{avail} = CaO - [1.274\ CO_2 + 0.7004\ SO_3 + 0.3(SiO_2 + Fe_2O_3 + Al_2O_3)]$$

The CO_2 content has the main influence on this calculation. The lowest CO_2 content will be reached in an RK.

3.2 *Reactivity according EN 459-2*

The definition of the reactivity of burnt lime will be done in the so-called wet-slag curve. The wet-slag curve serves to define the reaction velocity of burnt lime. The reaction of burnt lime (150 g) and distilled water (600 ml) leads to the hydration of lime and the emission of heat. The temporary temperature modification will be measured. With the help of the t_{60} value (time to reach a temperature of 60°C) burnt lime can be classified as soft, medium or hard-burnt lime.

Table 4. Classification of lime.

Lime	t_{60} (min)
Soft-burnt	1–3
Medium-burnt	3–8
Hard-burnt	>8

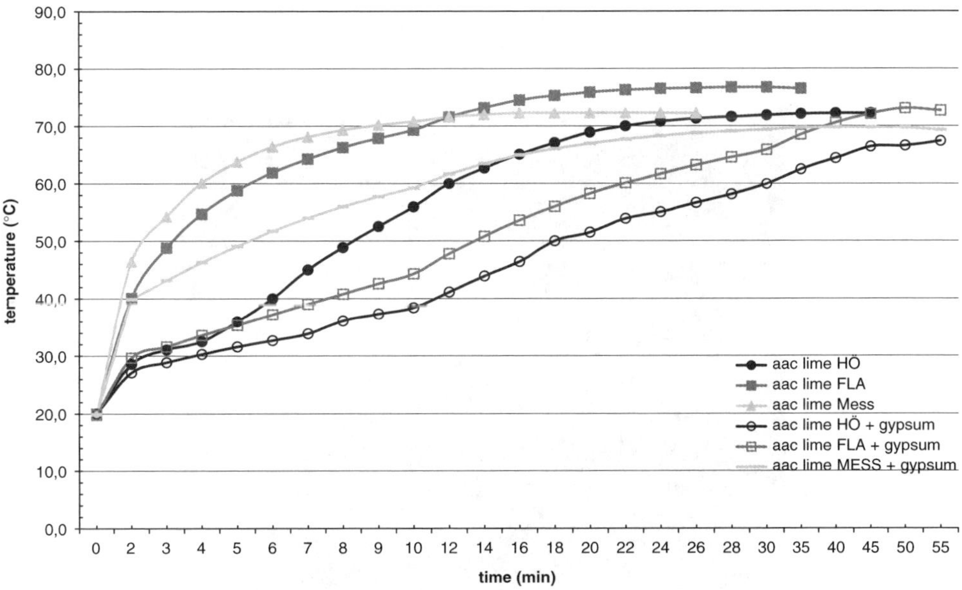

Figure 5. Wet-slag curves of different lime qualities for AAC production.

For the AAC producer it is very important to know not only the t_{60} but also the temperature after 2, 5, 10, 20 minutes and the maximum temperature and time. Figure 5 presents the wet-slag curves of the AAC lime from Rheinkalk. Furthermore, the wet-slag curve of these lime qualities under addition of gypsum is described.

The AAC lime HÖ is a really hard-burnt lime with a $t_{60} \approx 12$ min. It is well-known as a reference lime for the AAC industry. The lime qualities FLA ($t_{60} \approx 5.5$ min) and Mess ($t_{60} \approx 4.0$ min) are both medium-burnt limes. The reason for the difference between these lime qualities is the used kiln. Considering the behavior of these limes in a saturated gypsum water solution, the lime FLA from the RK shows a similar behavior as the hard-burnt lime HÖ.

This behavior is traceable to the calcinations of the limestone in the different kilns. Because of the smaller grain size of the used limestone for an RK and the low charging rate the lime is similarly sintered as a coke-burnt lime from an NSK. The AAC lime Mess is produced in an NSK with a central burner and pulverized lignite as fuel. This lime is only weakly sintered and presents itself as a normal medium-burnt lime. Figures 6–8 are photos by a surface electron microscope of the lime qualities described.

3.3 *Application of the described lime qualities in the AAC production*

All AAC lime qualities were or are in use in the AAC industry in Belgium, Germany and the Netherlands. At the end of March, the AAC lime production in plant Hönnetal was stopped. The

Table 5. Results of the wet-slag curve according to EN 459-2 and the wet-slag curve with gypsum.

Lime	t_{60} (min)	t_{max} (min)	T_{max}(°C)
AAC lime HÖ	12.0	28.0	72.2
AAC lime FLA	5.4	26.5	76.2
AAC lime MESS	4.0	10.1	72.4
AAC lime HÖ + gypsum	30.3	55.0	67.4 (M. was stopped)
AAC lime FLA + gypsum	21.8	50.0	73.1
AAC lime Mess + gypsum	10.6	40.9	69.8

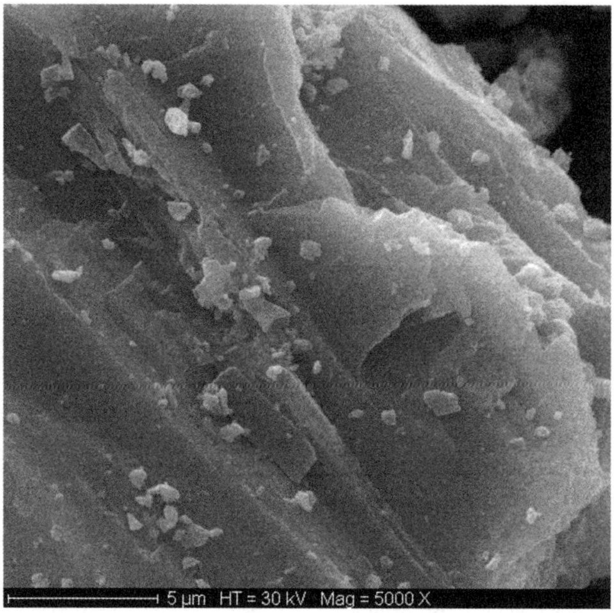

Figure 6. AAC lime HÖ, strongly sintered surface.

reason was the coke situation in Germany. Most of the former customers of AAC lime from Hönnetal are now working with AAC lime from the Flandersbach plant. Most of them only had to make little changes in their recipes. Because of the faster lime reaction the blowing-out of the hydrogen is also faster. This can lead to trouble in the green cake. On the other hand, the AAC producers obtain more available CaO for the process.

4 QUALITY CONTROL

The quality control for AAC lime includes three parts. Part one is the control of the running production. Part two is the control of the ready-milled AAC lime after loading the silo trucks and part three are special tests on the material according to quality contracts with the customers.

4.1 *Control of the running production*

As the RK is a very flexible kiln and can produce lime with different reactivities, it is necessary to have a fast method to define the right reactivity of the run-of-kiln lime. This method is called

Figure 7. AAC lime FLA, similarly sintered surface like AAC lime HÖ.

Figure 8. AAC lime Mess, slightly sintered surface, many small crystals.

coarse-grain titration. 50 g of pebble lime (grain size 2–10 mm) are stirred in 2 l of water with a temperature of 40°C. A titration with hydrochloric acid is performed (4 mol/l HCL) until the pH value reaches 7. The need of hydrochloric acid is measured. The need of the HCL after 2, 5 and 10 minutes gives the measure for the right reactivity of the lime. The kiln crew and the plant laboratory work with this method. This measurement will be done every hour for the AAC lime.

4.2 *Monitoring of the milled AAC lime*

The monitoring of the milled AAC lime takes place when loading the silo trucks. A sample of the lime is taken and analyzed in the laboratory. The tests are carried out according to EN 459-2. The main characteristics like chemical analysis, reactivity and grinding fineness are checked and the truck can leave the plant if the results are right. If the results do not meet the client agreements, the truck will blow out and reload.

4.3 *Special tests*

Rheinkalk has arranged special tests for the AAC lime with some clients. These tests are carried out once per week. For instance the measurement of the sediment volume of the AAC lime with and without gypsum will be checked. The results present hints to the reactivity, the sensitivity regarding gypsum and the uniformity of the lime.

5 CONCLUSIONS

AAC lime can be produced in Normal-Shaft Kilns with solid fuels like coke or anthracite and in Rotary Kilns with pulverized lignite and gas. The lime qualities produced demonstrate different reactivities.

For the choice of the AAC lime the following facts are significant:

- chemical analysis
- high available CaO content
- reactivity
- temperature behavior after 2, 5, 10, 20 minutes and maximum temperature
- t_{60}
- behavior of the lime in a saturated gypsum-water solution

The coarse-grain titration is a possibility for a fast control of the running lime production.

The quality control of the milled lime will be performed according to EN 459-2.

REFERENCES

Boynton, R.S. 1980. Chemistry and Technology of Lime and Limestone. JOHN WILEY & SONS, INC. New York

EN 459-1/2. 2002. Baukalk. Beuth Verlag Berlin

Honig, H., Mücke, F. 2004. Untersuchung von Hart- und Mittelbrandkalken für den Porenbetonprozess. Rheinkalk, unpublished

Joswig, M. 2004. Festbrennstoffe für Normalschachtöfen. Rheinkalk, unpublished

Schiele, E., Berens, L.W. 1972. Kalk. Verlag Stahleisen M:B:H:/Düsseldorf

Winkelrät, F. 2004. Verstromung von Niedertemperaturabwärme nach dem ORC-Verfahren. Rheinkalk, unpublished

Autoclaved Aerated Concrete – Limbachiya and Roberts (eds)
© 2005 Taylor & Francis Group, London, ISBN 0 415 38356 0

Mechanisms and countermeasures against cavity defectives in AAC during manufacturing

Y. Aono, S. Shibata & F. Matsushita
Sumitomo Metal Mining Siporex Co., Ltd., Research & Development Center, Japan

ABSTRACT: Cavity defective are of highest importance in AAC production. While, the mechanism of cavity has not been clarified for a long time. The defective simulation test with SEM observation showed that the cavity occurred in the foaming process. We investigated the foaming process using Al-powders and the cavity frequency can be classified in two types: the type-A and B. The type-B and showed the change in dh/dT that corresponds to the thermal expansion coefficient of the ideal gas. This probably means that the bubble in the AAC slurry is a closed system. The measurement of the internal pressure in the AAC slurry supports this conclusion. To avoid the occurrence of the cavity, the internal pressure by foaming gas should be released from the AAC slurry prior to the end of foaming process. The bubble in the AAC slurry should be in the open system. The characteristics of Al-powder suitable for the prevention of the cavity are summarized.

1 INTRODUCTION

In order to improve productivity and quality of the AAC, we must take suitable countermeasures for defectives of AAC products. Table 1 and Figure 1 show representative and typical defective types of AAC. Occurrence rate of the cavity is probably high in the AAC products that use Al-powder as a foaming detergent. Nevertheless, the mechanism of occurrence of the cavity had not been clarified for a long time. This paper presents the estimation of the timing of the cavity occurrence during the AAC production process, the investigation into the mechanism of the cavity occurrence, and a proposal of countermeasures.

2 ESTIMATION OF THE TIMING OF CAVITY OCCURRENCE

In order to estimate the timing of the cavity occurrence, the defective simulation test was performed in laboratory. The surfaces on artificially broken AAC samples were observed in SEM (Scanning Electron Microscope), and the results were compared with observation of real cavities.

Table 1. Representative defective types of AAC.

Defective type	Feature
Shadow	Vertically disturbed bubble that arises at the upper part of reinforcing steel.
Cavity	Horizontal void, which occurs at side or inside.
Crack	Many types of crack, which arises at anywhere.

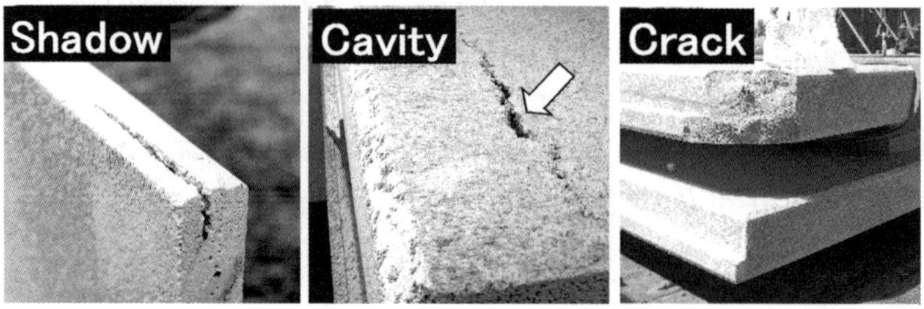

Figure 1. Photographs of typical defectives of AAC products.

Table 2. Mixture proportion of raw material.

Cement				Recycled materials			W/S*2
(%)	Silica	Lime stone	Gypsum	Before-AC*1	After AC	Al-Powder	(–)
26	39	6	6	13	10	0.065	0.60

*1: Autoclaving. *2: Total Water/Total Solid ratio.

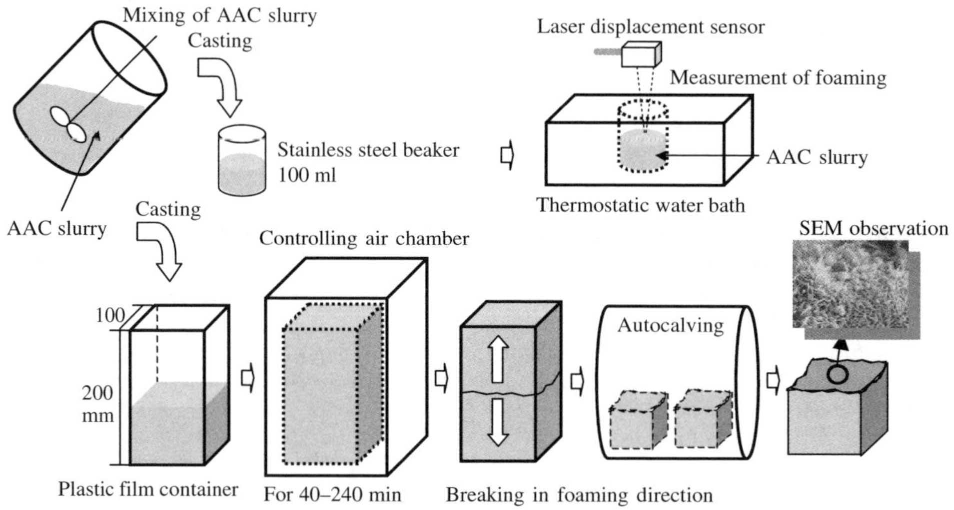

Figure 2. Conceptual scheme of defective simulation test.

2.1 *Experimental procedure of defective simulation test*

Table 2 shows the composition of the AAC slurry. Those materials except for Al-powder were mixed for 2 minutes at 45°C. Al-powder was added to the mixture, and was further agitated for 30 seconds to yield the AAC slurry.

Figure 2 shows the conceptual scheme of the defective simulation test. The AAC slurry was placed into a stainless steel beaker (100 cm^3) and in plastic film containers (100 × 100 × 200 mm). The stainless beaker was placed in the water bath, then adjusted to the AAC slurry temperature change shown in figure 3 measured in the real mold. The change in foaming height of the AAC slurry was measured using laser displacement gauge (KEYENCE LB300) in the stainless steel beaker for 120 min. The foaming height was converted into the foaming height rate, which made the final foaming height 100%.

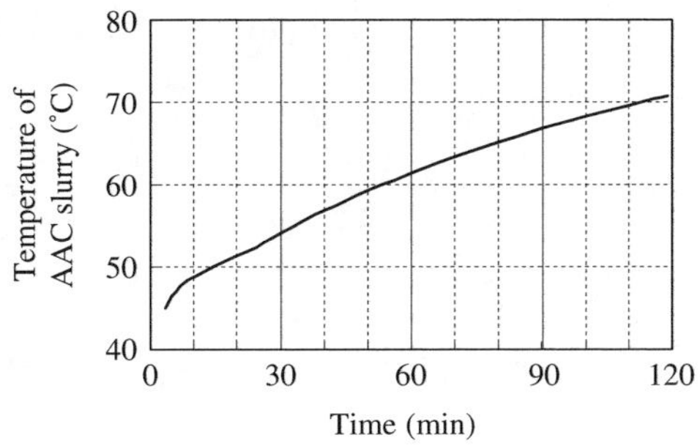

Figure 3. Temperature change in AAC slurry based on measurement result of real mold (7 m × 1.5 m × 0.7 m).

Table 3. Intentional breaking timing.

Sample (Ref.)	Breaking timing after casting (30 min)	Method of sample preparation (Not breaking by fluidity)
A	40 min	
B	60 min	Breaking in foaming direction
C	90 min	(See Fig 2)
D	120 min	
E	240 min	Breaking in foaming direction
F	240 min	Breaking in foaming direction, then joined together
G	After autoclaving	Breaking by hammer

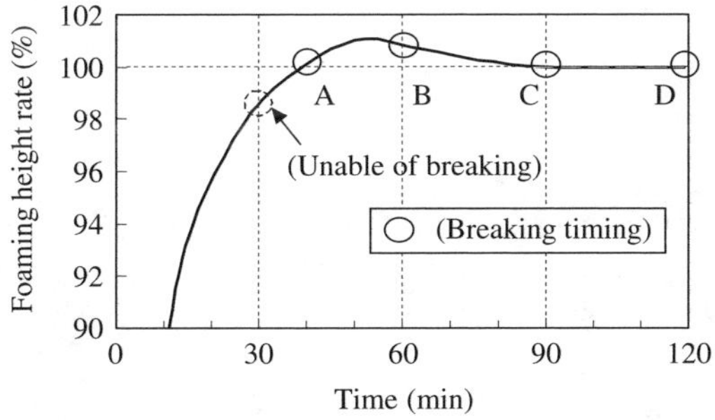

Figure 4. Change in foaming height rate of the defective simulation test. Internal temperature was controlled along with figure 3. Open circles mean breaking timing of AAC.

Simultaneously, the AAC slurry in the plastic containers were placed in the controlling air chamber, then also adjusted to the temperature change in figure 3. These were artificially broken with the plastic film containers in the foaming direction at the timing shown in table 3. All the broken samples were autoclaved at 180°C in saturated steam pressures, then the SEM observation of the surface on the sample was carried out.

2.2 *Identification of cavity occurrence timing*

Figure 4 shows the foaming height up to 120 min measured with a laser displacement sensor. We failed to make breaking samples at 30 min, because it collapsed by fluidity. Notation A (at 40 min) was the timing in which the foaming still continued, nevertheless we could break it without collapse. The foaming height reached maximum value at around 50 min, then it contracted 1%

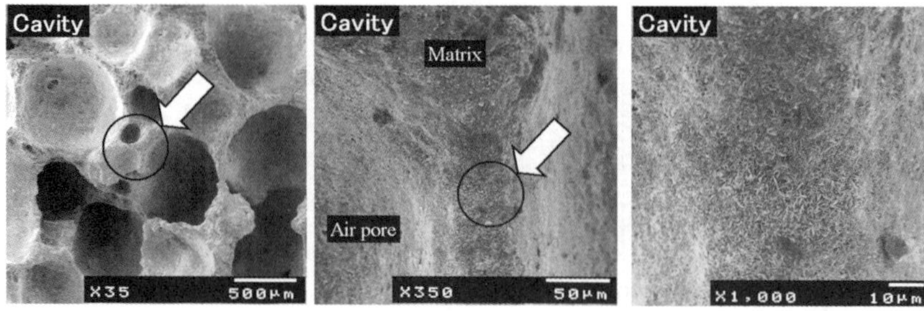

Figure 5. SEM micrographs of surface of typical real cavity. Air pores has been disturbed (×35). Border between matrix and air pore is smooth, round in shape (×350). Tobermorite growths are hardly observed (×1000).

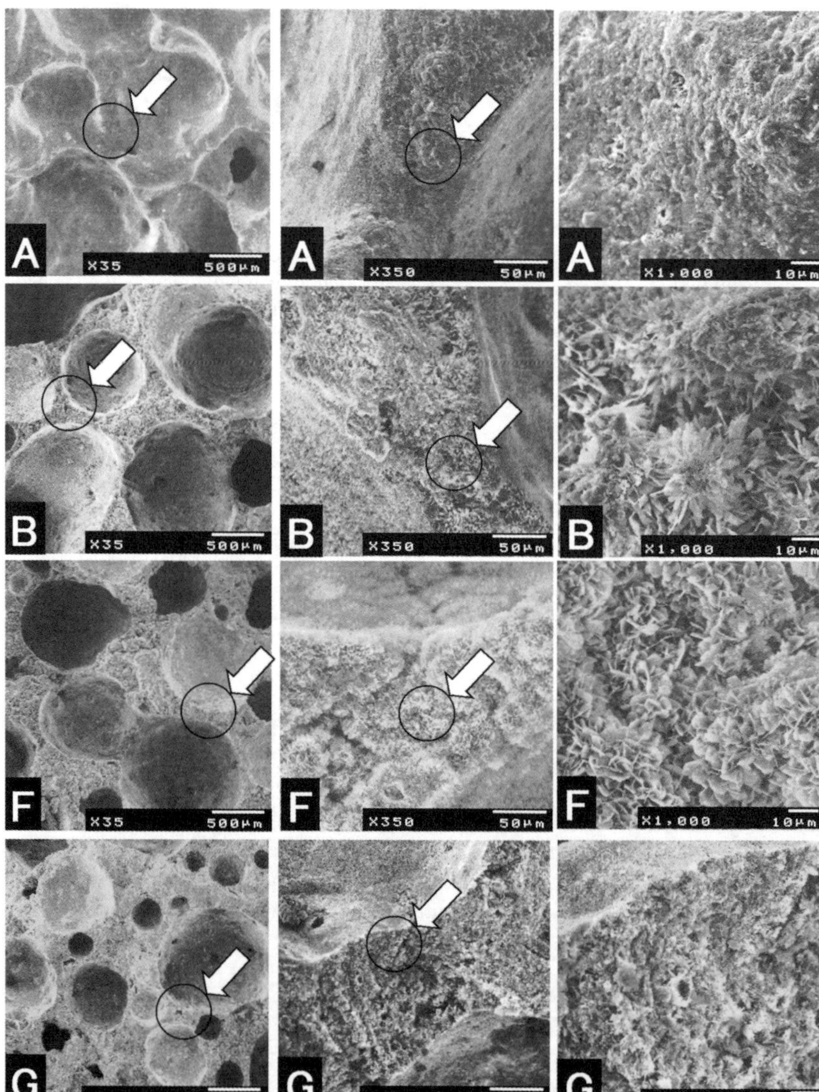

Figure 6. SEM micrographs of defective simulation test. A: Breaking at 40 min.

until 90 min that corresponded to the timing of the C. Notation B (at 60 min) was in the timing in which the AAC slurry was contracting. At the timing of the D, E and F (latter two are not shown in figure 4), the expansion or the contraction had stopped.

The border between matrix and air pore was so smooth and no tobermorite crystals were observed. B: Breaking at 60 min. The fracture surfaces were crisp, and grown plate-like tobermorite crystals were observed especially in the concavity of matrix. C, D and E were almost same. F: Breaking at 240 min then joined together. Plate-like tobermorite crystals grew well at any part of matrix. G: Breaking after autoclaving. The fracture surfaces was clear. Plate-like tobermorite crystals could be observed. Comparison with figure 1 and 2, the cavity probably occurs in the timing just after A and before B.

Figure 5 shows the SEM micrographs of surface on the typical real cavity. Air pores were out of shape. Border between matrix and an air pore was indistinct and showed round in shape. Crystal growth was hardly observed.

Figure 6 shows the SEM micrographs of surface of the defective simulation test sample. The micrographs of A 40 min after casting resembles with that of the typical real cavity of figure 5, however the air pores smoother than that of the A.

The air pores of B, F and G were not disturbed compared with the timing of the A or figure 5. Borders between matrixes and air pores can be distinguished clearly showing the clear edges. The micrographs of C, D and E were omitted as these were almost same as B at any magnification.

From these observations, the cavity seemed to occur during the timing A to the B, just before the end of the foaming process.

Though the both of the timing B (or C, D, E, broken in the foaming direction) and F (broken then put together) showed plate-like tobermorite crystal growth. Tobermorite crystal growth were observed only at the concavity in B, while for F these were observed entire surface of the matrix.

For G (breaking after autoclaving), we could not observe the growth of the plate-like tobermorite crystals even at the magnification of $\times 1000$. The differences in the timing of B (or C, D, E), F and G depend on the availability of space in which the tobermorite crystal can grow, as well as that of silicate ions during autoclaving. Figure 6 is very useful in determining the process in which defectives of AAC product may occur.

3 INVESTIGATION FOR MECHANISM OF CAVITY FORMATION

After determining that the cavity occurs at the end of foaming process, we observed the foaming process in the full scale molds at some production conditions by which the cavity occurs in a different rate. From these results, we proposed the mechanism of the cavity formation, and confirmed it in the laboratory test.

3.1 *Relationship between production condition and cavity*

We have experientially grasped some relationships between the production conditions of the AAC and the occurrence of the cavity. Table 4 shows typical relationships between them. The Al-powder type shows the largest influence on the occurrence frequency of cavity. Total Water/Total Solid ratio (W/S) and quantity of bubble stabilizer are also relevant.

3.2 *Characteristics of Al-powders*

Figure 7 shows H_2 gas generation of type-A and B powders. The H_2 gas generation rate of the type-A is slower than, that of type-B.

Figure 8 shows the particle size distribution of Al-powders of type-A and B. BET surface area is also shown in figure 8. Both of the particle size distribution are almost the same, but BET surface area of the type-A is twice as much as that of type-B.

Table 4. Production conditions and occurrence rate of cavity.

Production condition	Occurrence frequency of cavity
Al-powder type	Type-A ⋘ Type-B
W/S*	Low W/S ≪ High W/S
Quantity of bubble stabilizer	Less < Much

* Total Water/Total Solid ratio.

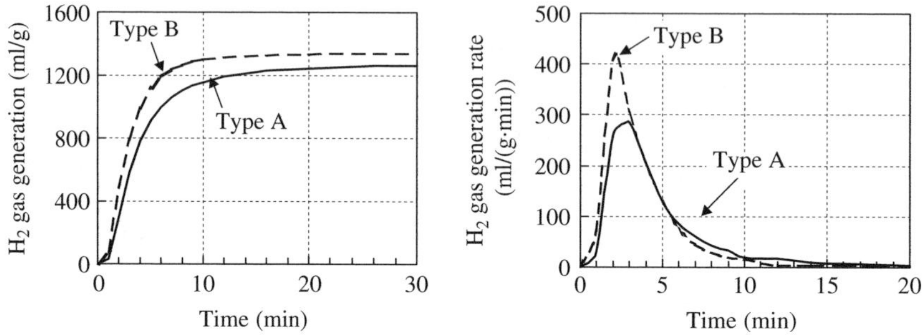

(a) H_2 gas generation (b) H_2 gas generation rate

Figure 7. H_2 gas generation of type-A and B. Solution: 1%-$Ca(OH)_2$, Al: 60.0 mg, Temp: 20°C.

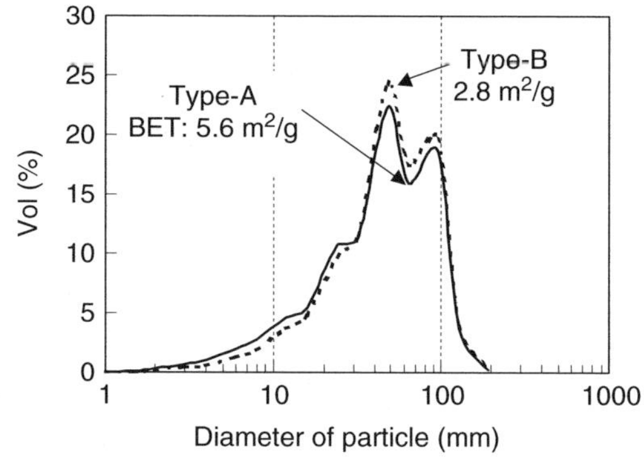

Figure 8. Particle size distribution of Al-powder. Equipment: SEISHIN SK Laser Micron Sizer Pro-7000, Solvent: Water, Pretreatment: Ultrasonic for 60 sec.

Figure 9 shows the SEM micrographs of Al-powders and shapes of air pore in AAC. Al-powder of the type-A seems to cohere, or wrinkle, while the type-B shows the comparatively smooth surface. The difference between them probably corresponds to BET surface area. The air pores of AAC using type-A shows more interconnections than that of the type-B.

3.3 *Investigation of foaming process considering production condition affecting cavity*

3.3.1 *Changes in foaming height*

We investigated the foaming process in terms of the production condition such as Al-powder types and W/S, relevant to the cavity generation as shown in table 4. The foaming height of the real mold was measured by the laser displacement sensor as mentioned section 2.1. Simultaneously, the temperature changes of the AAC slurry were measured at the middle of the real mold.

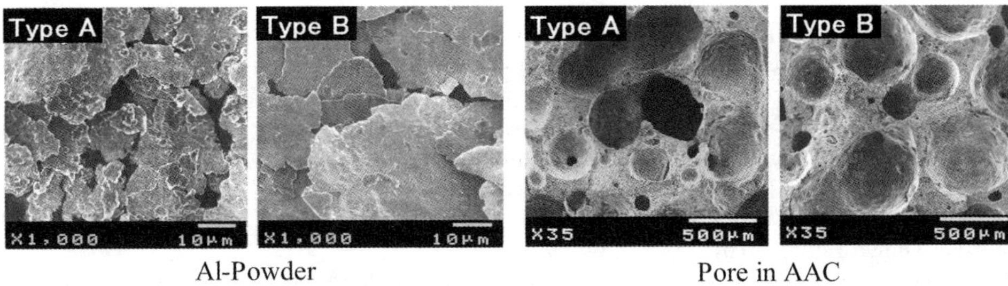

Figure 9. SEM micrographs of Al-powder and air pore in AAC.

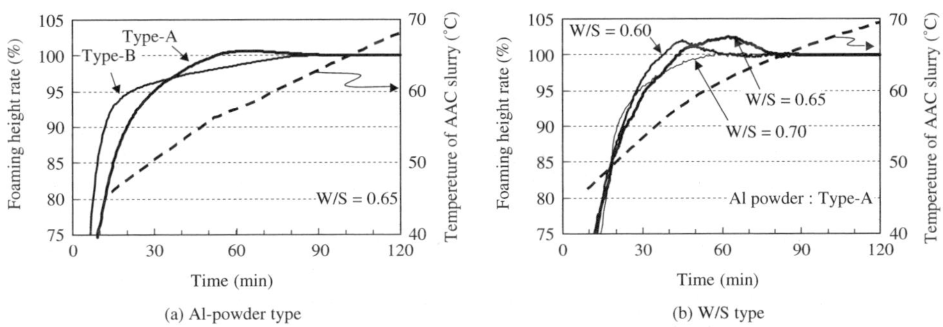

Figure 10. Change of foaming height rate of the real mold. (a): Al-powder types differ. (b): W/S differ.

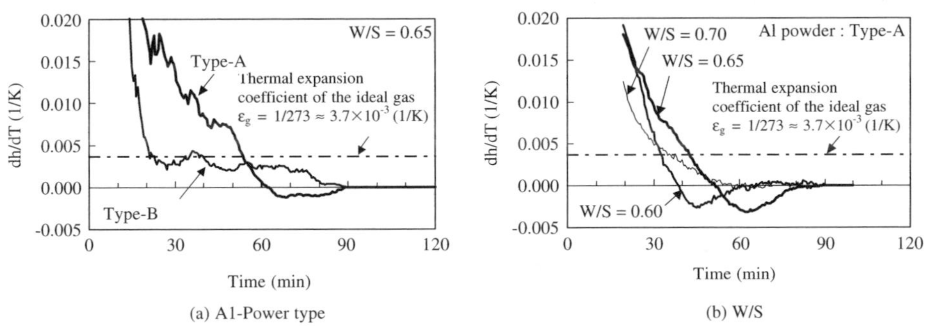

Figure 11. Change of dh/dT. (a): Al-powders differ. (b): W/S differ. Type-B showed the change that was equal to the thermal expansion coefficient of the ideal gas: $\varepsilon_g = 1/273 \approx 3.7 \times 10^{-3}$ (1/K).

Figure 7 shows in change the foaming height, of slurries with different Al-powder and W/S. The foaming height of type-B changed quicker by 30 min than that of type-A, then they were reversed afterwards, type-A became higher than type-B. The type-A reached the maximum value at 60 min, then it contracted 1% at 90 min. The type-B increased gradually with time until 90 min, and did not show the contraction.

The time of the maximum foaming height was shortened, as W/S became lower. The W/S = 0.70 did not show the contraction as well as the case of the type-B, these tendency correspond to the cavity occurrence rate.

3.3.2 *Derivative of foaming height with respect to temperature*

The result of the defective simulation test indicates that the cavity occurred at the end of the foaming process. In addition, there were the notable tendency concerning the cavity occurrence rate in figure 10. Then, we carried out the calculation of dh/dT, derivative of foaming height with respect to temperature, in order to analyze the behavior of AAC slurry at the end of foaming process. Figure 11 shows the change in dh/dT with time. Because AAC slurry was in a rigid mold, the volume expansion coefficient is equal to the change of the foaming height rate. The thermal expansion coefficient of the ideal gas (ε_g) is a reciprocal of absolute temperature; $\varepsilon_g = 1/273 \approx 3.7 \times 10^{-3}$ (1/K). We have found out that the dh/dT of the type-B was almost equal to the ε_g within 20 to 75 min. The dh/dT of the type-A has the negative minimum indicating the contraction.

The zero point of dh/dT probably means the setting of AAC slurry. As W/S increases, the negative minimum disappeared, the gradient of the dh/dT decreases, and setting time (reaching dh/dT = 0) delayed as shown figure 11(b).

When the dh/dT approaches to the ε_g shows that the bubble in the AAC slurry is being put in the closed system. While, the contraction behavior of the type-A (at W/S = 0.65, 0.60) shows that the bubbles in the AAC slurry are interconnected.

3.3.3 *Internal pressure of AAC slurry in real mold*

Verifying the above consideration, we measured changes in the pressure at the center of AAC slurry in the real mold, using a soil pressure transducer. Figure 12 shows the results of the pressure change in the AAC slurry for type-A and B. The pressure of both types of Al-powder were almost equal until 60 min. Then, the pressure of the type-A once lowered at 60 min, and slightly rose by 120 min. The lowering pressure point at 60 min of the type-A in figure 12 is probably corresponding to the contraction of the AAC slurry after 60 min in figure 10 or 11. This results support that the bubble of the AAC slurry with powder type-A was in the condition of the open pore system. The pressure of the AAC slurry with type-B powder increased steeply from 60 min to 120 min at least. It should be noticed that the pressure has been rising still steeply, even if the expansion of the slurry had stopped at 90 min, that will be discussed later.

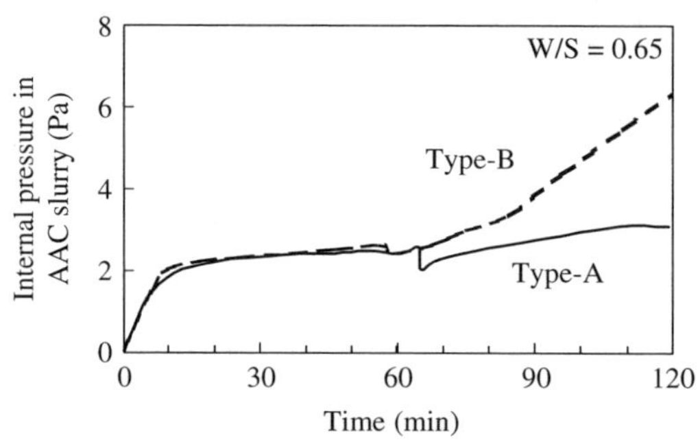

Figure 12. Change in internal pressure of AAC slurry in real mold for type-A and B.

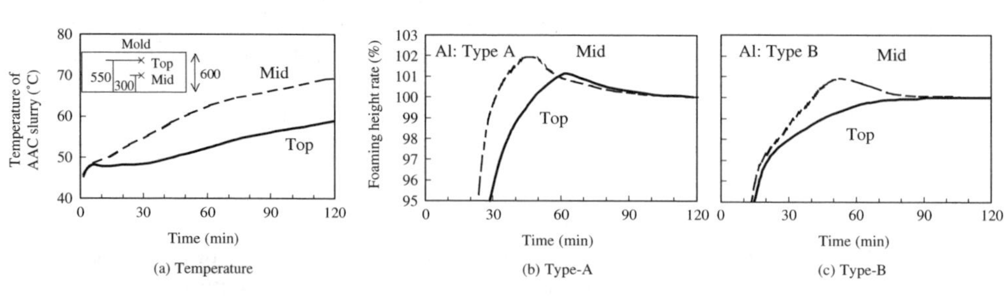

Figure 13. Temperature simulation test. (a): Temperature of Top and Mid measured in real mold for simulation. (b): Foaming height rate of type-A. (c): Foaming height rate of type-B.

3.3.4 *Laboratory test with type-A and B powders*

To clarify the differences of foaming behavior between powder type-A and B, we performed the laboratory scale test. Generally, there is temperature distribution in AAC slurry in the real mold, the temperature of upper side exposed to the outside air is likely to be lower than that of the middle part. The reaction of Al-powder is strongly affected by temperature, thus we performed the laboratory test, taking into account to temperature distributions that simulates the temperature changes in the upper side (Top: 550 mm from the mold bottom) and the middle part (Mid: 300 mm from the mold bottom). The test procedure and composition of the AAC followed the defective simulation test as described Section 2.1.

Figure 13 shows the results of temperature measurement. Temperature of the Top is lower than that of the Mid approx 10°C after 30 min. For the type-A, both of the Top and the Mid showed the contraction, while for the powder type-B, the Mid showed the contraction, but the Top did not.

These results probably corresponds to the result of figure 10 and 11 that the powder type-A showed the contraction, while the powder type-B did not. In addition, it also supports to the consideration that the foaming bubbles are in the open system for the slurry with type-A, but in the closed system for that with the type-B, especially at the Top side.

4 COUNTERMEASURES TO CAVITY

We conclude that the cavity is a result of accumulation of the foaming enable to escape to the outside of the AAC slurry. In order to avoid the cavity occurrence, the internal pressure by foaming gas should be released from the AAC slurry just before the end of foaming process. The type-A of Al-powder is more effective to avoid the cavity than type-B. Hence the most effective countermeasure is to adopt the optimum Al-powder. To select the optimum Al-powder, the characteristics of Al-powders is summarized in Table 5.

The decrease in W/S is also effective, since it could shorten the setting time of the AAC slurry of the Top side.

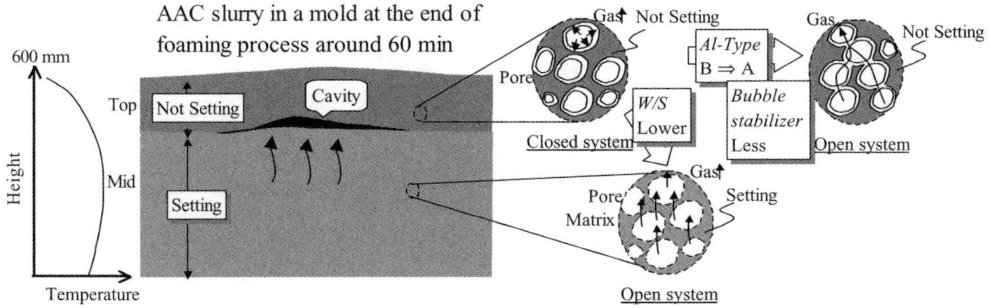

Figure 14. The conceptual scheme of cavity occurrence mechanism at the timing of the end of foaming process around 60 min.

Table 5. Optimum Al-powder as a countermeasure of cavity defective.

Character	Optimum	Undesirable
H_2 gas generation rate	Comparatively slow	Comparatively rapid
BET surface area	Large	Small
Shape of surface	Wrinkle	Smooth

5 CONCLUSIONS

(1) SEM observation of the surface of the cavity showed the characteristic shape that the interface between matrix and air pore was indistinct and round in shape. The timing of the occurrence of the cavity was at the end of foaming process.
(2) The cavity occurrence frequency is strongly affected by Al-powder type, and the type-A powders is better than type-B in preventing the cavity. The characteristics of the optimum Al-powder are slower H_2 gas generation rate, larger BET surface area, and surface of the particle is like wrinkle not smooth.
(3) We found out that the AAC slurry with type-B powder showed a volume change equal to the thermal expansion coefficient of the ideal gas. In addition, the internal pressure of the AAC slurry rapidly increased when using type-B powder.
(4) An AAC slurry with type-A powder showed the contraction of the AAC slurry regardless the temperatures distribution (the Mid part and the Top side), but the type-B did not show the contraction at a lower temperature that corresponds to that of the Top.
(5) We proposed the mechanism of the cavity occurrence based on the above investigation. The setting time of the Top is retarded compared to that of the Mid. The AAC slurry of the Mid had set in advance, and air pores are interconnected, allowing foaming gas to escape through the AAC matrix. The AAC slurry of the Top had not set yet, and air pores isolated. Foaming gas passing through the AAC matrix is blocked and accumulated beneath the Top, where the cavity is generated. In order to avoid the cavity generation, the internal pressure by foaming gas should be released from the AAC slurry just before the end of foaming process.
(6) Using the type-A powder air pores are likely to interconnect allowing foaming gas to escape through the AAC slurry, nevertheless which had not setting yet. The lowering of W/S can shorten the setting time of the AAC slurry of the Top side, and allows the foaming gas escape through the AAC matrix. These are the effective countermeasures against cavity.
(7) We also proposed the estimation of the occurrence timing of the other defective like cracks by SEM observation.

Autoclaved Aerated Concrete – Limbachiya and Roberts (eds)
© 2005 Taylor & Francis Group, London, ISBN 0 415 38356 0

The use of pulverised fuel ash for the manufacture of Autoclaved Aerated Concrete

R.A. Carroll & J.E. Guest
Hanson Thermalite Ltd., Birmingham, UK

ABSTRACT: Pulverised fuel ash is a major silica source used for the manufacture of autoclaved aerated concrete. Within the United Kingdom, it is the main raw material used to produce approximately two million cubic metres of building blocks each year. Technical and environmental benefits of using pulverised fuel ash for the manufacture of autoclaved aerated concrete are discussed. The characteristics of pulverised fuel ashes have changed over recent years, as a consequence of the economic and environmental factors operating at coal-fired power stations. These changes have so far been accommodated within the manufacturing process, without adversely affecting the properties and performance of autoclaved aerated concrete. Further changes can be expected, therefore more efficient utilisation of pulverised fuel ashes and the adoption of novel processing methods must be evaluated. Methods of processing pulverised fuel ash have been developed within the last decade, which have the potential to enhance performance and reduce variability.

1 INTRODUCTION

Formation of an autoclaved aerated concrete (AAC) matrix is achieved by reacting together finely divided calcareous and siliceous raw materials in saturated steam at temperatures above 100°C (hydrothermal conditions). The calcareous raw material is normally quicklime (calcium oxide), or a combination of quicklime and Portland cement. The siliceous component is often finely divided quartz, obtained from sand or sandstone. Alternatively, a raw material containing amorphous silica, or aluminosilicate glass, may be used. Pulverised fuel ash (PFA), the by-product of coal burning power stations, is a suitable siliceous raw material and is extensively used within the United Kingdom for the manufacture of AAC.

2 HISTORY OF AAC MANUFACTURE

Autoclaved aerated concrete was patented in 1924 by Dr Johan Axel Eriksson of the Royal Institute, Stockholm, Sweden. All the aspects of current manufacturing were covered in the invention. The patent described the preparation of a water-based slurry, containing calcareous and siliceous raw material, formation of a cellular structure by hydrogen gas generated from aluminium powder, and hydrothermal curing using high-pressure steam. Interestingly, Eriksson used slate ash, an aluminosilicate, as the silica source.

Manufacture of AAC began in 1929 at Yxhult, Sweden, using the trade name Ytong. Early production units appear to have used shale ash, PFA or blast furnace slag as the silica source. Expansion of production took place in continental Europe after the Second World War, particularly in Germany where conventional building materials were scarce. Generally, the siliceous raw material used was finely ground quartz sand.

Interest in AAC within the United Kingdom can be traced to a patent filed by John Laing and Son in 1948 (Starnes 1951). The document noted that PFA as a fine powder did not need grinding,

Table 1. Utilisation of PFA in the United Kingdom in 2003.

PFA use	Tonnes	Percent
AAC blocks	843,730	12.12
Concrete addition (BS3892 and EN450)	805,727	11.58
Fill and ground remediation	800,858	11.51
Grouting	307,019	4.41
Cement raw material	286,078	4.11
Non-aerated concrete blocks	238,855	3.43
Miscellaneous	204,699	2.94
Blended cement (EN197-1 and BS6610)	97,138	1.40

unlike quartz sand. The trade name Thermalite was registered in 1948 and a factory was built in Reading in 1951. This was the first AAC manufacturing plant in the United Kingdom and used PFA as its primary raw material. In 1957/1958 a larger factory was built next to the Hams Hall Power Station near Birmingham. The location was selected to take advantage of PFA readily available from the nearby coal-fired boiler units.

Celcon Limited was formed in England in 1949 and began developments in the early 1950's. A factory was built at Grays, Essex in 1958. The market for AAC within the United Kingdom expanded throughout the 1960's and 1970's. Factories were generally located either on the site of coal-fired power station or within close proximity. AAC factories using quartz sand were established in the UK, notably Durox and Tarmac in the 1960's and Quinnlite in 1995. The latest factory commissioned, owned by Thomas Armstrong Ltd, was constructed at Catterick in 2003.

Table 1 lists the most recent data on the use of PFA in the United Kingdom (Sear 2004). It is striking that AAC manufacture was the main use of PFA derived from British power stations, accounting for 12% of the total. Figures for the whole of Europe also list utilisation at 843,000 tonnes per annum, showing that AAC manufacture using PFA is largely confined to the United Kingdom (ECOBA 2004).

3 MANUFACTURING PROCESS

Generally, run of station ash is used for the manufacture of AAC. The fineness specifications of BS3892 Part 1 or EN 450 are less relevant for autoclaved products than for concrete cured at ambient temperature. It is possible to use relatively coarse PFA because the elevated temperatures and high alkalinity during autoclaving ensure rapid dissolution of the aluminosilicate particles.

The variability of PFA must be considered during AAC manufacture. In contrast, fine quartz sand can be derived from a high-purity source and ground to a particular fineness specification, thereby reducing many process irregularities.

Maintaining the correct water content of the AAC mix is important to control the early stages of the manufacturing process. If the water/solids ratio is too low the mix will be very viscous at casting. Expansion may be impeded and the mix will not reach the correct final height. Conversely, if the water/solids content of the slurry is too high, the cast AAC mix will have a low viscosity and the stability may be reduced. If the rapidly expanding mix has low viscosity and is not sufficiently cohesive, the hydrogen bubbles will coalesce to form large cells. This can be so extensive that the cellular structure breaks down completely. Mix stability is critical for the successful manufacture of AAC and is affected by raw material selection and the formulation used.

The carbon content of PFA, normally assessed by loss on ignition (LOI), has a significant effect on the casting and setting characteristics of AAC. High-carbon ashes, because of the large surface area and porosity of the carbonaceous particles, can have excessive water demands and require

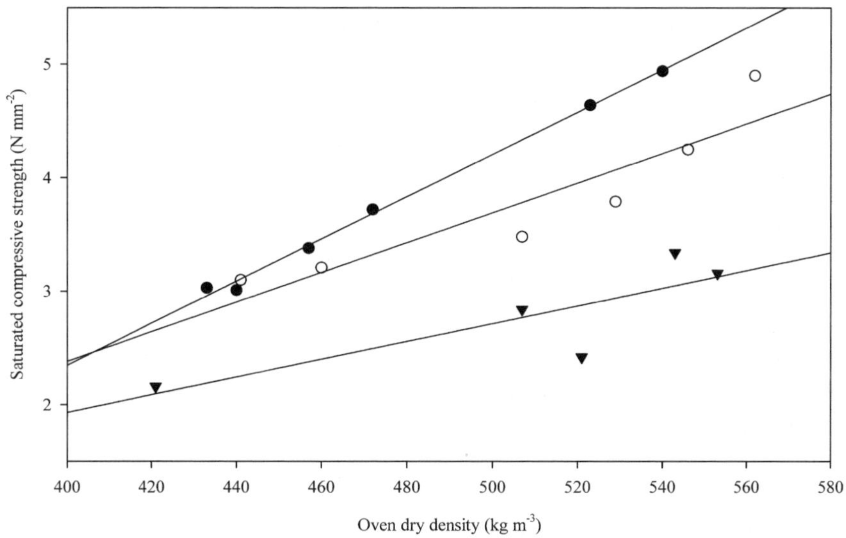

Figure 1. Strength/density relationship for AAC cubes made with three different US fly ashes produced in 1990 (Amos JE •, Fort Martin Unit 1○, Fort Martin Unit 2 ▼).

mixes with high water/solids contents. Instability within mixes in the early stages of manufacture can arise and collapse can occur. Wide variations in the carbon content of PFA can give extensive process difficulties.

The green strength developed within AAC must be sufficient to enable demoulding, cutting and loading of material into the autoclaves. Development of green strength is influenced by the water content of the mix and high-carbon ash can give a high water/solids ratios and prolong setting.

4 PERFORMANCE OF PFA

Most silica is present in PFA as spherical particles of aluminosilicate glass. There will be considerable heterogeneity between particles, but bulk elemental analysis is likely to be representative of the glassy material present. Small amounts of quartz, mullite and haematite will also be present, along with alkali salts. Rapid reaction of these glassy particles will occur in the early stages of autoclaving, liberating silicate ions, which react to form calcium silicate hydrates (CSH). Prolonged autoclaving transforms these into more crystalline CSH phases, such as tobermorite, which binds the AAC matrix together, thereby providing compressive strength.

The effect on compressive strength of AAC made with PFA from different sources is considered in Figure 1. Strength versus density plots were obtained for small-scale AAC cakes made in the Autoclave Laboratory of Hanson Thermalite, Hams Hall. An identical formulation and autoclaving cycle was used for each set. Elemental and mineralogical analyses of each ash are given in Table 2.

Amos JE ash gives AAC with the highest strength. The lowest strength is achieved with PFA from Unit 2 of Fort Martin, whereas ash from Unit 1 gives intermediate strength. Clearly the PFA sample with the highest silica content gives AAC with the greatest strength. The two Fort Martin ashes have similar silica and alumina contents, yet give specimens of different strength. The amount of iron within the glassy phase could affect reactivity during autoclaving. The results demonstrate that the performance of PFA sources should be evaluated in test mixes and cannot be predicted from their chemical composition.

Table 2. Analytical data for three US fly ashes produced in 1990.

Component %	Amos JE	Fort Martin Unit 1	Fort Martin Unit 2
SiO_2	55.1	49.6	51.1
Al_2O_3	32.9	30.2	29.7
CaO	0.90	2.07	2.33
Fe_2O_3	3.60	8.95	10.2
MgO	1.15	1.72	1.43
K_2O	2.63	1.99	2.08
Na_2O	0.28	0.46	0.54
TiO_2	1.79	1.30	1.35
SO_3	0.22	0.49	0.77
LOI	0.99	2.95	1.15
Quartz	5.4	4.0	3.2
Mullite	14.2	12.6	9.2
Magnetite	0.2	1.6	1.2
Haematite	0.4	0.8	0.8
Glass	78.8	78.2	84.1
D_{50} (μm)	20.4	26.5	22.5

5 BENEFITS

5.1 *Environmental*

The 'Ecopoints' scoring system is a method developed by the Building Research Establishment (BRE) to evaluate environmental Life Cycle Assessment (LCA) data for construction products. The Ecopoint score represents a measure of the overall impact of a particular product or process, taking into account factors such as energy use, mineral extraction, waste disposal and transport. One hundred Ecopoints are equal to the impact of one UK citizen per year. The higher the score, the greater the environmental impact.

Replacing the PFA in Thermalite with ground sand would cause the Ecopoints score per tonne to rise from 1.9 to 2.4. This represents a considerable 24% increase in environmental impact, due to the burden of primary aggregate extraction and grinding.

5.2 *Thermal conductivity*

Autoclaved aerated concrete produced with PFA as the silica source has a lower thermal conductivity than material of the same density made with ground sand. This is shown in Figure 2, in which tabulated values from EN1745 for AAC made with ground sand are compared with measured results for specimens made with PFA.

Within AAC made with ground sand, significant amounts of unreacted quartz remain after autoclaving. In material made from PFA, any unreacted particles are likely to be amorphous aluminosilicate glass. The rate of heat flow through the matrix containing crystalline quartz is higher than through the amorphous matrix, and therefore the thermal conductivity is increased.

6 CHALLENGES

During the 1980's and 1990's measures to reduce the main atmospheric pollutants associated with acid rain were introduced in the United Kingdom. Limits were introduced on the amount of sulfur and nitrogen oxides (SO_x and NO_x) emissions from coal-fired power stations.

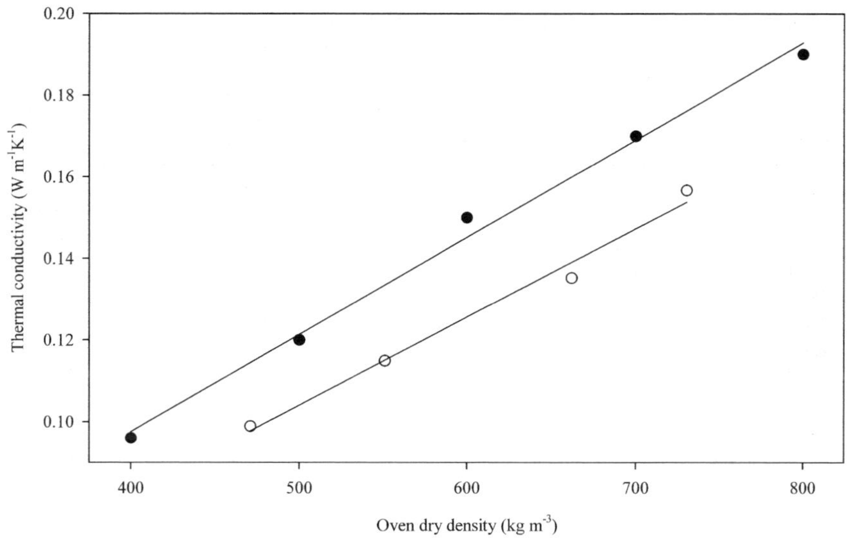

Figure 2. Thermal conductivity/density relationship for AAC (Tabulated mean values from EN1745 based on ground sand •, measured values for AAC produced with PFA ○).

Sulfur capture techniques were introduced at selected stations. Since the flue gases were treated, which did not interfere with combustion, no influence on PFA quality was observed. In contrast, the main measure to reduce nitrogenous pollutants, the installation of low NO_x burners, had a significant effect on the composition of the PFA produced. Low NO_x burners were designed to reduce the flame temperature in pulverised fuel boilers, thereby reducing the oxidation of atmospheric nitrogen. Reductions in combustion efficiencies occurred, with increased amounts of unburnt carbon in the PFA produced.

Changes in the coal sources used by British power stations also took place during this period. The amount of indigenous coal burnt reduced and large quantities of coal were imported to compensate.

Recent changes in the compositions of United Kingdom PFA are reviewed in Tables 3, 4 and 5. Chemical analyses of ashes from three major power stations are summarised for the decade 1978 to 1987 (Central Electricity Generating Board 1988) and compared with results for 2003 and 2004. In all cases there was an increase in the LOI value. The increases in LOI for Drax PFA and Ratcliffe PFA were relatively modest, but a large increase in the amount of unburnt carbon was evident in the ash obtained from Didcot Power Station. Reductions in the silica and alumina contents also occurred, probably associated with the increases in carbon. Other changes in the elemental composition were evident, such as iron, calcium and alkalis, but no consistent trend was observed and may be explained by variations in the coal burnt.

The reduction in silica content of each of the PFA source over the fifteen years was relatively modest. Minor adjustments to the formulation of AAC mixes would have maintained the CaO/SiO_2 ratio. Increases in the carbon contents have been more troublesome, since the water content of mixes have risen, affecting the early stages of manufacturing, from casting to cutting. However, these effects have been accommodated within the existing manufacturing process.

Future environmental changes have the potential to adversely affect PFA quality. The Large Combustion Plant Directive (LCPD), the National Emissions Ceiling Directive (NECD) and the Pollution Prevention and Control Regulations (PPC) are the key legislations likely to regulate United Kingdom coal-fired power stations for the next decade and beyond. The LCPD and NECD limit the amount of SO_2, NO_x, volatile organic compounds (VOCs) and ammonia released to the atmosphere. Further desulfurisation plant is likely to be installed and a new generation of low NO_x burners. Ammonia contamination of the ash could occur if systems such as selective catalytic reduction (SCR) are introduced to further reduce NO_x levels.

Table 3. Comparison of chemical analyses of Didcot PFA 1978–1987 and 2003–2004.

	Mean %		Range %	
Component	1978–1987 (n = 50)	2003–2004 (n = 8)	1978–1987 (n = 50)	2003–2004 (n = 8)
SiO_2	48.6	47.2	41.9–50.9	41.8–52.7
Al_2O_3	28.1	24.3	25.4–31.1	17.9–27.5
Fe_2O_3	10.9	5.0	7.2–15.0	2.7–10.1
CaO	4.6	4.4	3.3–8.9	2.2–7.7
MgO	1.6	1.9	1.0–2.2	1.6–2.3
TiO_2	1.1	1.3	0.8–1.3	1.1–1.9
K_2O	3.0	1.1	2.5–3.6	0.5–2.4
Na_2O	1.3	0.7	0.8–1.8	0.1–1.6
SO_3	0.8	0.7	0.6–1.6	0.5–1.1
LOI	5.0	12.4	2.6–6.7	8.5–19.0

Table 4. Comparison of chemical analyses of Drax PFA 1978–1987 and 2003–2004.

	Mean %		Range %	
Component	1978–1987 (n = 50)	2003–2004 (n = 16)	1978–1987 (n = 50)	2003–2004 (n = 16)
SiO_2	52.6	50.5	51.3–53.8	48.1–52.5
Al_2O_3	27.3	26.2	26.9–28.0	24.8–28.1
Fe_2O_3	10.8	8.9	9.0–11.5	6.2–11.3
CaO	1.5	2.3	1.2–1.6	1.7–6.0
MgO	1.6	2.2	1.5–1.9	1.7–2.9
TiO_2	1.0	1.0	1.0–1.1	0.9–1.4
K_2O	3.8	3.3	3.5–4.1	1.1–3.8
Na_2O	1.2	1.2	0.8–1.4	0.1–2.9
SO_3	0.6	0.7	0.2–1.0	0.4–1.4
LOI	2.8	3.3	1.0–5.5	2.1–5.4

Table 5. Comparison of chemical analyses of Ratcliffe PFA 1978–1987 and 2003–2004.

	Mean %		Range %	
Component	1978–1987 (n = 50)	2003–2004 (n = 11)	1978–1987 (n = 50)	2003–2004 (n = 11)
SiO_2	50.4	46.1	46.2–53.5	43.7–48.4
Al_2O_3	28.2	25.0	26.0–30.1	23.3–26.7
Fe_2O_3	8.9	11.7	7.1–9.5	10.1–13.1
CaO	2.4	3.3	1.7–3.3	2.7–5.0
MgO	1.6	2.3	1.4–1.9	1.8–3.2
TiO_2	1.1	1.0	0.9–1.2	0.9–1.1
K_2O	3.7	2.4	3.4–4.3	2.0–2.8
Na_2O	1.6	0.7	1.4–2.1	0.2–2.2
SO_3	0.6	1.0	0.4–0.9	0.6–1.3
LOI	4.5	6.0	2.9–8.3	4.1–7.8

7 STRATEGIES

It should be assumed that greater variability in the properties of United Kingdom run of station ashes can be expected. In particular, carbon contents are likely to increase. Measures to compensate for the greater variability and problematic water demands are considered.

7.1 *Blending*

It is possible to select and blend PFA from different sources. A high-carbon ash may be blended with a low-carbon raw material to achieve a mix with a satisfactory water demand. However, to be effective a considerable investment in silo storage may be necessary. Haulage costs may rise as ash is obtained from power stations more remote from the AAC factory. Chemical and physical characterisation of each ash sources is critical if blending is to be a successful strategy for AAC manufacture.

7.2 *Beneficiation*

Several methods of PFA beneficiation, to remove unburnt carbon, have been investigated. Techniques worthy of note include burn-out using microwaves or conventional ovens, oxidation of carbon using supercritical water, electrostatic separation and wet froth flotation or gravitational separation from slurries. Many attempts at beneficiation have been taken to only experimental or pilot plant scale. Several patents have been filed, but few processes have been commercialised as stand-alone systems.

The RockTron® PFA beneficiation process is a wet system, with plants registered at two United Kingdom power stations and one ash disposal site (RockTron Plc 2004). Progress Material's thermal carbon burn-out method is operational at two US power stations (Kepperler & Kirconnell 2005). However, the most established beneficiation process is the Separation Technologies LLC (ST) electrostatic system. Units have been in commercial operation since 1995 and now a total of twelve have been built in the US, Canada and the UK. The newest was installed this year at Didcot A Power Station, near Oxford.

Inside the separator the feed ash passes between parallel planar electrodes. The particles become triboelectrically charged by inter-particle contact. The positively charged carbon particles and negatively charged mineral particles are attracted to opposite electrodes and swept away on a moving belt (Bittner & Gasiorowski 2005).

Changes in the chemical composition of PFA, brought about by electrostatic beneficiation, are shown in Table 6. The elemental analysis of feed ash, with 14% LOI, is compared with a beneficiated product, in which the LOI has been reduced to 4%. The concentration of most elements increase. This is particularly notable for silicon and aluminium. However, the amounts of iron, calcium, sodium and sulfur decrease, showing that the levels of these elements are enhanced in the high-carbon fraction. The higher silicon, reduced iron and low carbon are likely to be advantageous for AAC manufacture. The water demand of PFA is greatly reduced after beneficiation. Figure 3 shows that a significant reduction in viscosity is achieved for an aqueous slurry of given solids content.

7.3 *Stockpiles and lagoons*

There are large quantities of PFA deposited in stockpiles throughout the United Kingdom. The UKQAA has estimated that 38Mt of raw material may be available (Sear 2004). Experiments and factory experience has demonstrated that ash recovered from these stockpiles is generally unsuitable for AAC manufacture.

Problems arise because of the chemical and physical changes that occur to ash particles exposed to water for periods of months to several years. Weathering processes have a profound effect on the PFA and severely limit its use as a raw material for AAC manufacture. The presence of water and compaction leads to physical agglomeration of ash particles (Figure 4).

Table 6. Analysis of fly ash before and after ST beneficiation.

Component %	Feed material	Beneficiated material
SiO_2	44.36	50.44
Al_2O_3	28.71	33.69
Fe_2O_3	2.47	1.93
CaO	3.91	3.52
MgO	1.93	2.13
TiO_2	1.34	1.44
K_2O	0.47	0.54
Na_2O	0.64	0.49
SO_3	0.89	0.67
MnO	0.03	0.04
Cr_2O_3	0.03	0.02
P_2O_5	0.86	0.85
LOI	14.36	4.26

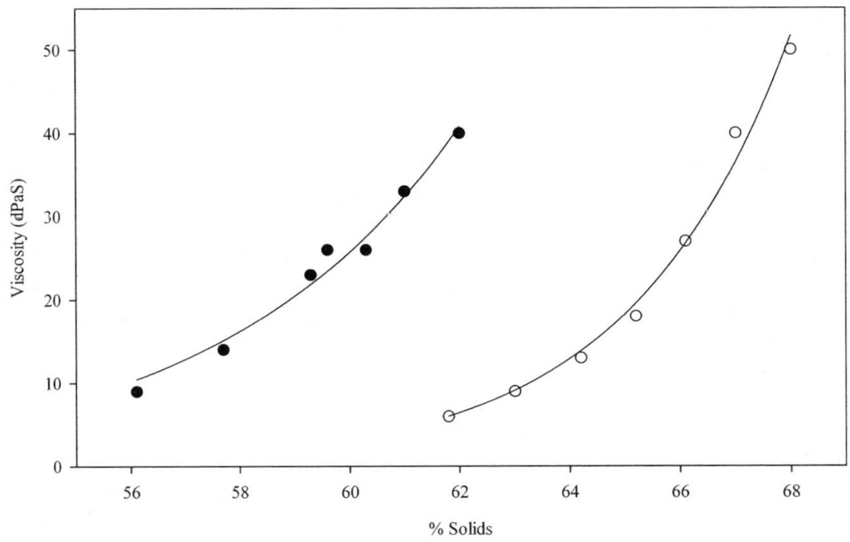

Figure 3. Viscosity/solids content relationship for slurries made with fly ash before and after beneficiation using the ST process (Unprocessed •, beneficiated ○).

Exposure to water and atmospheric CO_2 leads to the formation of phases such as calcite, gypsum and ettringite (Figure 4). Prolonged weathering can lead to some dissolution of the aluminosilicate glass, which constitutes many of the ash particles. An amorphous clay-like material is deposited on the surfaces. The formation of these various reaction products binds discrete ash particles into agglomerates. An impermeable layer forms on the ash particles. This impedes dissolution of silicate ions from the aluminosilicate glass during the autoclaving of AAC. Critically, the formation of calcium silicate hydrates, which bind the intercellular matrix together, depends on the availability of silicate ions. The practical consequence of this reduction in the hydrothermal reactivity of the PFA is the manufacture of autoclaved aerated concrete with low compressive strength.

Agglomeration of the ash gives lumps and "gritty" particles, making handling the raw material troublesome. It is difficult to disperse the PFA effectively and low quality mixes are achieved. An unsatisfactory cellular structure forms during initial setting and this leads to unstable mixes,

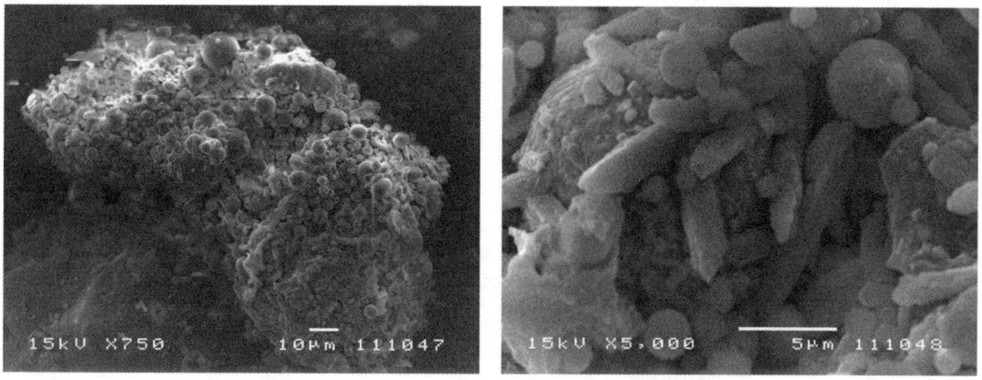

Figure 4. Scanning electron micrographs of stockpiled PFA, showing agglomeration (left) and growth of phases on particle surfaces (right).

which are prone to slumping or collapse. Hanson Thermalite has developed methods of processing stockpiles of weathered PFA to supply raw material for AAC manufacture.

8 CONCLUSION

For over fifty years pulverised fuel ash has been the main silica source used for the manufacture of autoclaved aerated concrete within the United Kingdom. PFA has technical and environmental benefits over the main alternative, ground quartz sand. There are challenges associated with the use of PFA, and these are likely to increase, mainly as the result of environmental legislation at coal-fired power stations. However, strategies such as blending and processing can be used to ensure that PFA continues to be a viable raw material for AAC manufacture in the future.

REFERENCES

ECOBA 2004. Production and utilisation of CCPs in 2003 in Europe. *http://www.ecoba.com/news/document/statistics_ECOBA_2003.pdf, accessed 19 May 2005.*

Environment Agency 2002. Discussion on the feasibility of a trading scheme for NO_x and SO_2 emissions from large combustion plant. *http://www.environment-agency.gov.uk/commondata/acrobat/so2_and-_nox_discussion_paper.pdf, accessed 13 June 2005.*

Bittner, J.D. & Gasiorowski, S.A. 2005. Triboelectrostatic fly ash beneficiation: An update on Separation Technologies international operations. *World of coal ash: science, applications and sustainability, Lexington, USA, 11–15 April 2005.*

Central Electricity Generating Board 1988. Summary of chemical and physical data for CEGB PFAs 1978 to 1987. *Report no. 8/88, CEGB by-products group laboratory.*

Eriksson, J.A., 1924, Förfaringssätt för framställning av gasbetong, *Swedish patent SE60091.*

Keppeler, J.G. & Kirconnell, S.F. 2005. Carbon burn-out: A state of the art in commercial ash beneficiation. *World of coal ash: science, applications and sustainability, Lexington, USA, 13–15 April.*

RockTron Ltd 2004. *www.rocktronplc.com, accessed 31 May 2005.*

Sear, L.K. 2004. *Pulverised fuel ash, cenospheres and furnace bottom ash production and utilisation data for 2003.* United Kingdom Quality Ash Association.

Starnes, P.E. 1951. Improvements in the production of cellular concrete blocks and the like, *British patent GB648,280.*

Autoclaved Aerated Concrete – Limbachiya and Roberts (eds)
© 2005 Taylor & Francis Group, London, ISBN 0 415 38356 0

Newly developed Pre-Hydration process of AAC manufacturing

Y. Aono, S. Shibata & F. Matsushita
Sumitomo Metal Mining Siporex Co., Ltd., Research & Development Center, Japan

ABSTRACT: The Pre-Hydration (PH) process during AAC manufacturing has been developed. In ordinary mixing method of AAC slurry, all raw materials are put into one mixer. In the PH process, quicklime and cement are separately introduced into two PH mixers where they are Pre-Hydrate. Then they mixed together in the casting mixer. The PH process results in the lowering of the viscosity of the AAC slurry and shortening in hardening time of the green cake. This newly developed PH process has already introduced into the production line. As a further potential application of the PH process, high temperature casting to shorten the hardening of AAC green cake was investigated. The lowering of the viscosity was attributed to the cohesiveness and dispersion of cement particle during PH.

1 INTRODUCTION

The control of viscosity of the AAC slurry is one of the important technology in the AAC production process. Increase primary viscosity of the AAC slurry causes the shadow damage, which is vertically disturbed bubbles that arise at the upper part of reinforcing steels (Aono et al. 2005). On the other hand, increase of water for lowering the viscosity causes hardening delay of the AAC green cake. The raw material that most greatly affects the viscosity is cement.

The effectiveness of the early hydration of cement on the viscosity has been studied in the field of the normal concrete by means of cement paste. The viscosity of cement paste is greatly affected by cohesiveness and dispersion of the cement particles in liquid phase. Furthermore, the cohesiveness and the dispersion are affected by the surface potential, namely zeta-potential.

The zeta-potential of the cement particle are affected by the mineral types in cement and the ion types and its quantity in the liquid phase. Intrinsic potential of the surface of C_3S shows minus due to SiO^- (Nägele, 1985). C_3S dispersed in water adsorbs Ca^{2+} on the surface and has the positive potential (Roy et al. 1980). Muhua and Roy (1987) proposed the model of the electric double layer on the surface of cement particle. They showed that Ca^{2+} is the potential-determining ion of cement particles, and the zeta-potential strongly affects to the viscosity of cement paste.

SO_4^{2-} affects the surface potential of C_3A. Tadaros et al. (1976) showed that SO_4^{2-} is adsorbed to Al^{3+} at the surface of C_3A, then zeta-potential changes from the positive to zero.

Nawa et al. (1986) studied the effects of gypsum types, $Ca(OH)_2$ content in mixing water and mixing method on the viscosity. In comparison of the two mixing methods, the single mixing (addition water at one time) and the dividing mixing (addition water dividing into two times), they showed that Ca^{2+} and SO_4^{2-} concentrations and the viscosity differed. Furthermore, they showed that the ettringite formation in the dividing mixing was more abounding than in the single mixing. They concluded that the viscosity was influenced by the zeta-potential and the cohesion due to the cross-link of the hydrates.

Like the above, viscosity and setting of the cement paste change with initial hydration environments, namely, ion type and its concentration in the liquid phase.

On the other hand, in the ordinary production process of AAC, all the raw materials such as cement, quick-lime, silica sand, gypsum and recycled material are put and mixed in one mixer at

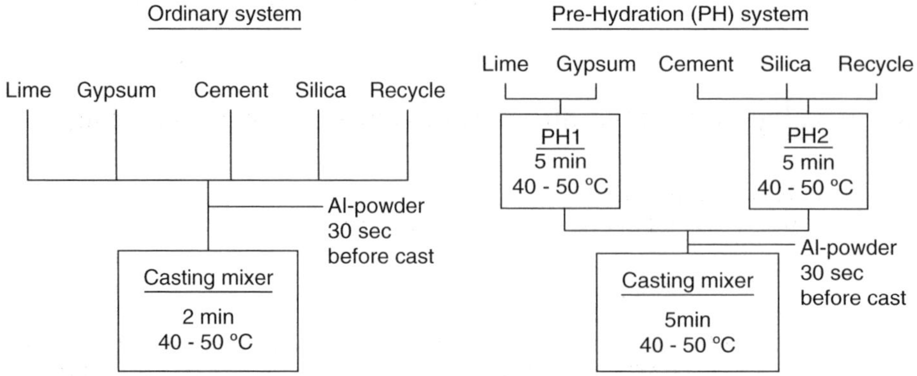

Figure 1. Conceptual scheme of ordinary process and Pre-Hydration process.

almost the same time. That is to say, the cement is in an environment where Ca^{2+} and SO_4^{2-} are coexisting.

Nevertheless, mixing method of AAC slurry has not been investigated until recently. Therefore we have studied the new mixing method of AAC slurry considering the environment of cement in the AAC slurry, and have developed what we call the Pre-Hydration (PH) process. The PH process is a method that the cement and the quick-lime are hydrated separately before mixing in the casting mixer. This newly developed PH process has been introduced into our factory in Japan.

In this paper, we outline the Pre-Hydration process adapted in our factory. For further possibility, we also explain the High-Temperature casting method of PH process.

2 PRE-HYDRATION PROCESS IN THE FACTORY

Figure 1 shows the conceptual scheme of the ordinary mixing process and the PH process. In the ordinary mixing process, all raw materials are put into one mixer, while in the PH process, quick lime and cement are separately mixed, PH1 (Pre-Hydration mixer 1) for quicklime and PH2 (Pre-Hydration mixer 2) for cement. Quick lime is mixed together with gypsum in PH1 for 5 min at 40–50°C. Simultaneously, Cement is mixed together with silica and recycle materials in PH2 for 5 min at 40–50°C. Then, slurries in PH1 and PH2 are mixed in the casting mixer for 5 min at 40–50°C. Al-powder is put into the casting mixer before 30 sec of pouring into the mold.

3 EFFECT OF PH PROCESS ON VISCOSITY AND HARDENING OF AAC SLURRY

The hydration environment of cement and quick lime are different between the ordinary process and the PH process. Therefore, viscosity and hardening of the AAC slurries are also different. In this section, we explain the experimental results in the effect of PH process to the viscosity and the hardness of the AAC slurry.

3.1 *Casting condition of AAC slurry*

Table 1 shows the mixture proportion of raw materials. In the PH process, the mixing water is shared by two mixers, PH1W (the mixing water for PH1 mixer) and PH2W (the mixing water for PH2 mixer), as shown in Table 1.

3.2 *Casting viscosity*

The casting viscosity was measured by the viscometer (RION Co., Ltd, Visco-testor VT-04F, NO.3 rotor and cup).

Table 1. Mixture proportion of raw materials.

Cement (%)	Silica	Limestone	Gypsum	Recycled materials Before-AC*1	After-AC*2	Al-Powder	W/S*2 PH1W*3	PH2W*4
28	37	4	5	12	14	0.065	0.12	0.50

*1: Autoclaving, *2: Total Water/Total Solid ratio = PH1W + PH2W, *3: Water for PH1 mixer/Total Solid ratio, *4: Water for PH2 mixer/Total Solid ratio.

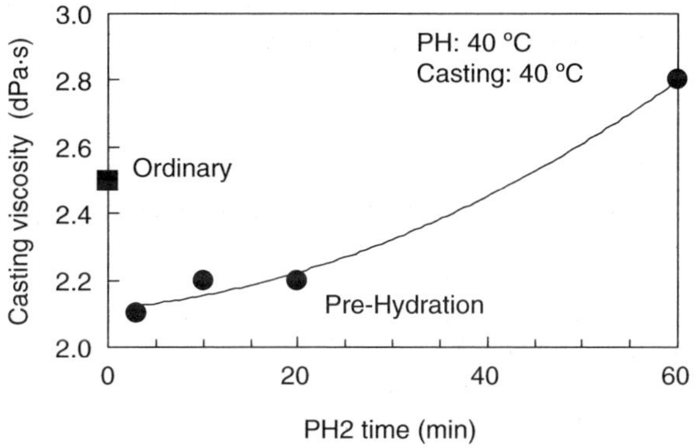

Figure 2. Casting viscosity of AAC slurry by ordinary process and PH process.

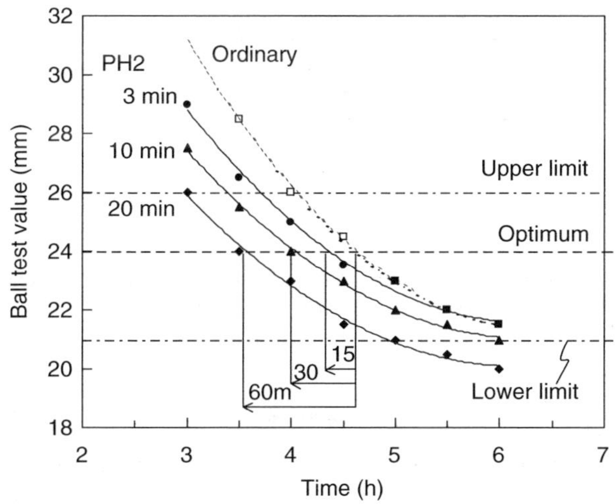

Figure 3. Hardening by the ball-test of the AAC green cake by ordinary process and PH process.

Figure 2 shows the casting viscosity of AAC slurry by the ordinary process and the PH process. The casting viscosities of the PH2 were lower than that of the ordinary process until 20 min. But, subsequently the casting viscosity of the PH2 sharply increased by 60 min. From this result, the optimum PH2 time at 40°C is earlier than 20 min.

3.3 *Hardening of AAC green cake*

Figure 3 shows the change of hardness by the ball test for the AAC green cake. The ball test measures the diameter of the depression when a steel ball (41 mm in diameter and 86 g in weight) is dropped from a height of 50 mm above the AAC green cake (Aono et al. 2005). The range of the hardness (the ball test value) applicable to cutting by the piano wire is from 26 to 21 mm, and the optimum is 24 mm.

The hardening of the AAC green cake increases, as the PH2 time of cement becomes longer. In the PH2 time for 3 min, 10 min and 20 min, the hardening quickened in 15 min, 30 min and 60 min, respectively. Considering the time when cement was in contact with water, the hardening times were corrected with the casting mixer agitation time. These results are summarized in Table 2. The PH2 for cement probably quickens the hardening of the AAC green cake.

Since the lowering viscosity by the PH2 could reduce the mixing water, or increase the casting temperature, PH process may have further potential of acceleration of hardening time. The latter, the high temperature casting with PH process will be described later.

3.4 *Effect of Pre-Hydration of quick lime*

The hydration of quick lime is very rapid without gypsum. In the PH1, quick lime coexists with gypsum, to avoid the rapid hydration of quick lime.

Figure 4 shows the reaction rate of quick lime with time. The reaction rate was measured with the thermogravimetric analyzer (RIGAKU TG-DTA TAS100), by the weight loss associated with the dehydration of $Ca(OH)_2$ around 400°C. The hydration of quick lime with gypsum advances slowly. Therefore control of production process of the AAC became easy and stable by PH1.

This phenomenon is probably explained by so called by the common ion effect where dissolution of the Ca^{2+} ion from the quick lime was prevented by Ca^{2+} ion from the gypsum.

Table 2. Change in hardening time by PH2 for cement.

	Hardening time by 24 mm (h : min)	Difference with the ordinary system (min)	Correction Time*[1] (min)
Ordinary	4:30	–	–
PH2 3 min	4:15	−15	−12
PH2 10 min	4:00	−30	−27
PH2 20 min	3:30	−60	−57

*1: The hydration time of cement is a sum of the PH2 time and the casting mixer agitation time; *(Correction time) = (Difference with ordinary) + (Difference casting mixing time: 5 − 2 = 3 min)*

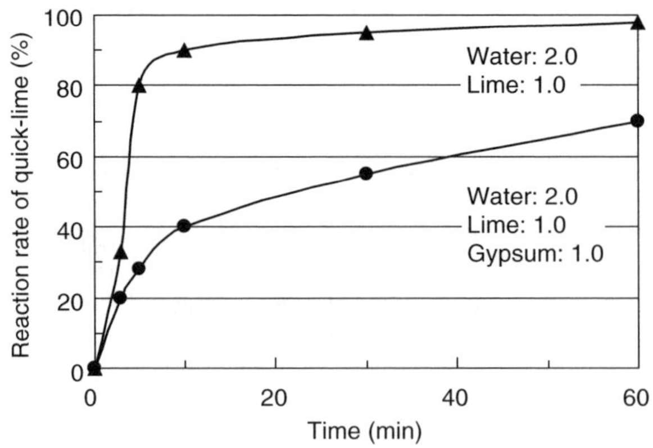

Figure 4. Change of reaction rate of quicklime with and without gypsum.

4 HIGH TEMPERATURE CASTING WITH PRE-HYDRATION PROCESS

4.1 *Concept of the HT-PH process*

In order to further improve the productivity, increase of the casting temperature has been investigated, since the viscosity is low in the PH process. We call this process High Temperature casting PH process (HT-PH). Figure 5 shows the conceptual scheme of HT-PT process, where the casting temperature is raised up to 50–60°C. However, the temperature of PH2 is preferably controlled as low as 20 or 30°C, and the mixing time of PH is depending on the temperature of PH2.

4.2 *Effect of the HT-PH process with and without control of PH2 temperature*

Figure 6 shows the viscosity, the hardening time (by ball test value of 24 mm) and the temperature of PH2 (a) without and (b) with control. The temperature control was carried out by a water-cooling jacket. Proportion of mixtures of the raw materials was the same as shown in Table 1. The time of PH1 for quick lime was fixed to 5 min, while the time of PH2 was 5 to 960 min. The casting temperature of the HT-PT was 60°C, and 45°C as a comparison of the ordinary process.

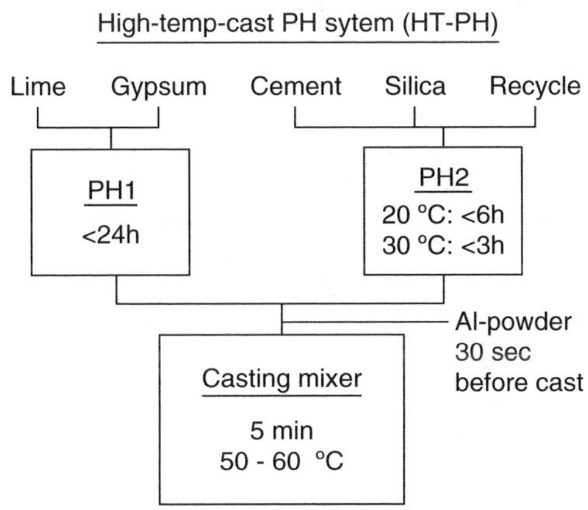

Figure 5. Conceptual scheme of High temperature casting Pre-Hydration process (HT-PT).

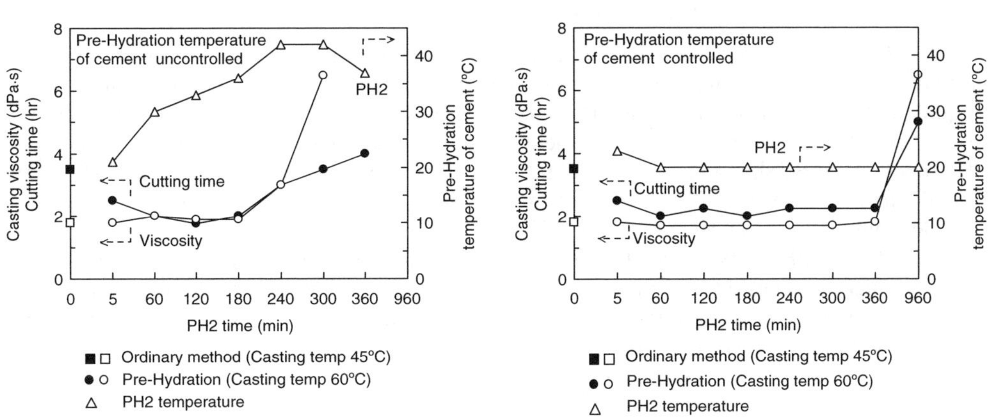

(a) PH2 temperature without control

(b) PH2 temperature with control

Figure 6. Viscosity, PH temp, and cutting time of HT-PH process.

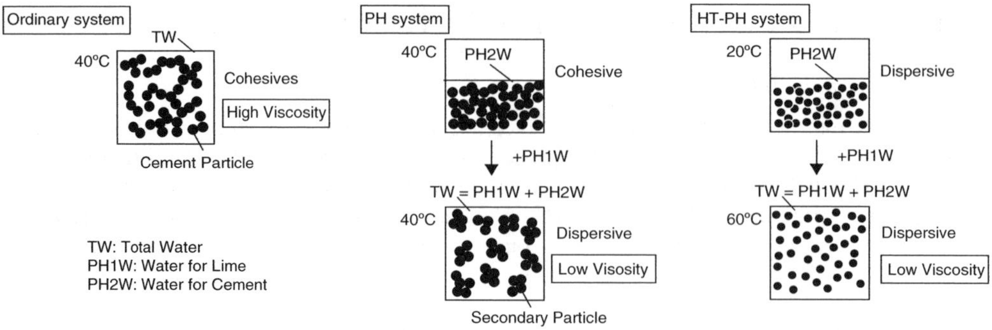

(a) Ordinary process (b) PH process (c) HT-PH process

Figure 7. Conceptual scheme on dispersion state of cement in the AAC slurry.

(1) Temperature of PH2 without control (Figure 6 (a))

The temperature of PH2 increased from 20°C to 45°C with the PH2 time and showed a maximum at 240–300 min due to the exothermic reaction by the hydration of cement. The casting viscosity of the AAC slurry within 5 to 180 min of PH2 time was the same as that of the ordinary process, even though the casting temperature was 60°C. After 240 min, the viscosity increased sharply. In comparison with the ordinary process, the cutting time was shortened by 1.0 to 1.5 h within the PH2 time of 5 to 180 min, but the difference was shortened for 240 min, it became equivalent for 300 min, and it was reversely retarded for 360 min. The rise of viscosity and the delay of hardening after the PH2 time of 240 min caused by the progress of cement hydration during PH2.

(2) Controlling temperature of PH2 (Figure 6 (b))

The temperature of PH2 was controlled at 20°C by water-cooling jacket. The casting viscosity was kept low until the PH2 time of 240 min, but it sharply increased at 960 min. The cutting time was shortened by 1.0 to 1.5 h within the PH2 time of 5 min to 300 min. It retarded at 960 min corresponding to the viscosity rise. In the temperature control, the progress of the hydration is in contrast suppressed according to the uncontrolled condition. In the HT-PH process, therefore, the PH2 temperature should be preferably controlled at 20°C, or the PH2 time should not be extended to the sharp viscosity rise.

5 DISCUSSION

5.1 *Lowering of viscosity in the PH process*

In the PH process, the viscosity lowers, and the hardening time shortens, for which change in cohesiveness and dispersion caused by surface potential may be responsible. However, the presence of quick lime, gypsum and recycling materials in the AAC slurry makes the reaction environment of the cement more complicated than cement paste, and it is not easy to clarify the effect of the PH process. Figure 7 shows the conceptual scheme on the dispersiveness of cement particle for the (a) ordinary process, the (b) PH process and the (c) HT-PH process. In the ordinary process, cement particle would be in a cohesive state. In the PH process, cement particle in the PH2 may be in a cohesive state, until PH1W is added to the PH2W, then cement particles would become dispersive with secondary particle, and the viscosity may decrease. The surface potential of cement particle is not taken into consideration, the reason for the acceleration of hardening of green cake is not explained in Figure 7, therefore, further examination is necessary.

5.2 *Effectiveness of HT-PH process*

The cement in the PH2 may loose its performance depending on the PH condition as shown in Figure 6. It is explained by the exothermic hydration reaction of the cement at different temperatures

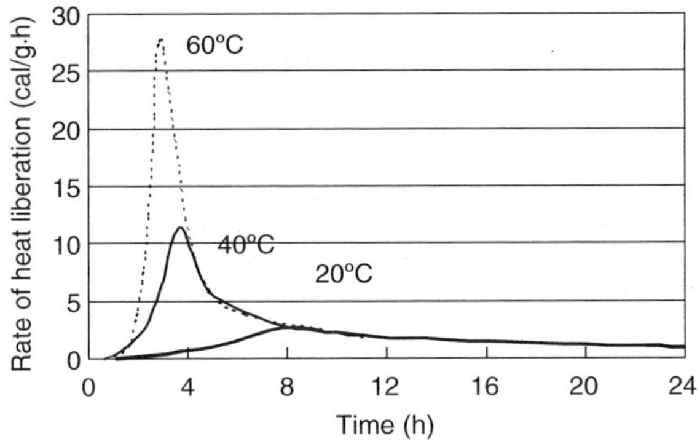

Figure 8. Exothermic hydration reaction of cement at different temperatures (Uchida et al. 1986).

in Figure 8 (Uchida et al. 1986). The rate of heat liberation at 40°C is more rapid than that of 20°C. In the PH2 temperature without controlling, the hydration of cement is accelerated with the temperature rise. By controlling PH2 temperatures, the hydration of cement is kept in a low level and the cement particle would become dispersive as shown in Figure 7 (c).

In the HT-PH process, the AAC slurry cast into a mold at 60°C. The hydration of the cement at this temperature is so rapid as shown in Figure 8 that the hardening of the green cake can be accelerated.

6 CONCLUSION

We have developed the Pre-Hydration (PH) process, in which quicklime and cement are separately hydrated. The PH process results in the lowering of the viscosity of the AAC slurry and attributed shortening the hardening time of the green cake. The lowering of the viscosity can be attributed to the cohesiveness and dispersion of cement particle during PH. However, the acceleration of hardening could not be explained with the dispersion state, and further examination is necessary.

This newly developed PH process has already been introduced into the real factory and works operates successfully.

Further potential of the High-Temperature PH process (HT-PT) has been investigated. HT-PH process allows the high temperature casting at 60°C without increasing viscosity. The hardening is accelerating for 1.5 h compared to the ordinary process at 45°C. In the HT-PT process, the preferable PH2 temperature is 20°C.

The PH process and the HT-PH process probably contribute for further improvement in the productivity of AAC.

REFERENCES

Aono, Y. Shibata, S. and Matsushita, F., 2005. Mechanism and Countermeasures against cavity defectives in AAC during Manufacturing, *4th International Autoclaved Aerated Concrete Conference-Innovation and Development*, (submitted)

Aono, Y. Shibata, S. and Matsushita, F., 2005. Effect of cement characteristics on the setting behavior of autoclaved aerated concrete, *4th International Autoclaved Aerated Concrete Conference-Innovation and Development*, (submitted)

Muhua, T. and Roy, D.M., 1987. An investigation of the effect of organic solvent on the rheological properties and hydration of cement caste, *Cement and Concrete Research*, 17(6); 983.

Nägele, E., 1985. The Zeta-potential of cement, *Cement and Concrete Research*, vol.15, pp 453–462.

Nawa, T. Eguchi, H. and Ooshima, H., 1986. Effect of early cement hydration on rheological behavior of cement, *CAJ proceedings of cement & concrete*, 40, 55–58 (in Japanese)

Roy, D.M. Daimon, M. and Asaga, K., 1980. Effects of admixtures upon electrokinetic phenomena during hydration of C_3S, C_3A and cement, *Proc. 7th Inter, Congress Chemistry of Cement*, Paris, Vol.II, Communications. II-242.

Tadros, M.E. Skalny, J. and Kalyuncu, R.S., 1976. Early hydration of tricalcium silicate, *J. Am. Ceram. Soc*, 59, 344–347.

Tadros, M.E. Jackson, W.Y. and Skalny, J., 1976. Study of the dissolution and electrokinetic behavior of tricalcium aluminate, *Colloid Interface Science* Ed. M.Kervker, Vol.IV, 211–223.

Uchida, K. Sakakibara, H. and Saitou, Y., 1986. Formulation of rate of heat liberation and prediction of temperature rise based on heat of hydration, *Concrete Journal*, 24(4); 105–113.

Autoclaved Aerated Concrete – Limbachiya and Roberts (eds)
© 2005 Taylor & Francis Group, London, ISBN 0 415 38356 0

Optimal autoclaving

M. Haas
Xella Technologie und Forschungsgesellschaft mbH, Brück, Germany

ABSTRACT: Properties of AAC like durability, freeze thaw resistance, shrinkage, equilibrium humidity are mainly determined by the autoclave process. Recent research in this topic showed how to optimize these properties during autoclaving and how to avoid damage to the material during the autoclave process.

Possibilities to reduce the energy consumption are shown too.

1 INTRODUCTION

1.1 *The CaO-SiO$_2$-H$_2$O system*

The simplified system, in which the reactions for the formation of tobermorite in AAC takes place, is formed by the three components CaO, SiO_2 and H_2O. The other components, which have a major influence on the properties of YTONG, are not considered in this system. They will be dealt with further on.

Figure 1 shows the main raw materials and hydrothermal phase formations in the three component system CaO-SiO_2-H_2O as well as the position of YTONG recipes (simplified). All in all, there are 20 ternary phases (containing all three components) in this system, of which, however, only 1.13 nm tobermorite and C-S-H(I) are of significance in production- relevant conditions. The stable phase in AAC mixtures is gyrolite but it would only develop via the intermediate stage xonotlite in extremely long curing times. The exact phase relations and stability ranges in this system are not fully understood yet. This is also due to the fact that there are many phases with a not exactly defined composition (e.g. C-S-H(I)). A lot of fundamental work still has to be done especially in terms of understanding and thus selectively influencing the reactions and the finished products of AAC production.

1.2 *Reactions during autoclaving*

It is only in the autoclave that the morphologically well-shaped 1.13 nm tobermorite, which is important in our process, is formed in the course of the following reactions steps:

1.2.1 *Dissolution of silicon from quartz and of calcium from slaked lime*

The solubility of silicon is determined by time and pH-value. The amount of dissolved Si^{4+} increases with rising ph-value.

1.2.2 *Reaction process of dissolved silicon and calcium with water into CSH-phases*

The formation of the CSH-phases takes place as more and more silicon (SiO_4^{4-}) and calcium (Ca^{2+}) dissolve. At approx. 200°C the solubility of $Ca(OH)_2$ is around 50 ppm, of SiO_2 around 300 ppm, of the CSH-phases, however, only 1–10 ppm. We conclude that the solubility product for the formation of C-S-H(I) and tobermorite is exceeded. This results a continuous precipitation of CSH-phases and a constantly renewed dissolution of $Ca(OH)_2$ and SiO_2, until one of the reactants has been completely used up (or the reaction is prematurely stopped).

Tobermorite has the formula $C_5S_6H_5$ ($5\ CaO \cdot 6\ SiO_2 \cdot 5\ H_2O$), i.e. a ratio CaO/SiO_2 of 0.83. Chemical reactions – especially when the formation of stable phases is not involved, which is the case with tobermorite – are more effective when the initial materials (in dissolved state) are approximately in the same proportion as in the product. This is the case for the range 180°C–190°C.

According to our recipes and curing conditions, 1.13 nm tobermorite does not represent the stable phase in contact with quartz. Under production-relevant conditions the reaction stops in this process step anyway. The further reaction to xonotlite and gyrolite is kinetically inhibited.

1.3 *Properties of steam*

The condensation of steam to water is a so-called "phase transition". As a rule, phase transitions are very intensive processes in terms of energy. This means that relatively small quantities of steam

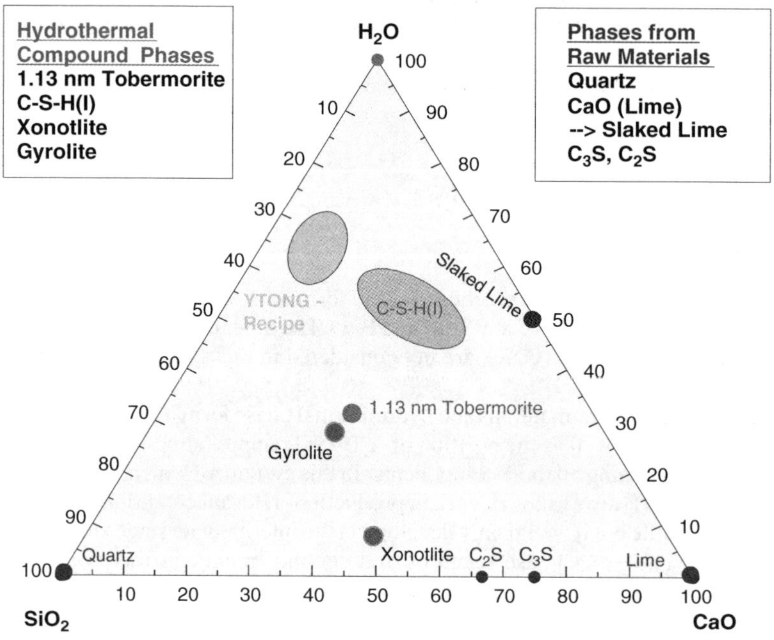

Figure 1. Raw material phases and hydrothermal phase formation in three-component system $CaO\text{-}SiO_2\text{-}H_2O$ as well as position of AAC recipes (from Zürn (1997), re-drawn according to Taylor (1965)).

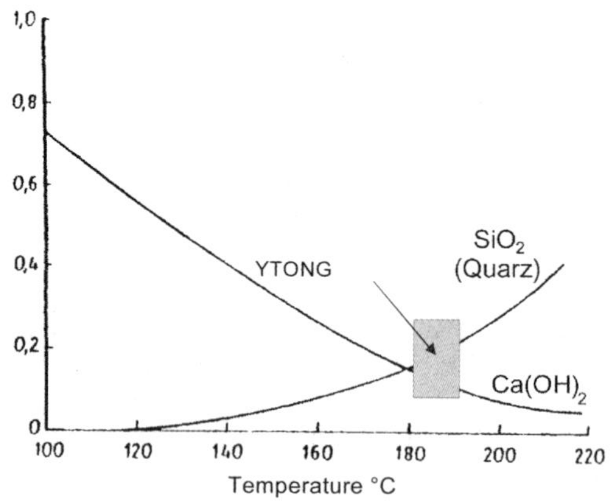

Figure 2. Solubility of $Ca(OH)_2$ and quartz as a function of the temperature (Schlegel (1982)).

are required to heat matters by means of steam, since steam can store a lot of energy. This will be shown in a few simple examples.

One of the essential findings described in this paper is that condensate is absorbed by AAC like by a sponge and – unlike on a steel surface – it will not run down the AAC surface. This means that the moisture of AAC will additionally increase during the pressure increase phase, which can cause different problems in the autoclaving of the material.

There is a clear dependence between pressure, temperature, volume (or density), enthalpy as well as a series of other thermodynamic factors in the phase transition from the liquid to the vaporous state, These interrelations are very well known and are summarized in so-called "steam tables" (Hill 1969).

In autoclaving of AAC, the process of steam condensation plays a most decisive role. Understanding this process, will allow you to understand all "problems" which may arise in connection with autoclaving of AAC.

1.4 *Condensation of steam on a steel plate*

Figure 3 illustrates the processes that take place when hot steam meets a cold steel plate (e.g. the autoclave. As the figures in the graph show, 1 kg of steam (temperature 100°C) is enough to heat up a steel plate of 100 kg by almost 50°C.

It is important to mention here, that in this process gaseous steam turns into liquid water (condenses). What is left is 1 kg of water with a temperature of 100°C and a correspondingly lower enthalpy.

Here, too, it shows how energy-intensive phase transitions are. The energy stored in steam with the evaporation of water (Chapter 2.2) is "abruptly" transferred to the steel plate. The heating of the steel plate and the relevant condensation of steam occur very quickly.

In other words: the condensation of steam has an excellent heat transfer. This heat transfer (heat transfer value $A \approx 10.000\ W/m^2K$) compared with other heat transfer mechanisms such as heat conduction or convection ($A \approx 10$–$100\ W/m^2K$) is better by a factor of 100 up to 1000.

Applied to an autoclave, the example in Figure 3 would mean that steam accumulates as condensate on the floor of the autoclave and from there it must be discharged with the aid of special condensate separators, since the water layer on the autoclave floor would impair the heat transfer between steam and the autoclave wall.

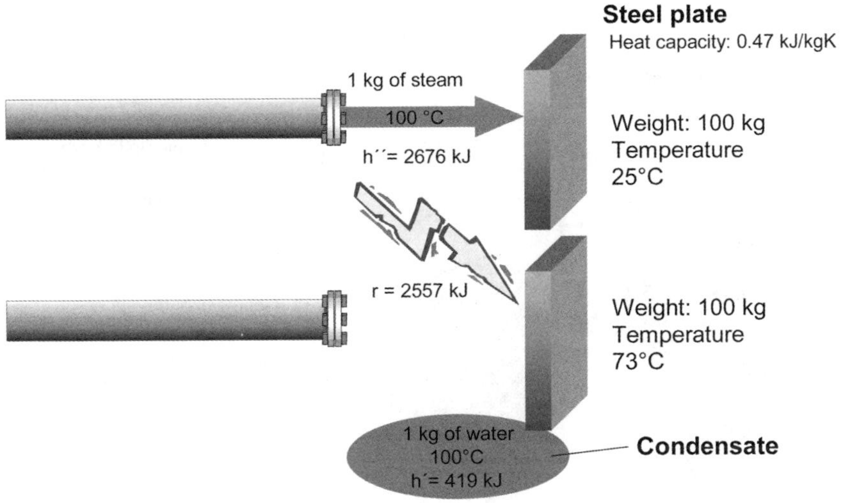

Figure 3. Condensation of steam on steel.

1.5 *Condensation of steam on AAC*

At first the conditions are not different from Chapter 1.4. Steam condenses to water and in this process it transfers its energy to AAC. Since the heat capacity of AAC, and above all the heat capacity of the water in AAC, is higher than that of steel, AAC will not be heated up so much by 1 kg of steam as steel will (approx. 12°C and 48°C).

The condensate will not run down the AAC surfaces, as it would with steel, and will not form a "puddle" on the floor of the autoclave – it will indeed be absorbed by AAC. The water content in AAC increases and its moisture rises. This finding is by no means obvious: it is the result of extensive measurements and calculations in steam and condensate audits. The fact that condensate is absorbed by AAC like by sponge is the cause of a series of problems in the autoclaving process. Here it is important to basically understand the process of condensation of steam on AAC.

1.6 *Shares of the steam quantity required in autoclaving*

Figure 5 shows the shares of the steam quantity required to heat both the steel of the autoclaves and of the curing trucks as well as the solid matter and the water content in the AAC cake during autoclaving. The major part of the steam (almost 3/4 of the total quantity) is required to heat AAC as indicated in Chapter 1.4 and 1.5.

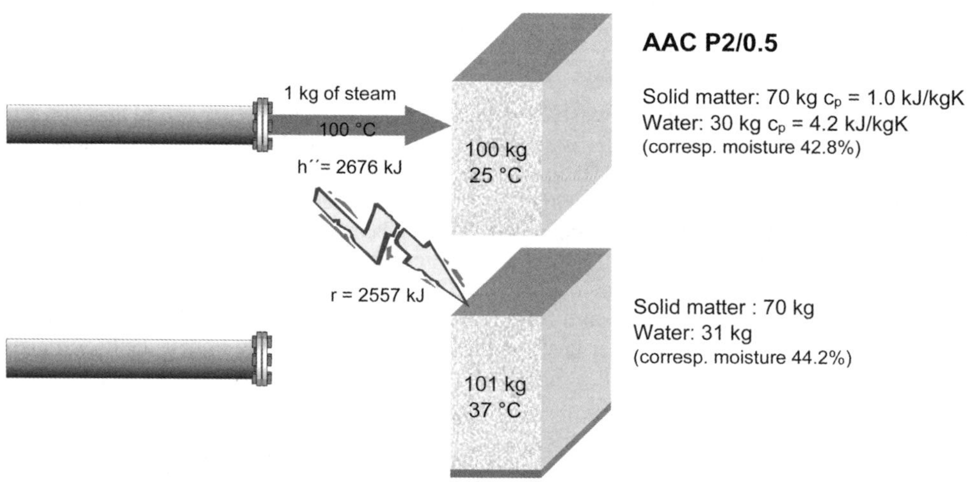

Figure 4. Condensation of steam on AAC.

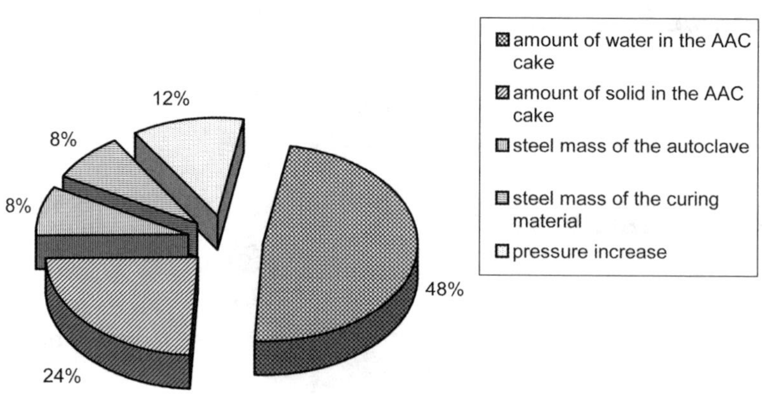

Figure 5. Shares of steel and AAC on steam consumption for autoclaving.

Almost half of the total steam quantity is required to heat the water in the AAC cake. This is not only because uncured AAC contains a lot of water (moisture is up to 50% of the dry mass in some factories), in the first place the cause of high energy consumption is the high heat capacity of water (approx. 4.2 kJ/kgK). Additionally, the condensate absorbed by AAC (Chapter 1.5) must be heated, too.

Therefore we must pay particular attention to the water in our cakes, not only because of the problems in condensate absorption by AAC but also in view of the steam quantities required and the relevant energy costs.

2 PRETREATMENT OF THE MATERIAL BEFORE AUTOCLAVING

2.1 *Introduction*

The pretreatment of AAC covers all the process steps from casting to the pressure increase phase. The objective of an optimal pretreatment is to prevent the material from cooling down before autoclaving and in the flushing and vacuum phases to remove the air inside the cake and the autoclave as much as possible. This creates the optimal conditions for the pressure increase and curing phases.

The main targets of pretreatment are:

- To avoid cooling down
- To reduce energy costs
- To prevent curing damage
- To shorten autoclaving time

We will discuss this interrelationship in general, because meanwhile there are a lot of technological solutions inside the AAC factories to reach these targets.

We recommend the use of a warming chamber in line to the autoclaves and an optimal control of the flushing and vacuum process. Warm cakes with a possibly low water content can be autoclaved in the smoothest, quickest and most cost-effective way. This is one of our fundamental findings.

3 THE OPTIMAL AUTOCLAVE CYCLE

3.1 *Introduction*

In this paper we will deal only with the phases pressure increase, constant pressure and pressure reduction.

An optimal autoclave cycle is characterized by the fact that the temperature of the AAC cake during the pressure increase and reduction phases is increased or reduced linearly. This paper describes how this can be implemented technically and which possibilities there are to approximate an optimal operation with the existing autoclave controls. We will also go into the influence of the exotherm on the condensate balance in an autoclaving process.

We will conclude with some tips and tricks about what should be basically observed in case of the (new) installation of autoclave control systems.

3.2 *Optimal autoclave cycle*

The main results of extensive measurements and investigations are the following:

The best possibility to heat up the AAC cake to the required reaction temperature of approx. 190°C is the *linear increase* (in the increase phase) and *linear reduction* (in the reduction phase) of the *temperature of the AAC cak* (not to be mistaken with the temperature of the autoclave).

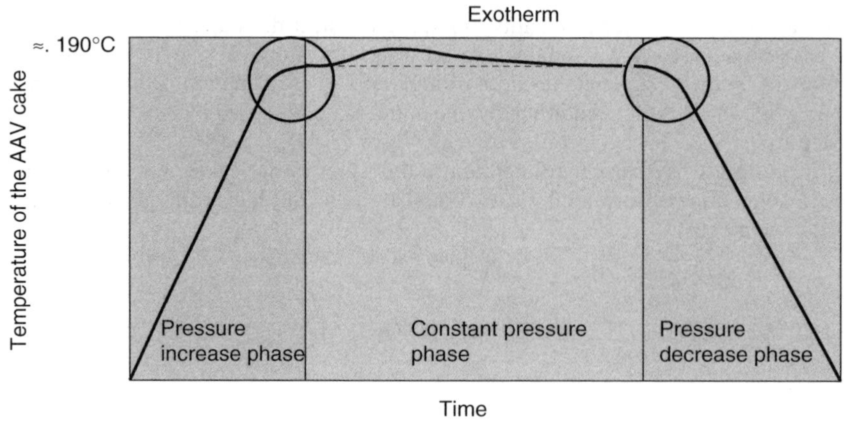

Figure 6. Qualitative representation of the cake temperature development in an optimal autoclave cycle.

Figure 6 shows a purely qualitative timely development of the curve in an optimal autoclave cycle. We have deliberately refrained from dimensioning the time and temperature axis. The graph is focused on the following essential points:

The ordinate axis indicates the temperature of the AAC cake (not the pressure or temperature of the autoclave).

Theoretically, the time interval on the abscissa axis can be chosen freely with the only exception that the constant phase should be at least 5 hours in practical applications.

The duration of the increase and reduction phases can theoretically be chosen as short as one wish. There are, however, several practical restrictions, which will be dealt with in detail in the following chapters.

3.2.1 *The increase phase*

In the increase phase the temperature in the AAC cake and in the autoclave is increased by steam fed from outside (and the relevant condensation of the steam in the AAC and on the autoclave wall). At the same time the pressure in the autoclave rises.

The objective of an optimal increase phase is merely to achieve a possibly linear, i.e. uniform, temperature increase in all parts of the AAC cake. In practice the increase phase takes between 70 and 180 min. This is due to the following reasons:

- Matter transportation processes during the condensation of steam.
  The (porous) structure of the cake constitutes a sort of "steam barrier" which prevents the steam from condensing in all parts of the cake at the same time. This effect is enhanced by residual air in the cake and by pores filled with water. This means: the higher the pore concentration in the cake and the drier the material, the shorter the increase phase can be.
- Technical limitation due to the piping.
  The faster the increase phase, the more steam quantity per time unit is required. There is a technical limitation here, however, given by the diameter of the piping and the capacity of the boiler.

3.2.1.1 Soft transition to the constant phase

Steam feeding at the end of the increase phase (i.e. upon reaching the reaction temperature of approx. 190°C) should not stop abruptly but it should be throttled slowly in order to reach a transition as illustrated in Figure 7. In practice this is achieved as the regulating valves often have settings of barely 1 minute. In most cases this is sufficient.

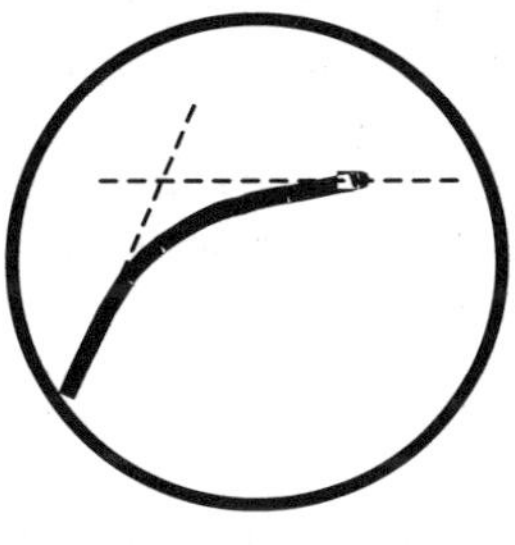

Figure 7. Transition from the increase phase to the constant phase.

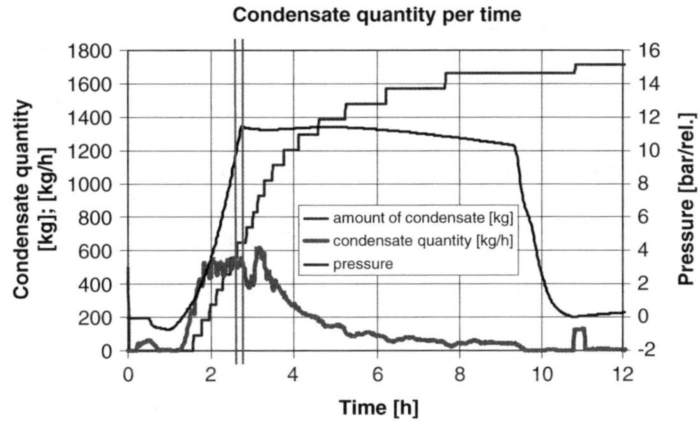

Figure 8. Quantity of condensate measured in a test autoclave.

In specific cases it may be necessary to specifically install this transition in the autoclave controls. This is due to the following reasons:

- Residual air in the autoclave is pressed into the core of the cake by the pressure increase during the increase phase.
- This air in the cake core prevents a simultaneous condensation of the steam in all parts of the cake. Hence the core will be heated up slower.
- The consequence is, that the cake core at the end of the increase phase is colder than the edge areas of the cake.
- If no further additional feeding of steam from outside occurs now, the steam in the autoclave condenses (with a time lag) in the parts of the cake that are still cold.
- This, on the other hand, brings about a pressure drop, which in unfavorable situations may lead to spalling of the cake.
- This should be prevented with the measure mentioned above. It is actually a pure safety measure. In any case it is always better to make sure that as low as possible residual air remains in the cake.

3.2.2 *The constant phase*

At a temperature of 191°C (13 bar) the time for the transformation of the mineral phases must be at least 4 hours. In practice, however, constant phases are virtually never less than 5 hours.

Again, the reason is that due to the air in the core of the cake, the material inside will be heated slower than the edge area. The additional hour therefore serves as time reserve to get all parts of the cake to the reaction temperature.

3.2.3 *The effects of the exotherm on the condensate balance*

Measurements of the condensate quantity, gave a first indication that condensate is totally absorbed by AAC in the pressure increase phase, thus increasing its moisture. Nevertheless condensate measurements are an excellent means to better understand the connections in autoclaving of AAC. In this chapter we display the effects of the exotherm on the moisture balance of AAC.

Figure 8 shows the quantity of condensate generated in a standard autoclaving process measured in a test autoclave. Based on the measurement of the condensate quantities, we will now show what happens in the single steps of autoclaving. The condensate quantity per time is important (kg/h).

- In the flushing phase only a small quantity of condensate is generated.
  The reason is that the flushing steam is absorbed by AAC and the autoclave in this example was hotter than 100°C. No steam has therefore condensed on the autoclave wall. The curing equipment (curing truck, curing plates), however, has a temperature of 20 to 60° C. The condensate measured in the flushing phase is therefore the steam which has condensed on the curing equipment.

- In the vacuum phase there is no condensate. As long as the autoclave is hotter than the core temperature of the cake (approx. 80°C) this is the way it should be.
- At the beginning of the pressure increase phase, too, no condensate is generated. Any quantity of steam streaming into the autoclave is absorbed by the AAC. In addition, our measurement of the condensate quantity does not work with underpressure. The small condensate quantity detectable as from 0 bar (ambient pressure) comes from the cold curing equipment.
- As from approx. 3 bar overpressure ($= 4\text{bar}_{\text{abs}} =$ approx. 144°C) the condensate quantity suddenly begins to rise quickly. The reason is the autoclave itself, which had a temperature of 140–145°C at the beginning of the cycle, must now be heated up, too. Consequently, steam that condenses on the autoclave wall is measured here.
- Until the end of the pressure increase phase the quantity of condensate per time unit remains fairly constant.
- At the beginning of the constant phase, the quantity of condensate per time unit first diminishes, but it rises again as the exotherm sets in. The condensate that is measured here comes from the AAC cake, since no more live steam is fed during the constant pressure phase.
- It can be seen, that due to the overpressure in the cake core developed by the exotherm, a lot of water is pressed out of the cake. This effect can be observed through portholes in the autoclave. The cake "sweats" downright.
- As the exotherm dies away, the quantity of condensate per time unit will get smaller and smaller in the course of the constant pressure phase. Due to the fact that water is "driven out" of the cake during the constant pressure phase, the material is "dried".
- Therefore the exotherm causes the fact that much more condensate is released during the constant pressure phase than during the pressure increase phase.
- In the constant pressure phase the cake gives off again a large amount of the condensate it had absorbed during the pressure increase phase.

An exotherm, therefore, apart from being a sign of a successful autoclaving, also contributes with its drying effect to avoid curing damage, such as spalling, at the beginning of the pressure reduction phase.

3.2.4 *The pressure reduction phase*

Like in the pressure increase phase, the lowering of the temperature of the cake should take place linearly. Apart from a uniform cooling down, it has a further technical advantage, which is explained later in this paper.

3.2.4.1 Soft transition from constant pressure phase to pressure reduction phase

Here, too, the aim is to achieve a soft transition from the constant pressure phase to the pressure reduction phase, as illustrated in Figure 9.

The reasons are the following:

- As described earlier, the AAC sucks itself full with condensate during the pressure increase phase. At the end of the pressure increase phase the material has a moisture of 70% and even more!

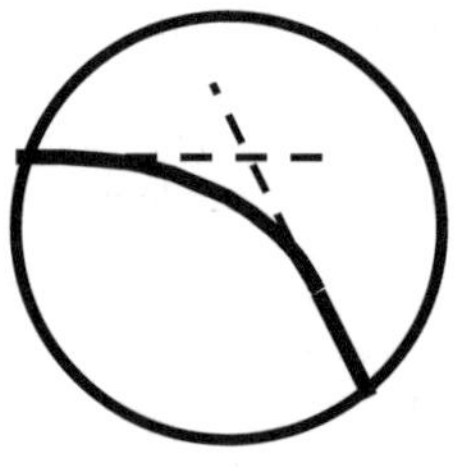

Figure 9. Transition from the constant pressure phase to the pressure reduction phase.

- Although a part of the condensate gets out of the cake during the constant pressure phase, the material still has a very high moisture content at the end of the constant pressure phase. In the worst case, the pores in the coldest spots of the cake (requiring the largest amount of steam to be heated up) may be virtually filled with water. The coldest spots of the cake are generally in the edge regions and close to the curing plate acting as a cooler.
- These pores filled with water form a "steam barrier". If the pressure is reduced too fast at the beginning of the pressure reduction phase, more water per time unit will evaporate in the cake than the pores can give off.
- This may cause the material to spall.

3.3 *Pressure development and steam quantities in the optimal autoclave cycle*

Figure 10 shows the calculated steam quantities as well as the temperature and pressure curves in an autoclave cycle in which a linear rise and fall of the AAC temperature have been adjusted. To simplify matters, the "soft" transition from the pressure increase to the constant phase and to the reduction phase was not considered for the calculation. These autoclave cycles have been calculated with a special computer program for the calculation of autoclave cycles, Figure 11 shows the advantages in terms of process technology of an autoclave cycle oriented on the cake temperature.

Figure 12 shows a linear rise and fall of the cake temperature as it is required for an optimal autoclave cycle. The temperature of the AAC cake is increased from approx. 70°C to approx. 191°C linearly, whereas the pressure rise and drop take place exponentially. The cause is the interrelation between temperature and pressure of saturated steam.

The continuous pressure and temperature drop during the constant pressure phase and the sudden increase after 250 min. are due to the fact that the computer program, with which this curve was calculated, although capable of calculating the heat losses of the autoclave, cannot take exotherms into account yet. The program at this point "feeds" more steam to keep the reaction temperature within determined preset limits. In practice this effect would be "smeared" by the transient behavior of the control valves and by the exotherm.

The curve of the steam quantity during the pressure increase phase is of particular interest. It rises with increasing temperature and reaches its maximum of nearly 13 tons per hour at the end of the constant pressure phase. The increase by leaps and bounds after approx. 40 minutes is due

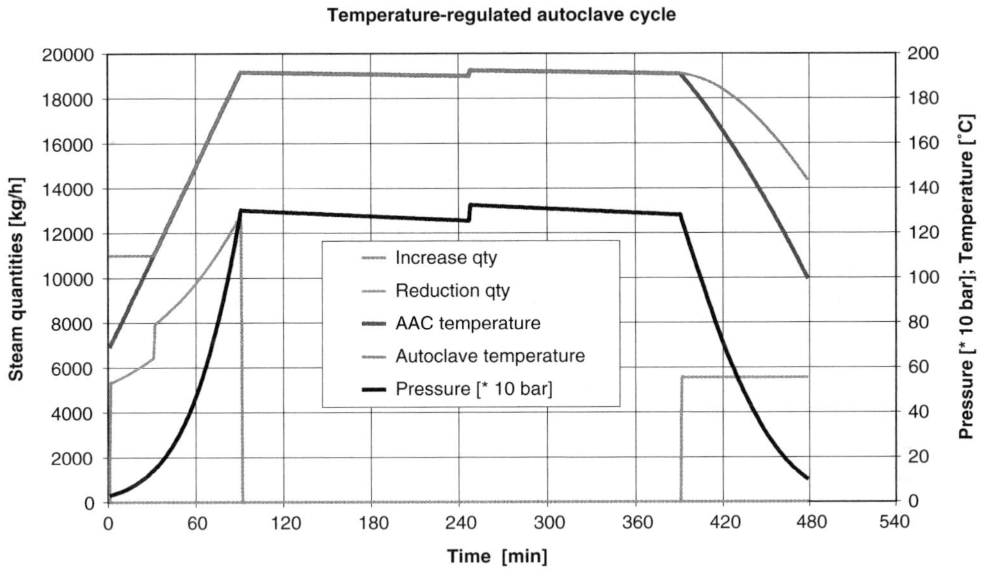

Figure 10. Calculated quantities of steam for an autoclave of 51.2 m length with 24 molds PPW2/ 0,4.

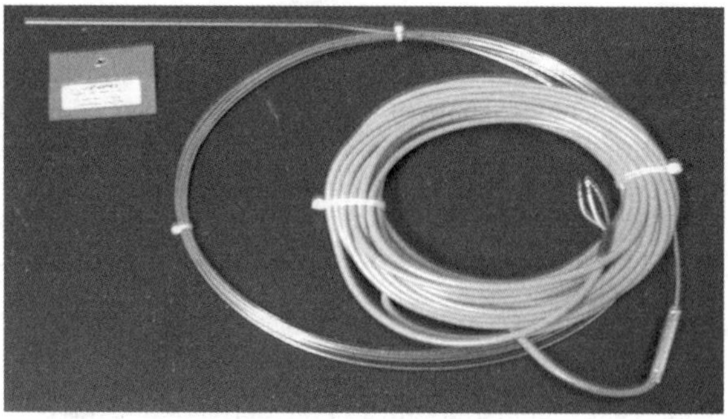

Figure 11. PT 100 resistance thermometers for the measurement of the cake temperature.

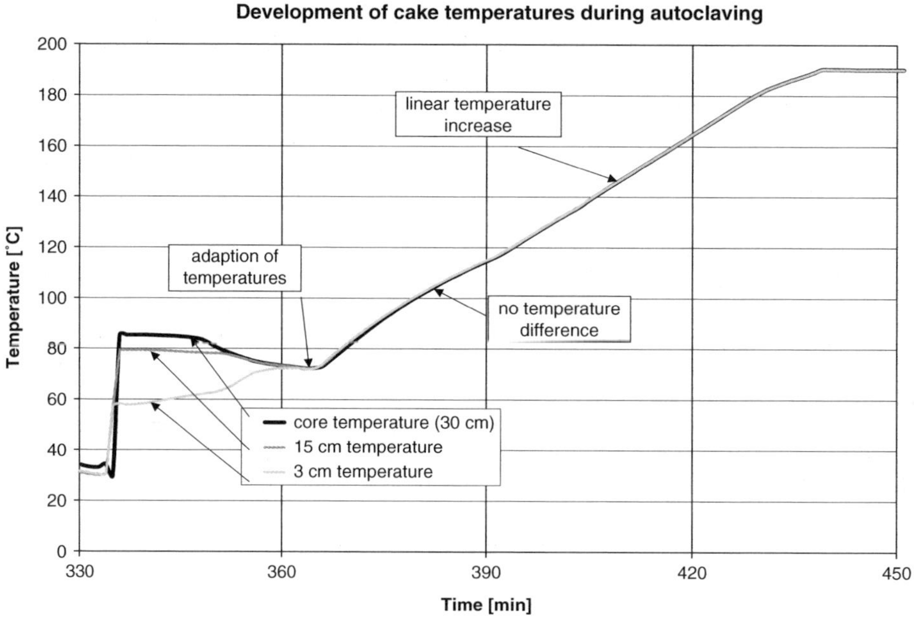

Figure 12. Ideal temperature development in the material during autoclaving.

to fact that the AAC at this point has reached the temperature of the autoclave and now additional steam is needed to heat the autoclave too.

This calculation was carried out for an autoclave of 51,2 m length in the Brück plant with 24 molds of P2/0.4 AAC material with a dry density of 389 kg/m^3.

This calculated autoclave cycle only starts with the end of evacuation at 0.3 bar$_{abs}$.

3.4 *Possible technical implementation of an optimal autoclave cycle*

3.4.1 *Installation of cake thermometers*

It is possible to measure the AAC temperature during autoclaving at different places of the cake (core, edge, etc.) and to regulate the process according to this factor.

However, we encounter some technical difficulties with this method:

- The thermometers are very sensitive, not particularly sturdy and susceptible to failure.
- Lead-in wires through the steel shell of the autoclave make a quick replacement of the thermometers difficult.
- Such thermometers are not suitable under the hard conditions of everyday production.

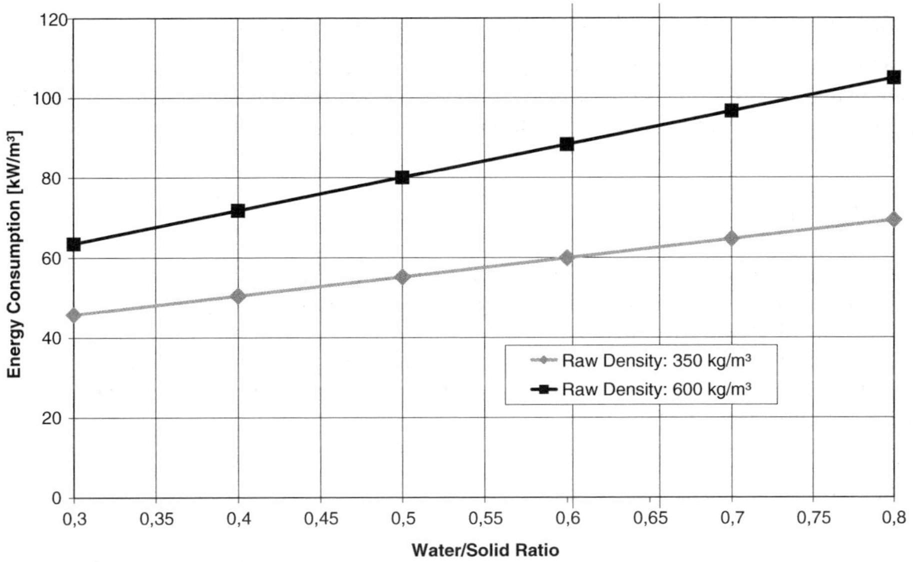

Figure 13. Energy consumption and water/solid ratio.

These cake thermometers, however, are used with great success in trouble shooting.

The only possibility we have at the moment to effect an autoclave cycle oriented to the cake temperature is indirectly by pressure or quantity regulation.

4 REDUCING ENERGY CONSUMPTION

The steam consumption depends on

- Mass and temperature of the autoclave and the curing material
- The raw density, the water/solid ratio and the temperature of the AAC cake
- The way the autoclave cycles are driven (e.g. using steam accumulators or overloading)
- The timing of the manufacturing process (e.g. avoiding of cooling down)
- The utilization rate of the factory.

Exemplary calculation are shown below.

4.1 *Influence of the autoclave temperature on the energy consumption*

The following assumptions were made:

- Mass of the autoclave: 72,4 to, including 100 m^3 AAC
- Autoclave temperature: 120°C ->60 kWh/m^3
- Autoclave temperature: 20°C ->67 kWh/m^3.

4.2 *Influence of the raw density on the energy consumption*

The same assumptions like above give the following results:

Raw density 350 kg/m^3: ->60 kW/m^3
Raw density 600 kg/m^3: ->92 kW/m^3.

4.3 *Influence of the water/solid-ratio on the energy consumption*

Figure 13 shows the influence of the raw density and the water/solid-ratio to the energy consumption. Assuming that the price for 1 kWh of steam is about 0,025 €, it can be easily seen that a difference in the water/solid ratio of 0.1 results in a difference in energy costs of 0,13 €/m^3.

5 CONCLUSION

- The autoclave control system should act on the base of the cake temperature.
- The cooling down of the cake or the autoclave definitely impairs the quality of the autoclaving process and of the finished material.
- Also the water/solid-ratio has not only an effect to the energy consumption but also to the other attendants like curing damages and sticking. Therefore handle your AAC cake like little babies: Keep them warm and dry!

REFERENCES

Hill, P.G., Keenan, J.H., Moore, J.G., Keyes, F.G. (1969), Steam Tables: Thermodynamic Properties of Water Including Vapor, Liquid & Solid Phases, John Wiley & Sons.

Schlegel, E. (1982): Grundlagen technischer hydrothermaler Prozesse. – Freiberger Forschungshefte, A 655.

Taylor, H.F.W. (1965): A review of autoclaved calcium silicates. – Int. Symp. Autoclaved Calcium Silicate Building Materials, London.

Zürn, S.G. (1997): Einfluss der Sandminerale auf die Bildung von Calciumsilikathydraten (CSH-Phasen), das Gefüge und die mechanischen Eigenschaften von Porenbetonprodukten, 226 p. – Logos Verlag, Berlin.

Autoclaved Aerated Concrete – Limbachiya and Roberts (eds)
© 2005 Taylor & Francis Group, London, ISBN 0 415 38356 0

Effect of cement characteristics on the setting behavior of Autoclaved Aerated Concrete

Y. Aono, S. Shibata & F. Matsushita
Sumitomo Metal Mining Siporex Co., Ltd., Research & Development Center, Japan

ABSTRACT: A cement suitable for the Autoclaved aerated concrete (AAC) was developed. Cements with different C_3S, fineness (Blaine), and hemihydrate (H)/gypsum (G) ratio (H/(H+G)) were produced experimentally. Using this cement, the effect of the cementitious property on the viscosity of the AAC slurry and the hardening of the AAC green cake were clarified. The H/(H+G) influenced viscosity, while C_3S and Blaine did not. Hardening quickened as C_3S was more abounding, Blaine value increased, and H/(H+G) decreased. The effect of H/(H+G) on slurry viscosity and hardening was regarded as the reactivity of C_3A and generation of the ettringite. The hydration temperature of the cement in AAC is higher than that of ordinary cement paste, and the rapid supply of SO_4^{2-} from hemihydrate, with a high solubility is required, since the hydration of C_3A is extremely active. The characteristic of the cement suitable for AAC is as follows; a) High-early-strengthen cement rich in C_3S is a base, b) the Blaine value of around 3900 cm^2/g is desirable, c) hemihydrate and gypsum are coexistent. The adaptability was confirmed by the factory scale test.

1 INTRODUCTION

1.1 *Background*

Though characteristics of cement for AAC are very important, no research has been yet reported. The authors aimed to develop a cement suitable for AAC manufacturing taking account of the cementations property in the foaming process and the hardening process of AAC.

The condition of the reaction of the cement in the AAC manufacturing is greatly different from that of ordinary mortar and concrete in terms of temperature, water/cement ratio, mixture system, etc., as shown in Table 1. Therefore, conventional methods of evaluating the cement hydration rate including cement setting test, cement strength test, and measurement of hydration heat from plain cement paste at 20°C are not fully applicable to cement for AAC.

Table 1. Comparison of curing conditions and mix proportions of AAC and normal concrete.

	AAC	Normal concrete
Major raw materials (wt%)	Cement 33, silicastone 53, lime 7, gypsum 7	Cement 16, fine aggregate 38, coarse aggregate 46
Water ratio	Water/Total solid 0.65 Water/Cement 2.0	Water/Total solid 0.09 Water/Cement 0.55
Curing temperature (°C)	Initial stage 40–50 Several hours later 70–80	Approx 10–30 (influenced by the ambient temperature)

Figure 1 shows the concept of the reaction profile of the cement required for AAC manufacturing. The required characteristic are as follows:

(a) The increase in viscosity during the foaming process (about 1 hour after the placing) is small. When the viscosity of the AAC slurry increases rapidly, the "shadow" (Figure 2) may be generated close to the reinforcing steels, and strength of the AAC panel may degrade.
(b) The hardening time (about 3–4 hours after placing) for the piano-wire cutting is short. The productivity decreases when the hardening time is too long.
(c) The time in which the cutting is possible, the cuttable time, is long (normal, 60 minute). Cutting with the piano-wire becomes impossible when the hardening advances rapidly, and the degree of freedom of production decreases.

1.2 *Development of cement for AAC*

For the characteristic of which the above are required, the following was evaluated:

(a) The viscosity of AAC slurry from placing to the end of foaming (about 60 minutes after placing).
(b) The curing time until the optimum hardness for the piano-wire cutting (about 3–4 hours after placing).
(c) The time keeping the hardness which the cutting is possible. For the development of the cement for the AAC manufacturing, the following two plans were considered:
   i) The curing time is shortened by using high-early-strength cement (HC) clinker with high tricalcium silicate (C_3S) content, while the viscosity during the foaming process of the

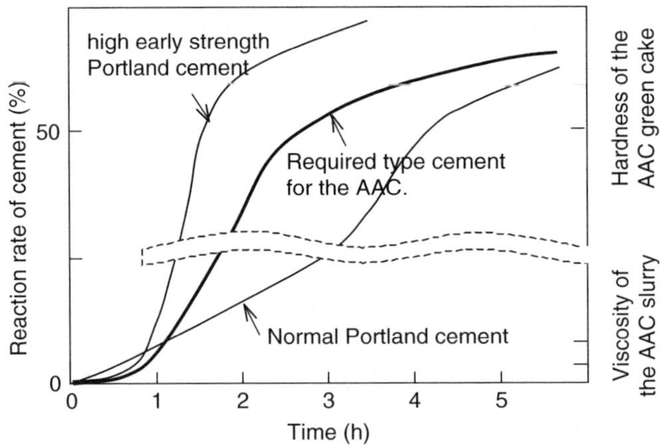

Figure 1. Reaction rate of the cement for the AAC (conceptual scheme).

Figure 2. Shadow of the AAC panel caused by a rapid increase in viscosity of the AAC slurry.

AAC slurry is lowered by lowering the fineness. By lowering the fineness, rapid hydration reaction is suppressed, and the cuttable time is ensured.

ii) The curing time is shortened by increasing the fineness, while the viscosity during the foaming process of the AAC slurry is held low using a moderate heat (MC) clinker with low C_3S content. By decreasing the C_3S quantity, the progress of the rapid hydration reaction is suppressed, and the cuttable time is ensured.

From an economical viewpoint in the production of the cement, it is disadvantageous with respect to cost to increase the fineness like (b). Thus, the plan (a) was developed.

In the meantime, Uchikawa et al. (1984) have reported on the gypsum type in the cement and the effect of hydration temperature on setting time of the cement paste. The setting time became longer in an order when using gypsum, hemihydrate, and anhydrite containing cement. The setting time was shortened with increase in curing temperature in a range from 10 to 30°C. However, the hydration environment of the cement in the AAC manufacturing are at a higher temperature, lime and gypsum coexistence, and high water/cement ratio, as shown in Table 1. Then, the evaluation of the AAC manufacturing was carried out with respect to the gypsum type in this study.

A factory-scale real operation test was also carried out after the characteristic of the cement suitable for AAC was ascertained in the laboratory test.

2 SAMPLE PREPARATION

Table 2 shows the type and sample number of cements subjected to the laboratory test. The parameters are fineness (Blaine value), hemihydrate/gypsum (H/(H+G)), and C_3S content (NC or HC clinker).

Table 3 shows the characteristics of cement used for the laboratory test. The C_3S content of NC and HC is 56% and 64% respectively.

Hemihydrate was prepared by heating and dehydrating natural gypsum used the actual cement manufacturing process. It was confirmed by powder X-ray diffraction that all gypsum became hemihydrate. The gypsum with different H/(H+G) was added to the clinker so that the SO_3 quantity in the cement became 3.0 weight %, and it was ground to a specified Blaine value using a test mill.

Quick lime and gypsum were used as calcareous materials other than cement.

Table 2. Type and sample name of cement for the laboratory test.

Clinker type (C_3S%)		NC (56)	HC (64)	HC (64)	HC (64)
H/(H+G)(%)	Blaine(cm^2/g)	3300	3300	3900	4300
0		–	–	4HC 3900-0	–
50		6NC 3300-50	1HC 3300-50	2HC 3900-50	3HC 4300-50
100		–	–	5HC 3900-100	–

Table 3. Characteristics of cement for the laboratory test.

Name	H/(H+G) (%)	Blaine (cm^2/g)	C_3S (%)	SO_3 (%)
1HC3300-50	50	3320	64.0	3.0
2HC3900-50	50	3880	64.0	3.0
3HC4300-50	50	4300	64.0	3.0
4HC3900-0	0	3910	64.0	3.0
5HC3900-100	100	3950	64.0	3.0
6NC3300-50	50	3310	56.0	3.0

Table 4. Mixture proportion of AAC.

				Recycled material			
Cement (%)	Silica stone (%)	Lime (%)	Gypsum (%)	Before-AC*1 (%)	After AC (%)	Al-powder (%)	W/S*2
25	40	5	5	13	12	0.06	0.65

*1: Autoclaving; *2: Total Water/Total Solid ratio.

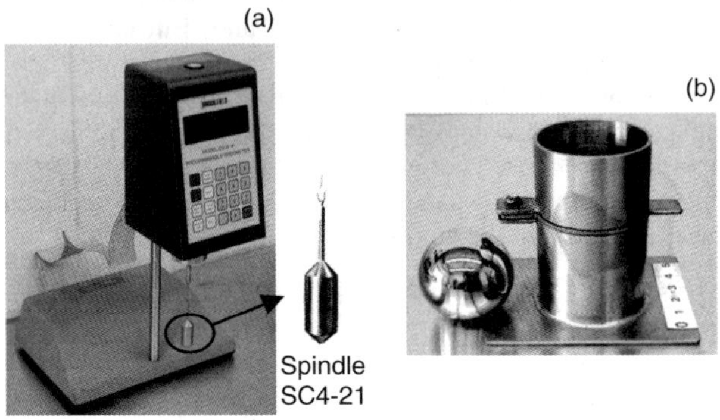

Figure 3. Apparatus for measurement of viscosity and hardness of the AAC. (a): The viscometer for the measurement of the viscosity during the foaming process of the AAC slurry (Rheocalc type HB3+ and spindle number SC4-21 of Brookfield CO., Ltd.), (b): The ball test equipment for the measurement of the hardness of the AAC green cake.

The natural quartz sand (SiO_2 93%, Blaine 3500 cm^2/g) ground by ball mill at the AAC factory was used.

Recycling material crushed AAC after autoclave curing, and used in addition to those carrying out in usual AAC manufacturing process.

2.1 *AAC slurry*

Table 4 shows the composition of the AAC slurry. Those materials except for Al-powder were mixed for 2 minutes at 45°C. Al-powder was added to the mixture, and was further agitated for 30 seconds to yield the AAC slurry.

3 TEST METHOD

3.1 *Viscosity*

For measurement of the viscosity during the foaming process, it was necessary that the rotor avoids the buoyancy effect caused by foaming. In addition, it is necessary to lower the rotational frequency as low as possible so as to not to destroy the bubble structure. For the viscometer which satisfied these requirement, Rheocalc type $HB3^+$ and spindle number SC4-21 of Brookfield CO., Ltd., were chosen (Figure 3(a)). It is confirmed that this viscometer can be operated at a very slow rotational frequency of 0.1 rpm. Moreover, the shape of the spindle exhibits very little buoyancy. A 400 cm^3 of AAC slurry was introduced into a stainless beaker (1000 cm^3), and the viscosity change during the foaming process for 60 minutes was measured. Simultaneously, the change in foaming height of the AAC slurry for 60 minutes was measured using laser displacement gauge (KEYENCE LB300). The foaming height was normalized with respect to the final foaming height of 100%.

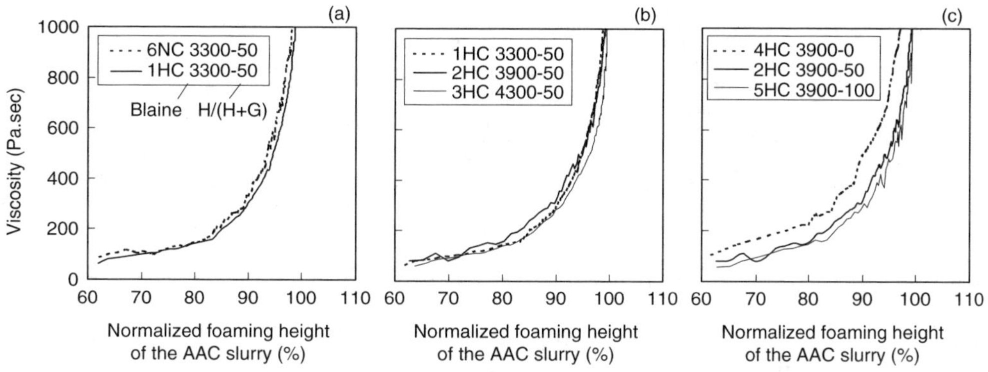

Figure 4. Viscosity change during foaming process of AAC slurry. (a): 6NC and 1H with different C_3S content. (b): 1HC, 2HC, and 3HC with different Blaine values. (c): 4HC, 2HC, and 5HC with different H/(H+G).

3.2 *Hardness*

The measurement of the hardness of the AAC green cake must be quick and convenient ascertain the timing of de-mold and cutting by the piano wire under continuous manufacturing. This is because the hardness change is much more rapid than standard mortar and concrete. Then the ball test, which measures the diameter of the impression by the steel ball (41 mm in diameter and 86 g in weight) dropped from a height of 50 mm above the AAC green cake has been adapted as shown in Figure 3(b). The hardness of the AAC green cake was estimated by the ball test at 15-minute intervals from 2 to 5 hours after the placing.

4 RESULTS AND DISCUSSION

4.1 *Viscosity of AAC slurry in forming process*

Since volumetric water ratio of the AAC slurry decreases by foaming, the viscosity of the AAC slurry during the foaming process increases. Simultaneously, the viscosity of the AAC slurry rises also by the hydration of cement and quicklime. It is necessary to lower the viscosity change of the AAC slurry during the foaming process to avoid the generation of shadows (Figure 2). Here, the characteristic effect of the cement parameters, C_3S content (NC and HC), Blaine and H/(H+G) on the viscosity of the AAC slurry is ascertained.

Figure 4(a) shows the viscosity change during the foaming process of the AAC slurry using 6NC (C_3S 56%) and 1HC (C_3S 64%). The change in the viscosity of 6NC and 1HC was almost equal. It was observed that the C_3S content did not influence the viscosity of AAC slurry during the foaming process.

Figure 4(b) shows the viscosity change during the foaming process of the AAC slurry using 1HC, 2HC, and 3HC with Blaine values. These viscosity changes were almost equal, and the Blaine value did not influence the viscosity of the AAC slurry in the foaming process.

Figure 4(c) shows the viscosity change during the foaming process of the AAC slurry using 4HC, 2HC, and 5HC with different H/(H+G). The viscosity change of 5HC (H/(H+G) = 100%) and 2HC (H/(H+G) = 50%) were almost equal, and 4HC (H/(H+G) = 0%) was even larger than that of the others.

4.2 *Hardness (ball test) of AAC green cake*

The range of the hardness (the ball test value) capable of cutting by the piano wire is from 26 to 21 mm, while the optimum ball test value is 24 mm.

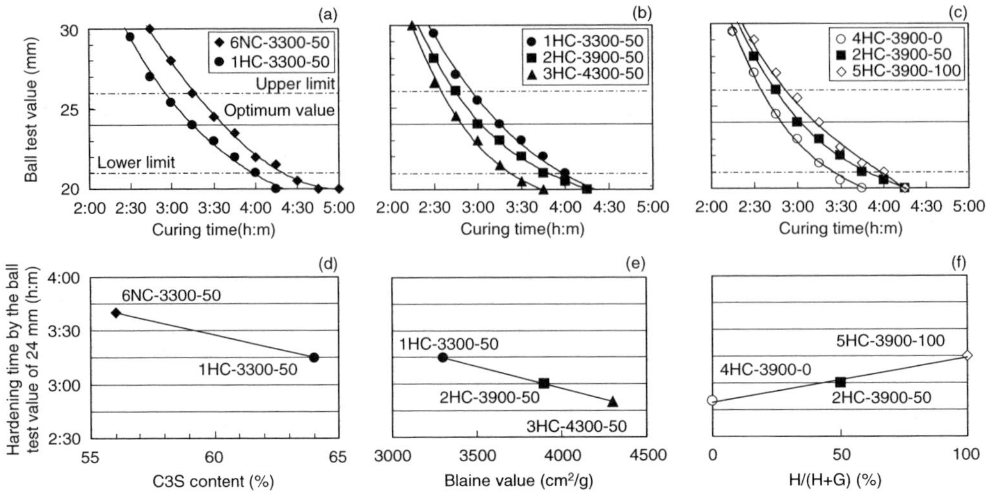

Figure 5. Results of hardening by the ball test. (a)–(c): Change in the ball test value as a function of curing time. (a): 6NC and 1H with different C_3S content. (b): 1HC, 2HC, and 3HC with different Blaine value, (c): 4HC, 2HC, and 5HC with different H/(H+G). (d)–(f): The relationship between parameters of cement and hardening time by 24 mm of the ball test value. (d): C_3S content, (e): Blaine value, (f): H/(H+G).

Figure 5(a) shows the change in the ball test value with time of 6NC and 1HC. The hardening of the AAC green cake of 1HC was more rapid than that of 6NC. Figure 5(d) shows the relationship between C_3S content and hardening time for 24 mm of the ball test value obtained from Figure 5(a). Hardening of the AAC green cake was more rapid, when the Blaine value was the same, as the C_3S content was higher.

Figure 5(b) shows the change of the ball test value with time of 1HC, 2HC, and 3HC. Hardening was quickened in an order of 3HC, 2HC and 1HC. Figure 5(e) shows the relationship between Blaine value and hardening time to the ball test value of 24 mm. Hardening of the AAC green cake is more rapid as the Blaine is higher when the C_3S content is the same.

Figure 5(c) shows the change of the ball test value with time of 4HC, 2HC, and 5HC. Hardening was quickened in an order of 4HC, 2HC and 5HC. Figure 5(f) shows the relationship between H/(H+G) and hardening time to the ball test value of 24 mm.

Hardening of the AAC green cake is more rapid as the H/(H+G) is lower. When the Blaine and C_3S content are the same

4.3 *Ideal cement for AAC manufacturing*

Figure 6 shows the comparison of hardening time (to the ball test value of 24 mm) and the range of cuttable time (within 26–21 mm) for each type of cement. With regards to the hardening process of AAC, 2HC is the most ideal cement for AAC, when curing time of within 3 hours and cuttable time of over 60 minutes are set as conditions. It is desirable that H/(H+G) is greater than 50% with regards to AAC slurry viscosity during the foaming process. Since cuttable time is shortened at Blaine value around 4300 cm^2/g and H/(H+G) = 0%, these values are not desirable. From these discussions, the ideal characteristic of cement for AAC manufacturing can be summarized as follows:

(a) HC rich in C_3S (64%) can be used as a base.
(b) The Blaine value around 3900 cm^2/g is desirable, which is lower than 4300 cm^2/g of general high-early-strength cement. Though this condition is suitable for our present plant, it can be adapted by changing only Blaine value, when the demands in hardening time are different.
(c) Hemihydrate and gypsum are coexistent, but, hemihydrate is more desirable, as it is more abundant.

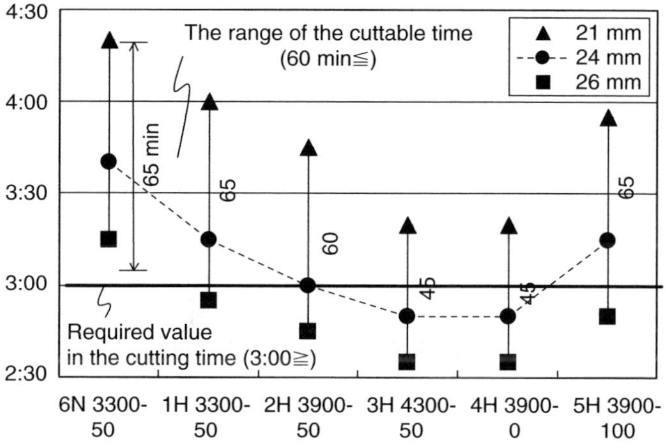

Figure 6. Comparison of hardness and the range of cuttable time attained by different type of cement.

Table 5. Representative example of characteristic of TBS cement.

Item	TBS
Clinker type (C_3S%)	64%
H/(H+G) (%)	50%≤
Blaine (cm^2/g)	3900

Table 6. Characteristics of cement for the factory scale test.

Name	H/(H+G) (%)	Blaine (cm^2/g)	C_3S (%)	SO_3 (%)
TBS	50	3880	62.3	2.9
NC	75	3220	49.6	2.3

We called this cement for AAC "the TBS cement" (TBS). TBS is an abbreviation of "Tei Blaine Soukyou" in Japanese. Tei and Soukyou mean low and high-early-strength, respectively. Full scale factory testing was carried out, and the adaptability of TBS discerned. Table 5 shows a representative example of the characteristics of TBS was confirmed.

4.4 *Factory scale test*

Table 6 shows the characteristics of TBS for factory-scale test. Using this TBS, the AAC manufacturing test was carried out at Sumitomo Metal Mining Siporex Co., Ltd. The characteristics of the conventional NC is also shown in table 6 for comparison. Other raw materials, placing temperatures, and evaluation items were the same as in the case of the laboratory test.

Figure 7 shows the viscosity change during the foaming process of the AAC slurry using TBS and NC. The viscosity of TBS and NC was almost equivalent, and the result of the laboratory test were reproduced.

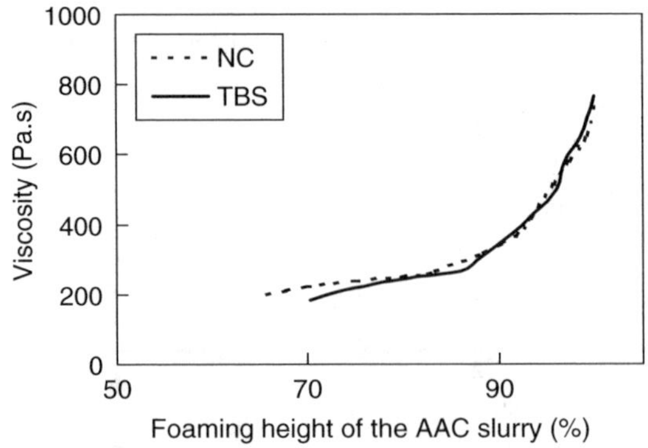

Figure 7. Viscosity change during the foaming process of the AAC slurry using NC and TBS in the factory scale test.

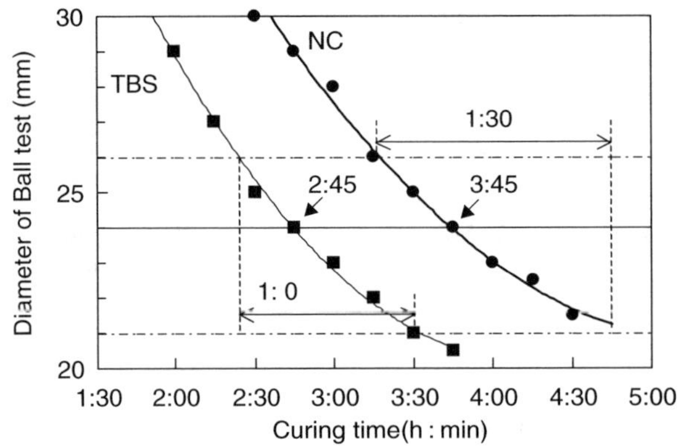

Figure 8. Change in the ball test value as a function of curing time using NC and TBS in the factory-scale test.

Figure 8 shows the change of the ball test value with time for TBS and NC. On the time of arrival to the ball test value of 24 mm was 3 h-45 min for NC, and was 2 h-45 min for TBS. In short, the hardening time to the ball test value of 24 mm was shortened for 1 hour (27%) by TBS compared to NC. For the cuttable time, NC and TBS were 1 h-30 min and 1 h-10 min, respectively. Though it shortened in TBS for 20 minutes, there are no problems during operation, since the requirement of over 1 hour has been ensured.

5 DISCUSSION

C_3S content (between 50% to 64%) and Blaine did not influence the viscosity of the foaming process, however H/(H+G) influenced significantly (Figure 4). The viscosity of the AAC slurry rose when H/(H+G) was 0% (namely, only gypsum). The setting time of ordinary cement paste tends to be retarded by gypsum differently. With regards to this, C_3A reactivity and SO_4^{2-} concentration in the solution seem to be related (Locher, 1980). The solubility of hemihydrate in water at ordinary temperatures is higher than that of gypsum. The setting time of standard cement paste is shortened because hemihydrate reprecipitates as gypsum by hemihydrate dissolving.

In the meantime, C_3A reactivity in the AAC slurry is higher than that for standard cement paste, since the temperature is higher as shown in Table 1. Therefore, in the AAC slurry, C_3A early hydration is active, and the calcium aluminate hydrate is formed easily. The precipitation of the ettringite on the C_3A particle surface is insufficient only in the presence of gypsum, for which the solubility is low in the cement, and the AAC slurry viscosity may rise.

Though the gypsum is also used as a raw material in AAC (shown Table 4), it does not seem to contribute to the generation of the ettringite on the C_3A surface. The grain size of gypsum is comparatively large (50–150 μm), as it is used without grinding. The viscosity change of the foaming process of AAC slurry is mainly related to the hydration of C_3A, and C_3S and Blaine value are responsible to the subsequent hardening process.

On hardening, it was proven that C_3S, Blaine value, and gypsum type influenced all the factors. The contribution of C_3S showed large influence on hardening, because the hardening of the AAC green cake quickened as the C_3S content. The effect of C_3S and Blaine value on hardening can be easily understood from the viewpoint of the cement chemistry. With regards to the gypsum type, the generation of the ettringite seems to be concerned, and the effect of gypsum type on the AAC slurry viscosity may be explained by the fact that the hardening quickened as H/(H+G) lowered. Interstitial material (C_3A, C_4AF) adheres to each mineral (C_3S, C_2S) unable of separating in simple substances during grinding of cement particles. It is supposed that the precipitation of the ettringite on the C_3S surface also became insufficient and that the induction period shortened.

6 CONCLUSIONS

(1) The following results were obtained in the study of the adaptability of the cement for AAC with different C_3S content rate, Blaine value, and H/(H+G).
- (a) C_3S content and Blaine do not influence the viscosity during the foaming process, however hemihydrate/gypsum does.
- (b) The viscosity of the AAC slurry increases when H/(H+G) is 0%.
- (c) The hardening quickens when C_3S is more abundant.
- (d) The hardening quickens when the Blaine value increase. On the other hand, the cuttable time shortens when the Blaine value increases.
- (e) The hardening quickens when H/(H+G) is decrease.

(2) Characteristics of the cement suitable for the manufacturing of AAC are as follows.
- (a) HC rich in C_3S is a base for accelerations the hardening process.
- (b) The Blaine value around 3900 cm^2/g is desirable, which is lower than 4300 cm^2/g of the general high-early-strength cement. The cuttable time increases as the Blaine value decreases, and the degree of freedom of production is ensured.
- (c) Hemihydrate and gypsum are coexistent, but hemihydrate is more desirable, as it is more abundant. The generation of shadow damage is suppressed by lowering the viscosity change during the foaming process.

(3) We called the cement for AAC the “TBS” cement. By conducting a factory test of TBS, the adaptability to AAC was confirmed.

REFERENCES

Locher, F.W. Richartz, W. & Sprung, S. 1980. Setting of cement Part II: Effect of adding calcium sulphate, *ZKG*, NO.6/80, 271–277.

Uchikawa, H. Uchida, S. Ogawa, K. & Hanehara, S. 1984. Influence of $CaSO_4 \cdot 2H_2O$, $CaSO_4$, $CaSO_4 \cdot 1/2H_2O$ on the initial hydration of clinker having different burning degree, *Cement and Concrete Research*, 14, 645–656.

Autoclaved Aerated Concrete – Limbachiya and Roberts (eds)
© 2005 Taylor & Francis Group, London, ISBN 0 415 38356 0

Non-linear static analysis of tall buildings

Santosh Kumar Rai & J. Prasad
Civil Engineering Department, IIT-Roorkee, Roorkee (U.A), India

ABSTRACT: In this paper an attempt has been made to conduct non-linear static analysis of a 30 storey framed building, followed by increasing the stiffness of the building by providing shear wall and a new arrangement i.e. diagonal shear wall panels. The diagonal shear wall panels are bay-wide and storey deep discrete panels to resist lateral loads. The non-linear static analysis of bare frame and shear wall/shear wall panels gives better understanding and more accurate lateral load evaluation of buildings as the progression of damage can be determined. With the trend, the non-linear static analysis leads to techno-economical solution for lateral load evaluation and retrofitting of buildings. Non-linear analysis of bare frame, shear wall and diagonal shear wall panels emphasizes on ductility, strength, stiffness and drift control to prevent possible damages followed by complete collapse.

1 INTRODUCTION

Three decade ago, engineers used simple hand calculations to determine design forces and moments, in many cases the word analysis was not mentioned even though it was being performed intrinsically. Engineers today typically use linear elastic static (first order) analysis to determine design forces and moments resulting from loads acting on a structure. First order analysis assumes small deflection behavior; the resulting forces and moments take no account of the additional effects due to the additional deformations of the structure under load. On the other hand second order analysis combines two effects to reach a solution; large displacement theory: the resulting forces and moments take full account of the effects due to the deformed shape of both the structure and its members, stress stiffening: the effect of element axial loads on structure stiffness, tensile loads stiffening an element and compressive loads softening an element. As structures become ever more slender and less resistant to deformation, the need to consider second order analysis and to be more specific, P-delta effects arises. As a result, Codes of Practice are referring engineers more and more to the use of second order analysis in order that P-delta and stress stiffening effects are accounted for when appropriate in design.

P-Delta is a non-linear (second order) effect that occurs in every structure where elements are subject to axial load [Sun et al, 2002]. It is a genuine effect that is associated with the magnitude of the applied axial load and the displacement. The magnitude of the P-delta effect is related to the (i) magnitude of axial load P, (ii) stiffness/slenderness of the structure as a whole and (iii) slenderness of individual elements. By controlling slenderness, the magnitude of the P-delta effect is often managed such that it can be considered negligible and then ignored in design; for instance, at the structure level by the use of more or heavier bracing (shear wall), at the element level by increasing member size.

In linear analysis of structures, the equations of equilibrium are based on the undeformed geometry existing before load application. This is sufficiently accurate for many practical cases. However, the frames with slender members may have large deformations, so that it is necessary to consider equilibrium in the real deformed configurations. This requires nonlinear analysis involving iteration, which can be done using Newton-Raphson's technique. The equilibrium equations are based on trial displacement values, whose accuracy is improved by iterations, so as to satisfy the equilibrium of the nodes or the members in their displaced positions or their deformed shapes. Nonlinearity caused by large deformations is referred to as geometric nonlinearity. A member subjected to large axial force combined with transverse loads or to transverse end translation or rotation represents a geometric nonlinear problem, often referred to as a beam-column problem, or casually, a P-delta problem.

2 NONLINEAR MODELING OF RC FRAMES

2.1 *Basic assumptions*

The following assumptions are used for modeling of a beam-column element. [Shuraim, 97]

(1) All elements are initially straight and prismatic. Plain cross-section remains plain after deformation.
(2) Materials have nonlinear elastic behavior and perfect bond exists between concrete and steel.
(3) The state of the boundary conditions is assumed to remain unchanged.

2.2 *Matrix displacement analysis*

When large deflections are present, the equations of force equilibrium must be formulated for deformed configuration of structure. This means that the linear relationship $\mathbf{P} = \mathbf{KU}$ between the applied forces $\mathbf{P}$ and the displacements $\mathbf{U}$ can no longer be used. To account for the effects of change in geometry as applied loading is increased we may obtain solutions for the displacements $\mathbf{U}$ by treating this nonlinear problem in a sequence of linear steps, each step representing a load increment. However, of the presence of large deflections, strain-displacement equations contain nonlinear terms, which must be included in calculating the stiffness matrix $\mathbf{K}$. The nonlinear terms in the strain-displacement equations modify the element stiffness matrix $\mathbf{k}$ so that

$$\mathbf{k} = \mathbf{k}_\mathrm{E} + \mathbf{k}_\mathrm{G} \tag{1}$$

where $\mathbf{k}_\mathrm{E}$ is the standard elastic stiffness matrix calculated for the element geometry at the start of the step and $\mathbf{k}_\mathrm{G}$ is the geometrical stiffness matrix, which depends not only on the geometry but also on the initial internal forces (stresses) existing at the start of the step. The elastic and geometrical stiffness matrices are calculated for each element and then assembled into the total stiffness matrix using the conventional procedures established for linear analysis.

$$\mathbf{K} = \mathbf{K}_\mathrm{E} + \mathbf{K}_\mathrm{G} \tag{2}$$

2.3 *Geometrical stiffness for beam/column elements*

The displacement distribution on a beam element is given by

$$\begin{bmatrix} u_x \\ u_y \end{bmatrix} = \begin{bmatrix} 1-\xi & 6(\xi-\xi^2)\eta & (-1+4\xi-3\xi^2)l\eta & \eta & 6(-\xi+\xi^2)\eta & (2\xi-3\xi^2)l\eta \\ 0 & 1-3\xi^2+2\xi^2 & (\xi-2\xi^2+\xi^3)l & 0 & 3\xi^2-\xi^3 & -\xi^2+\xi^3 \end{bmatrix} \begin{bmatrix} u_1 \\ u_2 \\ u_3 \\ u_4 \\ u_5 \\ u_5 \end{bmatrix} \tag{3}$$

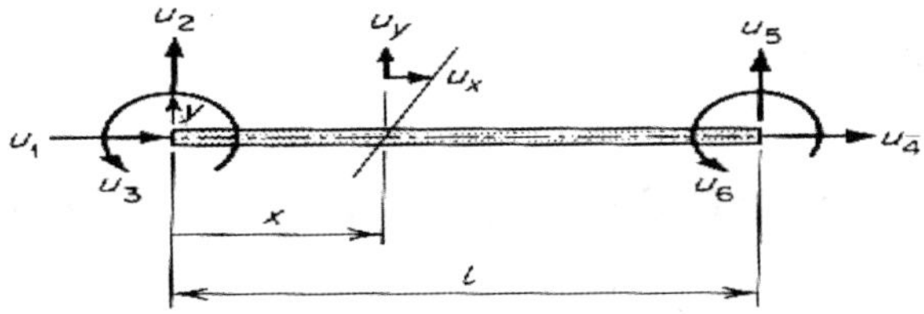

Figure 1. Beam/column element.

where $u_{1,\ldots\ldots}u_6$ are the element displacements shown in Figure 1.

In calculating the strain energy U_i we shall neglect the contributions from the shearing strains. Thus only the normal strains $\in_{xx}$ will be included. These strains for large deflections on a beam in bending are determined from

$$\in_{xx} = \frac{\partial u_0}{\partial x} - \frac{\partial^2 u_y}{\partial x^2} y + \frac{1}{2}\left(\frac{\partial u_y}{\partial x}\right)^2 \tag{4}$$

where y is measured from neutral axis of the beam and u_0 denotes the u_x displacement at $y = 0$. We use equation (4), and the strain energy U_i is given by

$$U_i = \frac{E}{2}\int_v \in_{xx}^2 dV$$

$$= \frac{E}{2}\int_v \left[\frac{\partial u_0}{\partial x} - \frac{\partial^2 u_y}{\partial x^2} y + \frac{1}{2}\left(\frac{\partial u_y}{\partial x}\right)^2\right]^2 dV$$

$$= \frac{E}{2}\int_{x=0}^{l}\int_A \left[\left(\frac{\partial u_0}{\partial x}\right)^2 + \left(\frac{\partial^2 u_y}{\partial x^2}\right)^2 y^2 + \frac{1}{4}\left(\frac{\partial u_y}{\partial x}\right)^4 - 2\frac{\partial u_0}{\partial x}\frac{\partial^2 u_y}{\partial x^2} y - \frac{\partial^2 u_y}{\partial x^2}\left(\frac{\partial u_y}{\partial x}\right)^2 y + \frac{\partial u_0}{\partial x}\left(\frac{\partial u_y}{\partial x}\right)^2\right] dx\,dA$$

The higher-order term $\frac{1}{4}\left(\partial u_y/\partial x\right)^4$ can be neglected in the above expression. Integrating over the cross-sectional area A and noting that sin y is measured from the neutral axis, all integrals of the form $\int ydA$ must vanishes, thus we get

$$U_i = \frac{EA}{2}\int_0^l \left(\frac{\partial u_0}{\partial x}\right)^2 dx + \frac{EI}{2}\int_0^l \left(\frac{\partial^2 u_y}{\partial x^2}\right)^2 dx + \frac{EA}{2}\int_0^l \frac{\partial u_0}{\partial x}\left(\frac{\partial u_y}{\partial x}\right)^2 dx \tag{5}$$

where I denote the moment of inertia of the cross section. We may note that the first two integrals in equation (5) represent the linear strain energy while the third integral is the contribution from the nonlinear component of strain. From equation (3) we obtain

$$\frac{\partial u_0}{\partial x} = \frac{1}{l}\left(-u_1 + u_4\right) \tag{6}$$

$$\frac{\partial u_y}{\partial x} = \frac{1}{l}\left[6\left(-\xi + \xi^2\right)u_2 + \left(1 - 4\xi + 3\xi^2\right)lu_3 + 6\left(\xi - \xi^2\right)u_5 + \left(-2\xi + 3\xi^2\right)lu_6\right] \tag{7}$$

$$\frac{\partial^2 u_y}{\partial x^2} = \frac{1}{l^2}\left[6\left(-1 + 2\xi\right)u_2 + 2\left(-2 + 3\xi\right)lu_3 + 6\left(1 - 2\xi\right)u_5 + 2\left(-1 + 3\xi\right)lu_6\right] \tag{8}$$

$$
\begin{aligned}
U_i &= \frac{EA}{2l}\left(u_1^{\,2} - 2u_1u_4 + u_4^{\,2}\right) \\
&+ \frac{2EI}{l^3}\left(3u_2^{\,2} + l^2u_3^{\,2} + 3u_5^{\,2} + l^2u_6^{\,2} + 3lu_2u_3 - 6u_2u_5 + 3lu_2u_6 - 3lu_3u_5 + l^2u_3u_6 - 3lu_5u_6\right) \\
&+ \frac{EA}{l^2}(u_4 - u_1)\left(\begin{aligned}&\frac{3}{5}u_2^{\,2} + \frac{1}{15}l^2u_3^{\,2} + \frac{3}{5}u_5^{\,2} + \frac{1}{15}l^2u_6^{\,2} + \frac{1}{10}lu_2u_3 - \frac{6}{5}u_2u_5 \\ &+ \frac{1}{10}lu_2u_6 - \frac{1}{10}lu_3u_5 - \frac{1}{30}l^2u_3u_6 - \frac{1}{10}lu_5u_6\end{aligned}\right)
\end{aligned}
\tag{9}
$$

We may note that even for relatively large deflections the quantity $AE(u_3 - u_1)/l$ may be treated as constant equal to the axial tensile force in the element. Hence introducing

$$
P = \frac{EA}{l}(u_4 - u_1) \quad \approx \text{Constant} \tag{10}
$$

and applying Castigliano's Theorem (I) to the strain energy expression (9). This results in the following element force-displacement equation

$$
\begin{bmatrix} S_1 \\ S_2 \\ S_3 \\ S_4 \\ S_5 \\ S_6 \end{bmatrix}
= \frac{EI}{l^3}
\begin{bmatrix}
\frac{Al^2}{I} & 0 & 0 & -\frac{Al^2}{I} & 0 & 0 \\
0 & 12 & 6l & 0 & -12 & 6l \\
0 & 6l & 4l^2 & 0 & -6l & 2l^2 \\
-\frac{Al^2}{I} & 0 & 0 & -\frac{Al^2}{I} & 0 & 0 \\
0 & -12 & -6l & 0 & 12 & -6l \\
0 & 6l & 2l^2 & 0 & -6l & 4l^2
\end{bmatrix}
\begin{bmatrix} u_1 \\ u_2 \\ u_3 \\ u_4 \\ u_5 \\ u_6 \end{bmatrix}
$$

$$
+\frac{P}{l}
\begin{bmatrix}
0 & 0 & 0 & 0 & 0 & 0 \\
0 & \frac{6}{5} & \frac{l}{10} & 0 & -\frac{6}{5} & \frac{l}{10} \\
0 & \frac{l}{10} & \frac{2l^2}{15} & 0 & -\frac{l}{10} & -\frac{l^2}{30} \\
0 & 0 & 0 & 0 & 0 & 0 \\
0 & -\frac{6}{5} & -\frac{l}{10} & 0 & \frac{6}{5} & -\frac{l}{10} \\
0 & \frac{l}{10} & -\frac{l^2}{30} & 0 & -\frac{l}{10} & \frac{2l^2}{15}
\end{bmatrix}
\begin{bmatrix} u_1 \\ u_2 \\ u_3 \\ u_4 \\ u_5 \\ u_6 \end{bmatrix}
\tag{11}
$$

In the equation (11) the first matrix is elastic stiffness matrix ($\mathbf{k_E}$) and the later one is the geometrical stiffness matrix ($\mathbf{k_G}$). Which again may be written symbolically as

$$
\mathbf{S} = (\mathbf{k_E} + \mathbf{k_G})\,\mathbf{u} \tag{12}
$$

3 PARAMETRIC STUDY

In this paper, an incremental displacement method is used as a solution technique. The relationship between element force and deformation can be represented by a piecewise linear curve as shown in Figure 2.

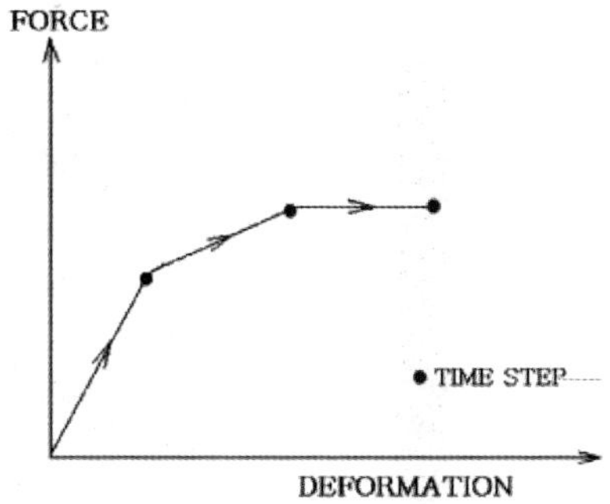

Figure 2. Incremental force deformation curve.

Table 1. Description of building.

S. No.	Particulars	Details/value
1.	Number of storeys	30
2.	Number of bay in X-direction	4
3.	Number of bay in Z-direction	6
4.	Spacing of frame in X-direction	5.0 m
5.	Spacing of frame in Z-direction	5.0 m
6.	Each storey height	3.5 m
7.	Size of all beams	300 mm × 500 mm
8.	Size of columns (from ground storey to twenty storey)	400 mm × 1000 mm
9.	Size of columns (from twenty one storey to thirty storey)	400 mm × 900 mm
10.	Thickness of shear wall/wall panels	300 mm
11.	Grade of concrete	25 MPa
12.	Grade of steel	500 MPa
13	Slab thickness	160 mm
14.	Modulus of elasticity of concrete	25000 MPa
15.	Poisson's ratio	0.18
16.	Live load	2 kN/m^2

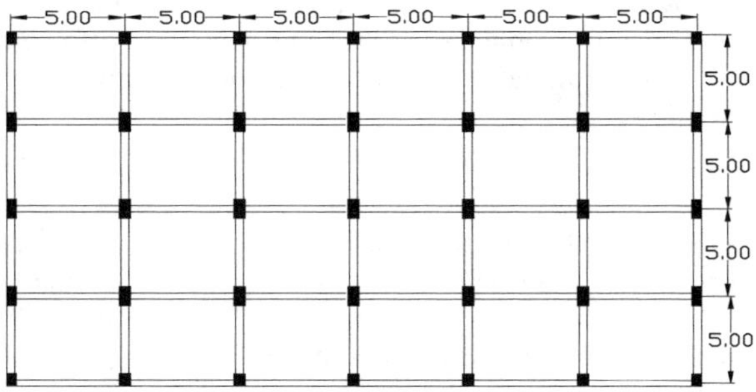

Figure 3. Floor plan of the building.

3.1 *Frame geometry and loading*

A typical 30-storey reinforced concrete building frame was chosen for analysis. The description of building is given in Table 1.

The building is assumed to be located in Seismic Zone IV as per IS: 1893–2000. The building consists of an assembly of cast in place reinforced concrete beams, columns and walls. The typical floor plan and elevation of the building are shown in Figure 3 and Figure 4 respectively.

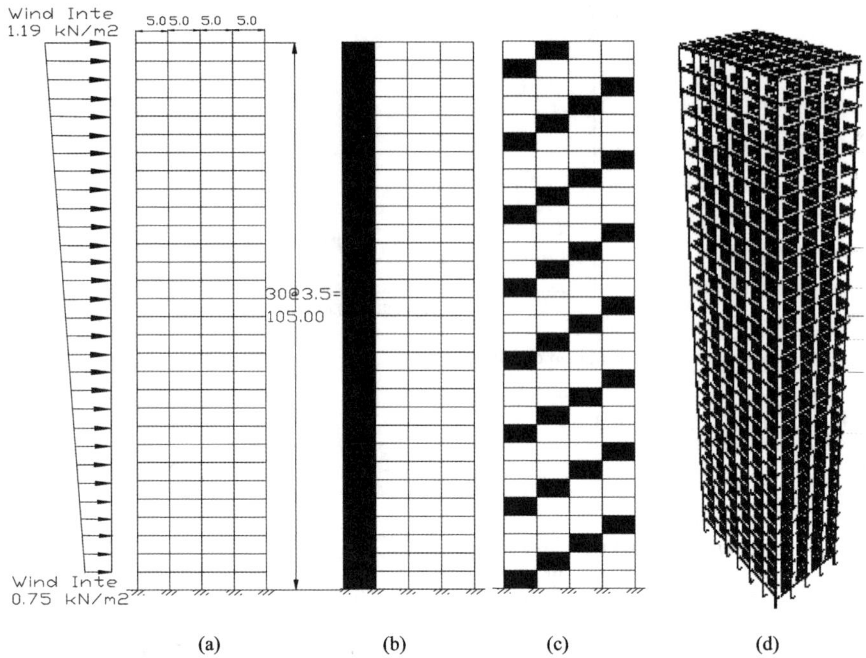

Figure 4. Front elevation of building in X-direction (a) Bare Frame (b) Frame with shear wall (c) Frame with diagonal shear wall panels (d) Isometric view of building.

The loading on building consists of gravity and lateral types that are assumed to be acting simultaneously, although the gravity loads, typically, should be applied first. The wind load is calculated as per IS: 875(3)-1987. The load combination is taken as per IS: 456–2000. The lateral loads are assumed to be acting at floor levels. Therefore, these loads are transferred to the columns and shear wall elements through the rigid floor diaphragms. Some of the assumptions and steps involved in designing the building frames are summarized here: (a) using standard elastic linear analysis and applying the gravity and lateral loads simultaneously, the required internal ultimate forces were computed; (b) the columns were designed for the axial and flexure; (c) the beams are designed for negative and positive moments and the same negative reinforcement was used at both ends of the beam when negative moment existed at both ends; (d) all the sections were designed according to the IS:456-2000 code.

3.2 *Drift analysis*

The 30-storeys building frame with different arrangements of beams, columns and shear wall/wall panels was modeled for analysis and design. In the first case, a bare frame is analyzed. In second and third cases, the shear wall and diagonal wall panels were analyzed. The analytical results obtained using STAADPro-2004 and SAP2000, software packages, have been presented in the form of graphs, indicating the trends of variable such as lateral deflection, interstorey drift and the stresses in the members.

3.2.1 *Lateral displacement*

The maximum deflection at the top of a building roof is restricted not to exceed a certain percentage of the building height as follows;

$d_H \leq 0.0007\ H$ Fintel

$d_H \leq 0.0010\ H$ Israel

$d_H \leq 0.0017\ H$ Yugoslavia

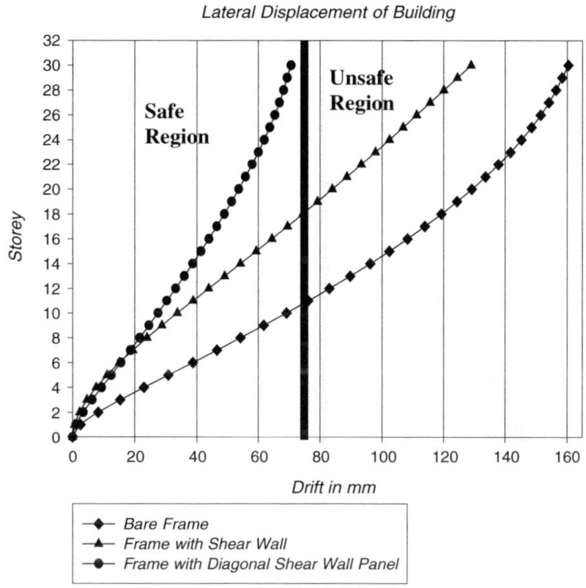

Figure 5. Lateral drift by linear analysis.

in which, d_H = the top floor deflection of a building at roof level and H = total height of the building. It is seen that the limitation on roof deflection is specified only in the earthquake codes of Israel and former Yugoslavia. Based on his observations, in the past earthquakes of Central America, Fintel's [1973,74] recommended a rather lower limit for the maximum roof deflection. As it will be seen in the example building, the roof deflection in linear and nonlinear analysis exceeds the above recommendations. In this paper, the value recommended by the Fintel for the maximum roof deflection is adopted and suggested for the restrictive roof displacement.

One may argue that the maximum roof displacement is not an important quantity, since the maximum interstorey drift limitation is already imposing a restriction on the lateral deflections at each floor level. This is true. But, the overall level of damage in a structure is also closely related to its global deflection measured at roof level. If the interstorey drift values at each floor level reach their maximum allowable limits, then the roof displacement will reach undesirable values. Therefore, an independent global limitation on the roof displacement is most desirable [Salse et al, 1973].

3.2.1.1 Bare frame

This represents the 30 storeyed rigid frame building without shear wall. In this case the maximum sway at roof level for linear analysis is 160.37 mm (0.0015 H) as shown in Figure 5. When nonlinear static analysis performed for the same building with same case the maximum sway at top is increased to 195.00 mm (0.0019 H), Figure 6. From Figure 6, we can see that there is 21.6 % increase in lateral displacement due to geometric nonlinearity.

3.2.1.2 Frame with shear wall

When lateral stiffness is increased by providing the shear wall to the frame system, the lateral deflection in linear analysis is 128.92 mm (0.0012 H) as shown in Figure 5. In nonlinear static analysis it is 144.38 mm (0.0014 H). In this case, the stiffness increases but is still higher than the recommendations given by the Fintel's. From Figure 6, we can see that there is 12% increase in lateral displacement due to geometric nonlinearity.

3.2.1.3 Frame with diagonal shear wall panels

A new system of bay-wide and storey-deep discrete panels arrangement is proposed here. In this case the shear wall is distributed in staggered manner in a diagonal fashion. This proposed system is

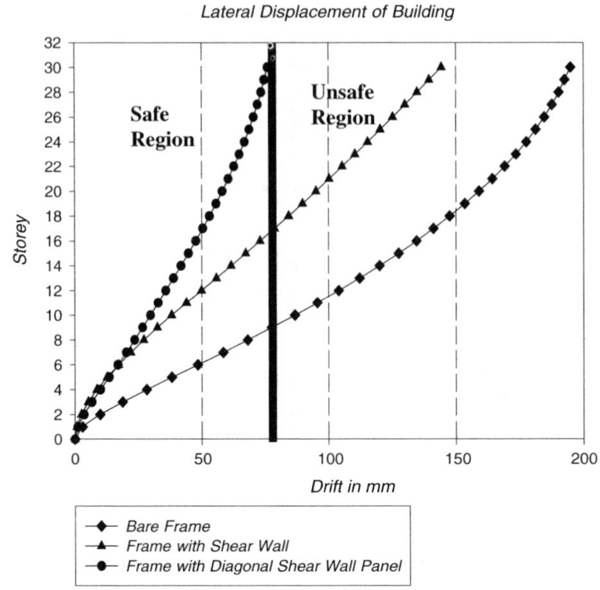

Figure 6. Lateral drift by nonlinear analysis.

called as diagonal shear wall panels. In this case the lateral displacement in linear analysis is reduced further to 70.54 mm (0.00067 H) and in nonlinear static analysis it is 73.65 mm (0.0007 H). From Figure 6, we can see that there is 4.4% increase in lateral displacement due to geometric nonlinearity, but the lateral displacement in case of linear and nonlinear analysis is well within the proposed limit (0.0007 H) as adopted in this paper.

3.2.2 *Storey drift*

The storey drift in any storey as per IS: 1893-2002, due to minimum specified design lateral force shall not exceed 0.004 h. where h = storey height. For reason of effective damage control and safety measures of the structures, in the example building, it is proposed herein, the upper limit of the storey drift to be taken as 0.0014 h.

Bare Frame: In this case the maximum storey drift for linear analysis is 7.88 mm (0.0023 h) as shown in Figure 7. When nonlinear static analysis performed for the same building with same case the maximum storey drift is increased to 10.05 mm (0.0029 h) as shown in Figure 8. From Figure 8, we can see that there is 27.54% increase of storey drift due to geometric nonlinearity.

Frame with Shear Wall: In this case the maximum storey drift for linear analysis is 5.12 mm (0.0015 h) as shown in Figure 7. When nonlinear static analysis performed for the same building with same case the maximum storey drift is increased to 5.84 mm (0.0017 h) as shown in Figure 8. From Figure 8, we can see that there is 14.06% increase in storey drift due to geometric nonlinearity.

Frame with Diagonal Shear Wall Panels: In this case the maximum storey drift for linear analysis is 3.12 mm (0.0009 h) as shown in Figure 7. When nonlinear static analysis performed for the same building with same case the maximum storey drift is increased to 3.44 mm (0.0001 h) as shown in Figure 8. From Figure 8, we can see that there is 10.26% increase in storey drift due to geometric nonlinearity, but its well within proposed limit (0.00014 h).

3.3 *Ductility*

The overall ductility [Hasan et al, 2002] of bare frame is initially 1.18, but when the shear wall provided, the overall ductility increased to 1.33 and when diagonal shear wall panels is provided the overall ductility comes maximum to a value 2.52. The storey ductility of the bare frame is initially

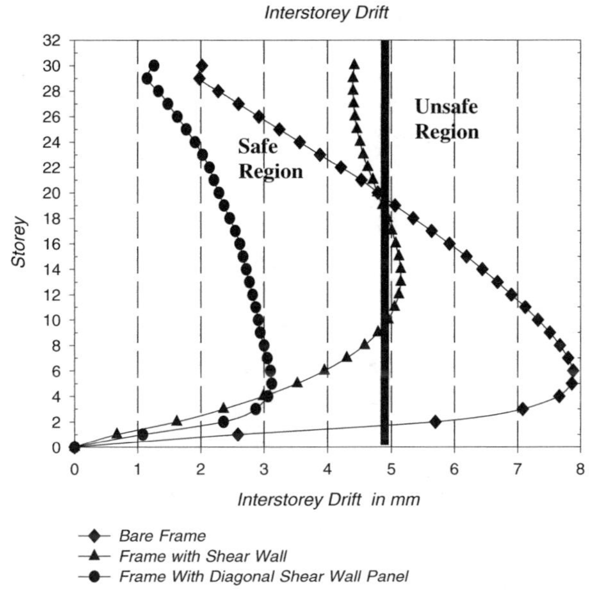

Figure 7. Storey drift by linear analysis.

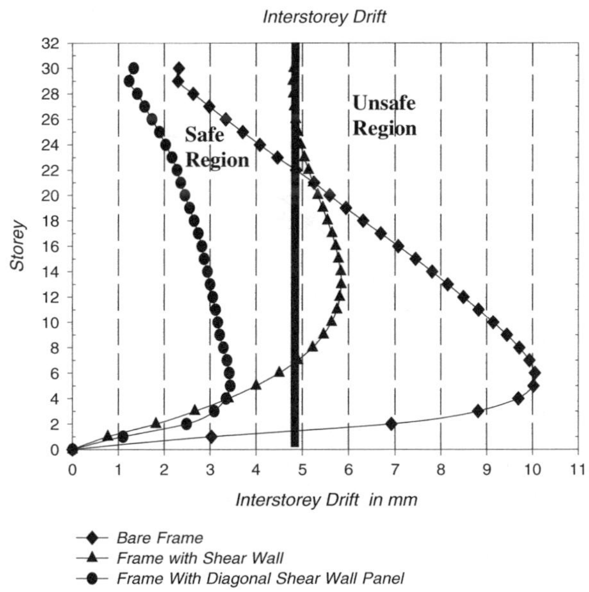

Figure 8. Storey drift by nonlinear analysis.

1.27, but when the shear wall provided the storey ductility increased to 1.93 & when diagonal shear wall panels provided, the storey ductility comes maximum to a value 3.19.

3.4 *Axial load – displacement curve for ground floor column*

In this the axial load Vs displacement curve for ground floor column has been plotted for linear analysis in combination with nonlinear analysis.

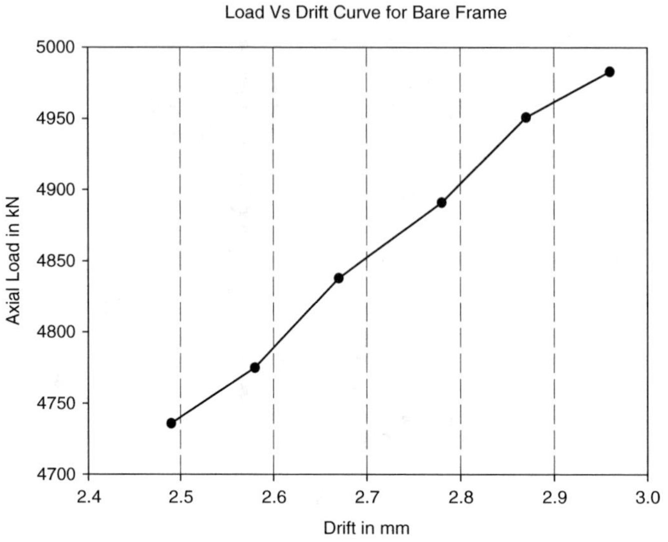

Figure 9. Axial load Vs drift curve for bare frame.

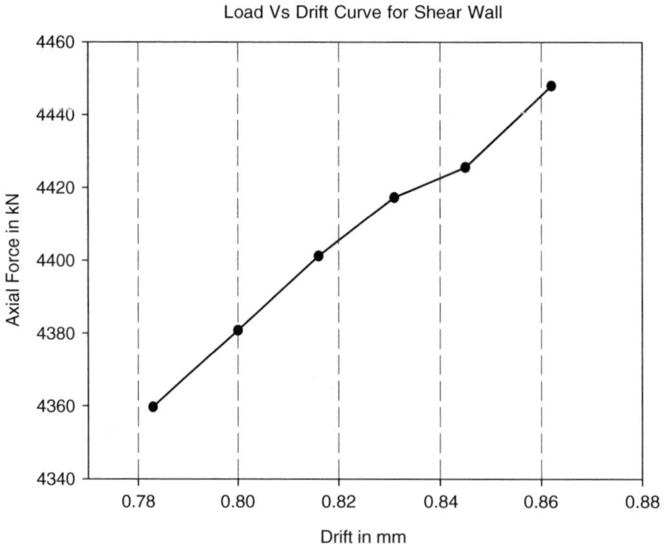

Figure 10. Axial load Vs drift curve for shear wall.

3.4.1 *Bare frame*

The axial load-displacement curve for ground floor column in bare frame is shown in Figure 9. In this for a unit displacement of column, the axial load comes out 525.5 kN.

3.4.2 *Frame with shear wall*

The axial load-displacement curve for ground floor column in frame with shear wall is shown in Figure 10. In this for a unit displacement of column, the axial load comes out 2.10 times greater in comparison with bare frame.

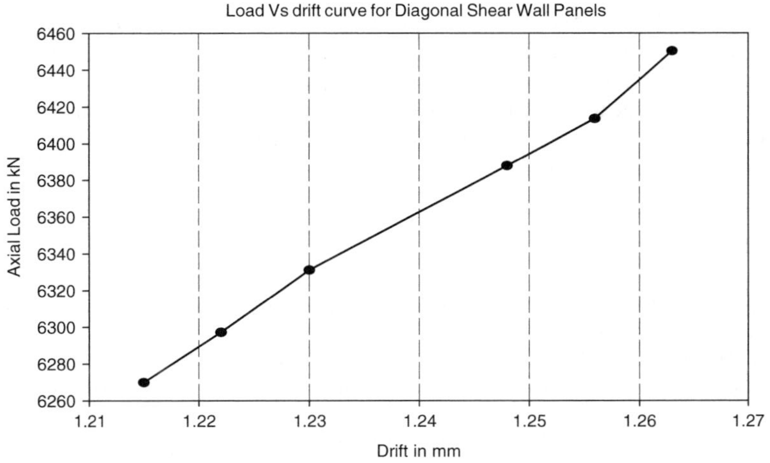

Figure 11. Load-Drift curve for diagonal shear wall panels.

3.4.3 *Frame with diagonal shear wall panels*

The axial load-displacement curve for ground floor column in frame with diagonal shear wall panels is shown in Figure 11. In this for a unit displacement of column, the axial load comes out 6.87 times greater in comparison with bare frame.

4 CONCLUSIONS

In this paper a comparison between linear analysis and nonlinear analysis of a 30-storey regular frame building has been presented. In this, at first instance the bare frame was analysed and then shear walls added to the frame and analysed. Finally a new kind of arrangement i.e. diagonal shear wall panels has been provided and then analysed. The following remarks are concluded based on parameters chosen in this investigation.

- Stiffness: The use of diagonal shear wall panels in bare frames increases the overall stiffness of the structure. In comparison with bare frame, by providing the shear wall, the overall stiffness of the structure is increased but it is lesser than the diagonal shear wall panels. Hence, the performances of the diagonal shear wall panels are better in nonlinear analysis also and provide maximum stiffness.
- Lateral displacement: Some time the lateral displacement in bare frame and frame with shear wall may be within permissible limit in linear analysis, but in nonlinear analysis its exceeds the limiting value. But by providing the diagonal shear wall panels arrangement the lateral displacement comes within limiting value in nonlinear analysis also, hence preventing the excessive damage to nonstructural elements.
- Storey drift: The performance of the diagonal shear wall panels is better and well within limit in case of storey drift in nonlinear analysis also, when compared with bare frame and shear wall.
- Ductility: The overall ductility of the frame is more in case of frame with diagonal shear wall panels.
- Strength: Significant reduction in the axial forces and shear wall base moment in case of the diagonal shear wall panels in comparison with bare frame and shear wall in case of linear and nonlinear analysis also.

It was vividly demonstrated in the example building discussed above that the lateral deflections are significantly reduced if the structural stiffness is increased by providing the shear wall or diagonal shear wall panels. The use of shear walls and diagonal shear wall panels may certainly increase the amount of concrete consumption per unit floor area. It is calculated in various example buildings, however, that the additional cost due to increase in the volume of vertical elements is only of the order of about 1% of the total cost of the building.

In comparison with shear wall and diagonal shear wall panels; the performances of the diagonal shear wall panels are always gives better performance with respect to lateral displacement, interstorey drift and member forces.

REFERENCES

Sun, C.H., Bradford, M.A. and Gilbert, R.I. 1993. Nonlinear analysis for concrete frame structures using the finite element method. Computer & Structure 48. 73–79.

Shuraim, A.B. 1997. Lateral Stiffness of Plane R.C. Frames. Computer & Structure 64. 771–782.

Fintel, M. 1973. Quake lesson from Managua: revise concrete building design? *Civil Engineering, ASCE* August, pp. 60–63.

Hasan, R. Xu, L. and Greierson, D.E. 2002. Push-over analysis for performance based seismic design. Computer & Structure 80. 2483–2493.

Salse, E.A.B. and Fintel, M. 1973. Strength, stiffness and ductility properties of slender shear walls, *Proc. 5th WCEE*, Rome, No. 109.

Fintel, M. 1974. Ductile shear walls in earthquake resistant multistorey buildings, *ACI J*, pp. 296–305.

Applied Technology Council, ATC. 1996. Seismic evaluation and retrofit of concrete buildings, *ATC-40*, Redwood City, Calif.

Chopra, A.K. 1995. Dynamics of structures theory and applications to earthquake engineering. New Jersey: Prentice-Hall.

Computers and Structures 2003. SAP2000: Integrated Software for Structural Analysis and Design, Computers and Structures, Inc. (http://csiberkeley.com), Berkeley, CA, U.S.A.

Earthquake resistant regulations, a world list, 1988. IAEE, Kenchiku Kaikan 3rd Floor, 5-26-20, Shiba, Minato-ku, Tokyo 108, Japan.

Gere, J.M. and Weaver, W.J. 1965. Analysis of framed Structures, New York: Van Nostrand Rainhold.

Ghali, A. and Neville, A.M. 1997. Structural Analysis: A unified classical and matrix approach, London: E&FN SPON.

IS:456-2000. Plain and Reinforced Concrete-Code of Practice, New Delhi: Bureau of Indian Standards.

IS:875-1987 (Part 1, 2, 3). Design Loads(Other than earthquake) for Buildings and Structures-Code of Practice, New Delhi: Bureau of Indian Standards.

IS:1893-2002. Criteria for Earthquake resistant design of structures (Part-1): General Provision and Buildings, New Delhi: Bureau of Indian Standards.

Stafford Smith, B., and Coull, A. 1991.Tall Building Structures: Analysis and Design, New York: John Willey and sons, Inc.

STAAD Pro.2004 (Build 1002). Research Engineers, USA.

Przemieniecki, J.S. 1968. Theory of Matrix Structural Analysis, London: McGraw-Hill book Company.

Autoclaved Aerated Concrete – Limbachiya and Roberts (eds)
© 2005 Taylor & Francis Group, London, ISBN 0 415 38356 0

Expansion and shrinkage behavior of green cake of Autoclaved Aerated Concrete

F. Matsushita, Y. Aono & S. Shibata
Sumitomo Metal Mining Siporex Co., Ltd., Research and Development Center, Japan

ABSTRACT: Expansion and shrinkage behavior of green cake of Autoclaved Aerated Concrete (AAC) during autoclaving process and those mechanisms are investigated in terms of hydration degree during pre-curing and of gypsum addition. Expansion and shrinkage behavior during heating-up period of autoclaving was divided into three regions; initial expansion approximately from 90°C to 130°C, shrinkage approximately from 130°C to 150°C and re-expansion approximately after 150°C. Expansion during autoclaving could be decreased by the sufficient formation of ettringite before autoclaving. It was suggested that gypsum addition has made expansion and shrinkage during autoclaving inactive.

1 INTRODUCTION

In AAC production, the pre-curing process to develop hydration of a porous green cake and the autoclave process to form a binder, tobermorite-1.1 nm, are essential to assure the good quality. After the expansion of green slurry by the reaction of aluminum powder and alkali, the developed porous green body was hardened by the hydration of calcic materials such as cement and lime during pre-curing process. The porous green body is shaped in panels or blocks by piano wires and then cured in an autoclave under saturated water vapour conditions approximately at 180°C. Expansion and shrinkage behavior during the autoclaving process of green body were studied by Ono & Ojiri (1968). The volume change during autoclaving should lead to a warp of specification and an initial stress in reinforced AAC (Koponen & Nieminen 1992). Augmented expansion by the addition of MgO was studied in both of heating-up and soaking periods of autoclaving (Ono & Ojiri 1968). Furthermore, most AAC engineer have learned by experience that cracking sometimes occurs during autoclaving process. However, the study of expansion and shrinkage behavior and those mechanisms for AAC green cake during pre-curing and autoclaving process are insufficient as compared with the importance.

In the cement chemistry, expansion associated with ettringite formation by the reaction of aluminate or ferrite phases with sulfate (Negro & Bachiorrini 1982) is well-known. Delayed ettringite formation accompanied by the expansion, which is strongly affected by the release and reaction of sulfate during long-term periods, is widely studied in recent years (Taylor et al. 2001). In addition, expansion associated with ettringite formation was reported to increase with increase in hydration temperature (Negro & Bachiorrini, 1982).

Therefore, it is estimated that sulfate and hydration degree should affect expansion and shrinkage behavior during autoclaving as well. In this study, the effects of hydration degree altered by casting temperatures, insulating and pre-curing time, and of gypsum addition on expansion and shrinkage behavior of AAC green cake during autoclaving are investigated.

2 EXPERIMENTAL

2.1 *Effect of hydration degree in pre-curing (A series)*

2.1.1 *Mixing of raw materials*

Quartz sand, ordinary portland cement, quick-lime, gypsum and recycled AAC powder as well as aluminum powder were mixed at a water/solid ratio of 0.695 as shown in Table 1. The gypsum/solid ratio was 6.1 wt%.

2.1.2 *Casting and pre-curing conditions*

The mixed slurry of the raw materials was cast in a mould under the conditions as shown in Table 2. The "Insulating" during pre-curing was done by covering the mould with the insulating materials.

2.1.3 *Characterization*

At the end of pre-curing, a small amount of green body was penetrated by acetone to stop hydration and was dried at 47°C for 24 hours. To examine hydration degree of cement and lime as well as consumption of gypsum, the dried specimens were analyzed by DSC (Rigaku Denki Co., Ltd, DSC8230D) with a programming rate of 20°C/min and a measuring temperature range from 20 to 700°C under the conditions of nitrogen gas flow. The targeted minerals were ettringite, gypsum and $Ca(OH)_2$, which were analyzed quantitatively in thermal unit of cal/g at around 90°C, 130°C and 370°C, respectively.

2.1.4 *Measurement of linear change during autoclaving*

After pre-curing, two horizontal notches with approximately 300 mm in distance were marked on the side of green cake. The initial distance of the notches (l_0) was measured precisely by a caliper. The green cake was cured in an autoclave at a temperature of 180°C for 8 hours and was cooled gradually until a room temperature. After the sufficient drying in atmosphere for about 1 week, the distance of the notches after autoclaving (l) was measured. Linear change during autoclaving was calculated with the equation (1).

$$\text{Linear change during autoclaving} = (l - l_0) / l_0 \times 100 \quad (\%) \qquad (1)$$

Table 1. Mixture proportions for specimens A series (wt%).

Quartz	OPC	Lime	Gypsum	Recycle	Al	Water/Solid	Gypsum/Solid (%)
35.9	25.9	4.6	6.1	27.5	0.1	0.695	6.1

Table 2. Casting and pre-curing conditions for specimens A series.

No.	Casting temperature (°C)	Insulating	Pre-curing time (hour)
A1	39.8	No	10
A2	40.0	No	9.8
A3	39.2	No	9.6
A4	40.4	No	9.5
A5	49.4	No	7.5
A6	35.1	Yes	7.3
A7	49.5	Yes	7.1

2.2 *Effect of gypsum addition (B series)*

2.2.1 *Mixing of raw materials*

Quartz sand, ordinary portland cement, quick-lime, gypsum and recycled AAC powder as well as aluminum powder were mixed at a water/solid ratio of 0.685 as shown in Table 3. The gypsum/solid ratios of specimens B1 and B2 were 4.0 wt% and 14.9 wt% respectively.

2.2.2 *Casting and pre-curing conditions*

The mixed slurry of the raw materials was cast in a mould at 45°C. A strain gauge with a thermocouple (Tokyo Sokki Co., Ltd, KM-100HB), which can compensate roughly the thermal expansion for concrete products and can be used at a temperature up to 180°C, was set in the center of the mould beforehand. The porous green body was pre-cured in a chamber, in which relative humidity was kept at 95% and temperature was controlled as shown in Figure 1.

2.2.3 *Measurement of linear change during heating-up period of autoclaving process*

After the pre-curing, the porous green body was removed from mould and introduced into an autoclave. The lead wire of the strain gauge embedded in the porous green cake was passed through a hole of an autoclave, which was sealed completely afterward, and was connected to a data-logger (Tokyo Sokki Co., Ltd, TDS-601). The atmosphere of heating-up period in an autoclave was controlled as shown in Figure 2. At first, vacuum of 0.5 atm was applied for 30 minutes. After that, the autoclave was heated up until 180°C within 4 hours. Linear change of porous green body during heating-up period was initialized at the end of vacuuming and measured through the heating-up period.

2.2.4 *Characterization of products at each temperature*

Five other specimens of the porous green cake without a strain gauge were prepared for characterization of the products in the same procedure as shown in 2.2.1 and 2.2.2. After the vacuuming at 0.5atm for 30 minutes, the autoclave was heated up until 100, 120, 140, 160 and 180°C and kept at each temperature for 3 hours as shown in Figure 2. After cooling down, a small amount of

Table 3. Mixture proportions for specimens B series (wt%).

	Quartz	OPC	Lime	Gypsum	Recycle	Al	Water/Solid	Gypsum/Solid (%)
B1	32.1	27.8	3.8	4.0	32.2	0.1	0.685	4.0
B2	32.1	27.8	3.8	14.9	21.4	0.1	0.685	14.9

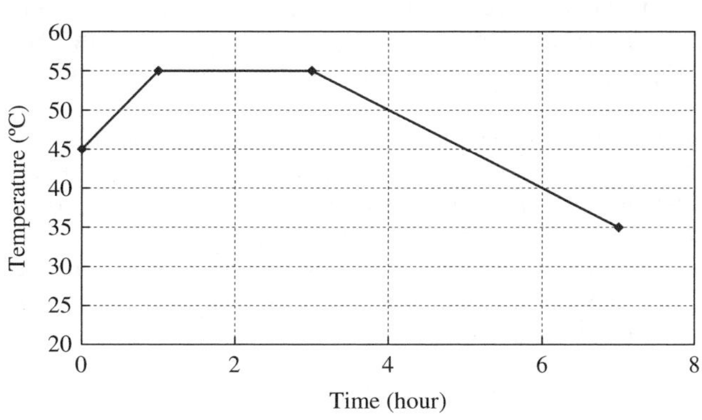

Figure 1. Temperature controlled in a chamber during pre-curing for specimens B series.

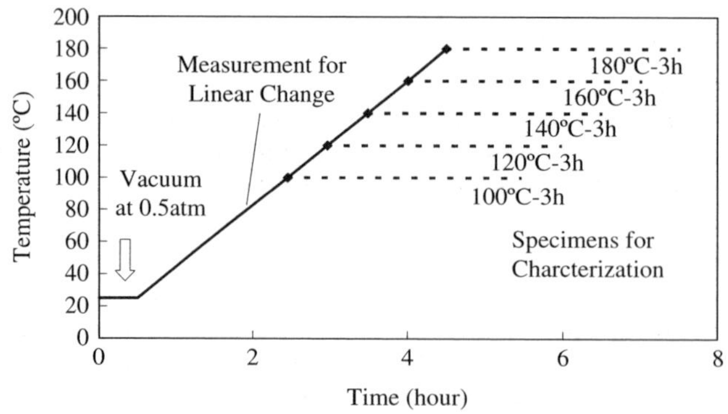

Figure 2. Atmosphere of heating-up period in an autoclave for specimens B series.

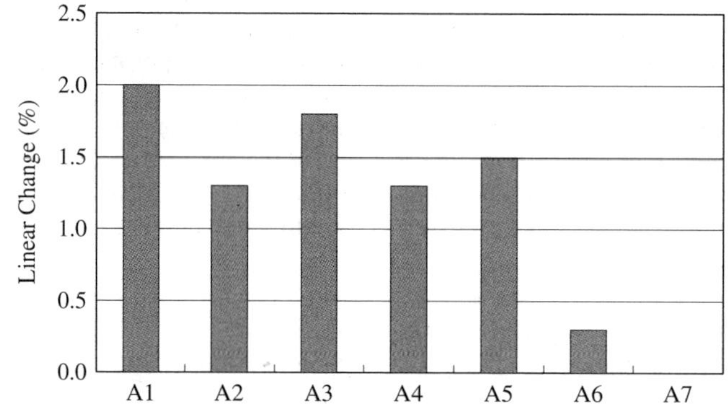

Figure 3. Linear change of specimens with different cast temperature due to autoclaving.

specimens was treated by acetone to stop hydration and was dried at 47°C for 24 hours. The formed products were analyzed by XRD (Rigaku Denki Co., Ltd, RINT-1000).

3 RESULTS AND DISCUSSION

3.1 *Effect of hydration degree in pre-curing (A series)*

Linear change after autoclaving were shown in Figure 3. Specimens from A1 to A5, which were not covered by the insulating materials, showed linear change more than 1%. On the other hand, specimens A6 and A7, which were covered by the insulating materials, showed a small amount of linear change.

Linear change after autoclaving as a functions of DSC peak intensities of $Ca(OH)_2$, ettringite and gypsum were shown in Figures 4, 5 and 6 respectively. Linear change after autoclaving had little correlation with DSC peak intensities of $Ca(OH)_2$, while it showed a correlation with those of ettringite and gypsum. It was suggested the expansion during autoclaving increased with increase in the amount of ettringite formation and decrease in the amount of gypsum remainder.

In the pre-curing period of porous green body, $Ca(OH)_2$ may form due to the hydration of quick lime and C_3S and/or C_2S as shown below:

$$CaO + H_2O \rightarrow Ca(OH)_2 \quad (2)$$

$$mC_3S\ (\text{and/or } C_2S) + nH_2O \rightarrow xC\text{-}S\text{-}H\ \text{gel} + yCa(OH)_2 \quad (3)$$

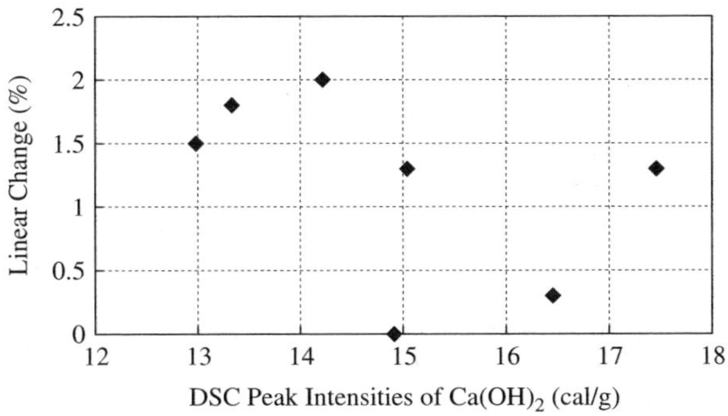

Figure 4. Linear change due to autoclaving as a function of DSC peak intensities of $Ca(OH)_2$.

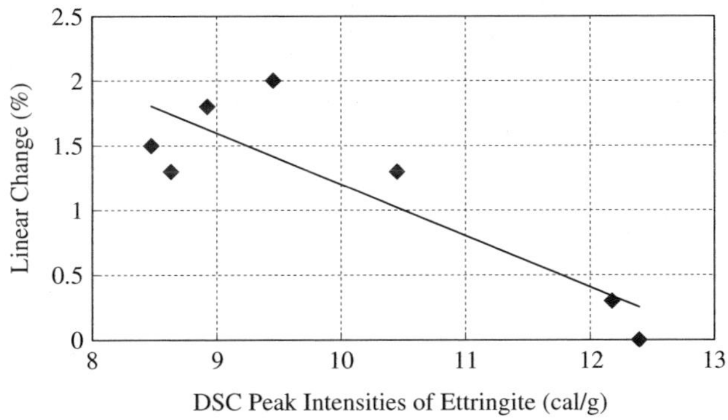

Figure 5. Linear change due to autoclaving as a function of DSC peak intensities of ettringite.

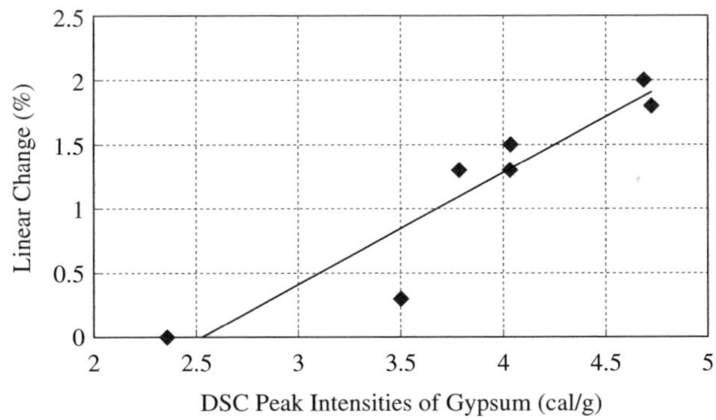

Figure 6. Linear change due to autoclaving as a function of DSC peak intensities of gypsum.

while, ettringite may form due to the hydration of gypsum and aluminate phase of cement as shown below:

$$C_3A + 3CaSO_4{\cdot}2H_2O + 26H_2O \rightarrow C_3A{\cdot}3CaSO_4{\cdot}32H_2O. \tag{4}$$

Therefore, it was suggested that expansion during autoclaving could be decreased by the sufficient formation of ettringite prior to the autoclaving, which can be accomplished by a rise of casting temperature, insulating and extension of the pre-curing time.

3.2 *Effect of gypsum addition (B series)*

Linear change during heating-up period of autoclaving as a function of temperature is shown in Figure 7. Expansion and shrinkage behavior during heating-up period of autoclaving may be divided into three regions; initial expansion approximately from 90°C to 130°C, shrinkage approximately from 130°C to 150°C and re-expansion approximately after 150°C.

With a small amount of gypsum addition (Specimen B1), the initial expansion lasted until 137°C and marked 0.28% at the maximum. Then, the sudden shrinkage was shown down to approximately 0% and at a temperature of 146°C, followed by the re-expansion to 0.35% at 180°C.

On the other hand with a large amount of gypsum addition (Specimen B2), the initial expansion lasted until 124°C and was as small as 0.15% at the maximum. Then, a slow and intermittent shrinkage was shown approximately down to −0.02% at a temperature of 152°C, followed by the re-expansion to 0.25% at 180°C, which was also smaller than that with a small amount of gypsum addition (Specimen B1). Therefore, it was concluded that gypsum addition was able to make expansion and shrinkage behavior during autoclaving inactive.

From the XRD results of specimens after cooling down (not quenching), the actual formed products in an autoclave could not be detected exactly. Therefore, conceptual scheme of the formed products after pre-curing and keeping at each temperature in an autoclave, which was supposed by the results of XRD, are shown in Figure 8.

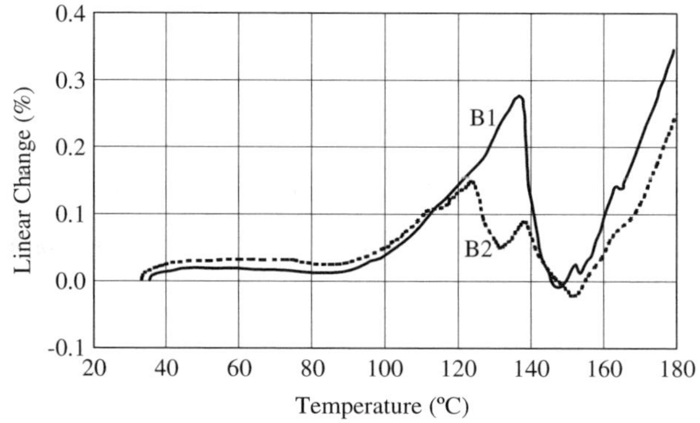

Figure 7. Linear change during heating-up period of autoclaving as a function of temperature for specimens B series.

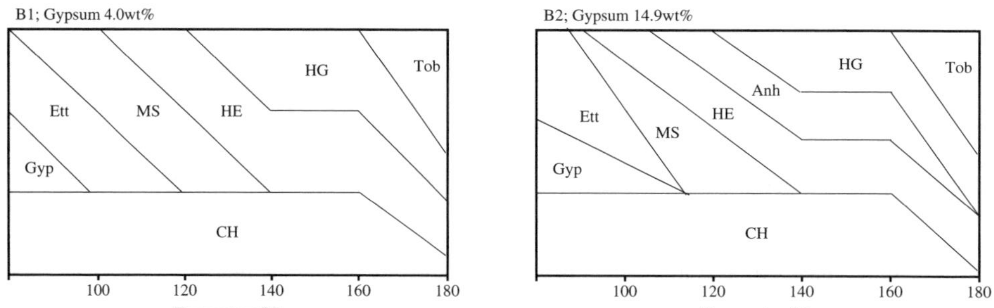

Abbreviation
CH:$Ca(OH)_2$
Gyp:$CaSO_4 \cdot 2H_2O$, Gypsum
Anh:$CaSO_4$, Anhydrite
Ett:$C_3A \cdot 3CaSO_4 \cdot 32H_2O$, Ettringite
MS:$C_3A \cdot CaSO_4 \cdot 12H_2O$,Monosulfate
HE:$Ca_{10}(SiO_4)_3(SO_4)_3(OH,Cl,F)_2$, Hydroxylellestadite
HG:C_3AlSH_4, Hydrogarnet
Tob:$5CaO \cdot 6SiO_2 \cdot 5H_2O$, Tobermorite-1.1nm

Figure 8. Conceptual scheme of formed products at each temperature for specimens B series.

With a small amount of gypsum addition (Specimen B1), $Ca(OH)_2$ and ettringite is formed and gypsum is remained from raw materials after pre-curing. At a temperature of 100°C, gypsum disappears and monosulfate forms. At a temperature of 120°C, ettringite disappears and hydroxylellestadite forms. At a temperature of 140°C, monosulfate disappears and hydrogarnet forms. At a temperature from 160°C to 180°C, $Ca(OH)_2$ disappears and tobermorite-1.1 nm forms. Takemoto & Kato (1968) reported that hydroxylellestadite formed in an autoclave under the condition of CaO/SiO_2 ratio more than 1. Therefore, it is supposed that solubility of quartz began at a temperature approximately from 120°C resulting in the formation of hydration products such as hydroxylellestadite, hydrogarnet and tobermorite-1.1 nm, nevertheless hydroxylellestadite should form because of the low solubility of quartz during the heating-up period of autoclaving.

The difference of formed products in specimen B2 with a large amount of gypsum addition than specimen B1 was that hydroxylellestadite may form earlier at 100°C and anhydrite may form at a temperature from 120°C to 160°C. Anhydrite is considered to be the re-crystallization of an extra amount of calcium sulfate released by the decomposition of ettringite and/or monosulfate as well as the formation of hydroxylellestadite at a temperature from 120°C to 160°C. From the formation of hydroxylellestadite at a relatively low temperature of 100°C, it is supposed that the hydration of aluminate phase of cement was accelerated even in an autoclave by the addition of gypsum.

The expansion and shrinkage behavior shown in Figure 7 will be considered with the results of formed products shown in Figure 8, at first for a small amount of gypsum addition (specimen B1). At a temperature smaller than 100°C, it was considered that linear change showed a little because of a little change in products. At a temperature more than 100°C, formation of both ettringite and monosulfate could be a cause of initial expansion, because it is common that ettringite has needle-like crystal shape and monosulfate has hexagonal platy crystal shape. At a temperature approximately 140°C, it is supposed that shrinkage happens suddenly due to the decomposition of such expansive products as ettringite and monosulfate. At a temperature more than 150°C, it is supposed that re-expansion happens due to the formation of characteristic card-house structure of tobermorite-1.1 nm with platy crystal shape. On the other hand, inactiveness of expansion and shrinkage due to a large amount of gypsum addition (specimen B2) is supposed to be the results of accelerated hydration of aluminate phases of cement before and during autoclaving, which was detected as the earlier formation of hydroxylellestadite.

3.3 *Relation between hydration during pre-curing and expansion and shrinkage during autoclaving*

In section 3.1, it was considered that expansion due to autoclaving can be decreased by the sufficient formation of ettringite before autoclaving. On the other hand in section 3.2, inactiveness of expansion and shrinkage behavior by addition of gypsum was considered to be the results of accelerated hydration of aluminate phases in cement before and during autoclaving. Therefore, it is concluded that expansion and shrinkage was closely related to the hydration of aluminate phases. With a sufficient pre-curing both by pre-curing-conditions and by a large amount of gypsum addition, expansion and shrinkage can be decreased.

4 CONCLUSIONS

(1) Linear change due to autoclaving had little correlation with DSC peak intensities of $Ca(OH)_2$, while it had some correlations with those of ettringite and gypsum. It was considered that expansion due to autoclaving can be decreased by the sufficient formation of ettringite before autoclaving,
(2) Expansion and shrinkage during heating-up period of autoclaving can be divided into three regions; initial expansion approximately from 90°C to 130°C, shrinkage approximately from 130°C to 150°C and re-expansion approximately after 150°C. It was suggested that gypsum addition made expansion and shrinkage during autoclaving inactive.

REFERENCES

Koponen, A. & Nieminen, J. 1992. Initial steel stresses in reinforced AAC units, In Wittmann, F. H. (ed.), *Advances in Autoclaved Aerated Concrete*: 187–194, Rotterdam, Balkema.

Negro, A. & Bachiorrini, A. 1982. Expansion associated with ettringite formation at different temperatures, *Cement and Concrete Research*, 12(6): 677–684.

Ono, K. & Ojiri, K. 1968. Some physical properties of aerated concrete under autoclave process, *Proceedings of the 5th International Symposium on the Chemistry of Cement, Tokyo, Part III*: 570–581.

Takemoto, K. & Kato, H. 1968. Hydroxylellestadite produced by hydrothermal reaction containing calcium sulphate, *Proceedings of the 5th International Symposium on the Chemistry of Cement, Tokyo, Part III*: 563–569.

Taylor, H.F.W., Famy, C. & Scrivener, K.L. 2001. Delayed ettringite formation, *Cement and Concrete Research*, 31(5): 683–693.

Autoclaved Aerated Concrete – Limbachiya and Roberts (eds)
© 2005 Taylor & Francis Group, London, ISBN 0 415 38356 0

The Keen-Crete plant, the *Keen* way to produce air*crete*

R. Beier & R.A. de Vos
Stork Bouwtechniek, Hengelo, The Netherlands

ABSTRACT: For more than 30 years Stork Bouwtechniek is a leading provider of robust, reliable and innovative machinery for the building material industry worldwide.

After 6 months of intensive customer research, severe engineering and thorough 3D simulations the new 'Keen-Crete' concept for producing AAC was launched in April 2005. An intelligent modular shaped concept for an AAC production capacity ranging from 200 m^3 to over 1.100 m^3 per day, offering high value for money. The benefits of this concept are the extreme flexibility in capacity and grade of automation, without a threshold on investment cost for the smaller plant capacities and superior product quality through the optimized *Stork* flat cake cutting system. This allows customers to start with an initially limited capital, to gain market share and increase capacity as the market develops.

Unlike other "low-cost" concepts, the Keen-Crete plant is based on the known high quality machinery, for which Stork has become known for over the years. It is the quality of the plant that determines the quality of the end product, which will be the key to our customers success.

1 INTRODUCTION

Stork was founded in 1868 and is nowadays a large group of companies with worldwide activities in various branches. Stork has a total of approximately 13.000 employees and an annual turnover of 1,82 billion EURO (2004). Stork is known worldwide for its excellent craftsmanship, engineering know-how and high quality of the products and services it supplies. In various branches Stork companies operate as worldwide market leaders.

Stork Bouwtechniek is a part of the Stork group. For more than 30 years, Stork Bouwtechniek is involved in design, manufacture and on site installation and commissioning of AAC production equipment, supplying to all of the world leading producers. Stork machinery and equipment has been installed worldwide. Stork is known for its high quality and reliable machinery and worldwide excellent after-sale services, as can be established by systems that have been supplied 30 years ago and that are still in successful operation.

2 AUTOCLAVED AERATED CONCRETE

The Autoclaved Aerated Concrete (AAC) material was developed in 1924 in Sweden and has become one of the most used building materials in Europe. The market for AAC is rapidly growing in many other countries around the world. AAC also known as Aircrete or Cellular Concrete is a lightweight, load-bearing, high-insulating, durable building product, which is produced in a wide range of various sizes and strengths.

AAC offers incredible opportunities to increase quality and at the same time reduce costs at the construction site. Due to its excellent properties, AAC is used in many building constructions, for example in residential homes, commercial and industrial buildings, schools, hospitals, hotels and many more.

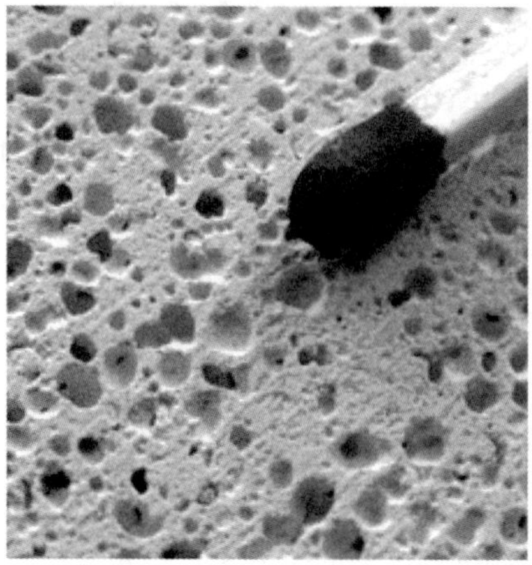

Photo 1. SEM picture AAC.

The advantages of AAC are:

- Large variety of sizes: AAC can be produced in a large variety of sizes, from standard bricks and blocks to large reinforced panels.
- High dimensional accuracy: As a result of its dimensional accuracy, AAC is extremely easy to install using thin bed mortar.
- Lightweight: AAC weighs approximately 50% less than other comparable building products.
- High compressive strength: AAC is a solid product with a high load bearing capacity. The entire surface area is used in structural calculations.
- Excellent thermal insulation: AAC has a very low thermal conductivity and therefore a very high thermal energy efficiency is achieved. This results in savings on heating as well as cooling costs.
- Great acoustic isolation: The porous structure of AAC provides a high acoustic isolation.
- High fire resistance: AAC has an extremely high fire rating of at least 4 hours and more.
- Termite resistance: Termites or insects cannot damage AAC.
- High workability: As a result of the excellent size/weight ratio, constructing with AAC is very rapid. Even though AAC is a solid building material, it can be cut, sawn, drilled, nailed and milled like wood, making it an extremely workable product.

AAC is produced out of a mix of quartz sand and/or pulverized fly ash (PFA), lime, cement, anhydrite, water and aluminium and is hardened by steam curing in autoclaves.

Autoclaved aerated concrete is actually hydrated calcium silicate ($Ca_5Si_6H_5$). The calcium (Ca) is obtained from quick lime (CaO), hydrated lime ($Ca(OH)_2$), cement (a complex of Ca, Al, Fe, Si) and gypsum ($CaSO_4$).

The silica (SiO_2) is obtained from silica sand, fly ash (PFA), crushed silica rock and/or stone. It is possible to obtain silica as a by-product from other processes, e.g. foundry sand or burgee from glass grinding, provided the levels of alkalis, or other impurities, are not too high.

3 GLOBAL MARKET

Due to its superior product qualities, AAC is rapidly expanding in the global market and especially applies to Asia, Russia, the Middle East and South-America. The Keen-Crete plant is designed with a focus on the major growing AAC markets.

These markets are looking for increasing production capacities, but for many investors it is important to first establish a market position. In many cases this requires for a smaller start,

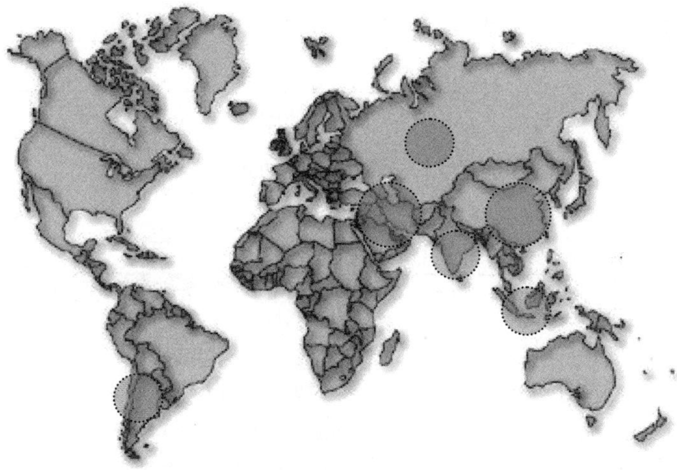

Figure 1. Main AAC markets with largest growth.

however still allow for the flexibility to increase capacity as market demand rises. The aim is to keep the initial investment minimal, without compromising future expansions.

Another aspect is that these markets for at large, still have comparatively low labour costs. This concludes that the investment in automation will only be considered if this offers financial benefits, ensures productivity or increases safety. Further we believe that with growing technical skills and considering the lower labour costs, only the key-production-machinery will be supplied by Stork. Other equipment can be produces and procures locally, using Stork's specifications and network, as well as engineering assistance. Besides further reducing the project investment costs, this also brings work to the country of the customer.

Other markets too, such as the United States of America, where AAC seams to be gearing up again, to try and gain a substantial market position. The Keen-Crete plant can play a role in succeeding with this goal. In the past, high investment costs required high capacity plants, a capacity that could not be sold in a developing market. This forced even the top international suppliers of AAC to rethink their position.

Other solutions to invest in "low-cost" plants resulted in lower end-product quality, making market penetration difficult. However with reduced starting capacities at reduced investment, without compromising end-product quality AAC will have a chance to succeed in the conservative building materials market. The flexibility to increase the degree of automation, will allow to cater for increased labour costs.

4 KEEN-CRETE[1], THE *KEEN* WAY TO PRODUCE AIR*CRETE*

The basic philosophy is a modular standard concept with a focus on the process critical components, to which Stork offers added value. As a result of close cooperation with the client, a maximum of components for the plant will be supplied from local sources, although a turn-key solution can be offered on request.

The development of the new Keen-Crete Plant is done with the use of proven machinery concepts, experienced procedures and with the aim to only improve the logistics to simplify the production flow.

[1] Application to register Keen-Crete as an trade mark has been filed and is awaiting final confirmation at the time of writing this paper.

Table 1. Overview of the Keen-Crete modular capacity steps.

Keen-Crete®	K-450	K-650	K-900	K-1100	
Maximum output capacity (m^3/day)	450	650	900	1100	Further expansion very well possible: customer defined.
Maximum output year 300 days) (m^3/yr)	135.000	195.000	270.000	330.000	

Upgradeable

4.1 *What makes this AAC plant different from existing systems?*

This is mixture of extreme flexibility in capacity and product range and the level of factory automation that can be included. Usually this kind of flexibility has an adverse effect on the investment cost, but the Keen-Crete factory is so easily extendable, that the factory for the lower capacity is not burdened with costs to offer the possibilities of an extension to a higher capacity. The extension of the factory is possible without rendering equipment superfluous or useless, in which in the initial stage has been invested.

Implementation of standardization where useful and participation in strategic partnerships with suppliers have resulted in a lean and mean concept, that still is providing the well-known Stork look and feel in robustness, in quality and reliability.

Unlike other "low-cost" concepts, the Keen-Crete plant is based on the known high quality machinery, for which Stork has become known for over the years. Improved logistical solutions allows the use of top-of-the-range equipment, especially for mixing, dosing and cutting, without substantial additional cost for the entire investment.

4.2 *How does this flexibility work?*

Every detail of the production line is taken into consideration and is engineered as a module with standard, well accessible, components.

Key elements for changing the capacity from a K-450 (suitable for a theoretical capacity of 200–450 m^3 per day of AAC) to a K-1100 are for instance:

- Adding a second mixer discharge and mixer track;
- Adding a mould circulation extension to obtain a one way flow as well as additional moulds and rising places;
- Adding additional weighing hopper for raw material handling, weighing and dosing;
- Adding a second unloading crane and sorting facilities;
- Adding autoclaves, buffer tracks, cooking frames and bogeys to increase autoclaving capacity (the main bottle neck in capacity of each concept).

While developing the concept, great care was taken to consider expansion solutions which will keep shut-down periods to a minimum. This will allow the customer to increase his output capacity by adding one or several modular expansion sets, without occurring considerable loss of production, thereby not running the risk of loosing market share to potential competitors.

4.3 *Basic characteristics of the concept*

The capacity is, with block production, from a theoretical minimum of 200 m^3/day of AAC (2 shifts) to a theoretical capacity over the 1.100 m^3/day (3 shifts). The process critical items, as for instance the cutting machine, are engineered to produce even more. Depending on a variety of variables (raw materials, mixing & dosing, autoclaves, moulds, etc.) it is possible to produce up to 1.500 m^3/day.

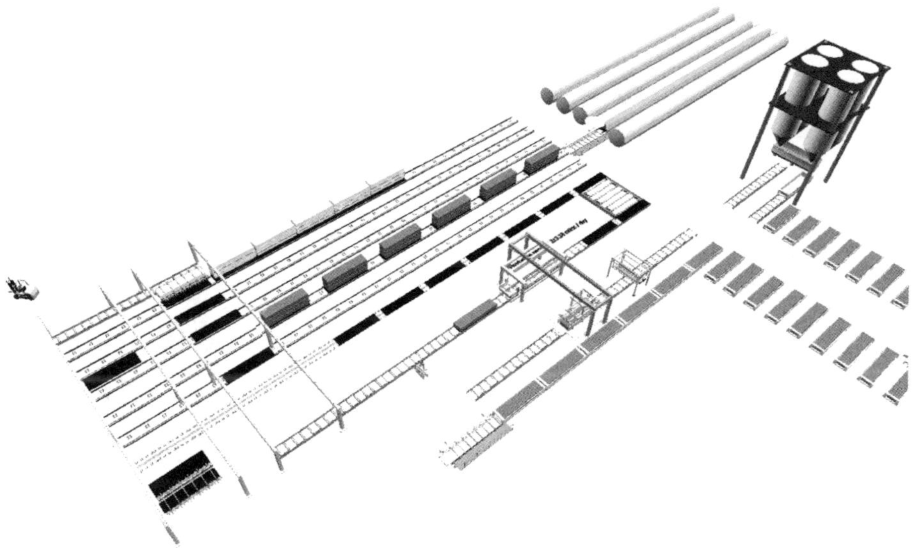

Figure 2. Keen-Crete plant concept tested with advanced plant simulation software.

The Keen-Crete plant has the future option for production of AAC reinforced panels of up to 6-meter length. In combination with the Stork flat cake cutting machine a high-speed cutting frame can be inserted for production of 'super-smooth' panels (a typically table top smoothness). One of the great advantages of these 'super-smooth' panels is the ability to apply wallpaper to the panel, without a need for plastering. This is another unique feature to the Keen-Crete AAC plant concept.

5 CONCLUSIONS

Since launching of the new concept offers has been well received by the market, attracting a great deal of interest from smaller investors already involved in the AAC business as well as from companies that are already involved in the building materials branch, however not yet in the field of AAC production. Also complete "start-ups" can now consider producing and selling high quality AAC.

The Keen-Crete concept turns the traditional disadvantage around that a system is expensive when supplied by a well-known supplier, who has over decades been involved in the high-end of the market, supplying to leading building material companies that have set the standard for AAC in the European market.

Without sacrificing on quality of the key machinery, the quality level of the product will remain high. The Keen-Crete user will enter into the AAC market with a solid product capable of achieving a high-end price, attributing to the worldwide expansion of AAC as a well received material in the building industry and propelling the possibilities to gain market share on local economic scale.

The modular logistic concept provides a reduction in investment costs and offers the possibility to subcontract a large share of the steel construction work as local content. For countries in which labour costs are low compared to European standards, saving possibilities may lay in reductions in the level of automation on a number of operations and monitoring.

The Keen-Crete concept is the result of many years of experience in the AAC market, a dedicated and continuous engineering effort to extend and improve the product range and the implementation of standardization in a modular concept for the benefit of our customers.

Autoclaved Aerated Concrete – Limbachiya and Roberts (eds)
© 2005 Taylor & Francis Group, London, ISBN 0 415 38356 0

Aircrete autoclaves by Terruzzi Fercalx

M. Sbrosi & E. Barberis
Terruzzi Fercalx Spa

ABSTRACT: TERRUZZI FERCALX SPA performs its activity on worldwide basis and provides industrial plants for several applications and fields of which the following are the most important:

- Aerated concrete industry
- Lime industry
- Pharmaceutical industry
- Aerospace industry
- Food industry
- Composite materials industry
- Chemical industry
- Steelmaking industry
- Aircraft industry
- Glass industry
- Impregnation industry
- Rubber industry

This paper briefly outlines quality aircrete autoclaved production by Terruzzi Fercalx Spa.

1 BACKGROUND INFORMATION

Terruzzi Fercalx Spa was generated by the merging of two companies of the Terruzzi Fercalx Group: Terruzzi Cristoforo Srl and Fercalx Spa, and is now organized in the following three divisions: Terruzzi division, Fercalx division and Terruzzi Foodtech division. The activity was started in the faraway 1897 by Terruzzi Daniele, the founder, pioneer in the manufacturing of carpentry and joinery, tank-wagons, boilers, autoclaves, innovative products of the mechanical industry.

Today Terruzzi Fercalx Spa, thanks to its organization gives to the customers the advantage to have access to comprehensive realization: process engineering, detailed design, manufacturing and

Figure 1. Production setup.

Figure 2. Ball mill for limestone.

Figure 3. Autoclave charging system.

supply of equipment, project management, erection and/or supervision to the erection and to the commissioning, training and after sales service. The above in order to provide a continuous and complete support through the whole installations life-cycle.

2 AUTOCLAVES

Terruzzi autoclaves have been realized for the most several operating conditions as herebelow mentioned:

- Diameter: up to 9.000 mm
- Length: up to 60.000 mm
- Vacuum/pressure: from 0.01 bar (absolute pressure) up to 28
- Temperature: up to 400°C.

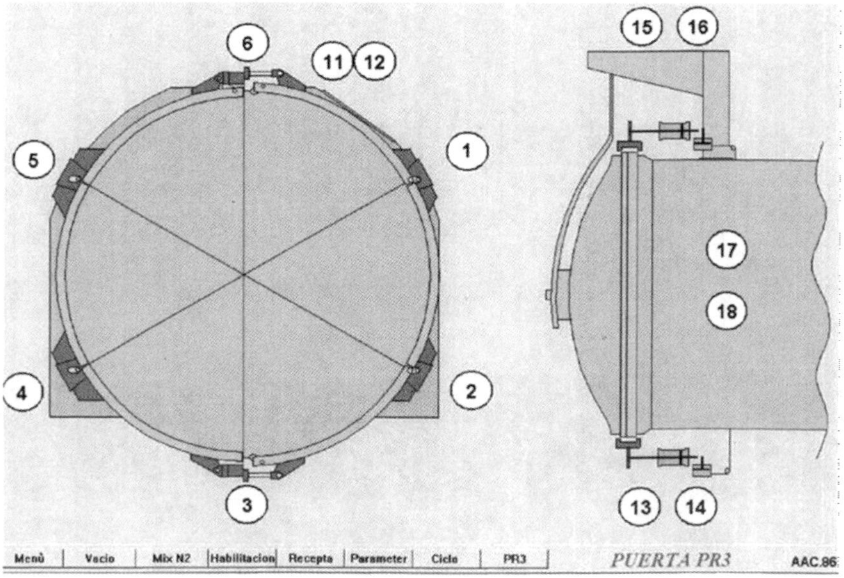

Figure 4. Fast locking system.

Figure 5. Fast locking system.

Specifically for aircrete applications the autoclaves have been realized as follows:

- Diameter: up to 3.250 mm
- Length: up to 60.000 mm
- Vacuum/pressure: from 0.1 bar (absolute pressure) up to 13
- Temperature: up to 200°C.

Figure 6. Autoclave body.

Of course the above dimensions and operating conditions are not a limit since the autoclaves can be sized according to specific customer needs.

3 MAIN FEATURES OF TERRUZZI FERCALX AUTOCLAVES

- Hinged, horizontal or vertical, with ball bearing support, with earth-discharged weight, designed in order not to stress the autoclave shell. The door can open/close by turning either aside or upwards.

  This system is widely used, but specific solutions are available for special applications. For these applications the tightness of the door is ensured by a special profile elastomeric gaskets, for operating conditions with service temperature up to 200°C and suitable to work under vacuum/pressure.

  Autoclaves working at higher temperature are equipped with a special cooling system suitable to keep low gaskets temperature to ensure their long life. Special and accurate machining of the contact flanges, is essential to ensure perfect tightness between the door and the autoclave body.
- The closing device used in the Terruzzi autoclaves, has the great advantage to prevent the gaskets from being stressed or rubbed.
- Door fast-locking system, consists of three forged and mechanically machined rings. The crowns of both the shell and the door are clamped up by the external hollow ring (the third) which is split into sections linked to each other through a hinges and guides system allowing it to close up and open through the control of oleodynamic pistons.

This system allows a perfectly tight enclosure of the first two rings into the third hollow ring during closing operation, as well as the release of the third ring to allow opening of the door.

Elastomer or vacuum gasket, self-tight, with special profile, suitable to withstand the maximum service pressure and temperature values of the autoclave. The gasket is set into a special, mechanically machined seat between the door ring and the shell terminal ring.

3.1 *Autoclave body*

Shell constructed with cylindrical rings longitudinally and cross electro welded to each other suitable for vacuum and the designed operating pressure. Convex bottom with elliptic profile worked out from a cold forged sheet and thermally treated whereas required by the standard specification.

3.1.1 *Bearing saddles, one stationary and the other skidding*

Saddles are sized to stand the weight of the autoclave structure and the material charged for the processing. Particularly strong is the stationary saddle (front one) which bears the weight of the door mechanism and of the related closing ring to avoid any mechanical stress deformation. The wide dimensions of the laid section allows a limited specific load on the foundations. Internal rails and supports, made of steel sections, positioned in the bottom of the autoclave, with gauge to be determined.

3.2 *Insulation*

The thermal insulation is installed on the autoclave body, made of high insulating ceramic and mineral fibre pads, duly anchored to the autoclave structure and coated with galvanized steel sheet.

Safety devices preventing the door from opening until the autoclave is completely depressurized, as well as the autoclave from pressurizing if the door is not perfectly locked. Three safety levels are set in series so that the door opening operation (or pressurizing with closed door) can take place only upon consent from all devices. The operation is controlled by a proper electronic logic.

They are:

- Safety switch: for minimum pressure: this device prevents the door opening until internal pressure has achieved the atmospheric value.
- Mechanical locking device: mechanic device which proves the perfect closure of the rings, pneumatically driven and controlled by the safety electric logic.
- Safety valve: the manual drive of this valve, both in case of closing as well as in case of opening, forces the operator to check the autoclave status in its interior.

Limit switches positioned both on the sectors of the split ring as well as on the other components of the safety closing device allow the release of the opening/closing procedure after approval of the electric logic of the safety device.

3.3 *Process and control equipment*

Terruzzi Fercalx realizes several solutions of process panels according to the specific requirements of the end users. Terruzzi Fercalx designed and elaborated its own processing software, highly flexible which can fit every specific customer's demands.

The hardware, realized in different solutions accordingly to the operative requirements of the plant and customer necessities, is mainly composed by:

- basic system with a simplified, semiautomatic control;
- intermediate system with dedicated PLC suitable to the automatic management of the process data;
- advanced system with PLC or DCS, TF supervision software, PC and colour printer.

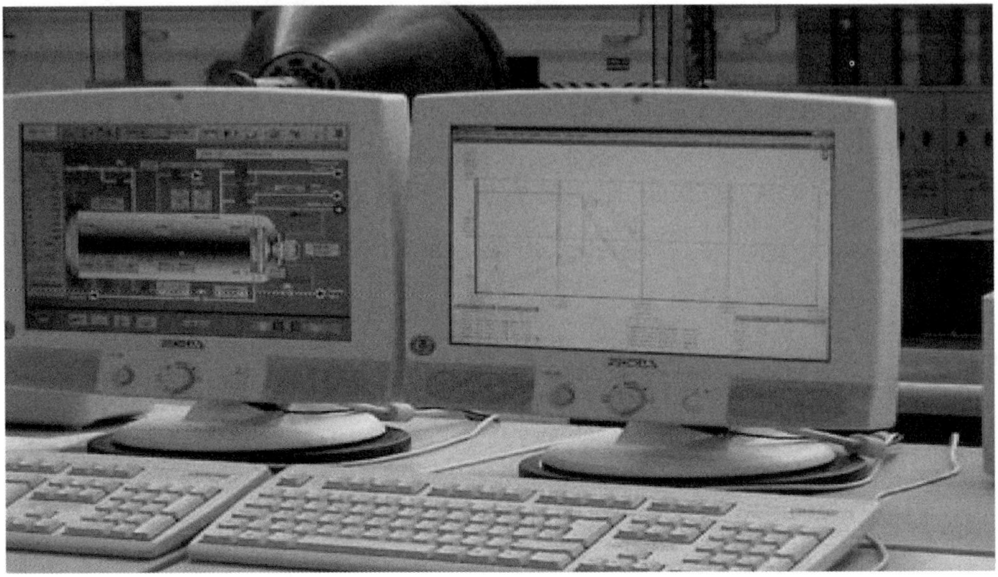

Figure 7. The plant automation is obtained by means of a PLC or DCS, that controls all the plant logical functions such as start-up and shut-down sequences, analogical and digital signals coming from field.

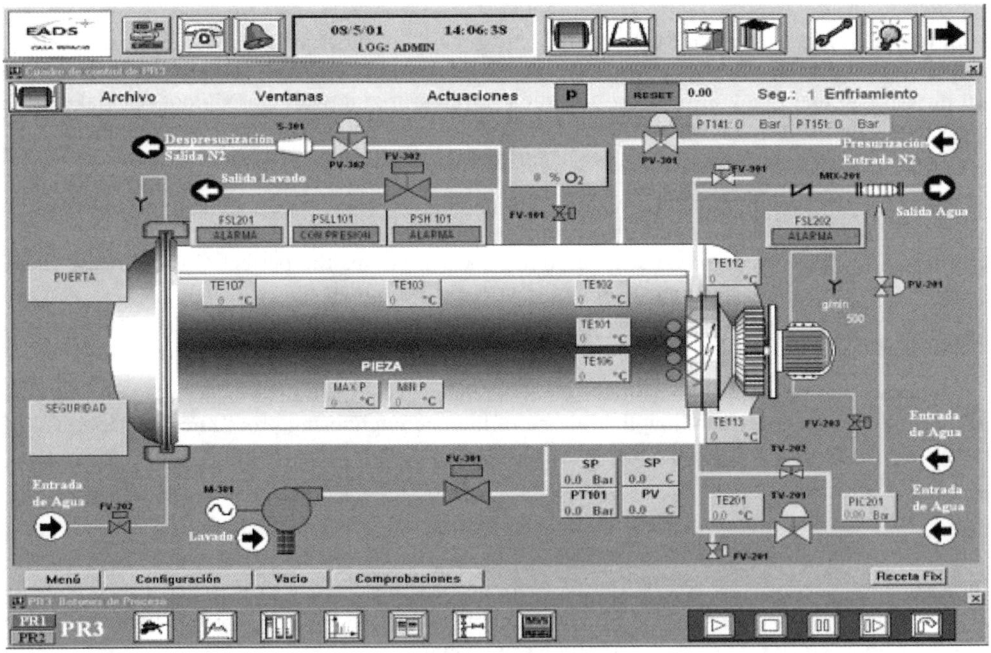

Figure 8. Overview.

Through the supervision graphic software it will be possible the visualization of the following functions:

– Colour plant synoptic, joined of a number of graphic pages with start-stop lighted signals of electric motors, alarms and main analogical values in real time.

Figure 9. Autoclave for aerospace industry.

- Historical trend in real time of the main variables of the plant.
- Alarms lists for blocks interventions.

It will be possible to fix the set-points of the various graphic video instruments and the recording of all the operation parameters and related trends on the personal computer for a max. period of 2 months. Only the totalized sizes are maintained for the whole year.

The control system, so managed by the TF supervision graphic software, can have two operation conditions:

- *Automatic operation*: In this condition the control is operated by PLC and visualized by personal computer, with the acquiring of the operating state of the equipment and machines, the verifying of block sequences and integration of the single regulation loops.
- *Manual operation*: In this condition the installation control is performed by keyboard with the possibility of:
  - Operation of every single end user.
  - Variation of the parameters of each regulator, operating on the related set-point.

Autoclaved Aerated Concrete – Limbachiya and Roberts (eds)
 ISBN 0 415 38356 0

A fast and accurate dosing easily achieved with a clever fluidization system

K. Saarinen & Jan-Peter Björklund
Raute Precision Oy, Plaster and Mortar Plants, The Netherlands

ABSTRACT: Even the best mixer can not ensure a good product if the batch is not dosed correctly. That's why dosing and weighing, together with the control system, is a very important part of the dry mix production process. With an excellent fluidization system it is possible not only to prevent bridging in the raw material silos but also to combine the fluidization with a dosing system using flaps. The advantage is an extremely fast yet very accurate dosing with a minimum of moving parts.

1 DOSING ACCURACY

To answer the demands from the market, Raute has developed a plaster and mortar plant design, which provides high efficiency and precision, as well as flexibility in adapting itself to the local raw material base and quick product changes.

Raute's plant design is based on a patented silo discharge and dosing system which makes it possible to build gravity-based plants without any screw conveyors for powder material transport. This provides many advantages like compact size of the mixing tower, easier raw material changes, faster and more accurate dosing and less moving parts. In this system, the dried sand and other raw materials are brought up to the top of the tower, and thereafter, all the material transfer takes place by gravity. The material is extracted from the silos by means of carefully controlled airflow. Horizontal transfer of material is minimised in all process stages.

Raute Precision has nearly 90 years' experience of supplying weighing, dosing and bulk materials handling processes. Raute's know-how is based on the continuous R&D work done in the company's pilot plant for testing of bulk materials, in its accredited mass and force laboratory and in the laboratory for development of plasters and mortars.

The weighing of the materials to be mixed is performed with Raute's high-precision process scales and filling and discharging devices. The weighing and dosing process is controlled by Raute WB-950 multi-component batch controllers, which combine very high accuracy with the fastest measurement and control. The WB-controllers control all weighing, dosing and discharging equipment directly and provide advanced calibration, diagnostic and maintenance functions to guarantee excellent batch quality. During the dosing of a particular component, the self learning batch controller controls the coarse/fine tuning with the butterfly valve and shuts off the feed at the appropriate time paying attention to the "in-flight material" factor. This factor is corrected/fine tuned by the batch controller always when the dosing result shows a deviation from the target. In this way, the dosing automation responds to changes in the raw material properties. This technique is not new, nor unique but together with a very uniform material flow, thanks to the fluidization technique, it is at its best.

1.1 *Optimum silo discharge*

1.1.1 *Economical dosing without screws*

A basic problem in the dry mix production plant is the tendency of fine materials to bridge, arch or build rat holes in the silos. For mixtures segregation can be a very bad problem. As a solution

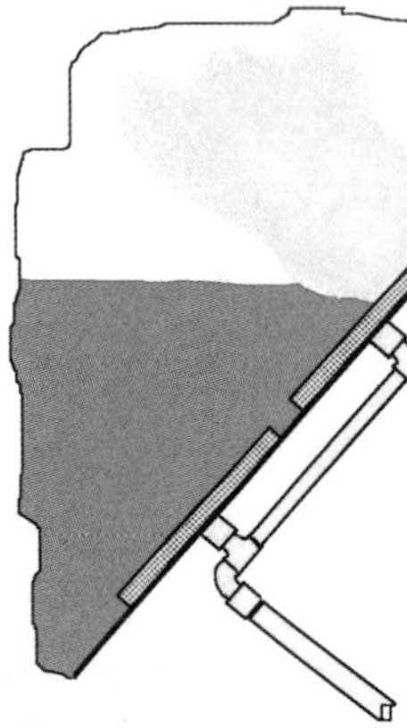

Figure 1. In traditional fluidization systems the fluidization is unstable or faulty because the powder resistance is the determining air restrictor.

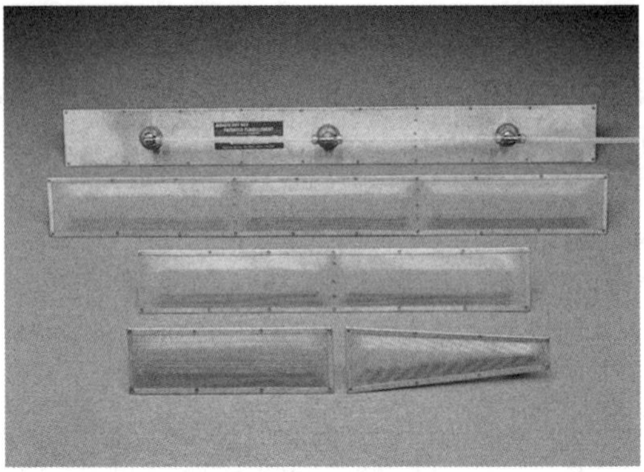

Figure 2. The elements only consume about 8 litres compressed air per minute at 1 bar pro nozzle. Because the fluidization is used only in dosing mode the energy demand is very low.

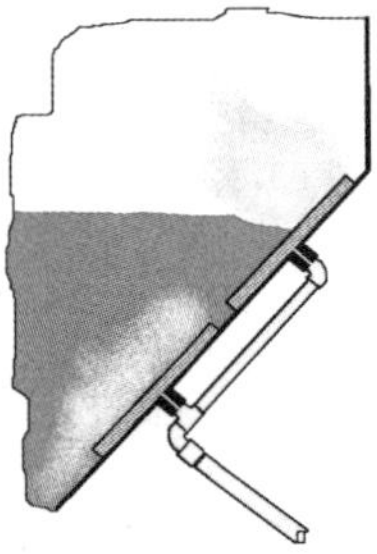

Figure 3. In Raute's patented fluidization system the nozzles keep the air stream steady regardless of the material resistance.

Figure 4. The controlled fluidization system keeps the material flow even. The fluidized material runs out of the discharge pipe almost like water.

to this, Raute offers a special fluidization technique, which is widely used by Raute's customers in building materials, chemical and metallurgical industry.

The core idea of any fluidization system is quite simple: the friction between the silo wall and the powder as well as between the particles is lowered by means of an evenly distributed air stream. The problem is usually to distribute the air in a controlled way, avoiding bubbling and chimney effects. It is also essential to get the air penetrating the settled compact material. The air should not be allowed to take the easiest way through weak layers or trough panels located above the material level.

The patented system keeps the flow even and prevents the creation of dust, excess pressure in the silo, and segregation in the material, and saves compressed air. The fluidization works with different types of raw materials such as fine limestone powders, cements, burnt lime and various dry powder chemicals.

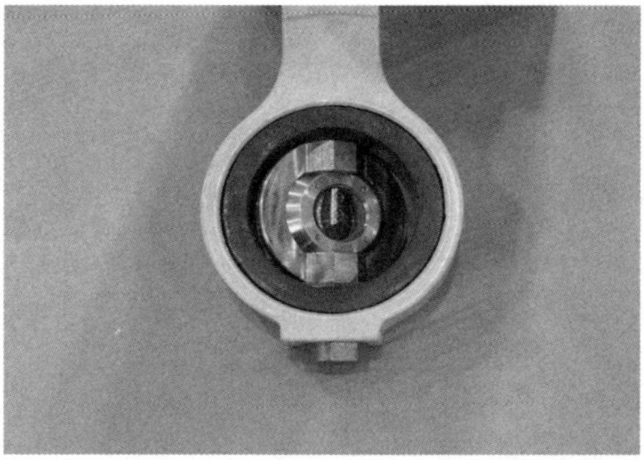

Figure 5. When combining the fluidization system, the dosing flap and the weighing technology with the advanced control system, very high accuracy – and also high capacity – can be achieved.

The fluidization is performed by means of specially designed fluidization elements on the wall of the conical part of the silo. The air stream through the nozzles of the elements is controlled in such a way that the amount of input air is virtually independent of the prevailing density of the material at the silo wall. In this way, a sustaining air cushion around the neck of the silo is achieved. The material in the tip of the silo is fluidized, and runs out of the discharge pipe almost like water.

To complement the specially designed fluidization elements, Raute has developed a double flap dosing valve to use with powder chemicals. In this valve, an additional, smaller built-in valve for fine feeding is placed inside of the main valve. The main valve is opened for coarser dosing. With the Raute fluidization system, the capacity of large diameter screw conveyors and the accuracy of small screws are achieved without any feeding screws.

2 WEIGHING AND DOSING OF MATERIALS

Dosing of materials is carried out by scales. In a factory there are normally scales for aggregates, binders and additives. The size of the scales depends on the amount to be balanced, e.g. the amount of aggregates like sand is 70–80% of a batch, which means several hundred kilos. Smaller scales are needed always for additives and sometimes for binders.

Considering the quality of dosing, we must distinguish the weighing accuracy from the dosing accuracy.

(a) Weighing accuracy
Weighing accuracy of a scale is defined by uncertainty in measurement. For that, a step value (d) is defined describing the scale resolution, being equal to one scale division on the scale display. In scales used in dry mix plants, the number of step values or scale divisions normally ranges from 1000 to 5000. The accuracy of scale (e) is illustrated in the chart below in terms of step value d. The weight, which loads the scale, is expressed as number of d values (n). However, the minimum acceptable load is 20d. In static weighing, the accuracy is in general 0,05–0,1% of the maximum capacity of the scale.

(b) Dosing accuracy
The dosing accuracy means the capability of the combination of scale and dosing device to feed a preset amount of material onto the scale or off the scale. The accuracy that can be achieved is affected by the properties of load cells, weighing instrument, mechanical devices and also of the material itself. In practice, the dosing accuracy in batch processes using a scale with 5000 divisions and with free flowing materials is 2 … 5 scale divisions. In the dry mix factories, a thumb rule is to have the factor 3d.

The Figure 6 presents examples of typical scales used in dry mix plants and their dosing accuracy. However, the factor e is not included in figures in the table for practical reasons. In

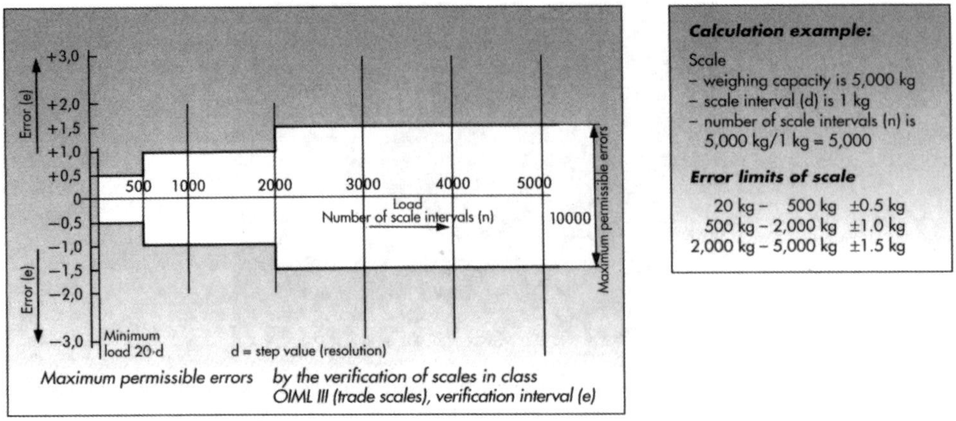

Figure 6. Achievements in practice.

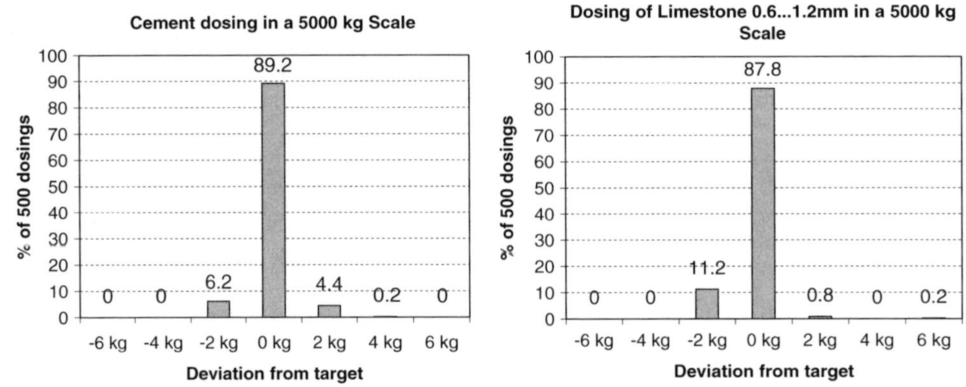

Figure 7. Minimum maintenance requirements.

practice, the dosing accuracy is followed by the scale display, and the weighing accuracy of scale is maintained by yearly calibration.

(c) *Requirements for dosing in practise and according to the standards*

For dosing accuracy, there are some regulations also in standards. The German standard for renders and masonry mortar requires for dosing of individual ingredients the accuracy of 5% of ingredient amount and 3% of total product amount.

The requirements seem to be quite wide. As a thumb rule in practice, 1% can be recommended for all raw materials except pigments for which the accuracy has to be 0,5% or better.

In the figures below there are example charts showing the deviation from the target for cement and limestone sand. The data is on line recordings collected by our client for 500 batches each. The scale capacity is 5000 kg with 2 kg divisions (steps).

All main raw materials were dosed in the same scale. Total batch sizes were between 4 and 5 ton each and the dosing time was not allowed to exceed 60 seconds for the whole batch (including all main raw materials). For a 4.5 ton batch this means that the *average* dosing speed (including coarse and fine feed and switching between several materials dosed after each other) has to exceed 75 kg/s.

Raute's gravity-based plant design with special silo discharge and dosing system reduces the requirements for maintenance and spare parts. Since no screw conveyors are required, the number of moving, maintenance-dependent parts is very small. Actually, the only wearing parts are the flaps i.e. the sealings. By keeping one spare flap of each size (only a few sizes normally) in stock the change is carried out easily. The flap taken out can then be fitted with a new sealing for future use.

The large number of raw materials and silos in a modern dry mix plant sets stringent requirements on the flexibility of product change. With a wide selection of products, it may sometimes be necessary to change raw materials in the silos. The swiftness and cleanness in changing are significant. Over the use of screw feeders, the direct dosing from the silo discharge offers a distinct advantage: the need for inter-product cleaning is radically reduced.

Raute's process control software, known as RAUTE Batchware™, also includes elements for service and trouble-shooting. The control system can be remote controlled even from external stations by modem or network connection. All necessary changes, covering programming, administrating and trouble-shooting, can be made by our experts via telephone, thus optimising the maintenance personnel's work and reducing service time and costs.

Theme Two: Product properties and performance

Autoclaved Aerated Concrete – Limbachiya and Roberts (eds)
© 2005 Taylor & Francis Group, London, ISBN 0 415 38356 0

UK housing and climate change: performance evaluation using AAC

C.A. Fudge
Autoclaved Aerated Concrete Products Association Limited, Sevenoaks, Kent, UK

J.N. Hacker
Arup, London, UK

ABSTRACT: This paper reports on a study that proposes a building design response to climate warming for dwellings. It draws on research work examining the impact of climate change on thermal comfort in buildings. With a significant proportion of the UK's expected new homes to be in the Southeast where climate change will be most pronounced, it will be important to ensure these homes are future proofed for climate change. The alternative is homes that may become prematurely obsolete; requiring the addition of mechanical cooling, increasing energy costs and future increasing greenhouse gas emissions.

This research illustrates the relatively good performance of AAC masonry construction, as part of a response that includes solar, ventilation and internal heat gain control. The paper reports on a computer simulation incorporating AAC masonry in response to changes in climate and the beneficial effect of thermal mass.

1 INTRODUCTION

Climate change has the potential to have a very major impact on buildings in the UK and the way in which they will be designed in coming years. The latest projections for climate change in the UK over the 21st Century (Intergovernmental Panel on Climate Change 2001) point to substantially warmer summers. The need for comfort cooling in summer is therefore expected to be greater than has been the case historically.

Arup have previously carried out a study for the UK Government to examine this question. Some of the results of this study were published as CIBSE TM36 (CIBSE 2005). The focus of the study was identifying architectural features that would enable reduction in mechanical cooling needs. Two key aspects identified were solar shading and 24-hour intelligent control of ventilation. It was also found that use of higher thermal mass building materials was generally beneficial, imparting greater thermal stability to a building and a reduction in peak space temperatures. The 'passive cooling' provided by use of thermal mass was predicted to keep internal temperatures several degrees below the external temperature on peak summer days, and moreover, several degrees below the space temperatures that would otherwise occur in lighter weight structures.

Use of passive measures to limit summertime mechanical cooling needs is an important part of international efforts to reduce the carbon dioxide emissions from fossil fuel use which are the major factor causing global warming. In this respect, dwellings are of particular importance, as in the UK they already account for around two-thirds of carbon emissions from the building stock. Wider use of air conditioning for summertime comfort cooling in dwellings in the UK would be likely to increase the carbon emissions from this sector considerably. Energy modelling studies in CIBSE TM36 indicate that even taking into account the reduced heating energy needed under a warmer climate, use of air conditioning in dwellings has the capacity to increase carbon dioxide emissions under present-day design conditions for London by around 20%, and by over 40% by the 2080s under the UK Government's climate change scenarios.

The CIBSE TM36 study considered four housing cases: a late 19th Century semi-detached house and three examples of a newly constructed 4-bedroom house – a so-called 'new-build house' – each meeting the current energy conservation insulation standards in UK building regulations – but with different levels of fabric thermal mass: lightweight, medium-weight and heavyweight. In each case the houses were also 'adapted' in a way that would improve their performance under the future climate change scenarios. The adaptation measures focused on controlling heat flows from solar gain and ventilation air and were intended to be generic and specific ways in which these measures could be implemented were not considered.

This paper presents an analysis of a house similar to the 'new-build house' case studies from CIBSE TM36, but constructed to a somewhat different specification using autoclaved aerated concrete blocks. The aim of the study has been to examine the effectiveness of the fabric thermal storage/mass cooling in a house constructed to this specification and in particular to examine whether the type of wall finishes used affect the overall thermal performance. The types of wall finish considered are: Case (A) lightweight plaster, and Case (B) plaster board on dabs, with a 10 mm air gap. These surface finishes are of interest because of their possible undesirable effect of insulating (isolating) the thermal mass of the AAC blockwork walls from the interior spaces. The predicted thermal performance has also been compared with the three TM36 cases of lightweight, medium-weight and heavyweight new-build houses to enable the new construction specification to be placed in the context of that study.

The building form and layout considered is the same as that considered in CIBSE TM36 and also related to previous reports. The construction details for the external and interior walls are somewhat different and are as outlined below.

- *General description:* Four-bedroom detached house built to the 2002 Building Regulations standards of fabric insulation. Total floor area of approx. 140 m^2, representative e.g. of a new-build 'executive home'.
- *Orientation:* The front façade of the building faces south.
- *Fenestration:* Double-glazed units. South façade glazed area is approximately 14%.
- *Insulation:* External walls: $U = 0.30\ W/m^2K$. Windows $U = 1.9\ W/m^2K$.
- *Air-tightness:* Background infiltration of 0.5 ac/h provided from trickle ventilators.
- *Heating system:* Gas central heating with thermostatic radiator valves fitted to all radiators. Heating set points of 22°C for living areas and 18°C for bedrooms.
- *Occupancy:* The house is considered to be occupied by two adults and two children. It is assumed that all occupants are out during most of the day from Monday to Friday and present to varying degrees during the day at the weekend. All occupants are present at other times.
- *Lighting:* The lighting level is assumed to be controlled by occupants according to available day-lighting. Lights come on when daylight falls below a fixed threshold in occupied rooms, except in bedrooms when occupants are asleep.
- *Internal gains:* Heat gains due to lighting and TV/Stereo equipment in the living room were included. Heat gains from kitchen appliances and hot water services were not included.
- *Ventilation and Shading:* The details of shading and ventilation provision and controls are dependent on the manner in which the house is being used. Two cases have been considered:
  - Unadapted: No modifications applied to take account of the changing climate: no shading; natural ventilation by opening windows in summer.
  - Adapted: Modifications applied to take account of the changing climate: shading applied in summer with only 10% transmittance; controlled mechanical ventilation supplying up to 6ac/h operating when space temperatures are above the heating set point and at least 3°C higher than the external temperature.
- *Weather data:* CIBSE Design Summer Years (DSYs) for London, Manchester and Edinburgh under '1980s' climate (1976–1995 period) and synthetic future weather years for climate time-slices 2020s, 2050s and 2080s under the UKCIP02 Medium-High emissions climate change scenario. Full details of the climate scenarios and weather data used are given in CIBSE TM36.

Table 1. Properties of materials considered for Cases A and B.

Material	Thermal conductivity (W/mK)	Density (kg/m^3)	Specific heat capacity (kJ/kg K)
• Low-density AAC block, 100 mm thickness	0.11	480	1.05
• Medium-density AAC block, 100 mm thickness	0.15	600	1.05
• Lightweight plaster	0.16	600	1.00
• Plaster board	0.16	950	0.84

- *Construction:* The house constructions considered have been:

External walls:	outer leaf – single course of brick (105 mm)/50 mm cavity/ 75–100 mm filled cavity (mineral wool or EPS)/100 mm low-density AAC block (thermal properties given in Table 1)/ one of two types of internal wall finish (Cases A and B specified below).
Ground floor internal walls:	100 mm medium-density Autoclaved Aerated Concrete block (thermal properties given in Table 1) with one of two types of wall finish (Cases A and B specified below).
Upper floor internal walls:	timber studwork walls with plasterboard finish.
Floors:	timber on upper storey, solid concrete on ground floor, all carpeted.
Roof:	Synthetic slate tiled roof.

- The two types of wall finished considered were: Case A: 13 mm of lightweight plaster, and Case B: Finish 2. 12.5 mm plasterboard on 10 mm dabs.

2 MODELLING

Dynamic thermal modelling of the three houses was carried out using ENERGY2, a program developed in-house at Arup over a number of years. The program has a finite difference representation of wall heat transfer, uses hourly weather data (with 4 time steps per hour) and has a 2-layer stratification model to represent buoyancy driven airflow. The program has been found to give predictions close to the median of a range of dynamic thermal simulation models considered in international software validation tests. The program was also used for the modelling studies reported in CIBSE TM36.

Each of the three house models was analysed using Design Summer Year (DSY) weather data for London, Manchester and Edinburgh from the '1980s' baseline climate period (1976–1995), as currently recommended by CIBSE for summertime thermal design, and synthetic 'morphed' future DSY weather data under the UK Government's UKCIP02 Medium-High emissions scenario 2020s, 2050s and 2080s climate timeslices.

Full details of the weather data, climate scenarios and 'morphing' procedure are contained in CIBSE TM36. Here focus will be given to the results for London as an example.

3 COMFORT BENCHMARK TEMPERATURES

There is no universally agreed standard for thermal comfort temperatures and overheating risk for buildings in the UK. Typical target temperatures for naturally ventilated offices are 25°C to 28°C.

For schools the Department for Education and Skills specifies that not more than 80 hours in the year should exceed 28°C. There are no published guidelines for dwellings and little research has been carried out, particularly the important issue of the effects of warm night-time temperatures on quality of sleep and health. However, in order to enable an objective quantitative assessment to be made, here illustrative benchmarks have been defined.

For living areas, discomfort threshold temperatures of 25°C and 28°C have been considered. For bedrooms, lower discomfort threshold temperatures of 21°C and 25°C have been considered on the grounds that people are more susceptible to thermal discomfort at night, when they expect temperatures to be lower and when they are asleep. In each case, the lower threshold temperature has been taken to be indicative of when occupants will start to feel 'warm', and the upper threshold temperature to indicate when they will start to feel 'hot'. It is considered that occurrences of both hot conditions or prolonged periods of warm conditions will motivate occupants to seek comfort cooling. In this respect, the degree to which the comfort targets are met will be quantified in terms of the percentage of occupied hours in the year for which the comfort temperatures are exceeded.

Hours of exceedance provide a measure of the length of time the temperature is over the discomfort thresholds, but not the magnitude. To convey this information, the maximum space temperatures reached over the course of the year will also be considered. Temperatures over 35°C are considered to be potentially hazardous to the health of healthy adults, but adverse health effects can be experienced in elderly people, children and the physically infirm at lower temperatures (ASHRAE Inc. 2001).

A third diagnostic than can be used to quantify the potential for summertime discomfort is the number of 'cooling degree hours' for internal space temperatures relative to a specified base temperatures. For a given room and hour of time, the cooling degree hours are the number of °C by which the room temperature exceeds the base temperature (zero if room temperature is below the space temperature). The total number of cooling degree hours is calculated by summing the cooling degree hours for each hour of the year across all occupied rooms. This diagnostic provides, then, a combined measure of both the duration and magnitude of the level of thermal discomfort relative to the base-temperature. It may be also considered to provide a simple indication of the cooling energy that would be required to keep the space temperatures at or below the base temperature. Here cooling degree hours relative to a base temperature of 27°C have been considered. This diagnostic was used in a recent study of overheating in well-insulated dwellings (Orme & Palmer 2003; note that in this study all hours were considered, whereas here only occupied hours are considered).

In each case all comfort diagnostics will relate to operative temperature (average of air and surface temperatures) rather than ambient air temperature. Operative temperature gives a more accurate reflection of actually perceived temperature as it takes into account the radiant cooling (or heating) effect of room surfaces.

4 RESULTS

The results from the ENERGY2 modelling have been used to assess two aspects of the thermal performance of the house: i) the extent to which the type of surface finish affects thermal performance (difference between Cases A and B) and ii) how the thermal performance compares to the three different types of New-Build house construction considered in CIBSE TM36.

With regard to the first question, a negligible difference in thermal performance was predicted between the two types of surface finish, measured against the different discomfort diagnostics described in Section 3. This result may be further understood by considering the admittance (Y-values) of the surfaces of the two finish types. These have been calculated using the standard CIBSE admittance calculation procedure (CIBSE Guide A) and are given in Tables 2 and 3 for the internal and external wall constructions, respectively.

The Y-values for both the internal and external wall constructions considered in Cases A and B are very similar. The difference in the Y-values between the two cases for both internal and external walls is generally less than 0.2 W/m^2 K. The similarity of the Y-values is consistent with the lack of difference found between the two cases in the ENERGY2 modelling. The reason for the similarity

Table 2. Admittance Y-values and U-values of external wall construction types.

Construction	Y-value (W/m^2 K)	U-value (W/m^2 K)
• Case A – lightweight plaster finish	1.95	0.28
• Case B – plaster board on dabs finish	1.76	0.27
• As Cases A and B – no surface finish	1.83	0.29
• As Case B but with plaster board having thermal properties equal to those of lightweight plaster	1.65	0.27
• As Case B but with inner leaf air cavity replaced with solid timber	2.18	0.27
• CIBSE reference values for: 105 mm brick, 25 mm air space, 25 mm EPS insulation, 150 mm AAC block (density 500 kg/m^3), 13 mm lightweight plaster	2.18	0.47

Table 3. Admittance Y-values and U-values of internal wall construction types.

Construction	Y-value (W/m^2 K)	U-value (W/m^2 K)
• Case A (ground floor) – lightweight plaster finish	2.26	0.94
• Case B (ground floor)– plaster board on dabs finish	2.08	0.74
• Cases A and B (1st floor) – plaster board on studwork wall	1.64	1.32
• As Cases A and B (ground floor) – no surface finish	2.01	1.11
• CIBSE reference values for: 105 mm brick, 50 mm air space, 150 mm AAC block (density 700 kg/m^3), 13 mm lightweight plaster	2.50	1.07

in the Y-values is the materials used in the surface finishes all have similar thermal properties to the underlying blockwork. Although the air cavity between the blockwork wall and plaster board in Case B has an insulating effect that would tend to reduce the Y-value of the wall, this insulating effect appears to be offset by the additional thermal mass added to the wall by the surface finish and, for the external walls, the plasterboard is also slightly heavier weight than the blockwork. To extrapolate these conclusions to other cases, we would expect that the thermal performance of the house would not be greatly changed if no surface finish were applied to the blockwork (although note that the two different surface finished would be expected to have a significant impact on Y-values if the underlying blockwork were heavier weight).

With regard to the second question, the AAC house has been compared to the CIBSE TM36 Lightweight (timber frame), medium-weight (lightweight aggregate concrete blocks) and heavy-weight (dense aggregate concrete blocks) house constructions. Broadly speaking, from the comparison with the Case A results and those from TM36 for New-Build house models for lightweight, medium-weight, and heavyweight construction, the following conclusions can be reached.

Ground floor: the thermal response of the living room in Case A is predicted to be closest to the medium-weight case in both the unadapted and adapted cases. While the AAC blocks used in the external walls are relatively lightweight compared to the CIBSE TM36 medium-weight case (450 kg/m^3 vs. 1400 kg/m^3) all the internal ground floor partitions in the CIBSE TM36 medium-weight house are timber/plaster board stud whereas in the AAC house all the groundfloor partitions are blockwork (600 kg/m^3) thereby providing additional thermal mass and compensating for the relatively lightweight external walls.

Table 4. Internal operative temperature cooling degree hours (summed across all spaces; occupied hours only) to a base temperature of 27°C for the two AAC house constructions and those considered in CIBSE TM36.

London DSY climate	Cooling degree hours (°C hr)					Ratio case A/ Light (%)
	Case A	Case B	Medium	Light	Heavy	
1989s	430	455	446	888	132	48.4
2020s	723	761	738	1419	241	51.0
2050s	1357	1414	1358	2389	554	56.8
2080s	2866	2952	2849	4356	1572	65.8

First floor: the thermal performance of the main bedroom in Case A is predicted to be midway between the lightweight and medium-weight cases, reflective of the relatively lighter weight character of the first floor construction. On the one hand, the lighter weight thermal characteristic lead to reduced overall hours of overheating at night, against the two comfort temperatures, considered, but on the other hand, higher instantaneous peak temperatures.

Further information on the overall thermal performance of the houses with respect to summertime overheating may be gained from considering the internal space temperature 'cooling degree hours' ti a base temperature of 27°C. The values for the different cases are shown in Table 4. The values for the two AAC cases are similar to those for the medium-weight CIBSE TM36 case but are significantly lower than the lightweight case, by around 35–50% (last column of Table 4). The relative difference decreases through the timeslices because the average climate gets warmer relatively to the 27°C baseline. The cooling degree hours for the heavyweight case are significantly lower than the other cases, but again the degree of difference decreases through the timeslices.

5 SUMMARY AND CONCLUSIONS

In summary, for the two AAC house cases considered there was little difference in predicted thermal performance between the two types of surface finish – lightweight plaster or plasterboard on dabs – indicating that the positive role of the blockwork in providing fabric energy storage for the building could be accessed in both cases. The overall thermal performance of the AAC houses compares favourably against the range of cases considered in CIBSE TM36, being closest to the medium-weight case. The overall level of overheating in the AAC houses measured by the number of cooling degree hours to a base temperature of 27°C were very similar to the CIBSE TM36 medium weight case, significantly less than in the lightweight house, but also significantly larger than in the heavyweight case.

REFERENCES

ASHRAE Inc 2001, *Fundamentals.* ASHRAE Atlanta.

CIBSE 2005, TM36 'Climate change and the indoor environment: Impacts and Adaptation'.

CIBSE 1999, Guide A- *Environmental Design.*

Orme, M and Palmer, J 2003 Control of overheating in future housing – design guidance for low energy strategies. Faber-Maunsell.

Intergovernmental Panel on Climate Change, 2001, *Climate Change 2001:The Scientific Basis,* Cambridge University Press (www.ipcc.ch).

Autoclaved Aerated Concrete – Limbachiya and Roberts (eds)
 ISBN 0 415 38356 0

Simultaneous determination of sorption isotherm and water diffusivity of Autoclaved Aerated Concrete

S. Tada
Texte, Inc., Tokyo, Japan

ABSTRACT: Autoclaved Aerated Concrete – AAC undergoes substantial influence of transport and phase change of water throughout its service life. Numerical solution of moisture balance equations can provide time-dependent moisture content profiles of AAC panels, but the coefficients in the equations have to be determined in experiment with considerable time and care. A rapid method to determine sorption isotherm and water diffusivity at the same time is presented in this paper. Humidity boundary conditions were applied to AAC specimens, and the resulting water adsorption was continuously recorded in a sufficiently short period. The diffusion equation was analytically solved and was curve-fitted to the adsorption rate, resulting in the rapid estimation of both equilibrium water content and water diffusivity at the same time. Results of sorption isotherm and water diffusivity showed good agreement with experiments. The adsorption rate measurement took 3 hours with a measurement interval of 60 seconds to determine an equilibrium moisture content and a water diffusivity at an ambient relative humidity.

1 INTRODUCTION

Change in moisture profiles of AAC with time is one of the most important factors affecting the service life control of AAC building elements. Attempts of simulating the moisture condition in AAC have been performed since 1960's (Hanson 1960, Sandberg 1971, Andersson 1985, Gawin 1993). The numerical simulation needs to solve the moisture balance equation where coefficients, such as moisture capacity and water diffusivity, govern the overall precision of the simulation results. The most reliable value of the coefficients can be obtained through experiments for which, however, considerable time and care had to be taken.

This study presents a completely new experimental technique capable of determining sorption isotherm, the primary data of moisture capacity, and water diffusivity simultaneously and very quickly. For simplicity, the scope of this study was limited to the isothermal water transport and to the hygroscopic moisture content range. Stepwise changes in relative humidity were applied to the AAC specimen and the adsorption rate was recorded continuously. The adsorption rate at a stepwise change in relative humidity was fitted to an analytical solution of the diffusion equation at the boundary condition resulting in the rapid estimation of equilibrium moisture content and water diffusivity at a relative humidity.

2 PRINCIPLE OF MEASURING METHOD

2.1 *Conventional methods*

(1) *Equilibrium moisture content*
Correct equilibrium moisture content can be obtained when relative humidities are accurately generated and adsorbed mass is continuously weighed. Standard methods of relative humidity generation include the saturated salt solution method, flow-division method that mixes dry and

saturated air and the method altering temperature or pressure or both in a closed chamber. The volumetric method that isothermally changes the vapor pressure in a vacuum system has been widely used in chemical engineering laboratories.

Haggymassy (1970) used a weighing bottle with saturated salt solutions under vacuum and determined isotherms of hardened cement pastes. The mean free path of water vapor decreases under a reduced pressure leading to an increase in diffusivity of water vapor in air. This reduces the time to reach an equilibrium and is effective also in preventing carbonation of cement based specimens. Yuasa et al. (1998) ground specimens into a powder and introduced stirring mechanism in a saturated salt solution container. However, the saturated salt solution method has a disadvantage of generating a limited number of relative humidities. This method, as well as the volumetric method, is a standard with its simplicity and is placed as the reference method of this study.

Commercial instruments based on the two-temperature method cannot be used for the determination of equiriblium moisture content because the lower limit of available relative humidity is as high as 30% at a temperature of 20°C. Availability of a saturator capable of generating saturated air at a temperature as low as 4°C is unknown and no example of measurement was found. Ahlgren (1972) built an apparatus based on the two-pressure method and determined a variety of isotherms for concretes, AAC, mortars, stones, bricks and various types of woods at seven points of relative humidities ranging from 20 to 98 %.

The flow-division method can generate a wide range of relative humidities. Because the conditioned air flows continuously, the specimen chamber can be stirred and need not be a closed space making it easy to monitor the specimen mass, for instance by an electronic balance, from outside of the system. Takahashi et al. (1998) studied measuring conditions and precision of flow-division humidity generator. It was shown that the overall precision is dominated by that of the mass flow controllers and their change in performance with time, and that the precision falls in a range of 1 to 2 percent when employing the approximated formula specified in Japanese Industrial Standard – JIS Z 8806.

In this study, a commercial flow-division humidity generator was calibrated by a dew hygrometer to have a flow-ratio table dedicated to the generator enabling of running at a precision more than ±1%, and the computer controlled electric balance was used for a fully automated measuring system. Thus the apparatus can maintain a precise humidity for a long time and, after the long period of adsorption rate monitoring, we may obtain a moisture content as close as possible to the equilibrium moisture content. Results in a practically permissible duration, however, may only be obtained on the basis of a completely new prediction model.

(2) *Water diffusivity*

If the liquid flux and moisture content distribution in AAC specimen are obtained at the same time, the water diffusivity can be calculated from the flux equation, for which two methods of measuring the liquid flux have been known.

Kooi (1974) and Matsumoto (1978) applied horizontal water input to a cylinder specimen to minimize the gravity effect, and the flux was determined from a moving menisci in a measuring pipette connected to a water tank. When the inflow reached a steady-state condition, specimen was cut into pieces and the moisture distribution was determined. Water diffusivity obtained with this method is of a narrow moisture content range and is limited only to an absorption process.

Jonell et al. (1960) executed a vertical water uptake test for several specimens and cut one of them into pieces consecutively at a specified period to determine moisture content distributions. The water fluxes were calculated from the overall mass increase and moisture contents of each cut pieces. This method, unnecessary to wait until a steady-state condition, is quicker and can cover wider moisture range when taking a longer period of cutting a specimen. However, the specimen may be limited to easy-to-cut materials and the result may scatter because the results of different specimens are combined in a single curve. The second method, having a problem in accuracy, has an advantage of simplicity without a need of equipments.

With an initial and a boundary condition meeting the Bolzmann transformation, introduction of Bolzmann parameter to the mass balance equation gives an ordinary differential equation that has a solution of $t^{0.5}$ type. Further integrating the solution gives the water diffusivity as a function of moisture content and the Boltzmann parameter. By measuring distributions of moisture content at each time, relations between moisture content and Bolzmann parameter can be obtained, and the water diffusivity can be determined with these experimental data. This method has a wide variety of applications to construction materials. Determination of moisture content distributions of a specimen during the test is often performed with non-destructive methods such as Gamma-ray attenuation method (Nielsen 1972).

Claesson and Gaffner (1977) developed a method that can determine water conductivity measuring neither moisture profiles nor water flux. A prism specimen is conditioned to have a specified moisture content, and all sides of the specimen are sealed except for one side, which is then subjected to drying to have an equilibrium moisture content. One end of the specimen is hung with a wire hocked to the bottom of an electronic balance, and the mass changes are measured continuously. After the steady state of water flow is established, relationship between the gradient of the change in mass dm/dt and the water diffusivity can be given as functions of the moisture contents before and after he test, specimen length, cross-sectional area and dm/dt.

These techniques can determine the water diffusivity of AAC at a wide range of moisture content but are often time-consuming to wait for a steady state flow.

2.2 *Principle of the dynamic method*

In the equation of continuity of water in porous media, water diffusivity D_θ is normally a function of moisture content θ. In this experiment, the equation is taken as linear because an adsorption experiment will be executed under a constant relative humidity. For a sphere with a radius r, the water balance equation is given by

$$\frac{\partial \theta}{\partial t} = D_\theta \left(\frac{\partial^2 \theta}{\partial r^2} + \frac{2}{r}\frac{\partial \theta}{\partial r} \right), \tag{1}$$

which can be converted to the one dimensional linear equation by substituting θ with u/r. The initial and the boundary conditions are that the surface moisture content of a sphere with a radius R, or relative humidity in equilibrium with the moisture content, is always constant and that the initial moisture content of the sphere in the beginning of adsorption is null,

$$\begin{aligned} &u=0,\ r=0,\ t\geq 0,\\ &u=R\theta_0,\ r=R,\ t>0,\\ &u=0,\ 0<r<R,\ t=0. \end{aligned} \tag{2}$$

With variables separation method, the following formula is obtained (Crank 1975),

$$m(t) = m_e\left[1 - \frac{6}{\pi^2}\sum_{n=1}^{\infty}\frac{1}{n^2}\exp\left(-n^2kt\right)\right], \tag{3}$$

where $m(t)$ is the total mass passed from the surface of a sphere in a finite time, m_e is the total mass passed after the infinitely long time and k is $D_\theta\pi^2/R^2$.

These initial and boundary conditions can be attained by providing the following experimental conditions: the specimen is kept under a 0% relative humidity for a sufficiently long time, and then the specimen is introduced in the specimen chamber of the humidity generator where a specified relative humidity is continuously supplied. In the Equation 3, an $n=5$ polynomial may be a good approximation at an early stage of adsorption while $n=1$ may be a good approximation at all the stage, including the latter stage of adsorption. Thus executing the adsorption experiment of a spherical specimen and determine a, b and c in the following equations by non-linear curve fitting at a sufficiently high correlation factor, it is possible to predict the equilibrium moisture content θ_e

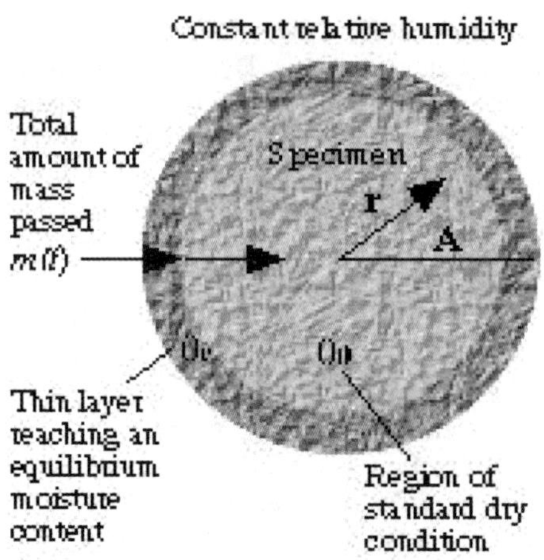

Figure 1. Water vapor adsorption system in a specimen particle.

by dividing a, corresponding to m_e in Equation 3, with the standard dry mass.

$$m(t) = a\left[1 - b\exp(-ct)\right] \tag{4a}$$

$$m(t) = a\begin{bmatrix} 1 - b\exp(-ct) - 0.25b\exp(-4ct) \\ -0.1b\exp(-9ct) - 0.0625b\exp(-16ct) \\ -0.04b\exp(-25ct) \end{bmatrix}. \tag{4b}$$

The equilibrium moisture content differs depending on adsorption or desorption, while only adsorption process was considered in this study.

The reason why the equilibrium moisture content appears in the rate equation can be explained as follows. The specimen is kept under constant null relative humidity for a sufficiently long time resulting in having uniformly null moisture content. When the ambient relative humidity changes into a constant value h and is maintained, a thin layer with a moisture content in equilibrium with the relative humidity h may appear immediately. The total amount of moisture entering in the specimen has to pass through this thin layer and reach the null moisture content region as shown in Figure 1. Rate of moisture transport is obviously controlled by the diffusivity of this region with an equilibrium moisture content corresponding to the relative humidity h. Thus, it is shown that the Equation 3 involves information of the equilibrium moisture content that cannot be obtained without waiting for the infinite lapse of time and of water diffusivity. The curve fitting program requires approximate coefficient values, as initial ones, to get optimum values after iterative calculations: an approximate convergence value of an adsorption for the parameter a, $6/\pi^2$ for b and 10^{-3} for c.

3 EXPERIMENTS

3.1 *Specimens*

Aerated concrete was a commercial product compatible to the Japanese Industrial Standards JIS A 5416 with a bulk density of 480 kg/m^3 and cutout dimensions of $100 \times 100 \times 30$ mm. Specimens were crushed with a hammer into pieces that all passed the 15 mm sieve and remained on the 10 mm sieve. Shape of the specimen may preferably be spherical as far as the Equation 3 is applied, but polyhedral shape was used aiming at a simpler specimen preparation. A 10 to 30 grams of specimen was in a stainless steel basket hanging from the bottom hock of the electronic balance and exposed to a relative humidity in the specimen chamber of the humidity generator.

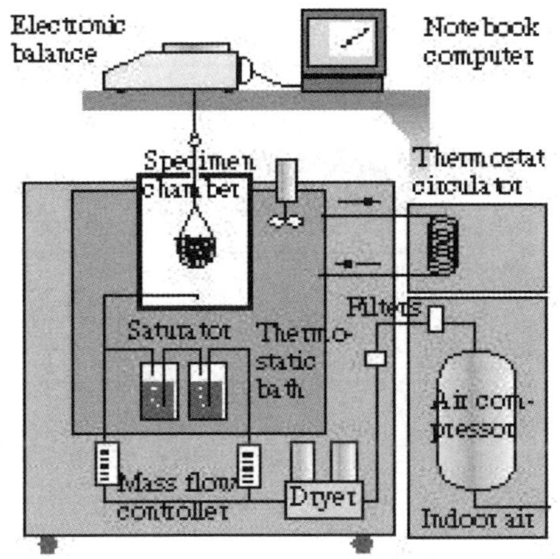

Figure 2. System diagram of the adsorption rate measuring apparatus.

Figure 3. Automatic testing apparatus.

As the standard dry conditions, the 105°C oven-dry method are normally used for AAC, however in this study, a method drying a specimen in the specimen chamber at the same temperature as that of subsequent adsorption was used. Dry-air of a dew point temperature of −42°C was continuously supplied 20 liters per minute. This drying method is milder than the D-dry method that uses a dry-air of a dew-point temperature of −76°C but can dry specimens under room temperatures to a level closed to that of 105°C oven-dry method. Furthermore, drying executed in the specimen chamber of the humidity generator makes it possible to proceed the subsequent adsorption experiment in a continuous manner.

3.2 *Test apparatus*

The test apparatus comprised three sets of a flow-division humidity generator and an electronic balance for mass determination, and all the balance were automatically controlled by a notebook computer. The configuration of the apparatus and the photo are shown in Figures 2 and 3.

The flow-division humidity generator, flowing 20 liters per minute, can produce a relative humidity at a precision of ±1% or higher in a relative humidity range from 0 to 95 percent. The nominal 0 percent relative humidity is a dry air having a dew point temperature of −42°C that corresponds to a relative humidity of 0.27 percent at a room temperature of 20°C. Because the dry air was generated with molecular sieve 5A dryer, atmospheric carbon dioxide can be removed to a level free from carbonation during adsorption experiment. The specimen chamber was shaped cylindrical, 200 mm in diameter and 250 mm in depth, and controlled at a constant temperature with a precision of ±0.5°C in a thermostatic water bath. The resolution of the electronic balance was 1 mg at the maximum weighing of 310 grams. Three electronic balances were supported with each holder independent from the humidity generators.

4 RESULTS AND DISCUSSION

4.1 *Equilibrium moisture content*

Specimens were kept in the standard dry condition of the specimen chamber, where each specified relative humidity of 5, 10, 20, 30, 40, 50, 60, 70, 80, 90, and 95 percent was introduced subsequently resulting in adsorption of water into specimens. Changes in mass with changes in relative humidity are shown in Figure 4. Each relative humidity corresponds to a curve consisting of 480 data points, once a minute for 8 hours. Equations 4a and 4b were used for curve fitting to the data and equilibrium

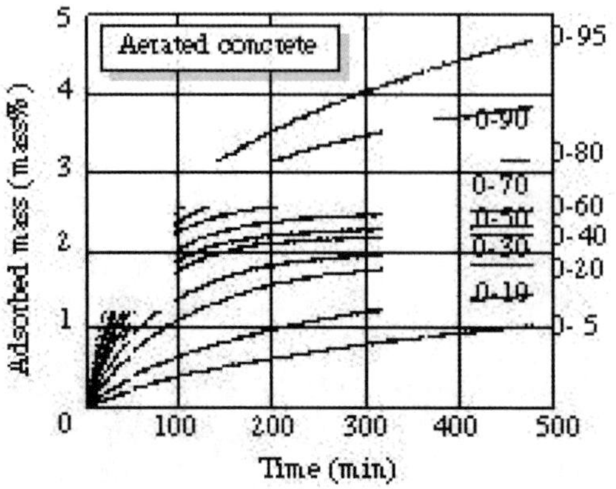

Figure 4. Adsorption rate at various relative humidity changes.

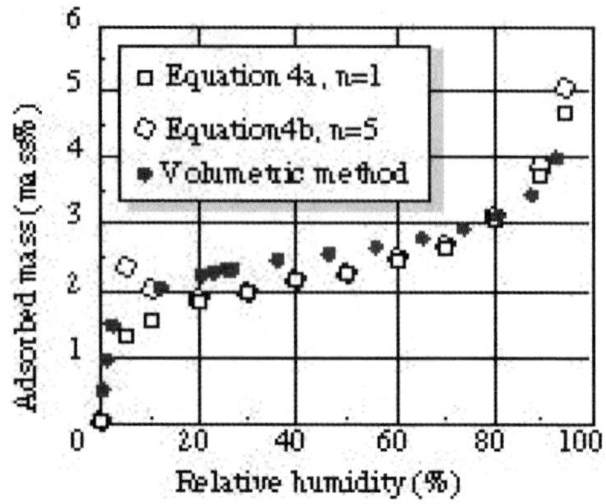

Figure 5. Sorption isotherms of AAC.

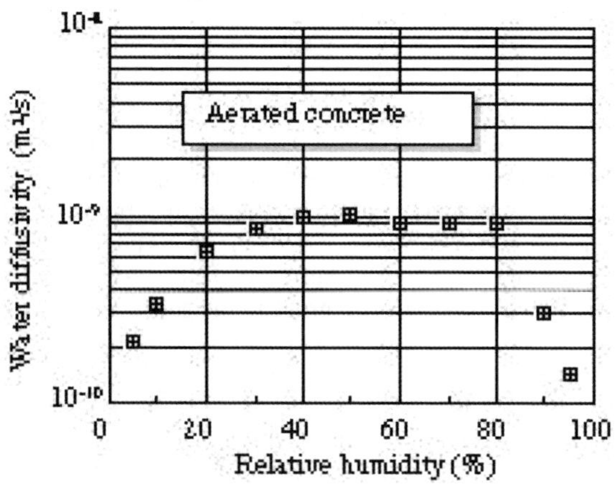

Figure 6. Water diffusivity of AAC estimated with dynamic method.

moisture contents were determined. The resulting sorption isotherms are shown in Figure 5. In each equation and for all curves, correlation coefficients of the curve fitting were greater than 0.994. In the mid-range relative humidities, the equilibrium moisture content predicted with Equation 4a and 4b were nearly identical, but differed at low and high humidity ranges. Especially at the low humidity range, equilibrium moisture content values predicted by Equation 4b became very unusual. Thus the Equation 4a was employed throughout the subsequent experiments.

The volumetric method using a commercial instrument was applied to AAC of the same brand. The result showed 0.3 percent higher equilibrium moisture content than that predicted using Equation 4 except for a high humidity range. This may be attributed to a large porosity of AAC, more than 80 percent, where adsorption proceeded in a short time and reach the final stage of adsorption in 8 hours resulting in violating the principle of the dynamic method that the thin surface layer controls the net moisture flow only at an early stage of adsorption, as shown in Figure 1.

4.2 *Water diffusivity*

Water diffusivity is obtained as a function of moisture content and varies in an order of 10^3 at high moisture content regions (Tada & Nakano 1983). The proposed dynamic method covered only low moisture content regions that correspond to equilibrium moisture contents of given relative humidities. The water diffusivity showed a peak at a relative humidity around 50 percent and was small at low and high relative humidities as shown in Figure 6. This may be attributed to the fact that

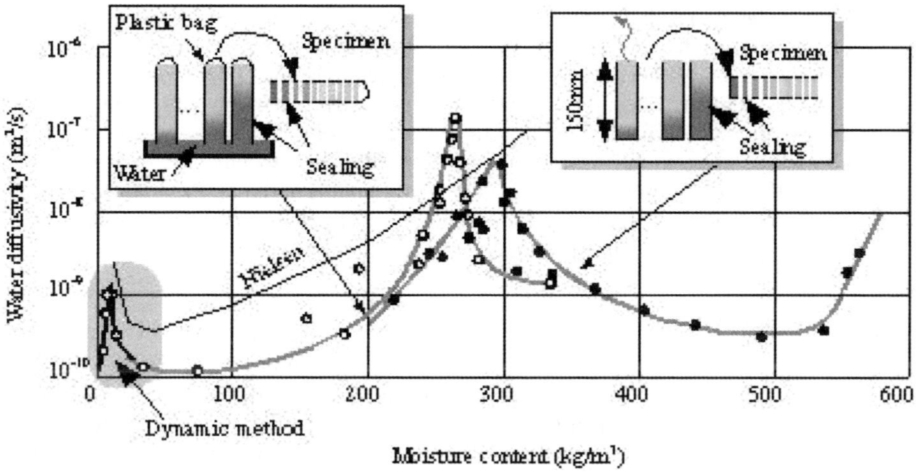

Figure 7. Water diffusivity of AAC determined with various methods.

adsorption to the solid surface may reduce moisture transport at a low humidity and liquid water may interfere vapor flow at a high humidity. In Figure 7, the results are also displayed with those obtained with the conventional methods and by Nielsen (1972) in a wide range of moisture content. Water diffusivity determined with different methods fell in nearly the same range but the proposed method may favorably be applied for a precise determination at lower moisture content regions.

4.3 *Possible causes of error*

(1) *Assumption in the basic equation*

Among possible causes of the error in this dynamic method, non-fulfillment of the assumptions in the solution of basic equation, such as the shape of specimen, may be included. Because the specimen was not a perfect sphere, the shape factor b in the curve fitting equation was not treated as a constant of $6/\pi^2$, but determined as a variable like a and c, resulting in the high correlation coefficient for all the adsorption data.

(2) *Experimental conditions*

Application of curve fitting Equations 4a or 4b corresponding to the stage of adsorption depends on the amount of specimen and the extent of the stepwise change of relative humidity, and cannot be determined uniformly. Measuring duration, measurement interval and the limitation of the curve fitting equations were studied in a preliminary experiment using aerated concrete. Long term adsorption experiments were executed with measuring intervals of 10 seconds, 1 minute and 10 minutes and data was collected until the latter stage of adsorption as apparently shown from the adsorption curves. The results were verified by the correlation coefficients and the compatibility to literature values. As a result, it was decided that measuring interval was 1 minute and the measuring time was 8 hours for aerated concrete and 24 hours for hardened cement pastes.

(3) *Conditions of test apparatus and specimen*

Changes in performance of the testing apparatus with time may also become a cause of error. Degradation of mass flow controller due to fine particles or oil mist and resulting decrease in accuracy of relative humidity generation may occur within several years. Changes in the atmospheric pressure, when a moving low pressure approaches for instance, may require corrections of the flow division table. Electronic balance may show a shift when an abnormal acceleration is applied by, for instance, earthquakes. It is also affected by the thermal drift and air current in the thermostatic chamber.

Lack of knowledge about the nature of specimen may lead to an error. When a cement-based material is subjected to change in pore structure during testing by, for instance, hydration or carbonation, the sorption isotherm changes to a great extent. Drying under a room temperature is preferable to have a standard drying condition regardless of the type of specimen, and the dryness of the conditioned air should be defined in terms of a dew point temperature. AAC can be dried at an oven-dry temperature of 105°C, but the long term drying may dehydrate the surface silanol group and revive the siloxane bonds resulting in the hydrophobic surfaces. Prior to the experiment, AAC was kept under 0 percent relative humidity for one day. Such drying conditions have not been well documented making it difficult to compare with the literature values.

5 CONCLUSIONS

Rate of water adsorption to AAC involves information of both the equilibrium moisture content and the water diffusivity. Isothermal linear balance equation was solved in terms of the total moisture gain and this analytical form was used as a curve-fitting scheme to the adsorption rate resulting in the early-stage prediction of the equilibrium moisture content and water diffusivity simultaneously. Results showed good agreement with those measured by the conventional methods. The adsorption rate experiment is simple and the time necessary for obtaining an equilibrium moisture content and a water diffusivity in a relative humidity was 8 hours. Departing from the accepted notion, the information of equilibrium moisture content involved in an adsorption rate is best manifested even at an adsorption stage very far from the equilibrium.

REFERENCES

Andersson, A-C., 1985. Verification of calculation methods for moisture transport in porous building materials, Document, Swedish Council for Building Research, D6:1985, 223p.

Ahlgren, L., 1972. Fuktfixering i Poroesa Byggnadsmaterial., *Report 36, Division of Building Technology*, The Lund Institute of Technology.

Claesson, J., and Gaffner, D., 1977. Fukt i Porosa Byggnadsmaterial, Report 1977.1, Lund Institute of Technology.

Crank, J., 1975. *The Mathematics of Diffusion*. 2nd Ed., Oxford Science Publications.

Gawin, D., 1993. Numerical solution of coupled heat and moisture transfer problems with phase changes in porous building materials, *Archives of Civil Engineering*, XXXIX (4): 393–412.

Haggymassy, J., Jr., 1970. Pore structure analysis by water vapor adsorption, *Ph.D. Thesis, Department of Chemistry, Clarkson College of Technology*.

Hanson, R., 1960. Moisture in light-weight concrete roofs, *Proc. Rilem Symp. Lightweight Concrete*, Goteborg, 405-431.

Jonel, P., Karamustafaoglu, V. and Tepfers, R. 1960. Determination of the coefficient of capillary flow for steam-cures lightweight concrete, *Proc. Rilem Symp. Lightweight Concrete*, Goteborg, 433–437.

Kooi, J. van der., 1974. Moisture transport in cellular concrete roofs, *Proc. 2nd Int. Symp. Moisture Problems in Building*, Rotterdam, 2.5.5.

Matsumoto, M., 1978. *A Study on Simultaneous Heat and Moisture Transport and Moisture Accumulation in Building Walls*, Thesis of Kyoto University. (in Japanese).

Nielsen, A. F., 1972. Gamma-ray attenuation used for measuring the moisture content and homogeneity of porous concrete. *Building Science*, 7, 257–263.

Sandberg, P. I., 1973. Byggnadsdelars fuktbalans i naturigt klimat, *Lund Institute of Technology, Report* 43, 245p.

Tada, S., & Nakano, S., 1983. Microstructural approach to properties of moist cellular concrete. in *Autoclaved Aerated Concrete -Moisture and Properties*, F. H. Wittmann (ed.), 71–88 Elsevier.

Takahashi C., Inamatsu, T., & Kitano H., 1988. Construction of flow-division humidity generator, *Transactions of SICE*, 24(6): 557–562, (in Japanese).

Yuasa, N. et al., 1998. Sorption isotherm and moisture conductivity of cover concrete, *Cement Science and Concrete Technology*, 52, 1042–1049. (in Japanese).

Autoclaved Aerated Concrete – Limbachiya and Roberts (eds)
© 2005 Taylor & Francis Group, London, ISBN 0 415 38356 0

The most important aspects of microstructure influencing strength of AAC

G. Schober
Ing.-Büro für Material- und Verfahrenstechnik, Sägmühlstraße, Olching, Germany

ABSTRACT: AAC consists of pores by a volume fraction of 70 to more than 85%, depending on apparent density of the product. It is obvious that porosity is a major issue determining the properties of aircrete. A survey of all elements of microstructure will be given especially about those elements, which mostly affect the strength. Air pores make up the main part of pore volume in aircrete. Besides the typical ball like air pores some additional elements in pore structure are produced during expanding process. First of all there are small cracks often connecting the pores and showing a preferred orientation. The mineral composition of the solids skeleton, size, form and interconnection of the different crystals determine microstructure and resulting properties of the solid part of AAC. It is the amount of tobermorite, which provides high strength. But micro pore characteristics and mineral composition can vary and therefore result in a higher or lower strength.

1 INTRODUCTION

AAC is a material, which is produced in qualities with densities between 100 and 800 kg/m^3. There is no other industrial product that covers such a range in apparent density. Up from 350 kg/m^3 in apparent density AAC can be used as load bearing construction material. The material lower in density is used for thermal insulation purpose. No difference in production technology is needed to achieve this diversity. Only the part of pore volume produced by chemical reaction, in most cases aluminium powder is used, has to be adjusted. This way porosity is the most important microstructure element in AAC.

To provide you with an impression about the volume fractions of porosity and solid material in AAC products some examples for different density classes are listed in Table 1.

The material skeleton contains all the micro pores. Due to high total porosity skeleton material has to be in a rather optimized state in any way. There have to be a lot of very small tobermorite crystals,

Table 1. Porosity characteristics of AAC products different in density.

Apparent Density kg/m^3	Macro Pores %	Micro Pores* %	Total Porosity
100	83	13	0.96
150	77	17	0.94
350	70	16	0.86
400	65	19	0.84
600	45	21	0.76
800	27	41	0.68

* Total volume of micro pores is mainly controlled by water/solid ratio of raw materials mixture, thus it is possible to shift in a certain range from micro to macro pores and vice versa.

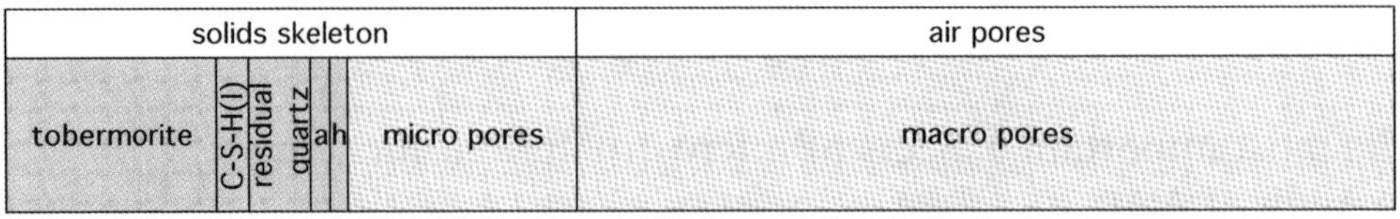

Figure 1. Schematic representation of volume parts in the structure of aircrete (a = anhydrite, h = hydrogarnet). Apparent density is about 400 kg/m^3 in this case, and raw materials are: pure quartz sand, lime, portland cement, anhydrite and water.

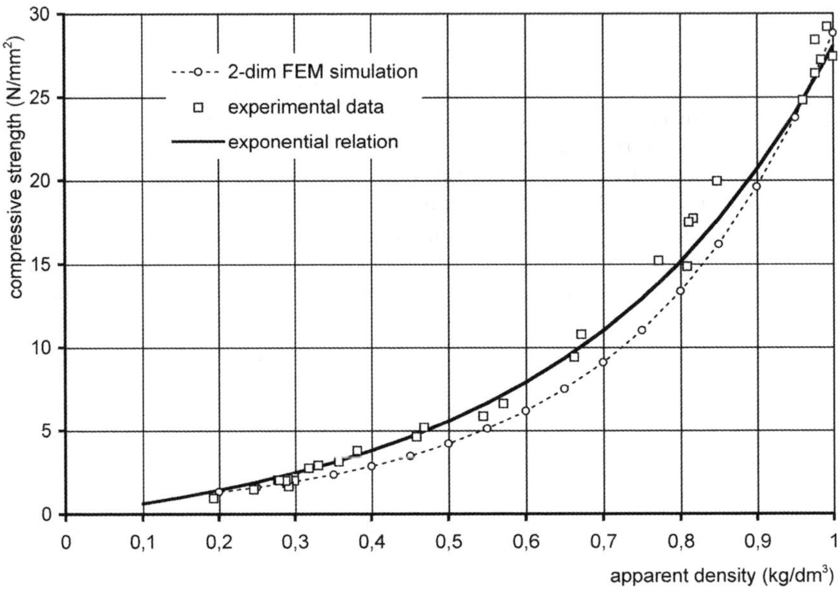

Figure 2. Relation between strength and apparent density of AAC over the range from 100 to 1000 kg/m^3.

well crystallized to create a binder which holds together all phases present. Besides tobermorite, which makes up between 50 to 80 vol.-% of solids one finds typically residual quartz and minor amounts of hydrogarnet and anhydrite. Figure 1 gives a representation of volume fractions of different microstructure components in AAC.

2 APPARENT DENSITY, MACRO PORE STRUCTURE AND STRENGTH

2.1 *Relation between strength and density*

An exponential law governs the relation between strength and apparent density. Over the range from 200 to 1000 kg/m^3 this has been proved by experiment and simulation.

The following expression was found to hold with a coefficient of determination of 0.99. It is a result from the simulation research work done in this field and from accompanying experiments (Schneider et al. 1998, Schneider 2001).

$$\sigma = \sigma_0 \times (\exp(3 \times D_{ap} \div D_{sk}) + D_{ap} \div D_{sk} - 1) \times \exp(-3) \quad (1)$$

where σ = compressive strength; σ_0 = compressive strength of not expanded but autoclaved material; D_{ap} = apparent density, D_{sk} = density of skeleton material.

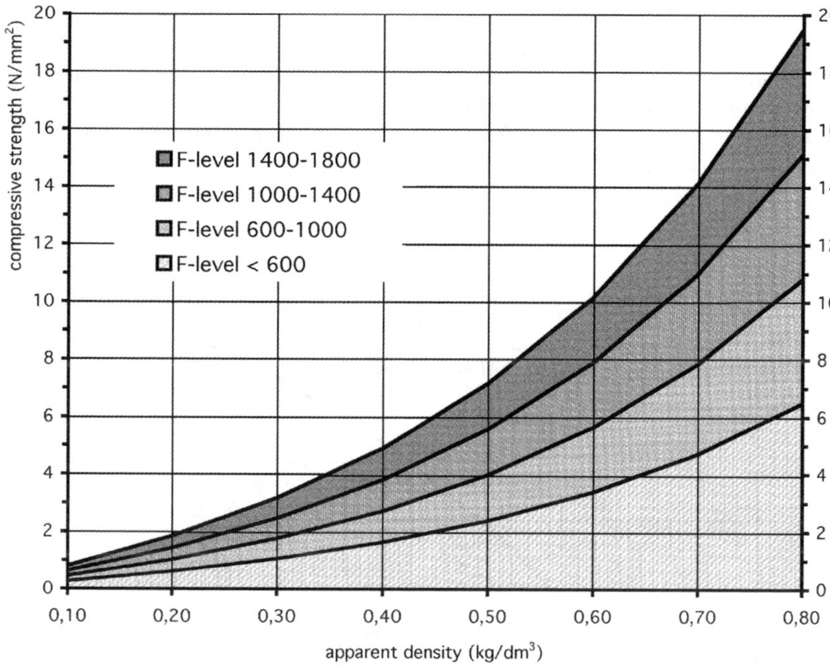

Figure 3. Strength density relations for AAC products, which differ in their quality levels. One curve represents one certain quality level in strength. These relations are also valid for foam concretes.

For different raw materials, properties of raw materials, formulations of AAC mixture and different autoclaving conditions one will get different curves. The following diagram shows a set of curves dividing the possible property combinations into quality classes.

If dry density values are not equal in one set of samples to be compared, it is possible to use this relation as a tool in quality control. A similar method is the so-called A-Zahl (A-value) widely used in Germany. But the A-value concept is only valid in the range from 400 to 750 kg/m^3. The figure of merit derived from the concept presented above may be called "strength level" or "F-level" and can be determined as follows when related to AAC:

$$F = \sigma \times 1000 \div (\exp(3 \times D_{ap}) + D_{ap} - 1) \tag{2}$$

where σ = compressive strength in N/mm^2; and D_{ap} = apparent density in kg/dm^3; it is assumed that σ_0 in equation (1) is about 21 N/mm^2 and D_{sk} around 1 kg/dm^3 respectively.

The "F-level" is a figure of merit for strength, which is independent of materials density.

In Figure 3 the possible property combinations are grouped into four classes of quality level regarding strength. One can use this to classify and compare raw materials situation and plant equipment conditions of different production sites for AAC.

2.2 *Anisotropy in strength of AAC*

It is a common known property of AAC that there is a difference in strength whether the load acts in rising direction or perpendicular to rising direction. This difference is not great but in some circumstances it is necessary to know about it. The following figure shows stress strain curves of two AAC samples out of the same lot, one compressed parallel to expanding direction and the other one compressed perpendicular to that direction.

The cause for this typical property of AAC has been found in a certain element of pore structure of the air pores. When the balls like air pores are generated, the slurry can move only in one direction, upwards. At the end of this expanding process the viscosity of this foam gets so high that fission occurs in the pore walls. All this tiny cracks, which are finally connecting neighboring air pores, are orientated horizontically. This cracks all straightened are responsible for anisotropy in strength and in some other properties, see Figure 5.

The anisotropy in strength can range from 5 to 25%. It is around 10 to 15% in most of the AAC products. All influences, which affect viscosity of expanding mixture, can increase this anisotropy if viscosity runs too high at the end of expanding movements.

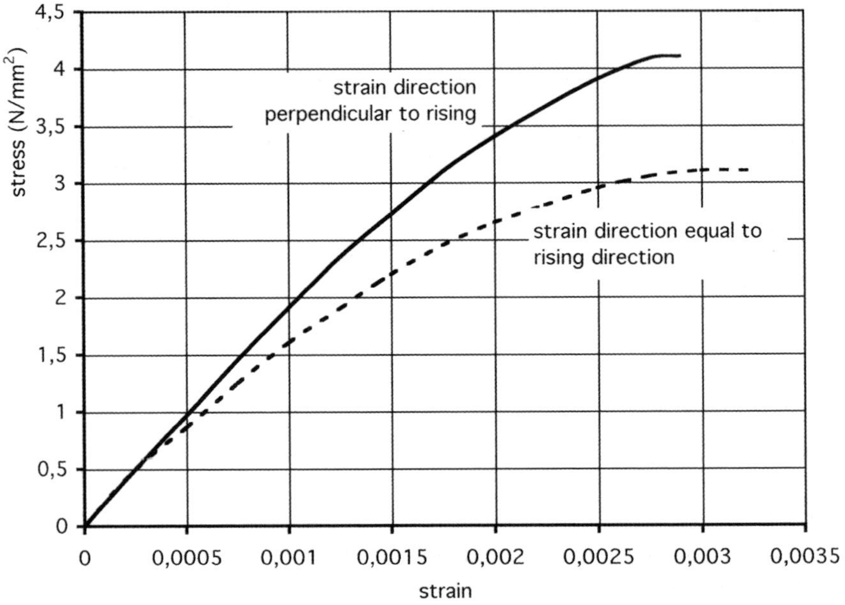

Figure 4. Stress strain relations of AAC cubes under compression in different directions relative to rising direction (expanding of the green cake).

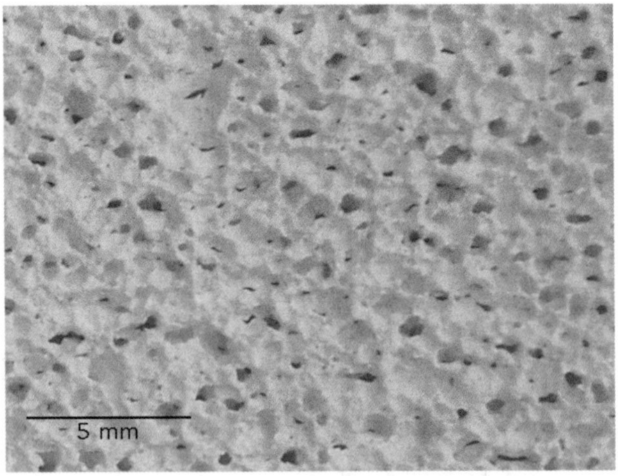

Figure 5. Photograph of AAC showing the micro crack structure produced during expanding process. Rising direction is from bottom to top, most cracks are orientated horizontically.

A suitable method to determine microstructure property related to micro cracks produced during expanding is measurement of gas permeability with respect to rising direction (Wägner et al. 1995).

Trials to avoid anisotropy in aircrete by using foam technology led surprisingly not to material without anisotropy. It turned out that the very small expansion due to temperature increase (lime and cement reactions) was sufficient enough to produce a similar effect as experienced with ordinary AAC making.

2.3 *Properties of pore size distribution of macro pores and their influence on strength*

The simulation of pore structures allowed to test a lot of characteristics of the pore size distribution of the air pores in AAC on their possible influence on strength.

The model in the simulation was a two-step one. Air pores were generated with certain properties of their size distribution, total porosity, minimum pore size, maximum pore size, minimum pore distance, mean pore size and type of size distribution. The solid part of the simulated pore structure was constructed by finite elements method (FEM). The solid material properties for FEM were Young's modulus, compressive strength and Poisson ratio. Additionally the solid part of the simulated pore structure was considered to be brittle and ruled by a Weibull distribution for its strength values. So another parameter the Weibull exponent (m) was included in this model. Most of the simulations were done on a two dimensional model. Details are described in Schneider et al. 1998 and Schneider 2001.

The results found in the simulation experiments can all be explained as effect of one single property of the air pores size distribution, and this is the minimum pore distance. In every case when two pores are very close together there exists a small volume element in the solid structure between these two pores, which comes under very high stress as soon as load acts on the sample. This volume element will be the weakest point in the structure and failure of the sample will start at this point.

As minimum pore distance is not independent of other pore structure properties the influence of mean pore size and broad or narrow size distribution in structures with the same porosity of air pores has been studied.

The influence of pore size proved to be directly proportional to strength as shown in Figure 6. This effect could be seen only when porosity was higher than 0.3. As soon as porosity is so high that the pores touch each other, this model doesn't hold any longer and therefore there is no answer in this case. The strength values as shown in Figure 6 can spread more or less according to minimum pore distance, which was allowed in the simulation runs. This does not change the positive trend for strength with pore size.

The character of pore size distribution turned out to be positive with respect to strength when it is a narrow one. Broad pore size distribution led to lower strength and also structures with a single pore size resulted in slightly lower strength than such with a narrow distribution of same mean pore size. These results too are only secondary effects of the minimal thickness of pore walls in the air pore structure of AAC.

3 SKELETON MATERIAL, PHASE COMPOSITION, MICRO PORES

3.1 *Solids*

The structure of the solid skeleton is made up by the binding material tobermorite as basic component. There is also a minor part of C-S-H(I) as a binder in most cases. All the CSH phases (Calcium-Silicate-Hydrate phases) in AAC are crystalline or semi-crystalline. No $Ca(OH)_2$ (portlandite) and C-S-H gel can be found in AAC of good quality.

The tobermorite crystals in AAC are very small in any case. What can be seen with SEM (scanning electron microscopy) always are only the biggest ones. Tobermorite forms during autoclaving and the crystals grow in form of laths, plates or bands. The single particles always crystallize with one

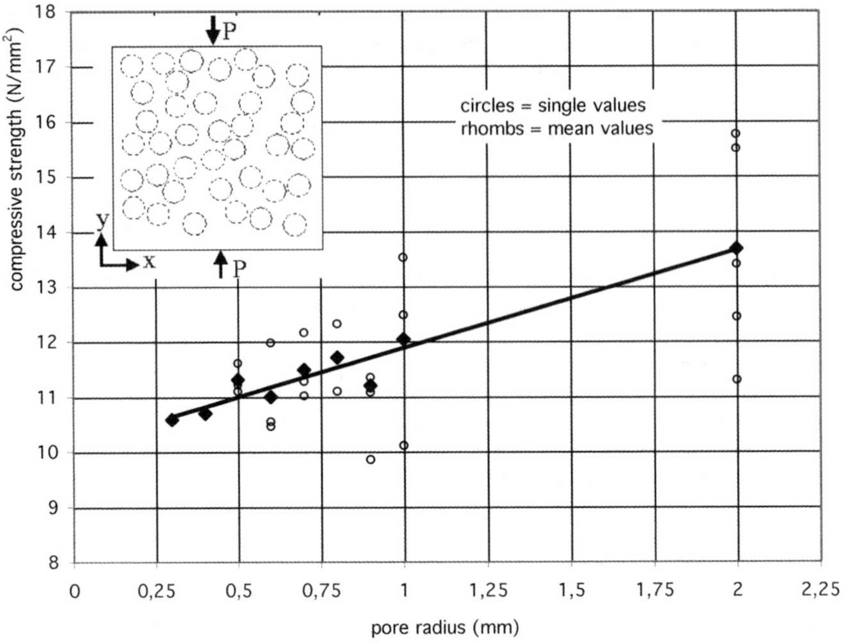

Figure 6. Strength of porous structures with different pore size of macro pores. Macro pore fraction is 0.4.

strongly preferred growing direction (silicate Dreierketten in the structure of tobermorite). Thus, intergrowing of these laths, plates or bands leads to the final strong microstructure, which becomes more and more dense during the autoclaving process.

There is no doubt that first of all the amount of tobermorite rules the strength of AAC products. All other components in AAC are incorporated in the tobermorite or C-S-H(I) mass. Figure 1 gives a schematic representation of the volume fractions for an example of AAC with 400 kg/m^3 in apparent density. Residual quartz, hydrogarnet and anhydrite are the typical accompanying phases in AAC. There may be some others according to which kind of raw materials are being used. If anyone of the minor components in AAC overruns about 10 to 15 mass-% strength will be reduced. This shows that none but CSH phases act as binder in AAC.

Formation of tobermorite microstructure is strongly influenced by raw materials, their properties and the autoclaving conditions. A major role concerning final strength are playing the intermediate sulphate phases, ettringite and hydroxylellestadite. The later acts as a precursor to tobermorite being formed in the final stage of autoclaving. Ettringite formation in the green cake seems to prepare the microstructure of the product, so that conditions for succeeding CSH intergrowing and reorganization of microstructure become more positive.

3.2 *Micro pores*

Most of the micro pores in AAC are remaining volume parts of space that first was occupied by water in the green cake and later on filled by reaction products. Only a small amount of the micro pores are new formed voids after dissolution of raw material particles e.g. quartz grains. Some of the residual quartz grains are not incorporated densely inside the CSH matrix. Quite often there can be found cavities around such quartz grains.

In the green cake mean pore size of micro pores is between 1 and 10 μm in diameter (Mitsuda et al. 1992). The final state after autoclaving shows mean pore sizes from 0.01 to 0.2 μm (Mitsuda

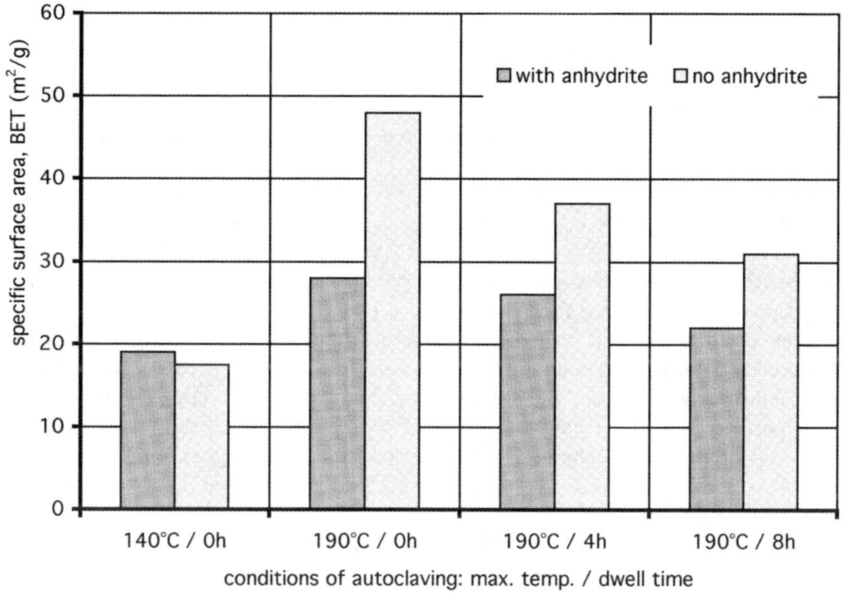

Figure 7. Specific surface areas of AAC autoclaved under different conditions. Results are given for two AAC sample series with different recipes, one with the other without anhydrite as a recipe component.

et al. 1992, Pospisil et al. 1992). In these cases measurement was done with mercury intrusion porosimetry method (MIP).

During autoclaving process, after a first CSH matrix has been formed with a maximum of frequency in pore size distribution lower than 0.1 μm, this maximum is going to shift slightly to somewhat higher values (Mitsuda et al. 1992). This effect is due to increase of tobermorite content. It means that the amount of crystalline CSH phases is growing and semi-crystalline CSH and intermediate reaction products are waning in the solid matrix. Determining the specific surface area with BET method shows this more clearly than MIP (Fig. 7). This change in microstructure is accompanied with an increase in strength.

What can be seen in Figure 7 as well is an effect of sulphates on microstructure. Anhydrite addition does not change the amount of CSH phases but it alters micro porosity. All the anhydrite added is present in the final product as anhydrite again, if autoclaving time was not too short. The positive effect on strength is due to formation of tobermorite in a way that leads to higher density of CSH matrix in AAC.

It is easy to change the amount of micro porosity in AAC by changing the water solid ratio (w/s) of the raw materials mixture. Such changes don't show a strong effect on strength. Trying to increase strength by reducing w/s results in higher viscosity of mixture and less stability of expanding process. Thus macro pore structure gets weakened and strength will be reduced.

4 DISCUSSION

An explanation to anisotropy in AAC that the air pores were egg shaped is still often told. But there is no evidence that the air pores are elliptical in shape and these ellipses would have a preferred orientation. The air pores are not all perfect spheres, but that is due to lack in space when the pores are almost touching each other. The only pores, which can be found with elliptical shape, are very large pores (entrapped air). They got their shape by streaming up some milli- or centimeters in the expanding slurry. If such pores were responsible for the difference in strength, then the effect would be the other way around, higher strength in rising direction and lower strength perpendicular to rising direction. The real causes for anisotropy are the small cracks in the walls of the air pores. They are produced unavoidably during expanding process and their occurrence is strongly influenced by viscosity of mixture.

Concerning influence of air pore size on strength some authors report that big air pores lower strength and they try to confirm this with reference to Griffith theory (Pospisil et al. 1992). In this view the results presented in Section 2.3, Figure 6 seem to be in contradiction to Griffith theory. The theory deals with a single crack or hole in an infinite plate. In this imagined plate there exists no other weak point besides the one hole, the edge of which is acting as starting point of failure. Therefore it is obvious that aircrete can not be described by this theory unless there were very big isolated holes in the ordinary pore structure of AAC, and this holes then would have to be more than twenty times in size than the mean air pore size is.

The weakest point in the AAC structure is not at the edge of one of the biggest air pores, it is a section between two pores. Therefore notice, not pore size is decisive in this case but pore distance, i.e. the minimal thickness of pore walls. Only as a secondary effect pore size of air pores has an influence. If porosity is held constant then an increase in pore size results in thicker walls between the pores and so strength should be higher in pore structures with bigger pores. This effect should be valid if porosity due to air pores is higher than 0.35 but lower than in a state where the pores have to touch each other.

In phase composition tobermorite crystals have been found to be the only ones of a lot of possible CSH phases to bring to pass a binder matrix with high strength. So all raw materials and conditions, which don't support formation of well-crystallized tobermorite in AAC, are not advantageous. The knowledge on physical interaction between binder and minor components in AAC microstructure is rather poor.

Micro pores in AAC are mainly pores of nano-scale. These are connected to binder phase, the tobermorite crystal mass. The part, which could be active in capillary transport, is low. Increasing w/s strongly would produce more capillary pores but lower strength. With thermal insulation products based on CSH this is a common way of processing. But this is not done in AAC production. As reducing w/s is limited due to process engineering restrictions other methods to obtain a denser CSH matrix have to be tried. Until now, only the addition of sulphates is a method, which is known to work.

5 CONCLUSIONS

The microstructure of AAC can be improved with respect to getting a high strength by trying to achieve the following:

- an air pore structure with less cracks in the air pore walls as possible
- a size distribution for air pores, which is rather narrow
- air pores as big as possible, if the air pore porosity is within the range from 0.35 to 0.75
- amount of tobermorite in phase composition as high as necessary
- intergrowing of CSH matrix as dense as possible
- less micro pores due to dissolution of raw material particles, e.g. cavities around quartz

REFERENCES

Mitsuda, T., Kiribayashi, T., Sasaki, K. & Ishida, H. 1992. Influence of hydrothermal processing on the properties of autoclaved aerated concrete. In F. H. Wittmann (ed.), *Proceedings of the 3rd RILEM International Symposium on Autoclaved Aerated Concrete, Zürich, Switzerland, October 14–16 1992*: 11–18. Rotterdam: Balkema.

Pospisil, F., Jambor, J., Belko, J. 1992. Unit weight reduction of fly ash aerated concrete. In F. H. Wittmann (ed.), *Proceedings of the 3rd RILEM International Symposium on Autoclaved Aerated Concrete, Zürich, Switzerland, October 14–16 1992*: 43–50. Rotterdam: Balkema.

Schneider, T., Schober, G. & Greil, P. 1998. Numerical modeling of the strength of highly porous aerated autoclaved concrete. In Schwartz, D.S. & Shih, D.S. & Evans, A.G. & Wadley, H.N.G. (eds), *Proceedings of the Materials Research Society, Vol. 521, San Francisco, California, USA, April 13–15 1998*: 21–26. USA: MRS Warrendale.

Schneider, T. 2001. Modellierung der Festigkeit poröser Keramiken. Dissertation, Technische Fakultät der Universität Erlangen-Nürnberg, Berlin: Logos Verlag.

Wägner, F., Schober, G., Mörtel, H. 1995. Measurement of gas permeability of autoclaved aerated concrete in conjunction with its physical properties, *Cement and Concrete Research* 25: 1621–1626.

Autoclaved Aerated Concrete – Limbachiya and Roberts (eds)
© 2005 Taylor & Francis Group, London, ISBN 0 415 38356 0

Developments in AAC constructions to meet the UK acoustic performance requirements

D.J. Harris
Autoclaved Aerated Concrete Products Association, Sevenoaks, Kent, UK

D. Baker
Robust Details Limited, Milton Keynes, Bedfordshire, UK

ABSTRACT: Since the introduction of Building Regulations within the UK, the usefulness of AAC as a suitable material to use in separating and associated flanking walls between dwellings to protect against unwanted noise from your neighbours has never been fully recognised. Changes to regulations in England and Wales in 2003 required new dwellings to be tested for their sound insulation before being sold. However, due to the unpopularity of such an approach, a new compliance route was introduced, intended to provide compliance without testing. This became known as Robust Details (RD). A Robust Detail construction requires a higher performance level compared to a construction that is designed using the testing option. This paper briefly describes the procedure necessary to obtain "RD" status and the results obtained, that indicates that an AAC cavity separating walls easily match the performance of the medium and heavier density masonry block walls of similar thickness.

1 INTRODUCTION

When the UK government revised their Building Regulations dealing with sound insulation (The Stationery Office 2003) they decided not only to increase the performance level but to change from the previous compliance route of building with one of the "deemed to satisfy" constructions presented in the Regulations, to a performance based approach.

This was due to the increasing number of complaints being received from occupants who were concerned that the dwellings being built to the existing regulations were not providing the necessary level of sound insulation to protect from the unwanted noise from their neighbours. The major reason for poor insulation was identified as poor workmanship and therefore it was decided to also introduce a pre completion testing regime (PCT), measuring a sample of new dwellings built to ensure that they meet the new higher performance level. This approach had already been implemented in Scotland as a result of a precedent set following a court case that stated "that a sound test could be used as a means to determine whether the workmanship was satisfactory". At least two building control bodies in Scotland subsequently introduced sample sound testing on sites in their area.(Smith 2001)As a result there has been a significant drop in noise related complaints. A test sample size of 1 in 10 dwellings within a group or sub group was proposed for all separating elements in new build constructions.

However, with an expected 160,000 dwellings to be built in England and Wales each year, and with a Government clearly inclined to see an increase in housing numbers, this meant that a large number of dwelling would have to be tested at a significant cost to the house builder. Perhaps unsurprisingly, the House Builders Federation (HBF) represented the industry's case that this would not only create uncertainty, delays and unnecessary costs but was also a poor use of funds. HBF submitted that it would be better to invest this money into the fabric of the building to obtain a higher level of sound insulation, ensuring that the property owner was the main beneficiary of this change in regulations. Therefore after discussions between the HBF and the UK government regulators it was agreed that the HBF would fund a program in which "Robust" construction would be developed, built and then tested.

2 ROBUST DETAILS

The basic requirements for a Robust Detail (RD) are that it should

- Be capable of consistently exceeding the performance standards given in the UK Approved Document relating to sound (AD E),
- Be practical to construct on site and,
- Accommodate acceptable levels of variations in workmanship.

Each Robust detail would include materials and construction details for the separating wall/floor and its key interfaces with other elements.

To qualify as a Robust Detail, and therefore be acceptable to the regulators in England and Wales as an alternative to Pre Completion Testing (PCT), the separating elements would have to achieve a mean level of sound insulation of at least 5dB better than the level set by the new Building Regulations. In addition there should be

- minimum of 30 wall or floor structures tested
- testing spread over a minimum of 4 sites
- no more than 8 test structures at any one site may be included
- at least 3 different measurement contractor companies should be spread over the 30 test sets
- the 30 test structures must not be constructed under the control of the same builder nor the same developer
- no measurement contractor company may carry out testing on more than 15 test structures.

The increased level of performance specified by the UK government in the revision of their Building Regulations were set at a minimum of 45dB $D_{nT,w} + C_{tr}$ for airborne sound and 62dB $L'_{nT,w}$ for impact sound. Therefore to achieve RD status the results of 30 test constructions would have to met a mean value of $\geq$50dB $D_{nT,w} + C_{tr}$ and any individual value $\geq$47dB $D_{nT,w} + C_{tr}$ whereas for impact sound insulation a mean values must be $\leq$ 57 $L'_{nT,w}$ must be met with individual values $\leq$60dB $L'_{nT,w}$.

This provided a unique opportunity for the UK Autoclaved Aerated Concrete (AAC) industry, as in the past the material, due to its lightweight nature, had never been recognised by the regulators as a material that would meet the adequate insulation required within separating or flanking walls. Accordingly, in the past, AAC or aircrete as it is known in the UK, was not included into the list of "deemed to satisfy" constructions specified within the regulations and that, in turn, it was often the case that Building Control organisations looked for test evidence to prove compliance. A basic tenet of the subject of building acoustics being that the airborne sound insulation performance of masonry is related to the element's mass, that is the higher the mass the higher (better) the sound insulation. As a result, for over 30 years, the UK AAC industry had to persuade developers to build aircrete blockwork separating walls between adjacent dwellings and then to measure the acoustic performance in order to demonstrate compliance with the regulations. Over the years, this process has generated a large database of hundreds of site test results incorporating AAC or "aircrete" as it is now more commonly known in the UK, in the flanking and separating walls.

In addition the industry had also undertaken to derive a mass law for AAC (RILEM 1993) that would further supported the field test results.

Under laboratory conditions controlled tests were undertaken to quantify the characteristics of AAC. This resulted in the following mass law relationship being derived specifically for aircrete masonry.

$$R = 22.9 \log(m) - 4.2 \text{ dB} \quad [3]$$

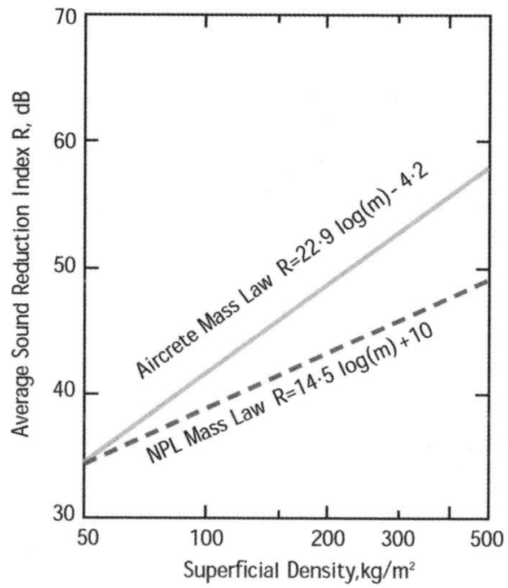

Figure 1. Comparison of the two mass laws.

where R is the average sound reduction index over a frequency range of 100 to 3150 Hz and m is the mass per unit area of the element (kg/m^2) The tests were carried out in accordance with the as then BS 2750:Part 3 (now BS EN ISO 140-3). The resulting relationship was then compared to the recognised empirical mass law derived by the National Physical Laboratory

$$R = 14.5 \log(m) + 10\,\text{dB}$$

which is still commonly used to assess the sound insulation of most materials. Figure 1 compares the two mass laws.

It can be seen that the aircrete mass law represents a superior performance compared with the generally accepted formula and also provides a useful method of predicting the sound insulation properties of a range of aircrete partitions.

3 OUTCOME

The industry then embarked on a program to meet these requirements within the short time period set by the UK Government, which required all testing to be completed within approx 6 months. At the end of this first phase 6 masonry, 2 timber and 2 steel framed separating walls, and 2 concrete, 1 timber and 1 steel separating floors achieved all the requirement to qualify as a Robust Detail and were issued as a Pattern book (Robust details 2004) for the industry to use. In addition an independent company was set up to administer the implementation and monitoring of the scheme.

Perhaps to the surprise of those except the AAC industry, within the 6 accepted masonry separating walls there was an aircrete construction comprising two leaves of a minimum of 100 mm aircrete blocks with a density of 600–800 kg/m^3, finished with a gypsum based board (nominal density 8 kg/m^3) mounted on dabs on a sand:cement render (nominal 8 mm) with scratch finish (fig. 2).

In Table 1 is a list of the RD separating wall plus their performance data and it can be seen the AAC construction E-WM-6, exceeded the 5 db extra requirement and produced a similar performance to that of much heavier concrete blocks, 1350–1600 kg/m^3 and 1850–2300 kg/m^3 and perhaps at last dismissing the previously held theory that the mass was the only way to achieve good acoustic insulation and that isolation is the primary factor of influence.

As with the AAC separating wall, the tests on all the other Robust detail masonry separating walls have indicated that AAC blocks can also be satisfactory used in the flanking walls to these

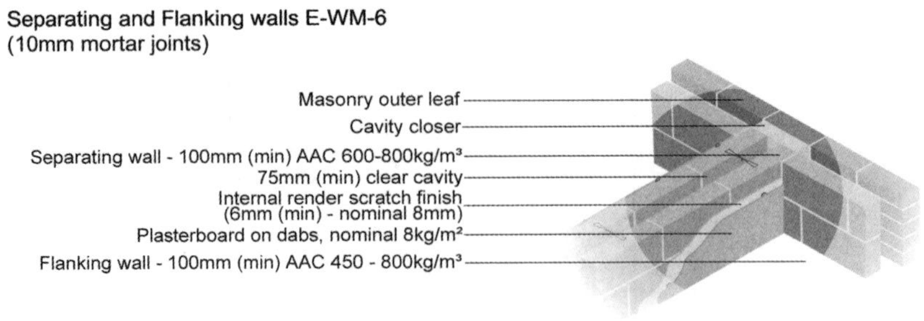

Figure 2.

Table 1. Robust Details-Masonry Separating Walls, including performance data.

Construction Type	Performance	
	Airborne Range dB	Airborne Mean dB
• E-WM-1 Masonry-Dense aggregate blockwork (wet plaster)	47–60	53
• E-WM-2 Masonry-Lightweight aggregate blockwork (wet plaster)	48–55	52
• E-WM-3 Masonry-dense aggregate blockwork (render & gypsum-base board)	47–57	51
• E-WM-4 Masonry-lightweight aggregate blockwork (render & gypsum-base board)	47–58	52
• E-WM-5 Masonry-Besblock "Star Performer" cellular blockwork (render & gypsum-base board)	47–57	52
• E-WM-6 masonry-AAC blockwork (render & gypsum-base board)	48–56	52
• E-WM-7 masonry-AAC thin joint blockwork (render & gypsum-base board)	50–58	54

Notes:

1. Mean performance figures were obtained from at least 30 field tests per Robust detail.
2. For airborne performance the higher the value the better the sound insulation.
3. The UK Building Regulations Part E performance standards:
   - Separating walls-airborne (min) 45 dB $D_{nT,w} + C_{tr}$.

RD walls, indicating that it not necessarily have a mass of at least 120 kg/m^2 in walls flanking to a separating element, as indicated within the regulations.

In January 2005 a second series of Robust Detail constructions were issued (Robust details 2005), which included a further aircrete separating wall construction using a thin layer mortar (fig. 3); the previous Robust AAC construction was using a 10 mm sand:cement mortar. The results of the tests for this AAC thin layer mortar construction indicated the best (highest, 54 dB $D_{nT,w} + C_{tr}$) performance level of all the blocks types (density) tested (see table1), again confirming that AAC can provide very good sound insulation solutions to protect occupants from intrusive noise. In addition to the thin layer AAC separating wall two Robust detail separating floors for use within apartments were approved (figs 4 & 5). With both constructions AAC blockwork was satisfactory used within the inner leaf of the external wall (flanking wall) and with one floor type, with a block density of 450 kg/m^3, again showing that lightweight AAC walls flanking a separating floor if built correctly can resist the transmission of sound and provide the required level of sound insulation.

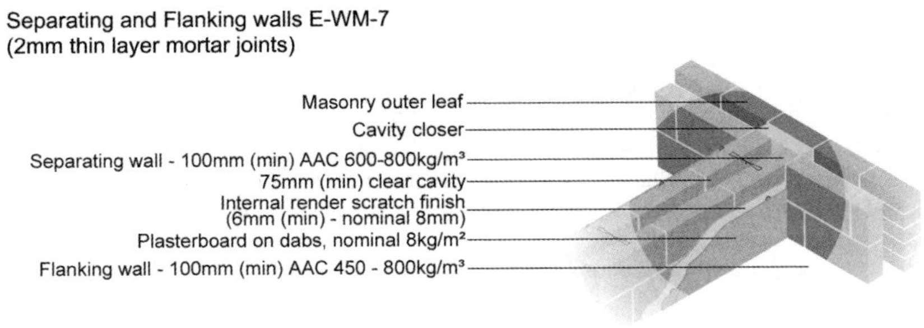

Figure 3.

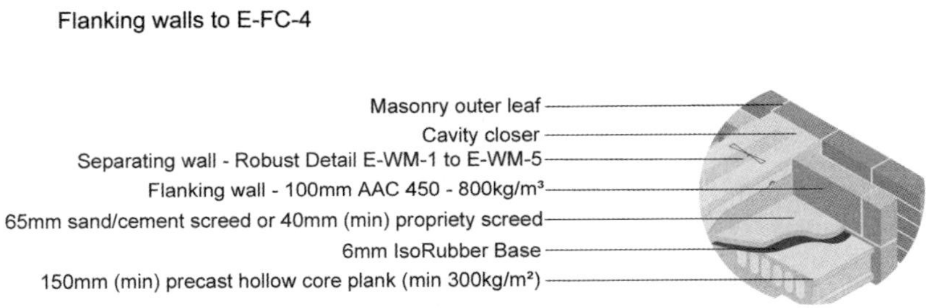

Figure 4.

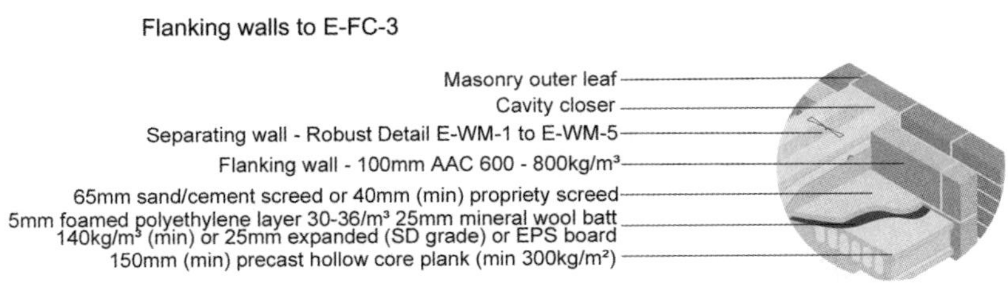

Figure 5.

4 CONCLUSIONS

The introduction of Robust details within the UK Building Regulations has at last proved to the construction industry, regulators and some acousticians that AAC with its unique structure consisting of millions of tiny individual bubbles can meet the required sound insulation set by the regulators and match the same performance levels, or even better, than similar thickness blockwork of higher density (mass).

The UK AAC industry has never sat back and relied on the assumption of mass alone to achieve the required sound insulation within dwellings but to ensure by laboratory and field sound testing that along with its other great attributes of excellent thermal insulation and fire resistance, the actual AAC constructions promoted do truly meet peoples expectations in providing the necessary sound insulation. Furthermore, with continued testing being carried out it is expected that further Robust detail AAC constructions will soon be available to meet market demands.

REFERENCES

RILEM Recommended Practice, Autoclaved Aerated Concrete, Properties, Testing and Design

Robust details Part E. Resistance to the passage of sound. Edition 1

Robust details Part E. Resistance to the passage of sound. Edition 2

R.S. Smith, R.K. Mackenzie, R.G. Mackenzie, T. Waters-Fuller. A Review of the Sound Insulation Performance in Scottish Domestic Construction. Report. Building Standards Division, Scottish Executive, Edinburgh, UK. May, 2001. (www.sbsa.gov.uk)

The Stationery Office, Building Regulations 2000, Approved Document E.2003

Autoclaved Aerated Concrete – Limbachiya and Roberts (eds)

A study on frost resistance of autoclaved high strength aerated concrete

Y. Hama
Muroran Institute of Technology, Hokkaido, Japan

T. Mimori & T. Tairaku
Kushiro National Collage of Technology, Hokkaido, Japan

T. Sanda
Misawa Homes Institute of Research and Development Co., Ltd., Tokyo, Japan

ABSTRACT: The frost resistance of high strength AAC with an oven-dry density of around 0.9 g/cm^3, which was developed to achieve high strength and high durability while making the most of AAC's characteristics including lightweight, was investigated. Having a good quality air-void system with a spacing factor of around 0.1 mm, high strength AAC was proven to be made resistant to frost damage similarly to normal concrete whose frost resistance is improved by inclusion of air voids. Since high strength AAC showed no significant scaling as found on conventional AAC or cracking deterioration during sustained one-side freezing testing, it will be resistant to freezing thawing action with little risk of deterioration under normal use conditions.

1 INTRODUCTION

Autoclaved Aerated concrete (AAC) is widely used as a construction material with excellent lightness, heat-insulating properties, fire resistance, placeablity, and dimensional stability. However, certain applications can make AAC prone to deterioration by frost damage due to its high water absorption. Care should therefore be exercised when using AAC in cold climates. Deterioration of AAC due to frost damage is classified into two types: scaling of surface layers and cracking at points of freezing temperatures (0°C). Frost damage of normal concrete is induced by freezing of water within capillary in concrete, while air voids in concrete mitigate the water pressure at the time of freezing. In the case of AAC, in which air voids prevail over capillary in volume, the air in voids is scarcely isolated, permitting water to enter in the voids, thereby causing scaling of surface layers impregnated with water. It is also said that, in an environment where water is allowed to permeate while the freezing point is maintained, water is drawn to points near 0°C, causing deterioration due to cracking (Senbu & Kamada, 1992). Similarly to AAC, lightweight concrete with an oven-dry density of around 1.0 g/cm^3 is reportedly prone to scaling due to freezing and thawing and cracking by sustained one-side freezing (Tabata et al., 1988).

However, in Japan, the recent enforcement of the Law for Promoting Quality Securement of Houses and the housing performance guarantee system has led to increased demand for housing materials having higher strength and durability than ever. New materials have therefore been under development based on AAC aiming for high strength and high durability while making the most of its characteristics, such as dimensional stability, fire resistance, and ease of placement.

This study intends to investigate the basic physical properties and frost resistance of prefoamed-type autoclaved high strength aerated concrete (hereafter referred to as high strength AAC). This AAC is produced by blending foams generated by an air-entraining admixture with the slurry and

Table 1. Mix proportions.

Series specimen	1				2	
	No. 1	No. 2	No. 3	No. 4	No. 5	No. 6
Water/Powders	0.40	0.40	0.40	0.40	0.45	0.45
Powders*						
Cement	44.0	44.0	44.0	44.0	37.5	37.5
Silica sand	40.0	40.0	40.0	30.0	28.5	28.5
Flyash	10.0	10.0	10.0	20.0	30.0	30.0
Quick-hardening admixture	6.0	6.0	6.0	6.0	4.0	4.0
Aramid fiber*	0.3	0.3	0.3	0.3	0.5	0.5
Air-entraining admixture**	0.00	0.38	0.75	0.75	0.50	1.00
Superplasticizer**	0.3	0.3	0.3	0.3	0.3	0.3

* wt% to all powers, ** wt% to cement.

incorporating fibers to increase the flexural strength and ductility, while using an air-entraining and high-range water-reducing admixture to strengthen the matrix.

2 EXPERIMENT PLAN AND PROCEDURE

2.1 *Experiment plan*

2.1.1 *Materials*

The materials included normal portland cement, silica sand, and fly ash with densities of 3.16 g/cm^3, 2.65 g/cm^3, and 2.20 g/cm^3, respectively, as well as aramid fibers, a quick-hardening admixture, air-entraining and high-range water-reducing admixture (superplasticizer, polycarboxylate type), and air-entraining admixture.

2.1.2 *Specimens*

The superplasticizer and air-entraining admixture were mixed with water in a mixer for 60 sec to generate foams. Powder materials including cement were then added and mixed for 2 min to 2 min and 30 sec. The mixture was placed in molds of specified sizes, demolded 1 hour later, and autoclave-cured under 10 atm at 180°C for 8 hours.

2.1.3 *Proportioning*

Tests were conducted in two series, and mixtures for each series were proportioned as given in Table 1. In Series 1, four types of specimens having different foam contents were prepared to examine the effects of the frost resistance of the strengthened matrix and the pregenerated foams on the frost resistance of concrete. For Series 2, two types of high strength AACs were prepared with a target oven-dry density of 0.9 g/cm^3 in consideration of the practical application together with conventional AAC (prefoamed type, hereafter referred to as P-ALC).

2.2 *Test procedure*

Basic physical properties of specimens, such as oven-dry density, compressive strength, and flexural strength, were measured. Frost resistance testing and air-void system measurement were also carried out. Compression and flexure tests were conducted on beam specimens measuring 4 by 4 by 16 cm in accordance with JIS R 5201. The frost resistance of AAC is generally evaluated in terms of volume loss by repeated freezing in air and thawing in water (ASTM C666 B method) in regard to scaling and in terms of the number of days to cracking by sustained one-side freezing in regard to

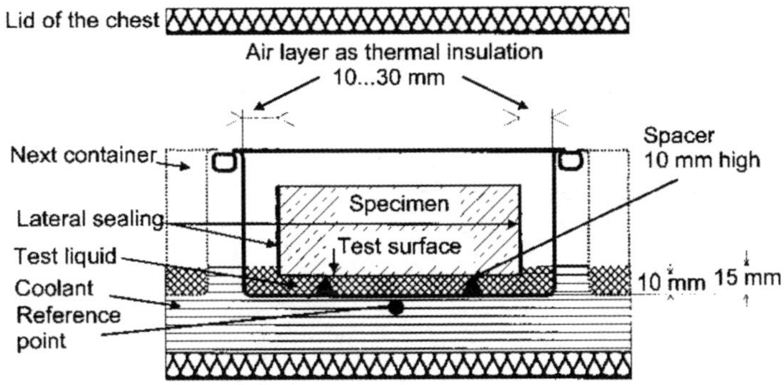

Figure 1. CIF test apparatus.

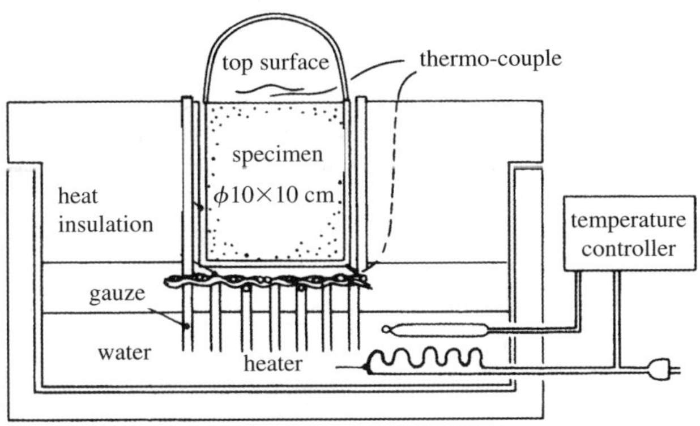

Figure 2. Sustained one-side freezing test apparatus.

cracking. In this study, however, changes in the scaling and dynamic modulus of both series were evaluated in accordance with the RILEM CIF test (Capillary suction, Internal damage and Freeze thaw test) (Setzer, 2001), while the resistance to cracking of Series 2 was evaluated by sustained one-side freezing test.

2.2.1 *CIF testing*

After a specified curing process, two facing sides and both ends of beam specimens measuring 10 by 10 by 20 cm were sealed with aluminum tape having butyl rubber lining to allow water suction only through the bottom for 7 days in a thermohygrostatic room at 20°C and 60% R.H. The specimens were then subjected to up to 100 one-side freezing and thawing cycles in a test apparatus as shown in Fig. 1 while maintaining water suction from the bottom. These cycles consisted of a high temperature phase of 20°C for 1 hour and a low temperature phase of −20°C for 3 hours, with transition periods at a temperature gradient of ±10 K/h in between (2 cycles/day). The measurement items included mass change, relative dynamic modulus, and scaling at the ends of the specified numbers of cycles. The scaling was measured by performing ultrasonic-cleaning on each container including the specimen for 3 min, filtrating the solution, drying the filter cake at 105°C for 24 hours, and weighing the dry delaminated flakes. The relative dynamic modulus was measured by the flexural oscillation method.

2.2.2 *Sustained one-side freezing testing*

Figure 2 shows the sustained one-side freezing test apparatus. A water tank containing water with a temperature kept constant at +20°C was set in a space under a cylindrical specimen 10 cm in diameter and 10 cm in length. Water was constantly supplied from the tank to the bottom of the specimen through absorbent wadding and cotton cloth, while the space on the top of the specimen was kept at −20°C. The changes in the mass were measured, and cracking was visually observed during a test period of 125 days.

2.2.3 *Measurement of air-void system parameters*

Measurement was made in accordance with ASTM C 457 (the linear traverse method) to calculate the air-void system parameters: air void content, average chord intercept of the air-voids, and spacing factor.

3 RESULTS AND DISCUSSION

3.1 *Basic physical properties*

Table 2 gives the basic physical properties of each specimen. The oven-dry densities of Specimen 1 containing no air-entraining admixture and high strength AACs in Series 1 are 1.47 g/cm^3 and 0.91 to 1.11 g/cm^3, respectively, whereas those of P-ALC and high strength AACs in Series 2 are 0.58 g/cm^3 and 0.90 to 0.95 g/cm^3, respectively. The oven-dry density is related to the compressive strength and the flexural strength in Figures 3 and 4, respectively. Both the compressive and flexural strengths linearly increase as the oven-dry density increases. High strength AAC with an oven-dry density of around 1.0 g/cm^3 attains a compressive strength of 20 to 30 N/mm^2 and flexural strength

Table 2. Basic physical properties.

Series	1				2		
specimen	No.1	No.2	No.3	No.4	No.5	No.6	P-ALC
Oven-dry density (g/cm^3)	1.47	1.11	0.96	0.91	0.95	0.90	0.58
Compressive strength (N/mm^2)	33.6	26.9	24.5	17.5	27.4	26.0	5.0
Specific strength (N/mm^2)	22.9	24.2	25.6	19.2	28.8	28.9	8.6
Flexural strength (N/mm^2)	8.1	6.3	6.1	5.0	5.4	5.3	1.0
Air-viod content (%)	7.5	25.0	35.8	38.3	27.3	36.0	46.6
Average chord intercept (mm)	0.555	0.187	0.189	0.168	0.127	0.149	0.274
Spacing factor (mm)	0.668	0.130	0.112	0.096	0.077	0.069	0.098

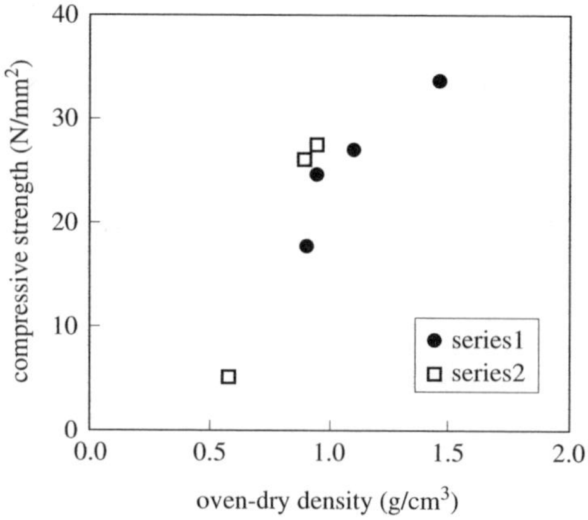

Figure 3. Relationship between the oven-dry density and compressive strength.

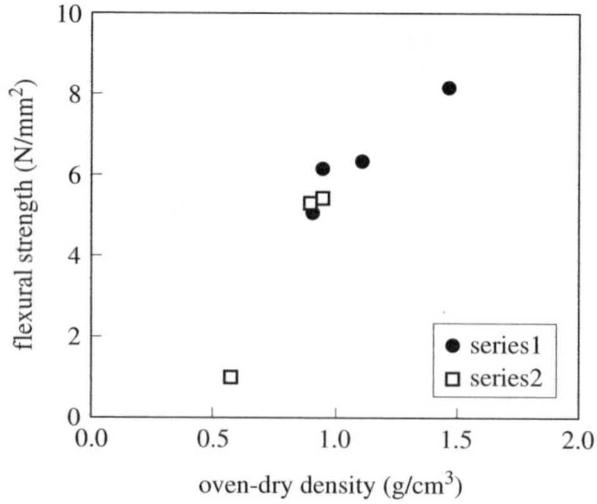

Figure 4. Relationship between the oven-dry density and flexural strength.

of 5 to 6 N/mm^2. Its specific strength is three times that of P-ALC, which has been known for its high specific strength.

Figure 5 shows the relationships between the oven-dry density and the air-void content, average chord intercept, and spacing factor. Whereas the air-void content of high strength AAC is scattered over a wide range of 25.0 to 38.3% depending on the oven-dry density, the scatters of its average chord intercept and spacing factor are relatively small. Its spacing factor is particularly concentrated around 0.1 mm, proving that high strength AAC has an excellent air-void system resistant to frost damage when evaluated similarly to normal concrete. Though the spacing factor of P-ALC is nearly the same as that of high strength AAC, its average chord intercept is greater than that of high strength AAC. On the other hand, the average chord intercept and spacing factor of entrapped air (7.5%) in Specimen 1 containing no air-entraining admixture are significantly greater than those of other specimens.

3.2 *CIF testing*

3.2.1 *Changes in relative dynamic modulus*

Figure 6 shows the changes in the relative dynamic modulus during CIF testing. In Series 1 for high strength AACs having different densities, the relative dynamic modulus of Specimen 1 drastically decreases from immediately after the beginning of freezing and thawing cycles, leading to a large crack on the top surface as shown in Photo 1 at 30 cycles. On the other hand, the modulus of other specimens containing foams scarcely decrease throughout the test period. It is therefore inferred that a good quality air-void system improves the frost resistance of high strength AAC similarly to normal concrete. In Series 2 in which high strength AAC is compared with P-ALC, the relative dynamic moduli of both high strength AACs scarcely decrease similarly to Series 1, whereas that of P-ALC shows a decreasing tendency from immediately after the beginning of freezing and thawing cycles.

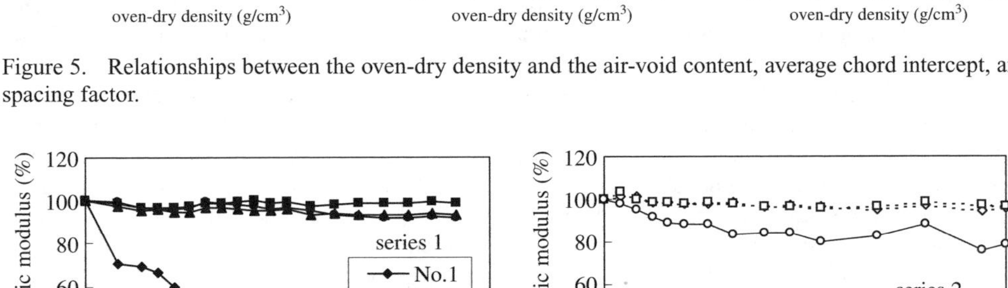

Figure 5. Relationships between the oven-dry density and the air-void content, average chord intercept, and spacing factor.

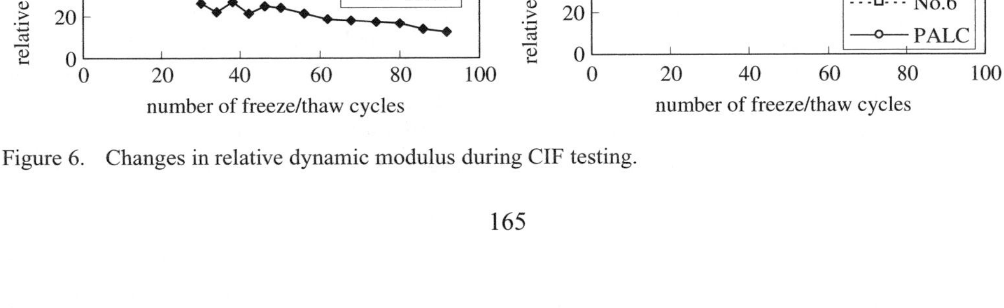

Figure 6. Changes in relative dynamic modulus during CIF testing.

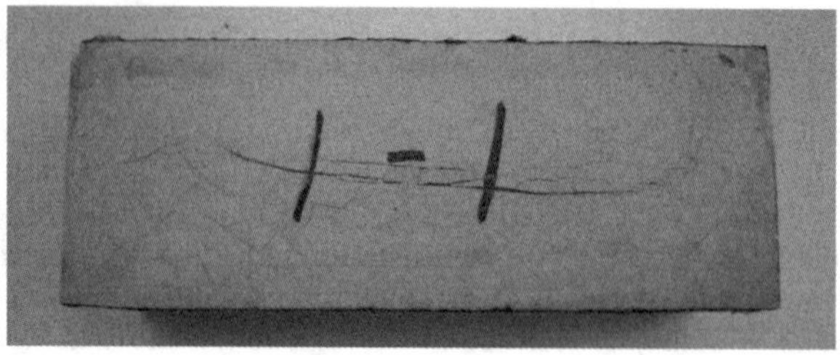

Photo 1. Large crack on the top surface of specimen No.1 by CIF testing.

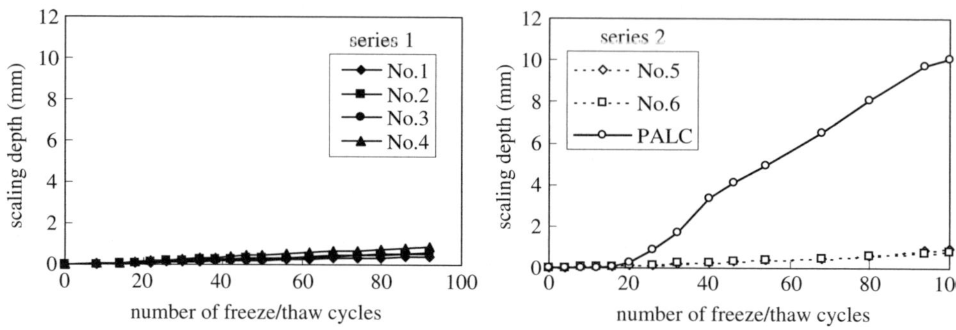

Figure 7. Changes in scaling depth during CIF testing.

3.2.2 *Changes in scaling depth*

Figure 7 shows the changes in the scaling depth during CIF testing. In Series 1, the scaling depths are generally marginal, being 1.0 mm or less, excepting that of Specimen 4 being slightly large. In Series 2, the scaling depths of high strength AACs are similarly marginal at 1.0 mm or less, but that of P-ALC rapidly increases beginning from around 20 cycles, exceeding 10.0 mm at the end of the test. Slight fluffing of fibers due to scaling was noted on high strength AACs.

3.2.3 *Changes in water absorption*

Figure 8 shows the changes in the water absorption during CIF testing. In Series 1, the water absorptions vary from one specimen to another at the beginning of testing, but their increases during the first 7 days through suction from the bottoms are nearly the same for all specimens at around 10% by volume. This can be attributed to the fact that water is absorbed mostly by capillary suction during this period through nearly the same matrices of Specimens 1 to 4 having the same basic mixture proportions excepting the air-void content. Under the subsequent freezing and thawing conditions, however, the water absorptions of Specimens 2 and 4 level off at 62 cycles (38 days), indicating saturation, while those of Specimens 1 and 3 significantly increase. In Series 2, while the water absorption behavior of the two types of high strength AACs are nearly the same, the increases in the water absorption of P-ALC are small during the first 7 days of suction from the bottom but are significantly large during the subsequent freezing and thawing cycles.

Since deterioration induced by freezing and thawing is closely related to water absorption properties, it is important to judge the possibility of reaching the critical degree of saturation (the threshold degree of saturation for deterioration) under the expected environmental conditions. Figure 9 shows the relationship between the water absorption and the relative dynamic modulus. Since the relative dynamic moduli of Specimen 1 and P-ALC decrease from immediately after the beginning of freezing and thawing cycles, their water absorption at the end of suction from the bottom, i.e., 36% and 25% by volume, respectively, are judged as their critical degrees of saturation. Water absorption is accelerated thereafter due to structural loosening concomitant with the progress of deterioration. No deterioration was observed on Specimens 2 to 6 of high strength AAC, though these should have

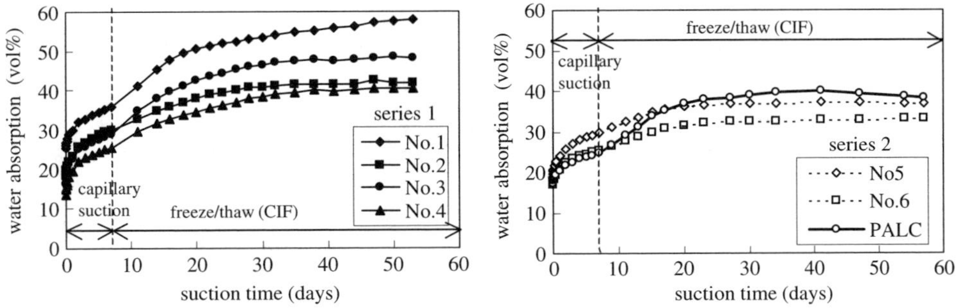

Figure 8. Changes in water absorption during CIF testing.

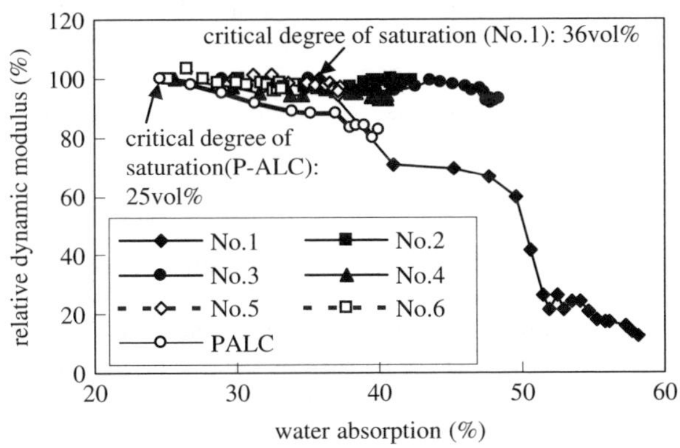

Figure 9. Relationship between water absorption and relative dynamic modulus.

been nearly saturated at the end of the test. Accordingly, high strength AAC is unlikely to undergo deterioration due to structural loosening, such as strength loss and collapse, by freezing and thawing in a normal use environment, though it is difficult to specify their critical degree of saturation.

3.3 *Sustained one-side freezing testing*

According to past studies, the water absorption of standard AAC having a density of 0.5 g/cm^3 with no water-repellent treatment increases to 30 to 40%, resulting in cracking of most specimens, by the end of 50 days of sustained one-side freezing testing[1]. In the present study, however, no cracking was observed in either of P-ALC and high strength AACs throughout the 125-day test period, proving their excellent cracking resistance. Figure 10 shows the changes in the water absorption during the sustained one-side freezing testing. Starting from water contents differing from one specimen to another, the water content of P-ALC tends to increase throughout the test period, whereas those of high strength AACs scarcely increase after 30 days. This tendency agrees with their absorption behavior in the CIF testing.

4 CONCLUSIONS

The frost resistance of high strength AAC with an oven-dry density of 0.9 to 1.0 g/cm^3, which was developed to achieve high strength and high durability while making the most of AAC's

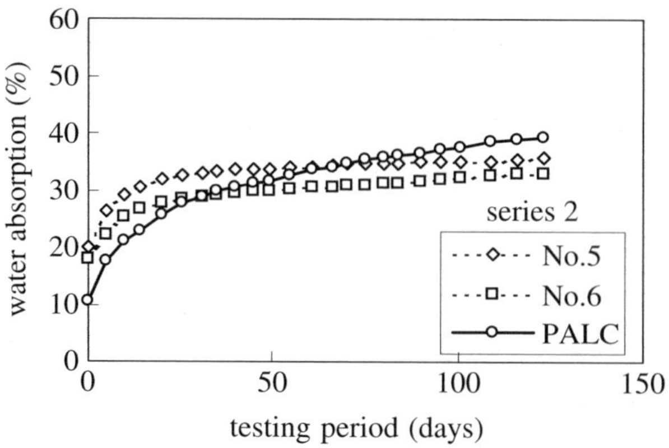

Figure 10. Changes in the water absorption during sustained one-side freezing testing.

characteristics including its lightweight, was investigated in this study, and the following results were obtained:

(1) Inclusion of foams enhances the frost resistance of high strength AAC similarly to normal concrete.
(2) The spacing factor of high strength AAC is around 0.1 mm, representing an excellent air-void system highly resistant to frost damage.
(3) No significant scaling as found with P-ALC was observed with high strength AAC.
(4) No cracking deterioration was observed by sustained one-side freezing tests.
(5) Under normal use conditions, it is unlikely that freezing and thawing should cause deterioration of high strength AAC due to structural loosening, such as strength loss and collapse.

REFERENCES

Senbu, O. & Kamada, E. 1992. Mechanism of frost deterioration of AAC. In Wittmann, F. H. (ed.), *Advances in Autoclaved Aerated Concrete*: 153–156

Setzer, M.J. 2001. Recommendation of RILEM TC 176-IDC: Test method of frost resistance of concrete "CIF-Test – Capillary saction, Internal damage and freeze-thaw test; Reference method and alternative method A&B", *Materials and Structure* 34: 515–525

Tabata, M. et al. 1988. Experimental studies on frost resistance of lightweight concretes. Summaries of technical papers of annual meeting AIJ (A-1): 245–246

Autoclaved Aerated Concrete – Limbachiya and Roberts (eds)
© 2005 Taylor & Francis Group, London, ISBN 0 415 38356 0

Development of a procedure for determining the freeze-thaw resistance of Autoclaved Aerated Concrete

D.J. Harris
H+H Celcon Ltd, Sevenoaks, Kent, UK

M.J. Robinson
Chairman of EAACA Working Group 1-Test Methods, Coleshill, Birmingham, UK

ABSTRACT: The introduction of the European Product Standards for masonry and reinforced AAC components required that manufactures should declare a number of essential requirements so as to meet the Construction Products Directive. In the absence of a recognized procedure to determine the freeze thaw resistance of AAC, the European Committee (CEN TC177) set up to develop the specifications for reinforced AAC products, requested the European AAC Association (EAACA) to develop a procedure to measure the materials durability. The aim was to develop a simple procedure that could be undertaken without the need for expensive and complex test equipment but still subjected the material to conditions likely to be experienced in service. Using a Standard freezing cabinet and 100 mm cubic samples, various test parameters were examined resulting in a procedure, which determines the volume (mass) loss of a sample with the option to also determine any loss in compressive strength.

1 INTRODUCTION

The European product standard for Autoclaved Aerated Concrete (AAC, or Aircrete as commonly known within the UK) masonry units, EN 771-4 and the draft product standard for prefabricated reinforced components, prEN12602, both require a manufactures declaration of their materials durability. In the absence of a harmonized test method the standards specified that the manufacture shall evaluate and declare the freeze/thaw resistance of units by reference to the provisions valid in the intended place of use of the masonry units or for reinforced components, the freeze/thaw resistance of AAC shall be verified by tests in accordance with test methods used in the place of use. In other words using the procedures that are in use within a national country that are accepted by their regulating organisations as indicating the materials resistance to freezing and thawing within its intended place of use. However, due to the different climate conditions and intended use of these AAC products; acceptance within another country than its manufacture may not be acceptable as a direct comparison of the materials performance was not possible.

The European committee therefore set up to develop the specifications for reinforced AAC products, CEN TC177, requested EAACA to develop a procedure to measure the materials durability by determining its resistance to freezing and thawing.

It should be noted that prEN12602 being developed by CEN TC177 also includes the manufacturing procedures and product characteristics for reinforced lightweight aggregate components but at the present time a similar freeze/thaw test method is not being developed.

The use of AAC blockwork for solid external walls in Scandinavia has been successfully used for many years without any distress to the material. This is a positive indication that AAC can resist freezing and thawing in the most extremes of conditions. Also in the UK, AAC has been successfully used for the construction of walls below ground level to support the above ground structure. Again in this application the material is regularly exposed to numerous freeze thaw cycles throughout the winter season.

The EAACA committee concluded that many freeze/thaw test methods have been developed in the past but considered that no single test could completely reproduce the conditions in the field for all individual scenarios. Nevertheless, any method should at least relate to the practical situations and give consistent results. Also, so as to ensure that all manufactures could undertake the testing, the method should be simple to carry out and without the need for expensive equipment.

2 VALIDATION PROCEDURE

In developing this test procedure the EAACA committee identified key areas that required investigating to ensure that the conditions for the action of freezing and thawing were achieved. At the same time ensuring that the material was subjected the normal stresses that would occur in service during the process of freezing and thawing.

There were two main criteria that needed to be considered during the validation procedure:-

a) that the all the material within the specimens reached a temperature below zero degrees centigrade in the freezing cycle and above zero degrees centigrade in the thawing cycle, and
b) the rate of freezing and thawing should be at a uniform rate to prevent sudden shock to the specimen. This would also allow parameters to be set when automatically controlled equipment was used for the testing.

The testing brief agreed by the EAACA Committee considered that a specimen size of 100 mm cubes was suitable for the test and that two different densities AAC samples (350 and 750 kg/m^3) should be used. The brief also considered that fine wire thermocouples were placed on the surface, at 10 and 30 mm from the surface, and at the centre of the test specimens. These should be connected to a data recorder to obtain continuous time/temperature monitoring whilst the specimens were subjected to freezing and thawing.

An important consideration of the Committee was to ensure that simple inexpensive equipment could be used to during the test. To comply with this remit a vertical commercial freezer was used to subject the specimens to the freezing temperature of -15 degrees centigrade, and a polythene tank, in which there was a reservoir of water above which the specimens could be placed on wooden bearers to maintain a relative humidity of $>95\%$, placed within a laboratory environment represented the conditions for thawing. The tank was covered and sealed during the thawing cycle to maintain the high humidity and prevent drying out of the specimens.

An initial series of tests measurements were carried out at the controlling laboratory to evaluate both conditions of freezing and thawing for a range of scenarios, and were undertaken on AAC samples with different density with different volume loadings of the freezer and thawing tank. This indicated the differences in the time that is required for the centre of the specimen to reach zero degrees centigrade but in addition concluding that by altering the degree of volume loading of the cabinets, different rates could be achieved.

Once the conditioning criteria for the freezing and thawing cycles had been set by the controlling laboratory, an equal number of samples of AAC were tested to the draft procedure by two different laboratories to ensure the test procedure was fully documented and easy to follow but at the same time evaluate the correlation of the results obtained under different testing regimes. Within this process a number of items were identified for inclusion to improve the reliability and hence reproducibility e.g. sampling scheme, distance of specimen above the water reservoir during thawing and the rotation of the specimens by 90 degrees after each cycle.

3 RESULTS

A summary of all the temperature profiles during the freezing and thawing cycles are presented in Tables 1 & 2.

Table 1. Test results.

Test No.	Density of cube monitored (kg/m^2)	State of freezer during the temperature monitoring	Time taken to reach 0°C, hh:mm		Minimum temperature after 7 hours, °C	
			Surface	Centre	Surface	Centre
1	750	Full	02:30	03:00	−2.5*	−1.2*
2	750	Single specimen	00:40	01:15	−13.5	−12.4
3	350	Single specimen	00:35	01:25	−15.6	−15.5

* The temperature was monitored for a total of 24 hours, at which point the temperature on the surface and in the centre of the cube was −6.2°C.

Table 2. Test results

Test No.	Density of cube monitored (kg/m^2)	Thawing conditiion	Time taken to reach 0°C, hh:mm		Maximum temperature after 7 hours, °C	
			Surface	Centre	Surface	Centre
4	750	Single Cube under Hessian	00:15	05:00	9.6	6.5
5	350	Single Cube under Hessian	01:00	05:50	10.7	8.7
6	350	Single Cube Thawing Tank	00:10	03:40	16.0	16.0
7	750	18 Cubes Thawing Tank	01:00	04:40	13.7	9.0

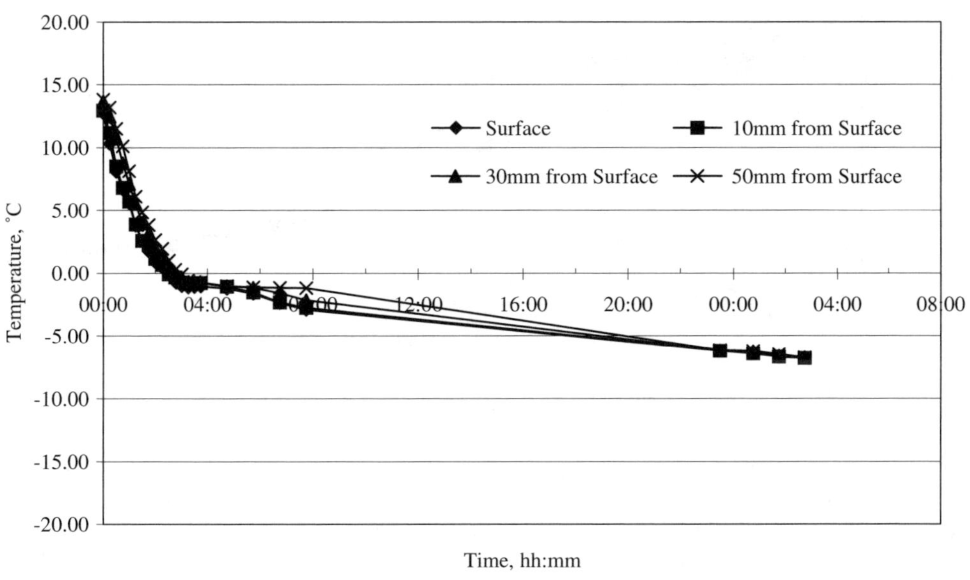

Figure 1. Freezing of a 100 mm AAC cube (750 kg/m^3) freezer loaded with 80 saturated 100 mm AAC cubes.

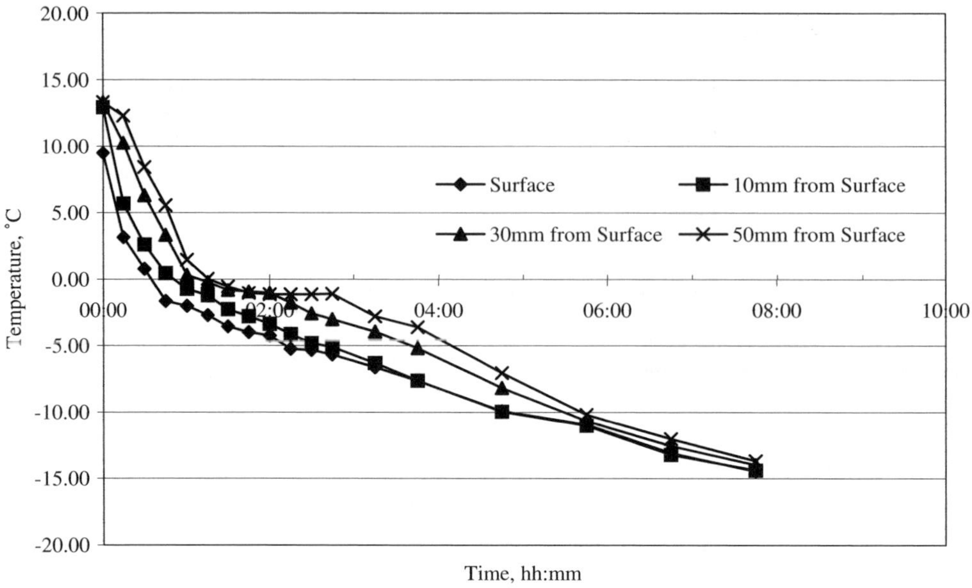

Figure 2. Freezing of a 100 mm AAC cube (750 kg/m^3).

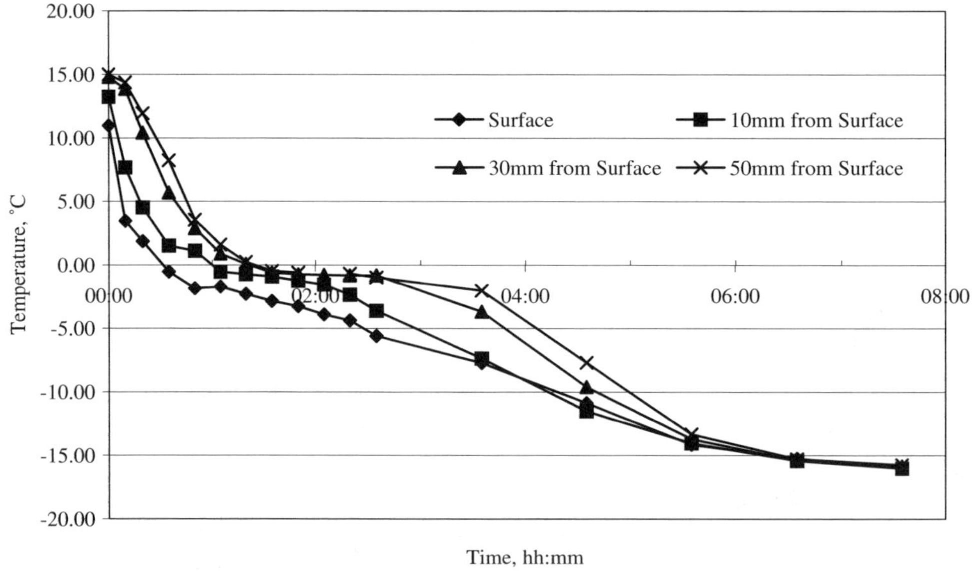

Figure 3. Freezing of a single 100 mm AAC cube (350 kg/m^3).

Figure 1 illustrates the temperature profile of a specimen of density 750 kg/m^3 during the freezing cycle with a total loading of 0.08 m^3 of AAC within a freezer of volume 0.571 m^3. It can be seen that the centre of the specimen reached zero degrees centigrade after 180 minutes. For the same material but with a single specimen loading (0.001 m^3), Figure 2, the centre temperature of zero degrees centigrade is reached in 75 minutes.

A single specimen loading for 350 kg/m^3 AAC reached zero in 85 mins, Figure 3.

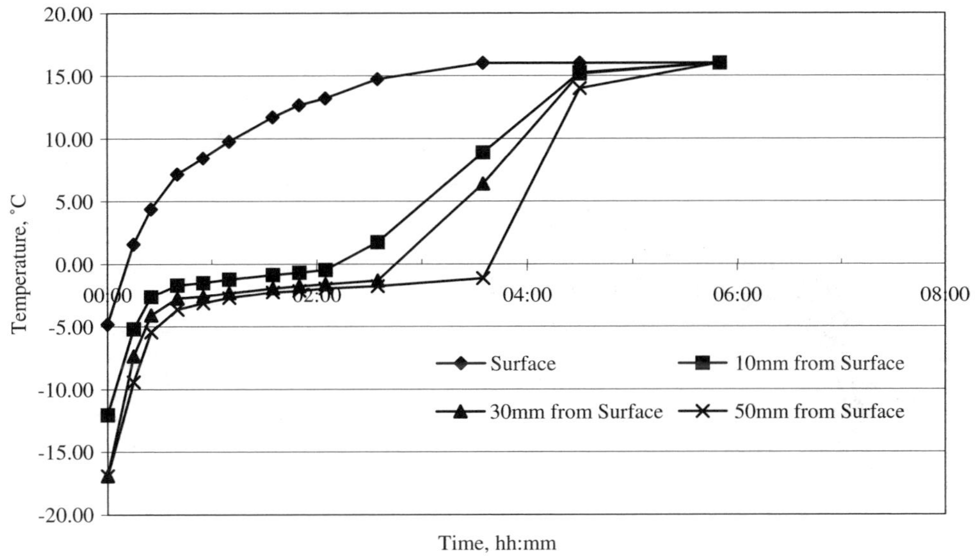

Figure 4. Thawing of a single 100 mm AAC cube (350 kg/m^3) within an enclosed tank.

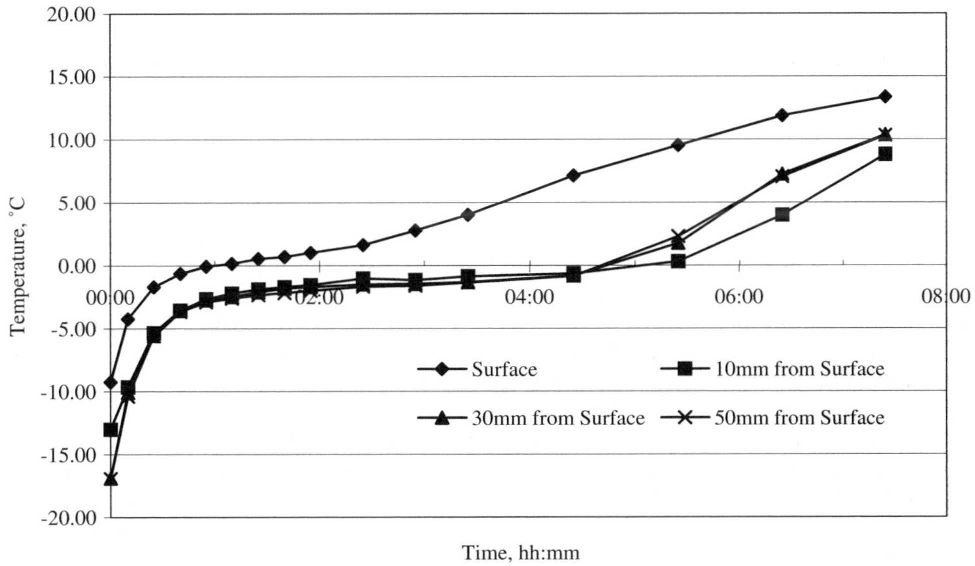

Figure 5. Thawing of a 100 mm AAC cube (350 kg/m^3) within an enclosed tank loaded with eighteen 100 mm AAC cubes.

Measurements during the thawing cycle indicated that for a single specimen (350 kg/m^3) within the thawing tank (volume 0.25 m^3) the centre reached zero degrees centigrade after 220 minutes (Figure 4) whereas a full load of 18 specimens of 750 kg/m^3 AAC (0.018 m^3) took 280 minutes (Figure 5).

Measurement of the relative humidity of the environment surrounding the specimens within the thawing tank was also undertaken using an electronic humidity probe to ensure that the required conditions are met. Levels of above 95% were recorded.

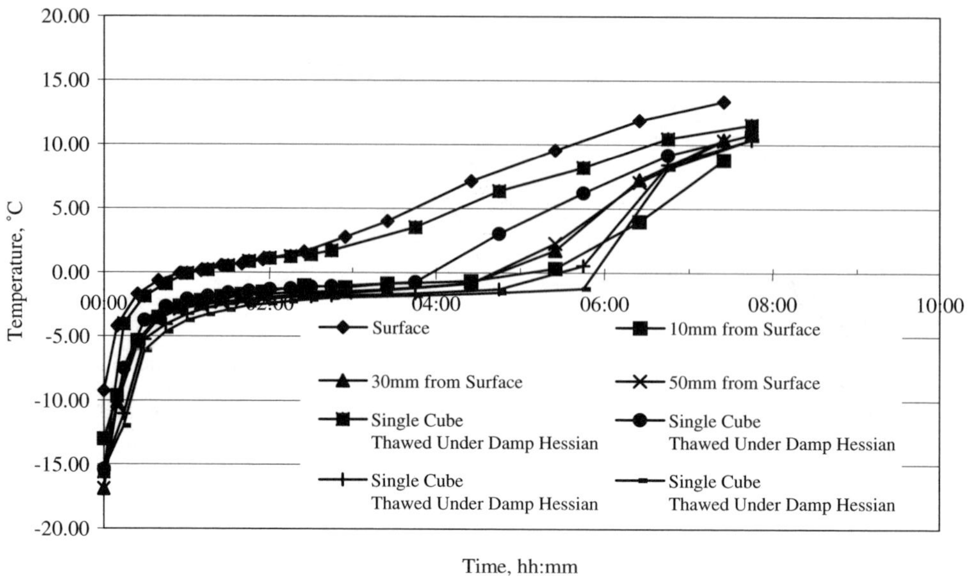

Figure 6. Thawing of a 100 mm AAC cube (350 kg/m^3).

A single specimen stored under damp Hessian during the thawing cycle was also measured indicating a similar rate of thawing to that within the thawing tank, Figure 6.

4 CONCLUSIONS

As a result of the work undertaken by EAACA Committee, a procedure for determining the freeze thaw resistance of AAC has now been submitted to CEN TC177 for their approval, the main sections of which are in appendix A. Once their comments have been agreed the document will then be circulated for public comment and it is hoped that the final version will form the basis of a procedure for determining the durability of AAC. A minimum of 15 freeze thaw cycles are required to be undertaken on each product but each member country will be able to specify the number of cycles required within their own country with reference to the materials intended use.

It was also considered that the significant parameter for regulators and end users to assess the materials performance to the resistance of freezing and thawing was to determine if any volume loss occurred during the test and was therefore considered as a normative requirement. As some member countries may also require to examine any loss in compressive strength loss due to freezing and thawing a procedure for determination of strength loss was added as an informative part of the Standard.

REFERENCES

Committee CEN /TC 177, Prefabricated reinforced components of autoclaved aerated concrete and aggregate concrete with open structure the Secretariat is held by DIN.
European Autoclaved Aerated Concrete Association, Dostojewskistraße 10, 65187 Wiesbaden, Germany.
EN 771-4:2003 Specification for masonry unit – Part 4 Autoclaved aerated concrete masonry units.
prEN 12602:2005 Prefabricated reinforced components of autoclaved aerated.

APPENDIX A

EAACA Recommendation

extract from EAACA R3
November 2004

1 DETERMINATION OF THE FREEZE-THAW RESISTANCE OF AUTOCLAVED AERATED CONCRETE

1.1 *Principle*

Prismatic test specimens are cut from products sampled from new production, immersed under water for 48 hours, removed and then placed in polythene bags or similar for 24 hours to equilibrate.

Specimens are then divided into two groups: main specimens and reference specimens. The main specimens are subjected to repeated freezing at -15 ± 2 °C for 16 hours followed by thawing in an environment of >95%RH for 8 hours. The reference specimens are stored in a climate chamber in conditions that prevent any drying of the specimens during storage.

After the specified numbers of freeze-thaw cycles, the volume (mass) loss of the main specimens is determined. The dry density and moisture content is also determined on the reference specimens.

1.2 *Test specimens*

The sample for the preparation of the test specimens shall be taken in such a manner that it is representative of the product to be investigated, and shall be cut from the unit as illustrated in Figs 1 & 2. The test specimens shall be cubes with an edge length of 100 mm, ±2 mm.

The test specimens shall be cut by means of a rotating carborundum blade or similar equipment. All surfaces shall be plane and clean.

The test set shall consist of equal number of main and reference specimens, (2.1).

The reference specimens are used for determination of the moisture content prior to undertaken the freeze-thaw tests and the dry density.

The main specimens are used for determination of the loss of volume (mass) and if required, the loss in compressive strength after the specified numbers of cycles of repeated freezing and thawing.

The numbering shall show the position of the test specimens in the material relative to the rise of the mass, and the direction of rise shall be marked on the test specimens.

Whenever possible, equal number of specimens shall be prepared from the upper third of the component, from the middle and from the lower third in the direction of rise of the mass during manufacture (see Figure 1). The specimens should be cut in pairs (reference and main) from adjacent areas of the unit to ensure a good comparison (see Figure 2).

1.3 *Conditioning of test specimens*

After their preparation, the test specimens shall be placed in underwater for 48 hours (20 ± 2)°C. The test specimens shall be stored for 24 hours with half of their thickness (50 mm) under water and for 24 hours totally under water. Remove specimens from the water and place in a Polythene bag or similar for 24 hours to allow the specimens to equilibrate without drying out.

The specimens shall then be divided into two groups, the main and reference group.

1.4 *Testing procedure*

The main specimens shall be loaded into the freezing chamber at the temperature of minus 15 ± 2°C in containers or placed on the net shelves. The distance between the main specimens, containers and the upper shelves shall be at least 50 mm. The air temperature in the chamber shall be measured in the middle of its working volume, in immediate closeness to the specimens. It is important to

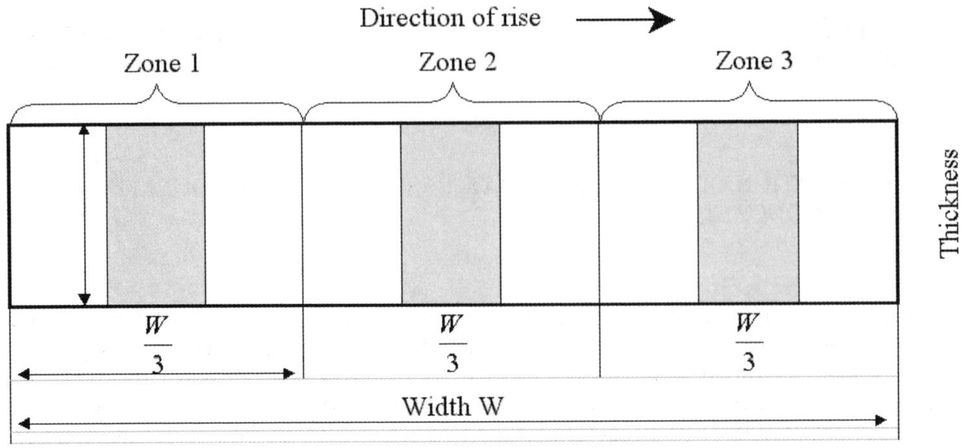

Figure 1. Sampling scheme.

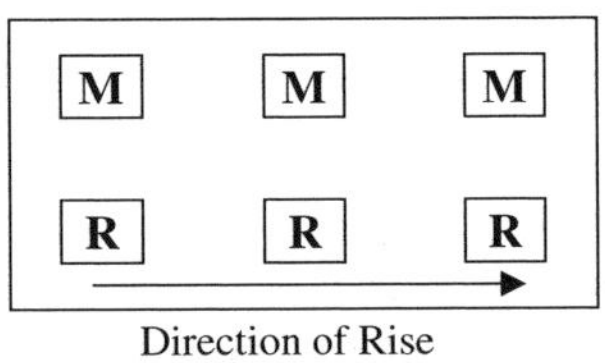

Figure 2. Plan; R = Reference specimen, and M = Main Specimen.

ensure that the specimens are not subjected to rapid freezing and thawing. The rate of freezing of the specimens shall be determined from measurement of the temperature at the centre of the specimen. The temperature at the centre of the specimens should reach zero °C within 3 ± 1 hours and should be determined before commencement of the test on a control specimen. The rate may be obtained by variation in the volume loading of the cabinet or by an automatic temperature control system. The specimens should be left in the cold cabinet for a total of 16 hours.

Thawing of the main specimens after removing from the freezing chamber takes place in the thawing cabinet (climate chamber) at a temperature of (20 ± 2)°C and relative humidity of >95%. The specimens should be positioned on wooden bearers with sufficient distance between each to ensure adequate air circulation and uniform thawing. The rate of thawing of the specimens shall be determined from measurement of the temperature at the centre of the specimen. The temperature at the centre of the specimen should reach 0°C within 5 ± 1 hours and should be determined on a control specimen prior to commencing the test. The rate may be obtained by variation in the volume loading of the cabinet or by an automatic control system. The specimens should be left to thaw for a total of 8 hours.

This completes one freeze-thaw cycle. The specimens should be rotated by 90° after each cycle.

In case of a forced pause at testing, the main specimens shall be kept in the thawed condition, in the climate chamber at a temperature of (20 ± 2)°C and relative humidity of >95%, so that drying of the specimens is excluded.

The reference specimens shall be kept during the period, corresponding to the same numbers of cycles as the main specimens, in the climate chamber at a temperature of (20 ± 2) °C and relative humidity of >95% or alternatively sealed in polythene bags so that the specimens do not lose moisture.

Note: The freeze thaw cycles are best programmed such that the placing of the specimens in the cold cabinet is undertaken at 17:00 and that of thawing at 09:00 to fit into the working day. If the specimens have to be left at weekends then the specimens should be stored in the cold cabinet over this period.

After 15 Freeze-Thaw cycles and at the completion of the test any visual damage to the specimens should be noted.

Note: Should a specimen split during the freeze-thaw test this should not be considered as frost damage. The specimen should be disregarded from the test result.

2 DETERMINATION OF MASS LOSS OF AAC TEST SPECIMEN DURING FREEZE-THAW TEST

2.1 *Number of specimens*

The test set shall consist of 12 test specimens: 6 reference specimens and 6 main specimens for determining mass loss only.

Reference specimens are used for determination of the moisture content prior to undertaken the freeze-thaw tests.

The main specimens are used for determination of the loss of mass after the specified numbers of cycles of repeated freezing and thawing.

2.2 *Determination of mass loss*

This calculation is based on the use of the loss in the dry mass of the main specimens. This is determined from the difference between the initial dry mass (calculated by assuming that the main specimens have the same initial moisture content as the reference specimens) and the final dry mass measured directly on the main specimens.

After the specified numbers of freeze thaw cycles both sets of test specimens shall be dried at ($105 \pm 5°C$) until constant mass and the mass measured. Care should be taken that no material is lost.

2.2.1 *Reference specimens*

M_i^r = measured initial moist mass of reference specimens immediately after removal from the polythene bag.

M_{id}^r = measured oven-dry mass of reference specimens (measured at n number of freeze thaw cycles).

2.2.2 *Main specimens*

M_i^m = measured initial moist mass of main specimens immediately after removal from polythene bag (before freeze thaw cycles).

M_{id}^m = equivalent dry mass of the main specimens immediately after removal from the polythene bag.

$$M^m_{id} / M^m_i = M^r_{id} / M^r_i$$

Therefore:

$$M^m_{id} = M^m_i \times (M^r_{id} / M^r_i)$$

Note: This assumption is correct when the main specimen and reference specimen are taken from the same AAC block or component and were located beside each other.

M_{fd}^m = measured oven dry mass specimen after n number of freeze thaw cycles.

M_L = percentage loss in dry mass.

$$M_L = 100 \times (M^m_{id} - M^m_{fd}) / M^m_{id}$$

$$= 100 \times (1 - M^m_{fd} / M^m_{id})$$

Autoclaved Aerated Concrete – Limbachiya and Roberts (eds)
© 2005 Taylor & Francis Group, London, ISBN 0 415 38356 0

Carbonation of in-service Autoclaved Aerated Concrete

F. Matsushita, Y. Aono & S. Shibata
Sumitomo Metal Mining Siporex Co., Ltd., Research & Development Center, Japan

ABSTRACT: Carbonation is one of the harmful factors of durability and service life of Autoclaved Aerated Concrete (AAC) panels. This paper deals with the carbonation and its influence on in-service AAC panels. Cracking, panel bending properties, compressive and tensile strength, drying shrinkage and carbonation degree were investigated for in-service AAC panels aged 4 to 33 years. When the carbonation degree was more than 50%, the AAC panels showed cracking and degradation in span-deflection ratio, compressive strength and drying shrinkage. Acrylic resin-finish, stable atmosphere and water-repellent additive may be desirable for the durability of AAC.

1 INTRODUCTION

Carbonation of AAC leads to deterioration such as changes of minerals and microstructures, the degradation of properties, cracking and the increase of drying shrinkage. For these reasons, many studies have been made on carbonation of AAC and deterioration (Asano et al. 1971, Sauman 1972, Dapkus & Stankevicius 1985, Iwasaki & Tada 1985, Sun et al. 1985, Ochiai 1993, Hanecka et al. 1997). However, various definitions of carbonation degree of AAC have been proposed and applied making it difficult to compare studies with each other. Besides these studies were not based on standardized durability criteria for AAC panels, which had yet to be proposed and prescribed. The authors studied the definition of carbonation degree of AAC (Matsushita et al. 2000) in order to investigate by a unified criterion in the same way as neutralization depth in concrete study. This paper deals with the durability and its criteria for normally used in-service AAC panels as determined by carbonation degree.

2 EXPERIMENT

2.1 *Samples*

2.1.1 *In-service panels*

Samples were in-service AAC panels with reinforcement steel bars aged 4 to 33 years taken from actual buildings under the warm climate in Japan. They were used as outer-walls and partition-walls. The usual warm climate in Japan is 0–35°C, 25–95% R.H. and 0.03 vol.% CO_2. Brand-new AAC panels made by Sumitomo Metal Mining Siporex Co., Ltd were used as control samples for the reference.

2.1.2 *Observation of surface finish and cracking*

The surface of outer-wall panels were non-finished, scratching finished of stucco and acrylic resin-finished. Finished samples had not been refinished at all in their service lives. The growth of the lattice-like cracking on the outer surface was observed.

2.2 *Test method*

2.2.1 *Carbonation degree*

The carbonation degree (Matsushita et al. 2000) was measured as shown below,

$$\text{Carbonation degree (\%)} = (C - C_0) / (C_{max} - C_0) \times 100 \quad (1)$$

where C, C_0 and C_{max} are the amount of carbon dioxide in the sample, in the control sample, and the theoretical amount of carbon dioxide needed to combine with the total reactive calcium oxide in the sample to form calcium carbonate, respectively. The reactive calcium oxide means all calcium oxide in tobermorite-1.1nm in the control-AAC, and was calculated with Equation (2).

$$\text{Reactive calcium oxide (mass\%)} = CaO\text{(mass\%)} - SO_3\text{(mass\%)} \times 56/80 \quad (2)$$

The amount of calcium oxide and sulfur oxide was analyzed by ICP, while that of carbon dioxide was analyzed as the heating weight loss from 600°C to 800°C by using TG-DTA. TG-DTA method was carried out under the conditions of nitrogen gas flow, temperature increasing rate of 20°C/min and the temperature range from 20°C to 1000°C.

2.2.2 *Compressive strength*

The test method of the compressive strength followed RILEM recommendation (Aroni et al. 1993). The specimens were cut from the original in-service AAC panels into the cubes with an edge length of 100 mm. A test set for each samples consisted of two specimens.

2.2.3 *Tensile strength*

The specimens were shaped from the original in-service AAC panels into the prisms with a cross section of 40 mm × 40 mm and the length of 160 mm. Attachments with universal joints were glued on both ends. Strain gages (Tokyo Sokki Kenkyujo Co., Ltd, PFL-10-11-3LT) were glued on the center of two surfaces facing each other. Tensile strength and strain was measured together. A setup of a sample with attachments and strain gages is shown in Figure 1.

2.2.4 *Panel flexural properties*

The test method of the panel flexural properties was based on RILEM recommendation (Aroni et al. 1993). The load was applied in two symmetrical equal points, acting at a distance of one-fourth

Figure 1. Setup of a sample with attachments and strain gages for tensile strength test.

of the clear span from the mid span. Cracking load and deflection were measured at the initiation of cracking except for cracks observed beforehand. Cracking load was normalized as a ratio of the test results to the design load and was calculated for each sample with the panel size and the reinforcement steel bars.

2.2.5 *Drying shrinkage*

The test method of the drying shrinkage was based on RILEM recommendation (Aroni et al. 1993). The specimens were sampled from the original in-service AAC panels and shaped into the prisms with a cross section of 40 mm × 40 mm and a length of 160 mm. A test set consisted of two specimens. Drying shrinkage was measured as the length change from the water saturated condition to that in the equilibrium state under 45% R.H. at 20°C.

3 RESULTS

3.1 *Carbonation degree and growth of lattice-like cracking*

Relationship between carbonation degree, 2 cm from the exterior surface of outer-wall panels and that of the maximum of partition-wall panels, and presence of lattice-like cracking in service life is shown in Figure 2. Carbonation degree increased during service life and was saturated approximately at 60%. In all samples with a carbonation degree from 50 to 60%, the lattice-like cracking was observed. Hereafter, carbonation degree will be used as an index, because it represents the chemical change of the matrix that increases with years.

3.2 *Compressive strength*

Relationship between compressive strength and carbonation degree is shown in Figure 3. For the samples of carbonation degree up to approximately 30%, compressive strength of approximately 5 N/mm^2 that is equal to the non-carbonated samples was maintained. However, compressive strengths were approximately 3 N/mm^2 at the minimum for samples of carbonation degree more than 50%.

3.3 *Tensile strength and stress*

Tensile strength and limit of tensile stress are shown in Figures 4 and 5. Tensile strength decreased with increase in carbonation degree, while limit of tensile stress was almost unchanged approximately 0.04% that is equal to the control-samples.

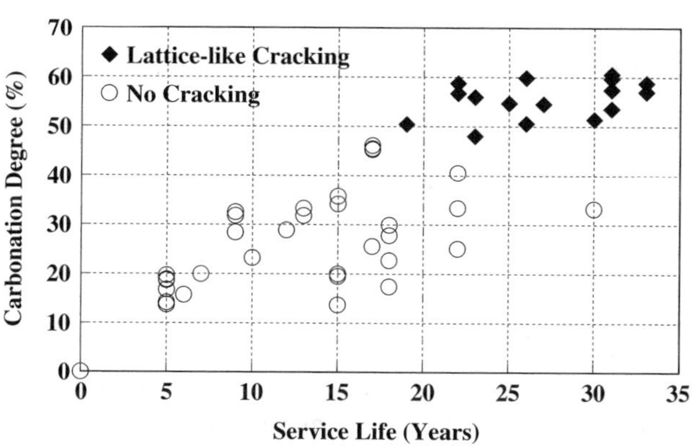

Figure 2. Carbonation degree and growth of lattice-like cracking during service life.

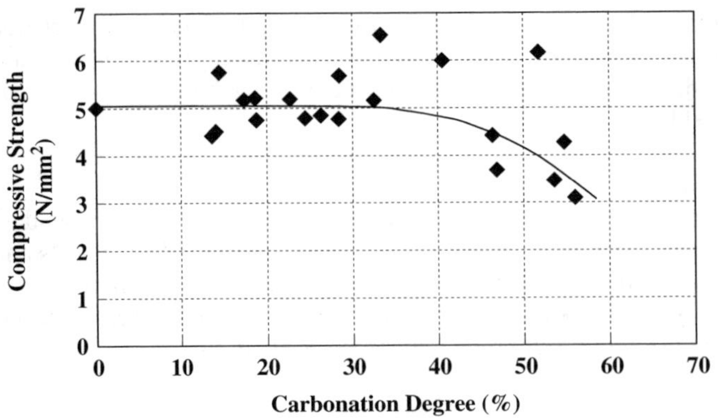

Figure 3. Compressive strength as a function of carbonation degree.

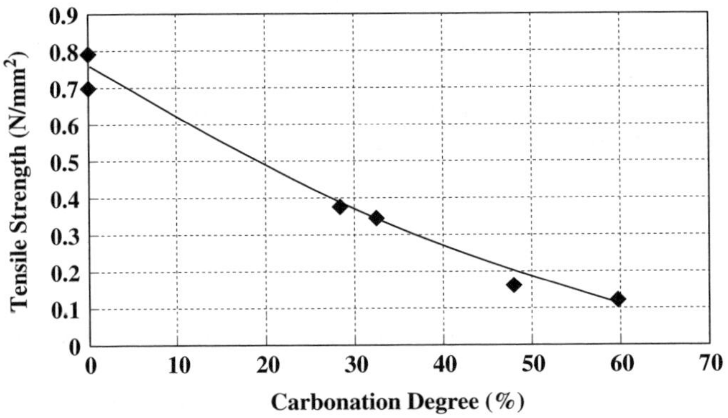

Figure 4. Tensile strength as a function of carbonation degree.

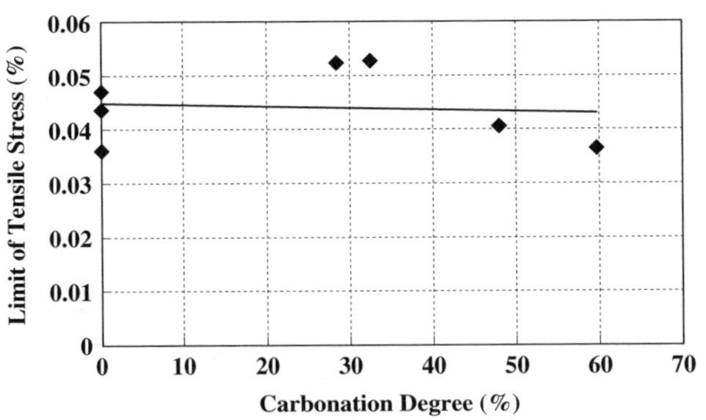

Figure 5. Limit of tensile stress as a function of carbonation degree.

3.4 *Cracking load and span-deflection ratio*

Normalized cracking load and span-deflection ratio are shown in Figures 6 and 7, respectively. Normalized cracking load was more than 1 for all samples, which meant cracking load was adequate as shown in the preceding study (Goodier & Matthews 1997). However, span-deflection ratio increased with increase in carbonation degree, particularly in region greater than 50%.

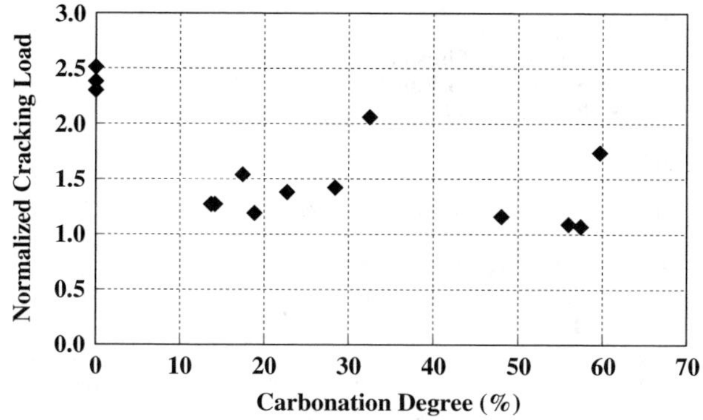

Figure 6. Normalized cracking load as a function of carbonation degree.

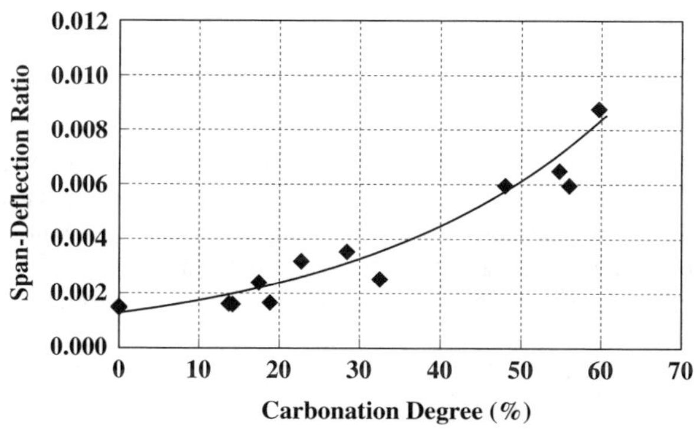

Figure 7. Span-deflection ratio as a function of carbonation degree.

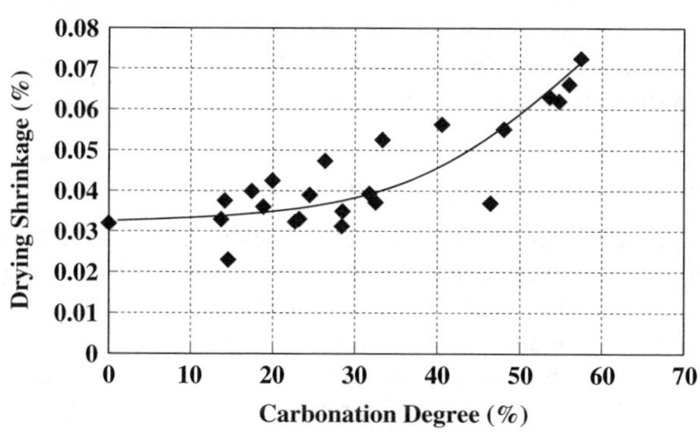

Figure 8. Drying shrinkage as a function of carbonation degree.

3.5 *Drying shrinkage*

Drying shrinkage as a function of carbonation degree is shown in Figure 8. Drying shrinkage increased with the increase of carbonation degree. Drying shrinkage of samples with a carbonation degree from 50 to 60% was approximately twice as large as that of the control-sample.

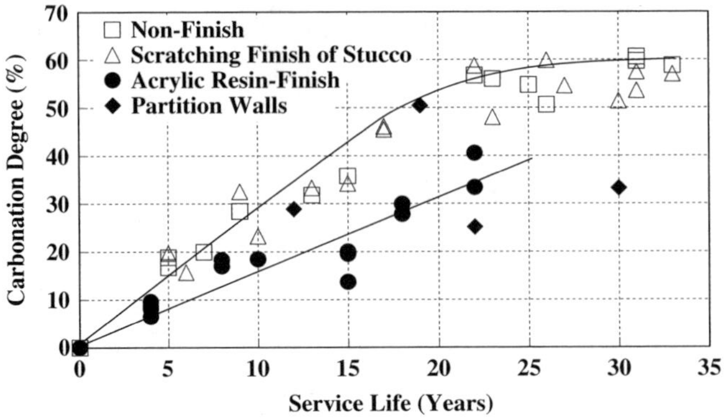

Figure 9. Carbonation degree for outer-wall panels by surface finish types and for partition-wall panels.

4 DISCUSSION

4.1 *Durability criteria*

When the carbonation degree of the sample was more than 50%, growth of the lattice-like cracking, decrease in compressive strength, increase in span-deflection ratio and increase in drying shrinkage were observed. This meant that the sample with a carbonation degree more than 50% is deteriorated. Thus, the author proposes a durability criteria for AAC as follows;

"When the carbonation degree of an AAC panel exceeds 50% as defined in this study, it may be considered as deteriorated."

It is possible to evaluate the durability of AAC by simply measuring the carbonation degree with a small sample.

4.2 *Deterioration due to carbonation*

It is obvious that the decrease in compressive strength and the increase in span-deflection ratio with the increase in carbonation degree were attributed to the growth of lattice-like cracking. It is also supposed that the increase in drying shrinkage with the increase in carbonation was due to the change in the main binding mineral from tobermorite-1.1 nm to a silica-gel.

4.3 *Durability improvement by surface finish and environment*

Surface finish is one of the principal factors affecting the durability of AAC because it has a large effect on the migration of carbon dioxide gas and moisture in AAC. Environment is quite different between outer-walls and partition-walls. Change in carbonation degree with time of outer-wall panels by types of surface finish and that of partition-wall panels is shown in Figure 9. Scratching finish of stucco had no effect on the durability improvement because the carbonation degree was equal to non-finished AAC panels. The acrylic resin-finished AAC panels were 1.5 times more durable than the non-finished ones because the carbonation degree of the acrylic resin-finished AAC panels were about two third of that of the non-finished. Several partition-wall panels showed relatively lower carbonation degree than outer-wall panels so that stable atmosphere and low water content was supposed to be desirable for durability of AAC.

4.4 *Durability of water-repellent AAC panels*

Water-repellent AAC panels have been produced more than 20 years in Japan. They contain a small amount of silicone-oil mixed with the starting raw materials at the beginning of manufacturing.

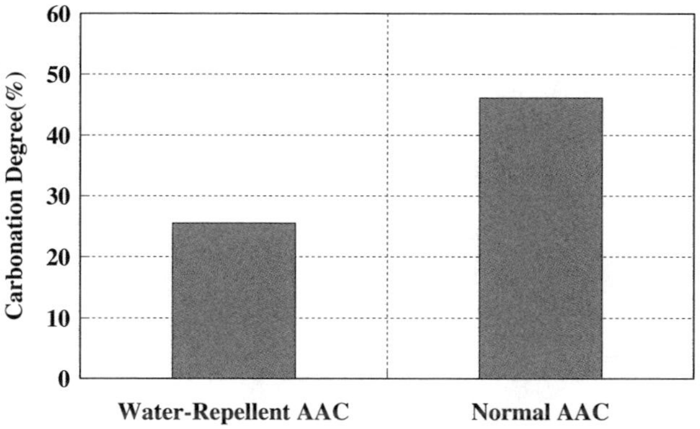

Figure 10. Carbonation degree of water-repellent AAC and normal AAC aged 18 years.

During the study, the authors happened to obtain water-repellent AAC panels and normal AAC panels 18-year old buildings of the same design in the same location. They were finished with a scratching stucco. Carbonation degree of them is shown in Figure 10. Water-repellent AAC panel was approximately twice as durable as the normal AAC panel because the carbonation degree of water repellent AAC panel was approximately half of that of normal AAC panel.

5 CONCLUSIONS

(1) When carbonation degree of sample was more than 50%, the growth of the lattice-like cracking, the decrease in compressive strength, the increase in span-deflection ratio and the increase in drying shrinkage were observed. Thus the durability criteria for AAC were proposed as follows: "When the degree of carbonation of an AAC panel exceeds 50% as defined in this study, the AAC panel may be deteriorated".
(2) Acrylic resin-finished AAC panels were 1.5 times more durable than non-finished AAC panels. Stable atmosphere and low water content in partition-walls may be desirable for the durability of AAC. Water-repellent AAC panels had twice as durable as normal AAC panels.

REFERENCES

Asano, S., Kamatani, Y. & Inoue, Y. 1971. On the carbonation of calcium silicate compounds, *Yogyo-kyokai-shi*, 79(9): 303–311. (in Japanese).

Aroni, S., de Groot, G. J., Robinson, M. J., Svanholm, G. & Wittmann, F. H. (eds) 1993, *RILEM Recommended Practice "Autoclaved Aerated Concrete; Properties, Testing and Design"*, London, E&FN SPON.

Dapkus, G. & Stankevicius, V. 1985. Cellular concrete carbonation, *Building Research and Practice*, 13(3): 184–187.

Goodier, A. & Matthews, S. 1997. In-service behavior of reinforced autoclaved aerated concrete planks, *Spec. Publ. Am. Concr. Inst.*, No.SP-170, Vol.2: 1197–1214.

Hanecka, K., Koronthalyova, O. & Matiasovsky, P. 1997. The carbonation of autoclaved aerated concrete, *Cement and Concrete Research*, 27(4): 589–599.

Iwasaki, M. & Tada, S. 1985. Carbonation of aerated concrete, *Proc. Int. Symp. Cem. Con., Beijing*, 3: 414–423.

Matsushita, F., Aono, Y. & Shibata, S. 2000. Carbonation degree of autoclaved aerated concrete, *Cement and Concrete Research*, 30: 1741–1745.

Ochiai, T. 1993. Carbonation of autoclaved light-weight concretes, *Gypsum & Lime*, 242: 22–31. (in Japanese).

Sauman, Z. 1972. Effect of CO_2 on porous concrete, *Cement and Concrete Research*, 2: 541–549.

Sun, G., Tang, D. & Zhao, Y. 1985. The carbonation of autoclaved aerated concrete and its hydration products, *Proc. Int. Symp. Cem. Con., Beijing*, 3: 331–342.

Autoclaved Aerated Concrete – Limbachiya and Roberts (eds)
© 2005 Taylor & Francis Group, London, ISBN 0 415 38356 0

The future of AAC – from a material scientist's point of view

M. Haas
Xella Technologie und Forschungsgesellschaft mbH, Brück, Germany

ABSTRACT: Autoclaved Aerated Concrete (AAC) is a porous, homogenous, isotropic, lightweighted building material having a lot of positive properties e.g. high compressive strength resp. to its weight, best thermal insulation, excellent working and handling, etc. The advantages have been continuously improved.

Because of its "built in properties", AAC has the best chances to fulfill the demands on building materials in the future. These are:

– Higher quality and improved values for the classical properties (strength, thermal insulation, sound insulation, etc.)
– Widening the range of applicable raw materials.
– Recycling of building materials.
– Unique international standards for building materials.

1 INTRODUCTION

1.1 *General development*

In the next 50 years there will be three leading technologies which will have great influence to our future [1]. These technologies are:

- Information technology
- Genom technology, and
- Material sciences.

Building materials like Autoclaved Aerated Concrete (AAC) belong of course to the last group. Building materials will not adept the same dramatic changes like semiconductors or nanosystems, but recently there are quite a lot of changes in the properties of AAC, which will allow to make some predictions about the future of this material.

Normally while trying to make predictions, we analyze the development of the past and extrapolate some properties to the future. The best examples for this method is Moore's Law. It means: Every two years there is a reduplication of the power of personal computers.

1.2 *The properties of Autoclaved Aerated Concrete (AAC)*

Like every other building materials for wall mounting AAC had and has to fulfill a lot of requirements:

- Compressive strength
- Thermal insulation
- Sound insulation
- Handling
- Form stability (e.g. shrinking, avoiding of cracks in the wall)
- Long term stability
- Resistance to environmental influences

- Ecological unhesitating (No unhealthy ingrediences, ability of recycling)
- Low production costs
- And so on.

In comparison to other building materials (e.g. bricks and hollow blocks), AAC is an homogenous, isotropic material, whose properties are approximately the same in every spatial direction. These properties can rather easy be controlled during the production process; respectively by the recipe and by the curing (autoclaving).

Therefore predictions about the development of AAC can be done with justifiable certainty.

2 THE PAST

The technical and physical properties of AAC have substantially improved over the past decades. A particularly notable change is the reduction of the dry density along with an increase in the strength within the individual quality classes. These improvements were possible as the material was constantly developed more and more by taking specific influence on the texture of the solid matter and on the pore structure. This development was characterized by two measures:

- An improved technology for the manufacture of AAC
- The application of new investigation methods.

2.1 *Dry density, compressive strength and thermal conductivity*

As a result of research and development work, it was possible to substantially reduce the dry densities required to reach the strength provided by the standard. Following the development of other technical construction properties such as the flexural tensile strength of the materials or also the modulus of elasticity (Young's modulus), similar trends can be observed. With the further development of the raw materials, recipes and process it was possible to reach the known technical properties with lighter and lighter products and hence to improve other properties such as thermal insulation.

Lower raw densities also naturally affect the other properties such as heat insulation and noise insulation. Figure 1 shows how the thermal conductivity values for products P2, P4 and P6 have changed since 1975. The calculated value of the thermal conductivity has almost halved during this time.

AAC has always remained a solid block with a homogenous distribution of the structural components. In other words, optimizations of technical properties were never reached by changing

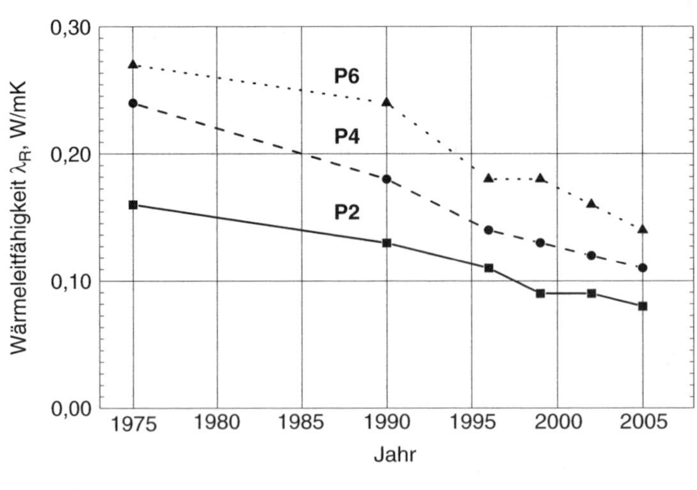

Figure 1. Development of the thermal conductivity of AAC of different quality classes (Thermal conductivity/Year).

the cavity pattern, the web thickness, the percentage of cavity, etc. but only by optimizing the porosity and the texture. This was reached by a continuous improvement of the texture and a better understanding of the microstructure of the material [2, 3, 4].

2.2 *The present*

In the last years, the building industry in Germany and some other West European Countries has suffered a dramatic decrease in market volume. This process is still lasting on and predictions for the development in the next future are very pessimistic (Figure 2). Therefore, most of the activities of the manufacturers of building materials had only one aim: Cost reduction! All activities in research and development (if there have been left any) were done under this aspect. The contribution of the research and development activities consisted (and consist) in the following measures:

- Reducing recipe costs
- Saving energy
- Reducing waste and substandard goods.

2.3 *Reducing recipe costs and saving energy*

Figure 3 shows the development of the raw materials and energy consumption for the production of AAC in percent. The red line is marking the increased activities in coherence with break down

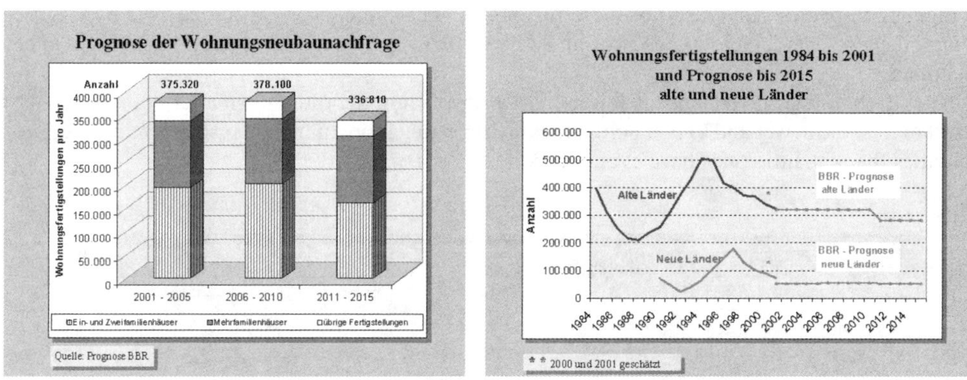

Figure 2. Prediction of the demand for new buildings and finishing of residental units in Germany.

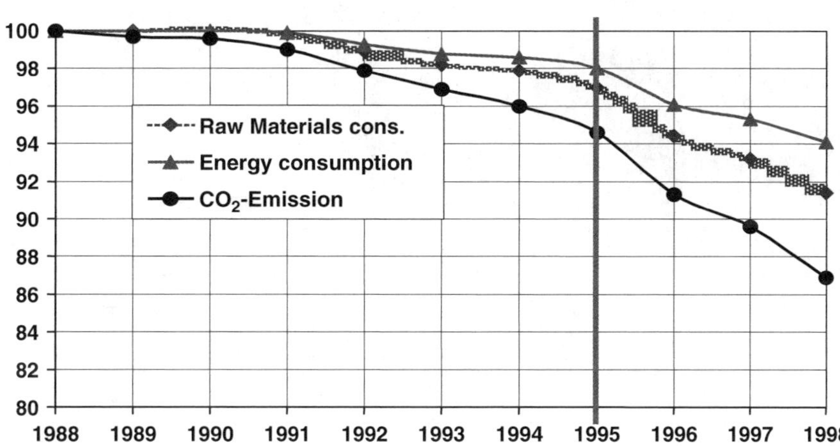

Figure 3. Development of raw materials and energy consumption as well as the CO_2 emission.

of the building materials market in Germany. The reduction of recipe costs was reached by the following measures:

- Reduction of the raw density by constant compressive strength.
- Use of cheaper raw materials with less quality without decreasing the quality of the final product.

2.4 *Reducing waste and substandard goods*

The production of high quality AAC is a rather complex and high sophisticated process which has not been understood in detail up to know. There are a lot of influence parameters (raw material properties, temperature, etc.) which affect the quality of the final product. To gain a better understanding, we have started to use a new kind of information technology to analyze these influences and their effects to the production process.

By investigating our process data with cluster analysis using so called "self organizing maps" (Figure 4) we discovered, that we are doing some serious mistakes by the calculation of our recipes. This mistake led sometimes to a higher amount of substandard goods.

2.5 *Company fusions of AAC manufacturers*

Due to the fact, that the market for building materials in (Western) Europe is still decreasing, a lot of little and mid companies of the building and building materials industry have been fusioned to bigger companies. As a result of these fusions and takeovers, over-capacities have been closed, central departments (including research and development) are reorganized and concentrated in new locations.

New, global, big players appear in the market displacing the old and smaller companies. As a result of this, the newer and bigger companies are investing more money in research & development activities than the little ones have ever been able to.

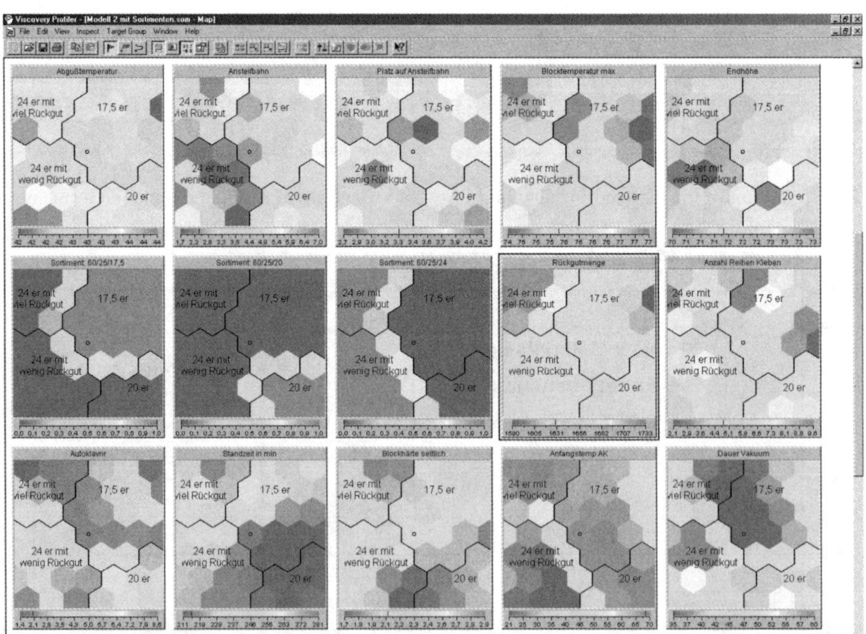

Figure 4. Cluster analysis of process data.

2.6 *The future*

2.6.1 *Improving the thermal insulation*

Greenhouse effect, CO_2-emission and increasing energy costs are the reason for the continuously increasing demands for improved heat insulation of buildings. These demands are reflected in national decrees or European or international standards (Figure 5).

2.7 *Technological aspects*

What consequences will the above mentioned demands for a higher quality of buildings have for AAC? One main topic in R&D is still the improvement of the thermal insulation of the material by keeping or increasing the compressive strength and the elasticity at the same time. This will have consequences to the product spectrum of AAC which is currently produced (Figure 6).

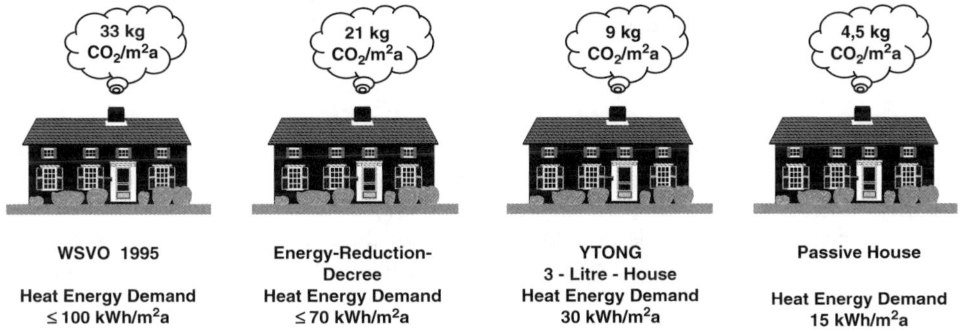

Figure 5. Development of standards and decrees for reducing energy consumption and CO_2 ejection.

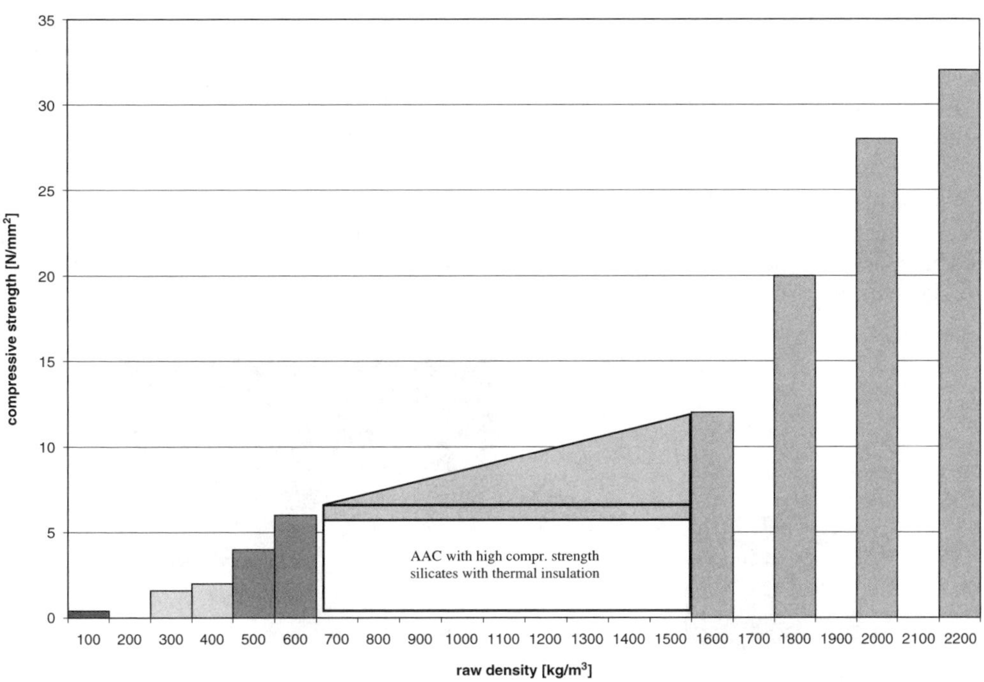

Figure 6. Actual and future raw density classes of AAC and silicates and their properties.

There are several gaps in the raw density classes. The raw density from 90 to 120 kg/m^3 presents our mineral building plate, which is only used for thermal insulation and has no static function. Another gap is between 700 and 1600 kg/m^3.

In future this gap will be closed. There will be a floating transient between pure thermal insulation materials and AAC with static properties for building walls. Even the differences between AAC and silicates will blur in future.

2.8 *Standardization*

2.8.1 *Demand for higher quality*

European committees are working on new standards for horizontal loads (earthquakes) and increased wind loads. Additionally demands from national institutes in static security and long term stability of building materials are also increasing.

Therefore there will be a need for building materials with better properties, as well as for new constructional solutions (for example new fixing technologies). This will change the way masonry is made in future.

2.8.2 *Internationalization of standardization*

Up to now, every country has its own building tradition and its own building standards. Even there are a lot of attempts inside and outside of the European Community to adapt the different national standards. However the results are less satisfying until now.

This will change, because it is of outmost interest of the building material producers, civil engineers, architects and building owners to have the same international standard for the same material worldwide.

2.9 *Environmental aspects*

2.9.1 *Waste free production*

Already now, AAC can be produced without any waste. Waste material from the production process is feed back into the mixer and recycled completely. Waste steam is used again by overloading between autoclaves or steam accumulators. Condensate from the autoclaves is used as mixing water in the recipe.

2.9.2 *Recycling*

The amount of additional, alternative raw materials which can be used in the building materials production will dramatically increase in the future, because the legislating for the handling of waste will become more severe.

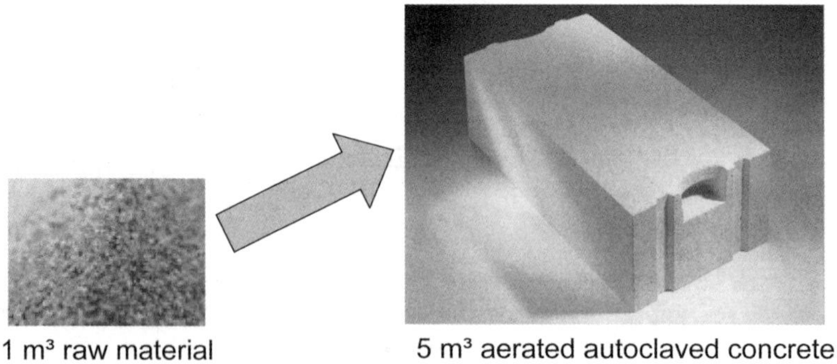

Figure 7. Using 1 m^3 of raw materials, 5 m^3 of AAC can be produced.

2.9.3 *Paucity of raw materials*

Some countries (e.g. China) are running short in special natural resources (e.g. clay). Considering the aspect, that almost all raw materials will run short sooner or later, AAC can take one of his greatest advantage: The optimal use of raw materials. For the production of 5 m^3 AAC, only 1 m^3 of raw materials (there from about 75% sand) is used (Figure 7).

3 CONCLUSION

I know, that this paper is aspecting only a few chances AAC has in the future. But this is the part; me and my coworkers are responsible for. Therefore the predictions made in this paper are based on a stable assuredness. Because:

The future can best foreseen, if you create it by your own.

LITERATURE

[1] Michio Kaku: Visions: How Science will revolutionize the 21st Century, Anchor Books/Doubleday

[2] Wittmann, F.H. (1983): Autoclaved Aerated Concrete, Moisture and Properties. Elsevier Scientific Publishing Company, Amsterdam

[3] Fehr, K.T. & Zürn, S.G. (2000): Mechanisms of calcium-silicate-hydrates formation under hydrothermal conditions. Proc. Joint ISHR & ICSTR, Kodi (IP), 278–281

[4] Zürn, S.G. & Fehr, K.T.: Kinetic study on the crystallisation of calcium-silicate-hydrates in steam cured building materials. Proc. 5th Int. Symp. Hydroth. Reactions, Gathinburg (USA), 302

Autoclaved Aerated Concrete – Limbachiya and Roberts (eds)
© 2005 Taylor & Francis Group, London, ISBN 0 415 38356 0

Carbonation shrinkage of Autoclaved Aerated Concrete

F. Matsushita, Y. Aono & S. Shibata
Sumitomo Metal Mining Siporex Co., Ltd., Research & Development Center, Japan

ABSTRACT: Carbonated autoclaved aerated concretes show no shrinkage at a carbonation degree approximately less than 20%. The ^{29}Si MAS NMR spectrum showed that, at a carbonation degree less than 25%, the typical double-chain silicate anion structure of tobermorite-1.1 nm was well maintained and interlayer Ca ions were exchanged with protons. It corresponded to the absence of carbonation shrinkage at a carbonation degree less than 20%. When carbonation degree exceeded from 25% to 50% and to 60%, double-chain silicate anion structure of tobermorite-11Å was decomposed and Ca ions in the Ca-O layers was dissolved, showing a possible mechanism of carbonation shrinkage.

1 INTRODUCTION

One of the most harmful factors affecting the durability of Autoclaved Aerated Concrete (AAC) is carbonation, in which tobermorite-1.1 nm ($5CaO \cdot 6SiO_2 \cdot 5H_2O$), the principal binding mineral of AAC, reacts with atmospheric carbon dioxide gas in the presence of moisture, and decomposed to silica-gel and calcium carbonate as shown in Equation (1).

$$5CaO{\cdot}6SiO_2{\cdot}5H_2O + 5CO_2 \longrightarrow 5CaCO_3 + 6SiO_2{\cdot}nH_2O + (5\text{-}n)H_2O \tag{1}$$

Carbonation leads to degradation such as the decrease in strength, the increase in deflection and the growth of lattice-like cracks (Goodier & Matthews 1997, Matsushita & Shibata 2000). They are mainly caused by the carbonation shrinkage.

In the microstructure change of AAC during carbonation (Matsushita et al. 1999), the planer particle shapes of tobermorite-1.1 nm and the associated inter-particle pores were not changed significantly, although the double-chain silicate anion structure of tobermorite-1.1 nm was decomposed to silica-gel-like structure and drying shrinkage increased with carbonation.

Numerical values of carbonation shrinkage of AAC were reported from 0.1 to 1.0% for laboratory carbonation though it varied according to the raw materials, carbon dioxide concentrations and ambient relative humidities (Fedin et al. 1970, Sauman 1972, Novikova et al. 1978, Nireki et al. 1990, Ochiai 1993).

However, the relation between carbonation shrinkage and change of tobermorite-1.1 nm crystal structure has not been reported at all. This chapter will investigate the mechanism of carbonation shrinkage of AAC carbonated under an accelerated condition with ^{29}Si MAS NMR spectrum and carbonation shrinkage of AAC carbonated under field conditions, which is extremely difficult to measure.

2 EXPERIMENTAL

2.1 *Specimens*

AAC blocks made by Sumitomo Metal Mining Siporex Co., Ltd. were used for laboratory carbonation. They were shaped in 40 × 40 × 160 mm and dried at 105°C for 2 hours. Subsequently, a pair of 20 mm-brass pins was fixed on both ends by an adhesive for the length change measurement. A near-surface part from 0 to 10 mm was used for determining the carbonation degree.

Field AAC wall panels (Field-AAC) under working conditions for 9, 26 and 33 years were used as field-carbonated samples. Carbonation degree of the specimens was approximately 25, 50 and 60%, respectively.

2.2 *Carbonation*

Laboratory carbonation samples (Labo-AAC) were pre-treated at a temperature of 20°C and relative humidity of 90% for 1 week. Subsequently, they were carbonated under conditions of CO_2 concentration of 3 and 0.3 vol.% at a temperature of 20°C and an ambient relative humidity of 90%.

2.3 *Carbonation degree*

The carbonation degree (Matsushita et al. 2000) was measured as shown below,

$$\text{Carbonation degree (\%)} = (C - C_0) / (C_{max} - C_0) \times 100 \tag{2}$$

where C, C_0 and C_{max} are the amount of carbon dioxide in the sample, in the control sample, and the theoretical amount of carbon dioxide needed to combine with the total reactive calcium oxide in the sample to form calcium carbonate, respectively. The reactive calcium oxide means all the calcium oxide in tobermorite-1.1 nm in the control-AAC, calculated using Equation (3).

$$\text{Reactive calcium oxide (mass\%)} = CaO\text{(mass\%)} - SO_3\text{(mass\%)} \times 56/80 \tag{3}$$

The amount of calcium oxide and sulfur oxide was analyzed by ICP, while that of carbon dioxide was analyzed as the heating weight loss from 600°C to 800°C by using TG-DTA. TG-DTA method was carried out under the conditions of nitrogen gas flow, temperature increasing rate of 20°C/min and the temperature range from 20°C to 1000°C.

2.4 *Carbonation shrinkage*

Shrinkage strain δ due to carbonation can be given by,

$$\delta(\%) = [\,(l - l_0) / l_0\,] \times 100, \tag{4}$$

where l and l_0 were a length and the initial length after pre-treatment under conditions of 20°C and 90% R.H. for 1 week. Measurement was always done in the carbonation chamber under conditions of 20°C and 90% R.H.

2.5 *^{29}Si MAS NMR spectrum*

^{29}Si MAS NMR spectrum were recorded by the JEOL Ltd. JNM-Λ-400WB with an observation frequency of 79.42 MHz, a repeating time of 7.0 s, a contact time of 6.4 μs and a cumulative number of 3000 to 12000 scans. The specimens measured were control-AAC and Labo- and Field-AAC at carbonation degrees of 25, 50 and 60%.

3 RESULTS AND DISCUSSION

3.1 *Carbonation degree*

Carbonation degree as a function of treated time of Labo-AAC is shown in Figure 1. The higher the carbon dioxide concentration was, the faster the carbonation degree increased. Carbonation degree saturated approximately at 60% after 35 days under a CO_2 concentration of 3 vol.% and after 120 days under a CO_2 concentration of 0.3 vol.%.

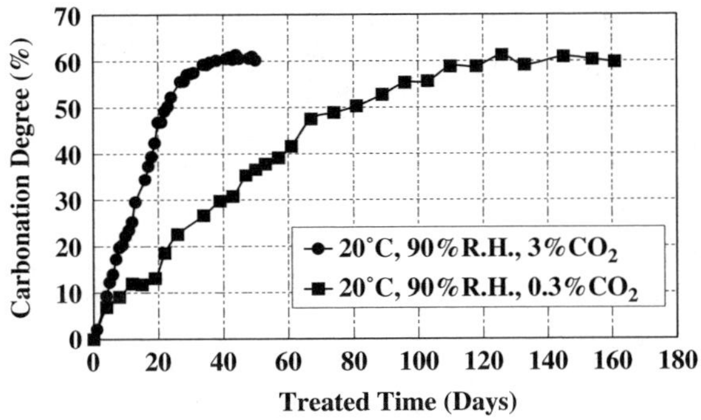

Figure 1. Carbonation degree as a function of treated time.

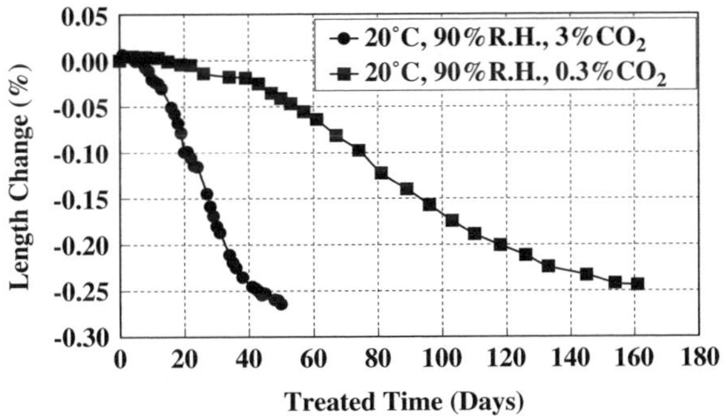

Figure 2. Length change during laboratory carbonation as a function of treated time.

3.2 *Carbonation shrinkage*

Carbonation shrinkage as a function of treated time for Labo-AAC is shown in Figure 2. Carbonation shrinkage reached approximately 0.25% after 40 days and 160 days when CO_2 concentrations were 3 vol.% and 0.3 vol.%, showing that higher CO_2 concentration led to quicker shrinkage. The maximum value of carbonation shrinkage was approximately 0.27% under 3 vol.% CO_2 and approximately 0.25% under 0.3 vol.% CO_2.

Besides the decomposition of tobermorite-1.1 nm, calcium carbonate was deposited during carbonation. The reason of the absence of swelling during carbonation was attributed the deposition of calcium carbonate into the inter-particle pores (Matsushita et al. 1999).

Length change during carbonation as a function of carbonation degree of Labo-AAC is shown in Figure 3. Regardless of the carbonation condition, the relation between carbonation degree and carbonation shrinkage showed nearly the similar tendency. No carbonation shrinkage occurred when carbonation degree was less than 20%. When carbonation degree ranged from 20% to 50%, carbonation shrinkage occurred gradually and reached approximately 0.1%. When carbonation degree exceeded 50%, carbonation shrinkage increased remarkably and reached approximately 0.25%.

3.3 ^{29}*Si MAS NMR spectrum*

3.3.1 *General remarks*

^{29}Si MAS NMR spectra of control-AAC and Labo- and Field-AAC at carbonation degrees of 25, 50 and 60% are shown in Figure 4 (a) to (g). The ^{29}Si MAS NMR chemical shifts by peak separation are shown in Table 1.

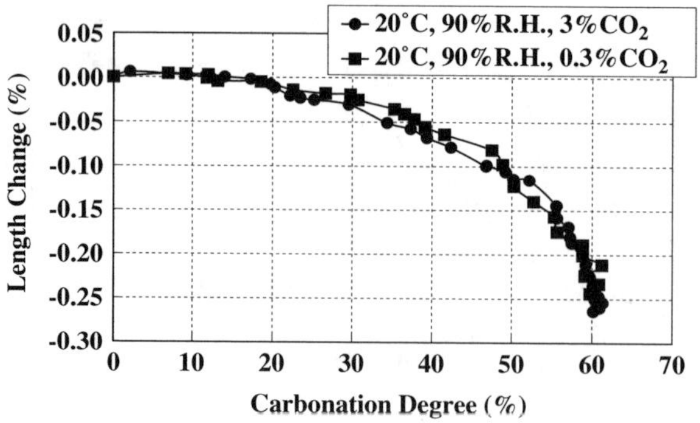

Figure 3. Length change during laboratory carbonation as a function of carbonation degree.

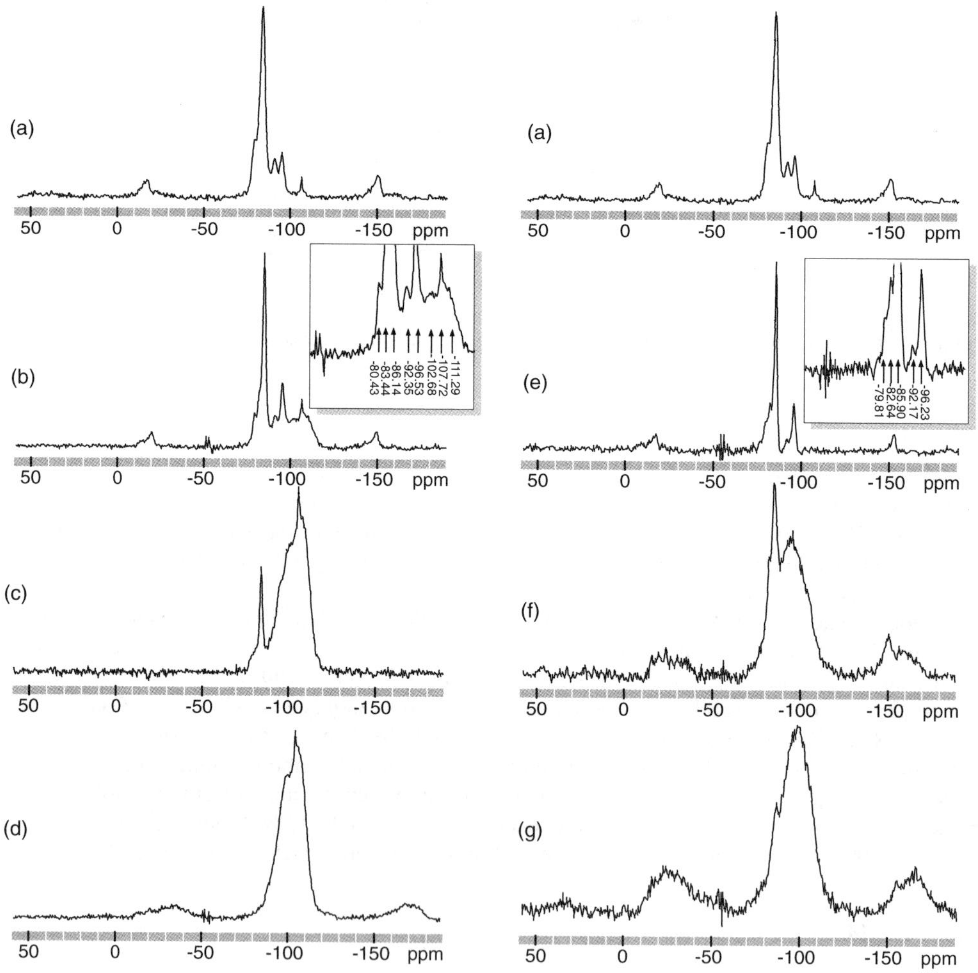

Figure 4. ^{29}Si MAS NMR spectrum; (a) Control-AAC, (b) Labo-AAC, Dc = 25%, (c) Labo-AAC, Dc = 50%, (d) Labo-AAC, Dc = 60%, (e) Field-AAC, Dc = 25%, (f) Field-AAC, Dc = 50%, (g) Field-AAC, Dc = 60%.

Table 1. ^{29}Si MAS NMR chemical shifts.

ppm from TMS (relative intensity)	Q^1	Q^2 Q^2_p	Q^2_i	Q^2_{Ca}	Q^3 1Al	0Al	Q^4 1Al	0Al	Q^4 Quartz
Control	−80.6 (28)			−85.3 (100)	−92.0 (16)	−96.0 (17)			−107.4 (4)
Labo Dc = 25%	−80.4 (15)		−83.4 (51)	−86.1 (100)	−92.4 (23)	−96.5 (42)	−102.7 (54)	−111.3 (48)	−107.7 (14)
Labo Dc = 50%		−82.8 (5)		−86.1 (21)		−98.0 (60)	−102.3 (90)	−111.1 (100)	−107.7 (15)
Labo Dc = 60%							−102.8 (100)	−110.3 (68)	−107.5 (5)
Field Dc = 25%	−79.8 (12)	−82.6 (30)		−85.9 (100)	−92.2 (3)	−96.2 (28)			
Field Dc = 50%		−82.9 (13)		−86.0 (48)		−96.9 (100)	−103.9 (60)		
Field Dc = 60%				−87.2 (29)		← −98.9 → (100)			

On the whole, changes of silicate anion structures of Labo- and Field-AAC were similar. When carbonation degree was less than 25%, ^{29}Si MAS NMR spectra were almost unchanged, which implied the double-chain silicate anion structure of tobermorite-1.1 nm was well maintained, however at peaks around −100 ppm a small change was observed in Labo-AAC (Figure 4 (a), (b) and (e)). On the other hand in the same progress of carbonation degree of 25%, ^{29}Si MAS NMR spectrum of specimen with a carbonation degree ranging from 25% to 50% changed considerably (Figure 4 (c) and (f)). All the more, when carbonation degree ranged from 50% to 60%, ^{29}Si MAS NMR spectrum changed remarkably in spite of an only 10% advance of the carbonation degree (Figure 4 (d) and (g)). During this process, the double-chain silicate anion structure of tobermorite-1.1 nm (Q^2 and Q^3) was supposed to be rearranged to a 3-dimentional silica-gel like structure (Q^4).

3.3.2 *Control-AAC*

For control-AAC, signals at −80.6, −85.3, −92.0, −96.0 and −107.4 ppm were observed corresponding to Q^1, Q^2, Q^3(1Al) Q^3(0Al) and Q^4, respectively (Wieker et al. 1982, Komarneni et al. 1985). Q^4 signal at −107.3 ppm corresponded to α-quartz, one of the raw materials of AAC. The value of Q3/Q2 of the untreated specimen was approximately 1/3, although the theoretical value is 1/2. Therefore the bridging tetrahedra was supposed to comprise approximately 75% of Q_3 and approximately 25% of Q_2. We call the Q_2 as bridging tetrahedra "Q^2_b" in this paper.

3.3.3 *Carbonation degree of 25%*

When carbonation degree reached 25%, signals at −83.4 ppm, −102.7 ppm and −111.3 ppm appeared in the Labo-AAC, and a signal at −82.6 ppm appeared in the Field-AAC. The signal at −111.3 ppm corresponds to Q^4 of silica-gel (Grimmer et al. 1982), which differs from Q^4 of α-quartz (−107 ppm). Sato et al. (1992) and Sasaki et al. (1996) concluded from ^{29}Si MAS/CPMAS NMR spectrum that the signal around −82 ppm corresponded the protonated silicate ions in the silicate chain. Klur et al. (1997) attributed −82.1 ppm to a bridging tetrahedra bonded to two protons "Q^2_p", −83.9 ppm to a bridging tetrahedra bonded to one proton and one calcium ion "Q^2_i" and −85.3 ppm to a silicate tetrahedra coordinated to the calcium plans "Q^2_{Ca}".

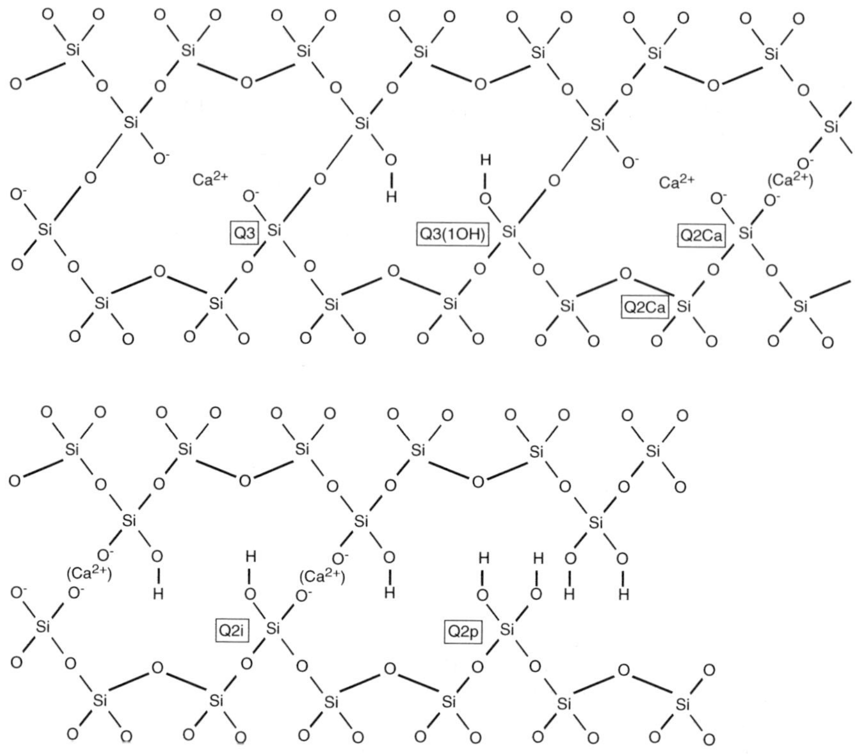

Figure 5. A tobermorite structure model including $Q^3(1OH)$, Q^2_b, Q^2_p, Q^2_i and Q^2_{Ca} bonds.

In short, Q^2_{Ca} called by Klur et al.(1997) means the ordinary Q^2. Therefore, −83.4 ppm for Labo-AAC and −82.6 ppm for Field-AAC corresponded to Q^2_i and Q^2_p, respectively. A signal at −102.7 ppm corresponds to a bridging tetrahedra bonded to one proton and one silicate ion, "$Q^3(1OH)$" (Cong et al. 1993). A tobermorite structure model including $Q^3(1OH)$, Q^2_b, Q^2_p, Q^2_i and Q^2_{Ca} bonds is shown in Figure 5. Appearance of Q^2_p, Q^2_i and $Q^3(1OH)$ including protonated silicate ions during the early carbonation periods signified the exchange of calcium ions by protons in interlayer spaces of tobermorite-1.1 nm.

3.3.4 *Carbonation degree greater than 50%*

When carbonation degree reached 50%, Q^1 disappeared, Q^2 around −82 ppm to −86 ppm became relatively lower and Q^3 and Q^4 appeared or became higher in both carbonated specimens. In Labo-AAC, Q^2_i(−83.4 ppm) disappeared, a slight signal of Q^2_p(−82.8 ppm) appeared and the highest signal changed from Q^2 to Q^4. In Field-AAC, $Q^3(1Al)$(−92.2 ppm) disappeared, $Q^3(1OH)$(−103.9 ppm) appeared and the highest signal changed from Q^2 to Q^3.

When carbonation degree reached 60%, $Q^3(1OH)$ and Q^4 were detected in Labo-AAC. On the other hand, the wide signal centered between Q^3 and Q^4 was mainly detected in Field-AAC. Therefore, double-chain silicate anion structure of tobermorite-1.1 nm was finally decomposed, shrunk and polymerized to a silica-gel-like structure at the final stage of carbonation.

3.4 *Changes in silicate anion structures and carbonation shrinkage*

During early carbonation periods at a carbonation degree less than 25%, double-chain silicate anion structure of tobermorite-1.1 nm maintained well and protonated silicate ions in the silica chain

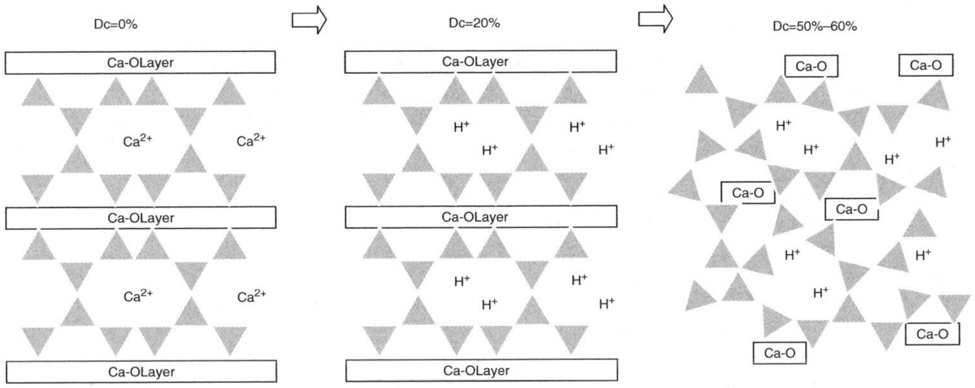

Figure 6. A schematic model of calcium silicate structure change and carbonation shrinkage.

were present in both types of carbonated AAC. Komarneni et al.(1989) studied the ion exchange properties of Ca ions in the interlayer space of tobermorite-1.1 nm. The Ca ions in the interlayer and the Ca-O layer occupied 20% and 80% of Ca atoms in tobermorite-1.1 nm respectively (Taylor 1964). In the case of Labo-AAC, no carbonation shrinkage occurred when carbonation degree was less than 20%. Therefore, the absence of carbonation shrinkage until the carbonation degree of 20% may be attributed to the retaining of the main structures of tobermorite-1.1 nm, the double-chain silicate anion structure and the Ca-O layer as shown in Figure 6.

When carbonation degree ranged from 25 to 60%, Ca ions leached from the Ca-O layer and the double-chain silicate anion structures was decomposed, shrunk and polymerized to a silica-gel structure. This is the most likely mechanism of carbonation shrinkage of AAC, a tobermorite-based material as shown in Figure 6. It can also support the relation that both of the progress of carbonation shrinkage and the change of silicate anion structure during carbonation degree ranged from 50% to 60% were more remarkable than that during carbonation degree ranged from 20 or 25% to 50%.

In Field-AAC, the measurements of carbonation shrinkage for several decades were extremely difficult. In this study, changes of silicate anion structures in Field- and Labo-AAC turned out to be very similar. Therefore, carbonation shrinkage of Field-AAC as a function of carbonation degree was supposed to be the same as that of Labo-AAC.

4 CONCLUSIONS

(1) In Labo-AAC under CO_2 concentrations of 3 and 0.3 vol.%, the relation between carbonation degree and carbonation shrinkage was nearly the same. Carbonation shrinkage was not observed when carbonation degree was less than 20%. Carbonation shrinkage reached approximately 0.1% and 0.25% at a carbonation degree of 50% and 60%, respectively.

(2) When carbonation degree was less than 25%, ^{29}Si MAS NMR spectrum suggested that the double-chain silicate anion structure of tobermorite-1.1 nm was maintained well and calcium ions were exchanged by protons. It was suggested that the reason of the absence of carbonation shrinkage until the carbonation degree of 20% was the retaining of the double-chain silicate anion structure and the Ca-O layer, that comprise main structures of tobermorite-1.1 nm.

(3) When carbonation degree ranged from 25 to 60%, Ca ions leached from the Ca-O layer and the double-chain silicate anion structures were decomposed, shrunk and were polymerized to a silica-gel-like structure. This is the most likely mechanism of carbonation shrinkage of AAC.

(4) Carbonation shrinkage of Field-AAC as a function of carbonation degree was supposed to be the same as that of Labo-AAC, because changes in silicate anion structures of Field- and Labo-AACs were very similar.

REFERENCES

Cong, X.-D., Kirkpatrick, R. J. & Diamond, S. 1993. ^{29}Si MAS NMR spectroscopic investigation of alkali silica reaction product gels, *Cement and Concrete Research*, 23(4): 811–823.

Fedin, A. A., Chernyshov, E. M. & Ponomareva, E. M. 1970. Stability of silicate aerated concrete during carbonatation, *Tr. Probl. Lab. Silikat. Mater. Konstr., Voronezh. Inzh. -Stroit. Inst.*, Vol. 4: 31–43. (in Russian).

Goodier, A. & Matthews, S. 1997. In-service behavior of reinforced autoclaved aerated concrete planks, *Spec. Publ. Am. Concr. Inst.*, No.SP-170, Vol.2: 1197–1214.

Grimmer, A.R., Starke, P., Wieker, W. & Maegi, M. 1982. High resolution solid-state silicon-^{29} NMR of silica gels, *Z. Chem.*, 22(2): 44. (in German).

Klur, I., Pollet, B., Virlet, J. & Nonat, A. 1997. C-S-H structure evolution with calcium content by multinuclear NMR, In Colombet, P. et al. (eds.), *Nuclear Magnetic Resonance Spectroscopy of Cement-Based Materials*: 119–141.

Komarneni, S., Roy, R., Roy, D. M., Fyfe, C. A., Kennedy, G. J., Bother-by, A. A., Dadok, J. & Chesnick, A S. 1985. ^{27}Al and ^{29}Si magic angle spinning nuclear magnetic resonance spectroscopy of Al-substituted tobermorites, *Journal of Materials Science*, 20: 4209–4214.

Komarneni, S. & Tsuji, M. 1989. Selective cation exchange in substituted tobermorites, *J. Am. Ceram. Soc.*, 72(9): 1668–1674.

Matsushita, F., Shibata, S., Aono, Y. & Kamada, T. 1999. Microstructure of autoclaved aerated concrete subjected to carbonation, *8th International Conference on Durability of Building Materials and Components*: 159–169.

Matsushita, F. & Shibata, S. 2000. Carbonation degree as durability criteria for autoclaved aerated concrete, *5th CANMET/ACI International Conference on Durability of Concrete*: 1123–1134.

Matsushita, F., Aono, Y. & Shibata, S. 2000. Carbonation degree of autoclaved aerated concrete, *Cement and Concrete Research*, 30: 1741–1745.

Nireki, T., Kabeya, H., Hirama, N. & Inoue, T. 1990. Durability of aerated lightweight concrete panels with surface coating systems, *Proc. 5th International Conference on Durability of Building Materials and Components*: 721–727.

Novikova, L. N., Ustimovich, A. B., Stankevich, Z. V. & Minchukova, T. L. 1978. Effect of artificial carbonization on the properties and structure of newly formed phases in cellular concrete with reduced bulk density, *Stroit. Mater.*, No.6: 32–33. (in Russian).

Ochiai, T. 1993. Carbonation of autoclaved light-weight concretes, *Gypsum & Lime*, 242: 22–31. (in Japanese).

Sasaki, K., Masuda, T., Ishida, H. & Mitsuda, T. 1996. Structural degradation of tobermorite during vibratory milling, *J. Am. Ceram. Soc.*, 79(6): 1569-1574.

Sato, H. & Grutzeck, M. 1992. Effect of starting materials on the synthesis of tobermorite, *Mat. Res. Soc. Symp. Proc.*, 245: 235–240.

Sauman, Z. 1972. Effect of CO_2 on porous concrete, *Cement and Concrete Research*, 2: 541–549.

Taylor, H. F. W. (ed.) 1964, *The Chemistry of Cements*, Chapter 5: 185-189, London and New York, Academic Press.

Wieker, W., Grimmer, A.-R., Winkler, A., Magi, M., Tarmak, M. & Lippmaa, E. 1982. Solid-state high-resolution ^{29}Si NMR spectroscopy of synthetic 14Å, 11Å and 9Å tobermorites, *Cement and Concrete Research*,12: 333–339.

Autoclaved Aerated Concrete – Limbachiya and Roberts (eds)
© 2005 Taylor & Francis Group, London, ISBN 0 415 38356 0

Experimental tests into the sustained load strength of autoclaved aerated concrete

E. Raue & E. Tartsch
Bauhaus University Weimar, Germany

P. Langer
Xella Technologie – und Forschungsgesellschaft mbH, Germany

ABSTRACT: In connection with a new set of standards for the analysis and design of structural members made of autoclaved aerated concrete (AAC), the question arose as to how to account for the destabilisation process of the material due to sustained loading. In order to be able to make clear statements on a short-term basis about the sustained loading strength of AAC, it would therefore be useful to define a threshold value based on the short-term stress–strain-relationship. The authors of the paper present some results of tests concerning this problem.

1 INTRODUCTION

Experimental tests of sustained load strength involve a large expenditure of time, since it must be guaranteed that certain loads can be resisted for a theoretically infinite time span. Only a few tests into the sustained load strength of AAC are known, and experimental results sometimes differ ([9], [12]). The destabilisation process of AAC was analysed in [3], [4]. The focus of the following paper will be on the macroscopic appearance of structural changes at the level of sustained load strength.

Tests were carried out at the Bauhaus University Weimar on 40 unreinforced cylindrical specimens made of AAC, applying uniaxial compression load with varying load history. This makes it possible to verify threshold values with regard to temporal independence. The experimental tests can be divided into three parts. One group of specimens was taken to test the compressive strength of the AAC using different loading rates. The second and most substantial part of the tests constitute investigations into sustained load strength. Stress was kept constant at certain load levels, ranging from 0.84 to 0.95 of short-term ultimate to subsequent failure. Finally, a specimen was tested under increasing low-cycle loading.

In the context of the analysis of the test data, three-dimensional textural responses are taken into account. This is why the analysis is based on the stress and strain tensor. Uniaxial loading is considered as a special case of multi-axial loading.

The research work was supported by the Forschungsvereinigung Porenbeton e.V.

2 TEST PROCEDURE

The tests were carried out under conditioned atmosphere (20°C, 65% relative humidity), using unreinforced cylindrical specimens made of AAC. Two different types of AAC were tested (Table 1, see also [2]). The uniaxial compressive loads were applied in load-controlled form, perpendicular to the direction of expansion.

Three different types of loading were used (Figure 1). First, the short-term strength of both series was tested using a constant loading rate of 0.0152 N/mm^2 per second. The tests into the sustained load strength then followed. The load levels were based on the short-term compressive

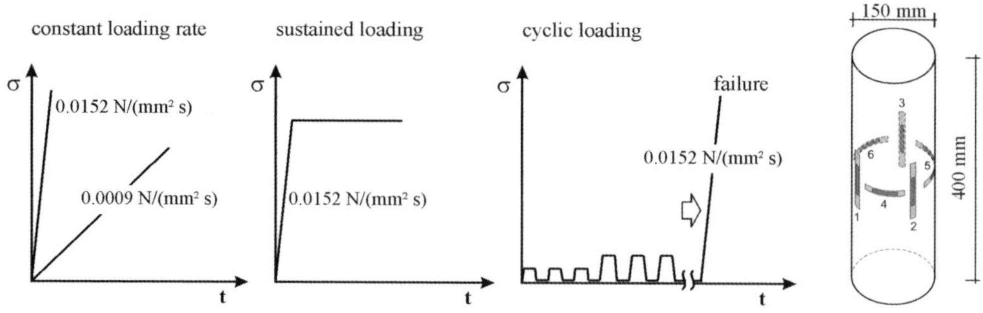

Figure 1. Stress–time-relationship, specimen with strain gauges.

strength. Specimens which did not fail were unloaded and reloaded at a higher load level. Preloaded specimens are indicated as part of the test results.

These tests are followed by one with increasing low-cycle loading. Within every cycle, load was kept constant for 180 seconds at basic (1 kN) and maximum level. This was carried out three times at each load level. Five percent of the short-term ultimate was chosen as the load increment between load levels. For all these tests, a loading rate of 0.0152 N/mm^2 per second was chosen. Finally, one further test of short-term strength was carried out on a specimen of series 1, this time using a lower loading rate of 0.0009 N/mm^2 per second. During the tests, force, deformation and time were monitored in parallel to each other at a rate of 5 times per second. The measurement of deformation was carried out on both a longitudinal and a transverse basis. For strain measurement, 6 strain gauges were used per specimen (Figure 1).

3 TEST RESULTS AND ANALYSIS

For the analysis of test data, it is assumed that deformations measured with strain gauges are representative of the specimen. It is also assumed that the stress state is uniaxial in the centre of specimen with regard to length. This means that influences due to specimen shape are disregarded. Both theoretical [6], [7] and practical tests [5], [8] justify this assumption, given slenderness ratio of the specimen.

Furthermore, it is assumed that the mean value calculated from the three transverse strain gauges represents the real deformation. This means that anisotropic material behaviour due to expansion during the production process (e.g. [1]) is disregarded.

3.1 *Short-term strength (constant loading rate)*

The results in Table 1 make clear that mean values of the short-term strength of the tested series differ due to material mixture. In addition to this, it is clear that a lower rate of loading leads to a significant decrease in the ultimate stress, as expected. In some of the following figures, stresses are given as a ratio of given stress and short-term strength in percent. These are based on the short-term strengths with a fast rate of loading.

Figure 2 shows longitudinal and transverse strains and the computed Poisson ratio for increasing stress. The Poisson ratio is 0.2 for lower load levels in the case of both material mixtures and different loading rates. This ratio increases sharply at higher loads. It can be assumed that, for low load levels, longitudinal and transverse deformations develop proportionally to each other as the load increases. Proportionality is not given for loads that exceed c. 40–50% of short-term compressive strength. For higher load levels, the Poisson ratio is dependent on both stress level and time. No material constant is therefore represented.

Table 1. Parameters of specimens and results of tests into short-term strength.

Series		1	1	2
Manufacturer		Hebel, Emmering	Hebel, Emmering	Ytong, Schrobenhausen
Geometry		cylinders, diameter d = 150 mm, length l = 400 mm		
Strength class		4.4	4.4	4.4
Gross density class		0.6	0.6	0.6
Dry density	[kN/m^3]	5.8	5.8	6.7
Loading rate	[N/(mm^{2}s)]	0.0152 (fast)	0.0009 (slow)	0.0152 (fast)
Number of specimens		10	1	10
Average ultimate stress	[N/mm^2]	6.41	5.84	5.48
Standard deviation	[N/mm^2]	0.08	–	0.18

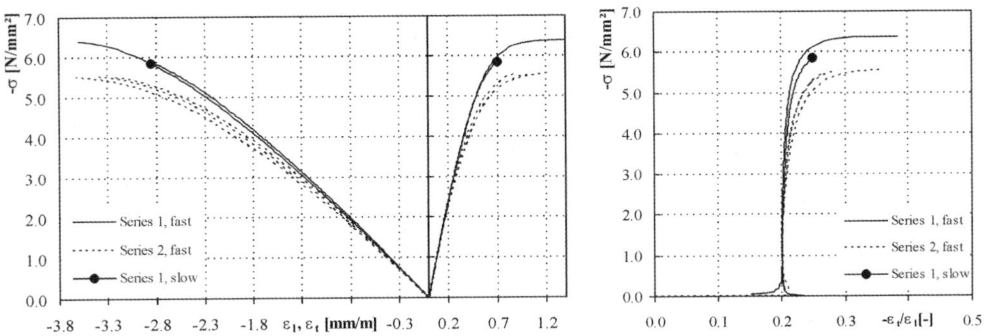

Figure 2. Longitudinal and transverse strains, Poisson's ratio for various material mixtures and loading rates (fast, slow).

The following analysis is based on the decomposition of strains and stresses, divided into two parts respectively – axiator (hydrostatic part) and deviator [10]. The hydrostatic stress σ_H and the deviatoric stress σ_D are derived from the principal stresses $\sigma_1 = \sigma$ and $\sigma_2 = \sigma_3 = 0$ as follows:

$$\sigma_H = \frac{1}{\sqrt{3}} \cdot (\sigma_1 + \sigma_2 + \sigma_3) = \frac{1}{\sqrt{3}} \cdot \sigma, \tag{1}$$

$$\sigma_D = \frac{1}{\sqrt{3}} \cdot \sqrt{(\sigma_1 - \sigma_2)^2 + (\sigma_2 - \sigma_3)^2 + (\sigma_3 - \sigma_1)^2} = \sqrt{\frac{2}{3}} \cdot |\sigma|. \tag{2}$$

According to the stresses, strains can be decomposed in a similar fashion. The dilatation ε_H and the deviatoric strain ε_D (distortion) are derived from the principal strains $\varepsilon_1 = \varepsilon_l$ and $\varepsilon_2 = \varepsilon_3 = \varepsilon_t$ as follows:

$$\varepsilon_H = \frac{1}{\sqrt{3}} \cdot (\varepsilon_1 + \varepsilon_2 + \varepsilon_3) = \frac{1}{\sqrt{3}} \cdot (\varepsilon_l + 2 \cdot \varepsilon_t), \tag{3}$$

$$\varepsilon_D = \frac{1}{\sqrt{3}} \cdot \sqrt{(\varepsilon_1 - \varepsilon_2)^2 + (\varepsilon_2 - \varepsilon_3)^2 + (\varepsilon_3 - \varepsilon_1)^2} = \sqrt{\frac{2}{3}} \cdot |\varepsilon_l - \varepsilon_t|. \tag{4}$$

where ε_l and ε_t are the longitudinal and transverse strain respectively. The associated rigidities bulk modulus K and shear modulus G are defined as

$$K = \frac{\sigma_H}{3 \cdot \varepsilon_H}, \qquad G = \frac{\sigma_D}{2 \cdot \varepsilon_D}. \tag{5}$$

Figure 3 shows deviatoric strains and shear modulus (tangent modulus) for an increasing degree of loading. Because of the continuity of the curves, no distinctive changes in material behaviour can be seen prior to specimen failure.

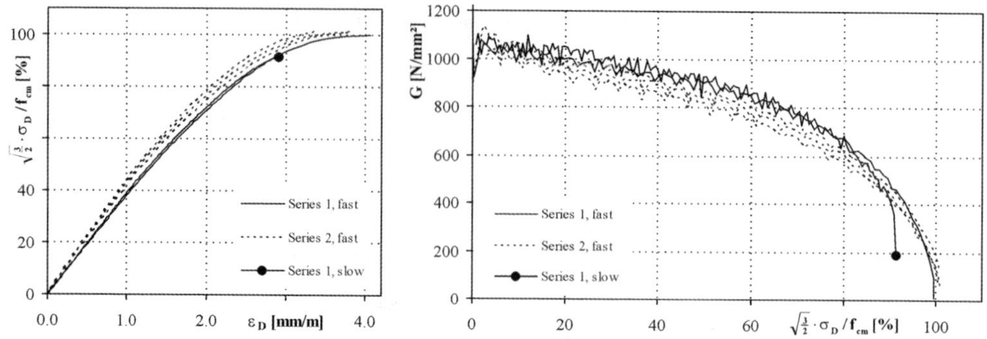

Figure 3. Deviatoric strains, shear modulus.

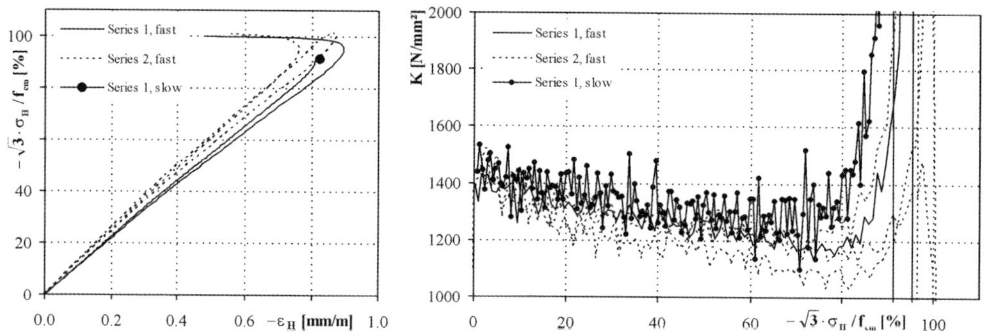

Figure 4. Dilatation, bulk modulus.

According to Figure 4 the specimen volume initially decreases with increasing loading. Some specimens show a slight volume increase immediately prior to failure. As far as a minimum in volume could be observed prior to failure, it is found at very high load degree of c. 0.95 f_{cm}. For this reason, it is not representative of the level of sustained load strength.

The bulk modulus shows that, for both possibilities of deformation development (i.e. with or without minimum in volume prior to failure) an extreme point exists in the function. Dependent on material mixture and loading rate, the minimum bulk modulus can be found to be between c. 0.6–0.8 f_{cm}.

3.2 *Sustained load tests*

The test results of both series are shown together because no differences in time-dependent strengths emerged. Time measurement for sustained loads began when the appropriate load level was reached, as the inset in Figure 5 illustrates. Thus time for applying loads (5–7 mins) is not included. Figure 5 makes clear that, throughout the tests, failure did not occur at stresses below 87% of the mean short-term ultimate.

Figure 5 is supplemented by the regression curve of all specimens that failed. The regression curve was assumed to have the form

$$\frac{-\sigma}{f_{cm}}(t) = a + \frac{1-a}{1+b\cdot t} \quad . \tag{6}$$

Whereas $-\sigma/f_{cm}$ is the ratio of applied sustained stress to mean short-term compressive strength, t is the time and a, b are coefficients.

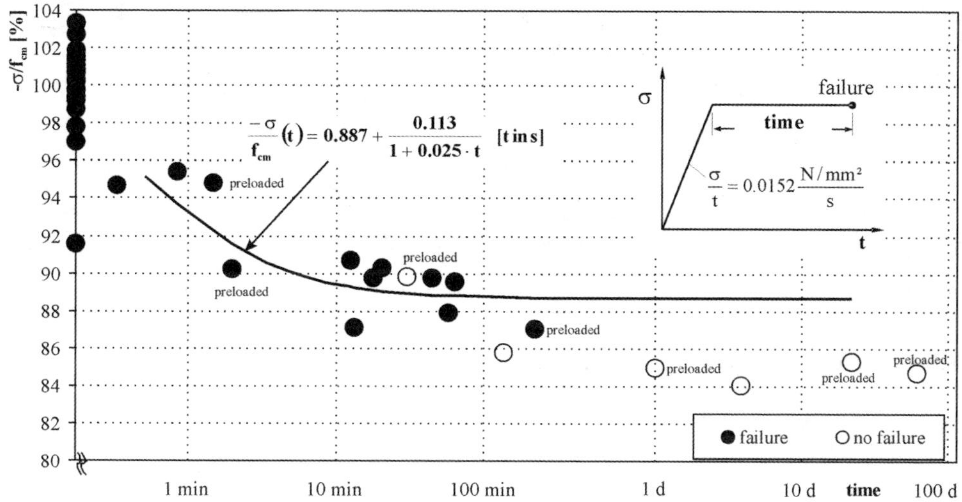

Figure 5. Relation between normalised stress and sustained load time.

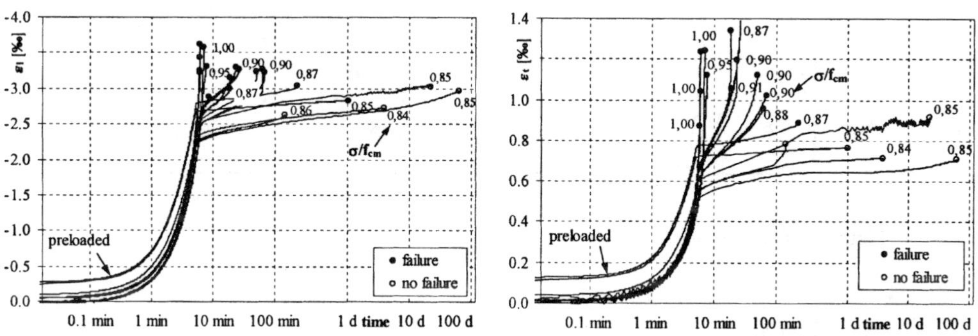

Figure 6. Longitudinal and transverse strains as a function of time.

The coefficients a and b were computed by using the least square method. For a failure probability of 0.50, the equation is

$$\frac{-\sigma}{f_{cm}}(t) = 0.887 + \frac{0.113}{1 + 0.025 \cdot t} \qquad [t \text{ in s}]. \tag{7}$$

The regression curve indicates the sustained load strength for the tested AAC to be somewhat above 85% of the short-term strength.

The functions of longitudinal and transverse strains in Figure 6 show that strain rates do not decrease after reaching sustained load level at high stresses.

Figure 7 shows maximum deviatoric strains for all specimens that failed to be close together. Graphs of preloaded specimens are characterised by a different deformation progress. This statement also applies to the dilatation. Furthermore, the curves of preloaded specimens make clear that dilatation can be regarded as elastic. Only deviatoric strains show inelastic elements of deformation. According to the short-term tests, failure can be characterised by both increasing and decreasing volume.

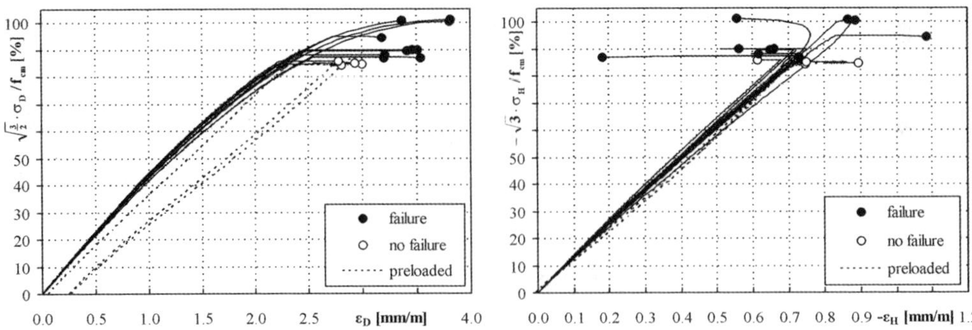

Figure 7. Deviatoric strains and dilatation, series 2.

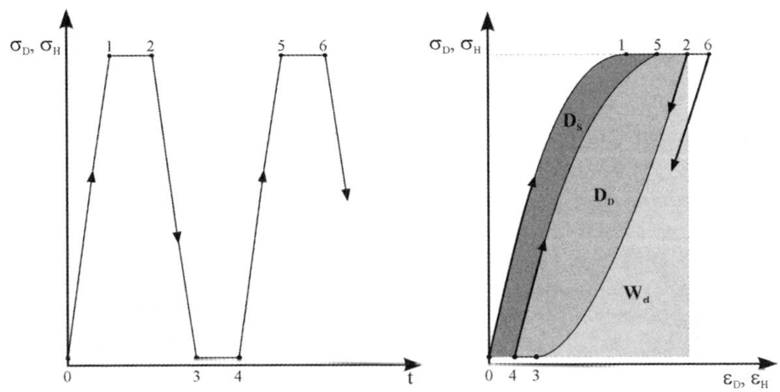

Figure 8. Partition of energies.

3.3 *Low-cycle loading*

In Figures 9 (a), it emerges that longitudinal and transverse strains increase disproportionately to load. For purposes of illustration clarity, the deformations at the moment of reaching a new load level are supplemented by data points.

The Poisson ratio, deviatoric strains, dilatation, shear modulus and bulk modulus in Figures 9 (a), (b) and (c) are similar in appearance to the curves of the short-term tests.

New information can be derived by studying energy dissipation. Deformation energy can be computed, based respectively on the sections of decomposed stress and strain tensor. Figure 8 illustrates the partition of the energies. The total energy can be divided into dissipated energy D_S, stored energy D_D and elastic energy W_{el} [11].

Figure 9 (d) shows summation of dissipated energy for distortion and dilatation. The figure makes clear that nearly no energy is dissipated while stresses are lower than 40–50% of the short-term ultimate. Textural responses can be regarded as elastic. Thus stresses within the serviceability limit state do not lead to any relevant textural changes. Only at higher load levels is energy dissipated. At c. 0.8 f_{cm}, the energy dissipated for dilatation reaches a maximum. For stresses exceeding the sustained load strength, a change of sign (positive/negative) can be seen.

It is to be noted that the absolute values of dissipated energy are in reality somewhat smaller than the computed values. This is due to visco-elastic deformations. Because of the short time when basic load is applied between the cycles, the resilience has yet to be completed. Part of the stored energy is thus assumed to be dissipated. The real values of dissipated energy are obtained when it is possible to eliminate this insufficiency. Creep models are therefore needed for both longitudinal and transverse time-dependent deformations.

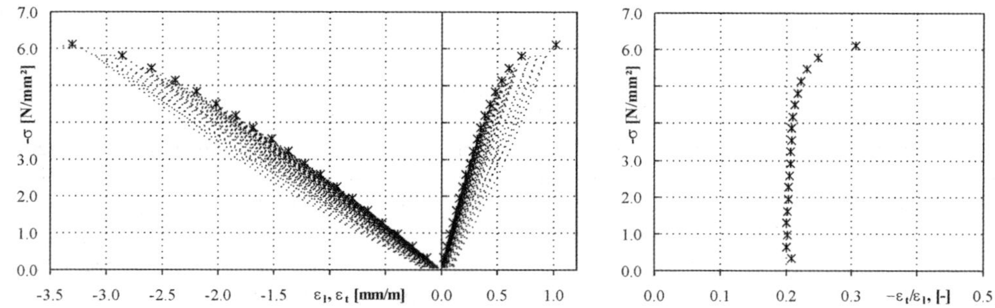

(a) Longitudinal and transverse strains as function of stress, Poisson's ratio

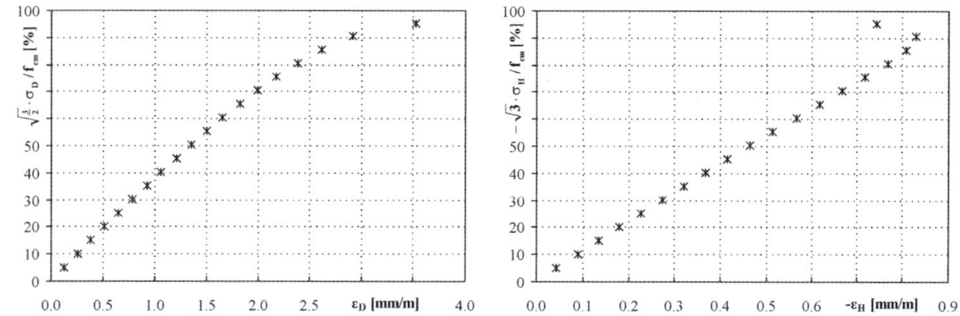

(b) Deviatoric strains and dilatation, primary cycle

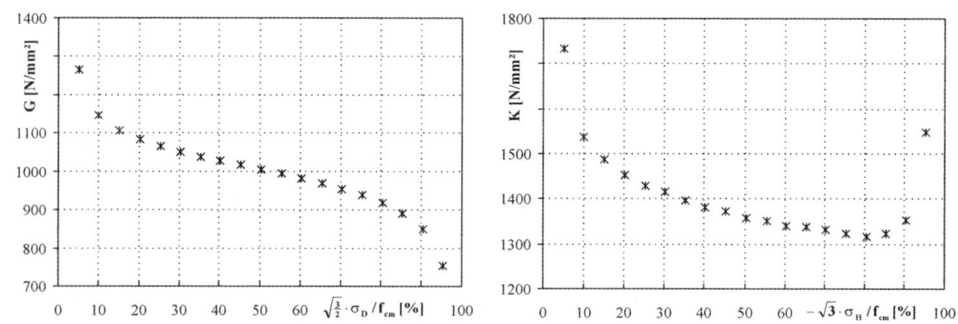

(c) Shear modulus and bulk modulus, primary cycle

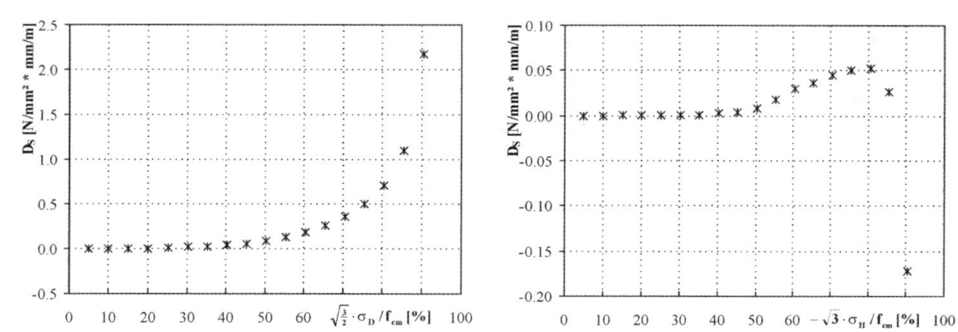

(d) Accumulated dissipated energy for distortion and dilatation

Figure 9. Results of tests with low-cycle loading.

4 CONCLUSION

The experimental tests have shown that the sustained load strength of autoclaved aerated concrete can be assumed at stresses somewhat above 85% of short-term strength.

According to the analysis of the test data, the decomposition of the stress tensor and strain tensor respectively was of special advantage. It could be shown that dilatation can be regarded as elastic for stresses which do not exceed the sustained load strength. Furthermore, it emerged that stresses below c. 50% of the short-term strength do not lead to significant energy dissipation. Thus stresses within the serviceability limit state do not result in relevant textural changes. Therefore, they do not affect the long-term behavior and the load-bearing capacity respectively.

REFERENCES

[1] Aroni, S. et al. (1993). *Autoclaved Aerated Concrete. Properties, Testing and Design.* RILEM Technical Committees 78-MCA and 51-ALC. London: E & FN SPON.

[2] Raue, E. & Tartsch, E. (2004). *Experimentelle Untersuchungen zur Dauerstandfestigkeit von Porenbeton (Experimental investigations into the sustained load strength of AAC.).* Abschlussbericht zum Forschungsvorhaben. Bauhaus-Universität Weimar.

[3] Raue, E. & Tartsch, E. (2005). *Modellierung des Kurzzeitkriechverhaltens unter niederzyklischer Druckbelastung (A model for the short-term creep behaviour of AAC under low-cycle loading.).* Revitalisierung von Bauwerken. Schriften der Bauhaus-Universität Weimar, Heft 117, 103–110.

[4] Raue, E. & Tartsch, E. (2005). *Experimental Results of fatigue and sustained load tests on Autoclaved Aerated Concrete.* Journal of Civil Engineering and Management. Vol. XI., Issue 2, June 2005. Vilnius, Lithuania. (in press)

[5] Schickert, G. (1980). *Schwellenwerte beim Betondruckversuch (Threshold numbers for concrete under compression).* Bericht der Bundesanstalt für Materialprüfung Berlin. Deutscher Ausschuss für Stahlbeton, 312. Berlin: Ernst & Sohn.

[6] Schleeh, W. (1975). *Theorie und Praxis bei der Druckfestigkeitsprüfung (Theoretical and practical investigations into the compressive strength of concrete.).* Beton, Heft 4, 132–138.

[7] Schleeh, W. (1975). *Die Spannungszustände in den Versuchskörpern zur Ermittlung der Betondruckfestigkeit (The stress states in test samples for the determination of the concrete strength).* Beton- und Stahlbetonbau 8, 194–201.

[8] Sell, R. (1970). *Festigkeit und Verformung von Gasbeton unter zweiaxialer Druck- Zug- Beanspruchung (Strength and deformation of AAC under biaxial loading.).* Deutscher Ausschuss für Stahlbeton, 209. Berlin: Ernst & Sohn.

[9] Sell, R. & Zelger, C. (1969). *Versuche zur Dauerstandfestigkeit von Leichtbetonen (Experimental tests into the sustained load strength of lightweight concrete.).* Deutscher Ausschuss für Stahlbeton, 207. Berlin: Ernst & Sohn.

[10] Skrzypek, J.J. & Hetnarski, R.B. (1993). *Plasticity and Creep. Theory, Examples and* Problems. London: CRC Press.

[11] Spooner, D.C. & Dougill, J.W. (1975). *A quantitative assessment of damage sustained in concrete during compressive loading.* Magazine of Concrete Research, 27, H. 92, 151–160.

[12] Wittmann, F. & Zaitsev, J. (1974). *Verformung und Bruchvorgang poröser Baustoffe bei kurzzeitiger Belastung und Dauerlast (Deformation and deterioration mechanisms in porous materials under short-term and long-term loading).* Deutscher Ausschuss für Stahlbeton, 232. Berlin: Ernst & Sohn.

Autoclaved Aerated Concrete – Limbachiya and Roberts (eds)
© 2005 Taylor & Francis Group, London, ISBN 0 415 38356 0

Improvement of carbonation resistance of Autoclaved Aerated Concrete by water-repellent additive

F. Matsushita, Y. Aono & S. Shibata
Sumitomo Metal Mining Siporex Co., Ltd., Research & Development Center, Japan

ABSTRACT: AAC containing water-repellent additive showed twice as much carbonation resistance as the normal AAC under accelerated carbonation conditions. No particular difference in chemical composition, BET specific surface area, pore size distribution by mercury porosimeter and XRD patterns were observed between water-repellent AAC and the normal AAC. Equilibrium moisture content of water-repellent AAC was a little lower than that of the normal AAC. However, it was found that, in cooling DSC, the exothermic peak intensities of water-repellent AAC were remarkably smaller than those of the normal AAC. The reason of the carbonation resistance of water-repellent AAC was attributed to the hydrophobic hydration of adsorbed water by the hydrophobic groups originated from the silicone-oil.

1 INTRODUCTION

The degradation of Autoclaved Aerated Concrete (AAC) is caused mainly by carbonation and frost damage. Carbonation is a chemical decomposition in which Tobermorite-11Å ($5CaO{\cdot}6SiO_2{\cdot}5H_2O$), the principal binding mineral of AAC, reacts with atmospheric carbon dioxide gas in the presence of moisture and decomposed to silica-gel and calcium carbonate. Carbonation leads to degradation such as the decrease of strength, the increase of deflection and the growth of lattice-like cracking (Goodier & Matthews 1997, Matsushita & Shibata 2000).

Frost damage of AAC has been studied in terms of the degradation phenomena such as cracking and scaling (Senbu & Kamada 1992) and particularly in terms of moisture content (Tada 1992) such as the critical degree of saturation method (Fagerlund 1978).

Thus moisture has a serious influence on the degradation both in carbonation and in frost damage. Consequently the outer-wall of AAC buildings and the joint between AAC panels or AAC panel and window flames are usually treated with waterproof treatments. For the material of AAC, water-repellent treatment has been applied by several methods such as esterification with alcohol (Nakamura et al. 1987), impregnation of water glass (Nakamura et al. 1988) and low temperature DC plasma treatment (Nakamura & Ketani 1990). Adding a small amount of silicone-oil in the green slurry has been a practical method of water-repellent treatment for AAC since 1979 in Japan (Japanese patent 1980).

The molecular structure of the normal silicone-oil, polydimethyl-siloxane, in which hydrophobic groups are all methyl groups is shown in Figure 1. The species added to water-repellent AAC were normal silicone-oil and alkyl-modified silicone-oil in which some methyl groups were substituted by alkyl groups. High density AAC with the water-repellent property by adding silicone-oil has been developed for a frost resistant capability (Hama et al. 1992). Whereas we found out the carbonation resistant capability of water-repellent AAC containing silicone-oil by the inspection of existing buildings made of AAC panels (Matsushita and Shibata 2000). The water-repellent AAC containing silicone-oil showed twice as much carbonation resistance as that of the normal one as shown in Figure 2. Consequently, we have studied the mechanism of carbonation resistance of the water-repellent AAC containing silicone-oil in terms of minerals, chemical structure and moisture characteristics.

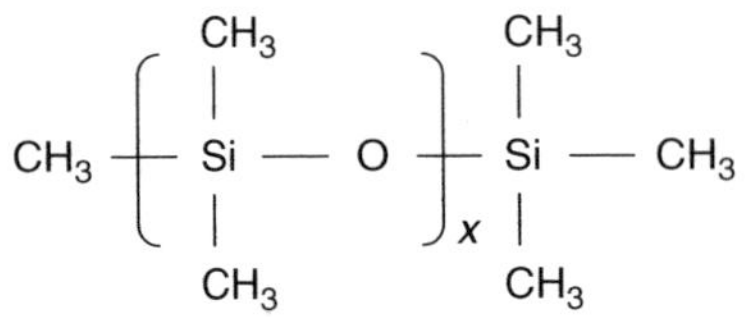

Figure 1. Molecular structure of the normal silicone-oil.

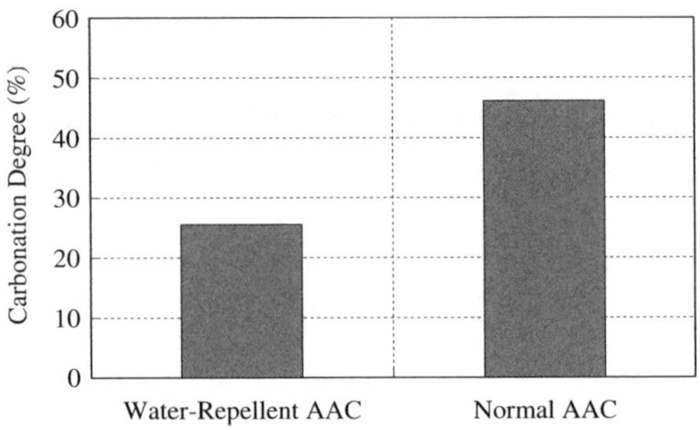

Figure 2. Carbonation degree of water-repellent AAC and normal AAC aged 18 years.

2 EXPERIMENTAL

2.1 *Specimens*

AAC blocks made by Sumitomo Metal Mining Siporex Co., Ltd were used. A green slurry was prepared by mixing quartz-sand, cement, lime, aluminum powder and water. A 0.20wt% of normal silicone-oil were added to the green slurry for the water-repellent AAC. The green slurry was cast into the mould, followed by an expansion due to the reaction of aluminum powder with alkali and subsequently hardened by the hydration of lime and cement forming a porous green body. The porous green body was cured sufficiently under the saturated vapor pressure at approximately 180°C for approximately 6 hours in an autoclave.

2.2 *Carbonation*

Samples were shaped in 10 mm × 40 mm × 100 mm, dried at 105°C for 2 hours and then pre-treated at 20°C and 90%R.H. for 1 week. Subsequently, they were carbonated under conditions of carbon dioxide concentration of 3% at a temperature of 20°C and an ambient relative humidity of 90%.

The amount of combined carbon dioxide after carbonation and calcium oxide were measured to determine the carbonation degree (Matsushita et al. 2000). TG-DTA was used for analysis of the amount of combined carbon dioxide except for the amount of adsorbed carbon dioxide gas and carbonated salt other than calcium carbonate. The amount of combined carbon dioxide was measured as a weight loss during heating from 600 to 800°C corresponding to the decomposition of calcium carbonate. The amount of calcium oxide was analysed by ICP. The carbonation degree (Dc) can be given by

$$\text{Dc}\ (\%) = [(C - C_0) / (C_{max} - C_0)] \times 100 \qquad (1)$$

where C, C_0 and $C_{\max}$ are the amount of combined carbon dioxide in a sample, that of an untreated AAC and that when all calcium oxide transformed to calcium carbonate.

2.3 *Characterization*

Chemical properties of specimens were analysed as a chemical composition of CaO, SiO_2, Al_2O_3, Fe_2O_3, SO_3 and Ig-Loss. N_2-BET specific surface area, pore size distibution by mercury porosimetry and XRD patterns were obtained. Silicone oil was analyzed by IR spectrometer with the KBr method after the treatment of an ultrasonic penetration of fine-ground specimen by chloroform, a centrifugal separation and continuous drying.

Table 1. Respective salt in saturated salt solutions for controlling relative humidity.

R.H./%	65	75	85	95	98
Salt	$NaNO_2$	NaCl	KCl	KNO_3	K_2SO_4

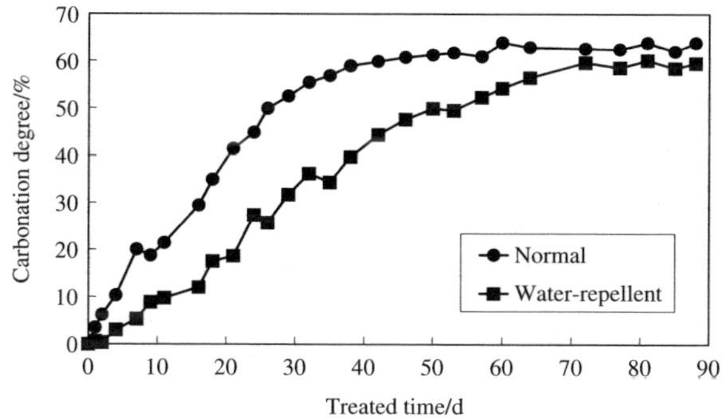

Figure 3. Carbonation degree as a function of treated time.

2.4 *Moisture property*

Water vapour sorption isotherms were measured by automatic sorption apparatus BEL Japan BERSORP-18. Cooling DSC from a room temperature to the liquid nitrogen temperature were executed for specimens conditioned to have a constant weight in the relative humidities of 65%, 75%, 85%, 95% and 98% controlled by the respective saturated salt solutions as shown in Table 1.

3 RESULTS AND DISCUSSION

3.1 *Carbonation degree*

Carbonation degree as a function of treated time of the normal AAC and water-repellent AAC is shown in Figure 3. Carbonation progress of water-repellent AAC was approximately a half of the normal AAC, which indicated that water-repellent AAC had the twice as large carbonation resistant property as that of the normal AAC. This result shows good agreement with the field AAC panels of the 18-year existing building (Matsushita & Shibata 2000). Carbonation came to the end at a carbonation degree of approximately 60% for both AAC.

3.2 *Characterization*

Chemical composition, as shown in Table 2, was almost the same reflecting the same composition of the raw materials. N_2-BET specific surface area was 20.4 m^2/g and 22.0 m^2/g for the normal and the water-repellent AAC, respectively. Pore size distribution by mercury porosimeter, as shown in Figure 4, was almost the same. XRD patterns, as shown in Figure 5, was also almost the same for both AAC showing existence of tobermorite-11Å, quartz, anhydrite and bassanite. IR spectrum, as shown in Figure 6, indicated the existence of silicone-oil in water-repellent AAC after the autoclaving (Bonanno & Griffiths 1995).

Table 2. Chemical compositions of normal AAC and water-repellent AAC.

	/mass%					
	CaO	SiO_2	Al_2O_3	Fe_2O_3	SO_3	Ig, loss
Normal	27.5	52.9	2.56	1.47	4.52	10.40
Water-repellent	27.4	54.0	2.44	1.73	4.84	9.78

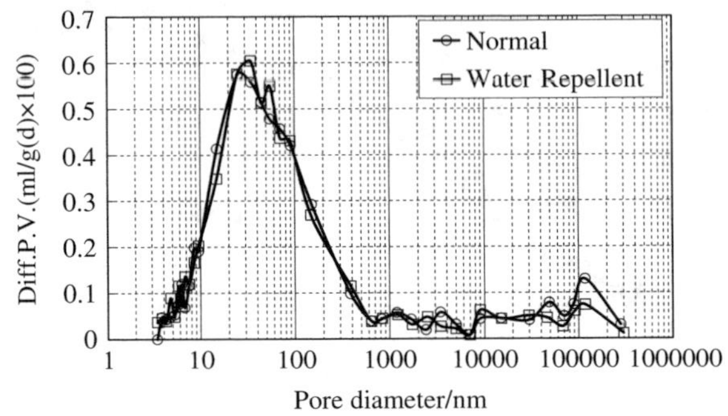

Figure 4. Pore size distribution by mercury porosimeter.

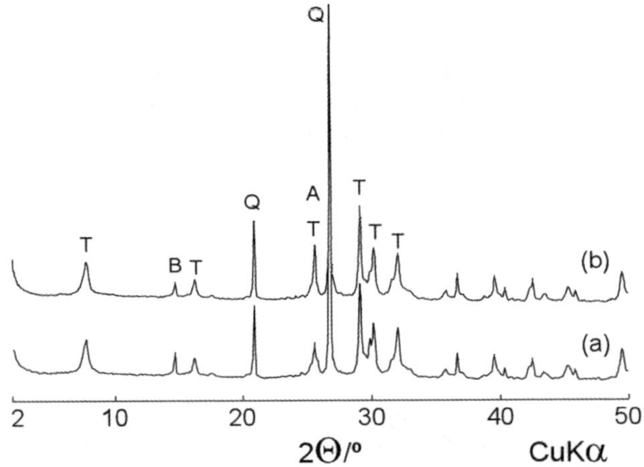

Figure 5. XRD patterns of normal AAC and water-repellent AAC (a: Normal AAC, b:Water-repellent AAC, T: Tobermorite-1.1 nm, Q: Quartz, A: Anhydrite, B: Bassanite).

3.3 *Moisture property*

3.3.1 *Water vapour sorption isotherms*

Water vapour sorption isotherms are shown in Figure 7. At a relative humidity of 80% or less, moisture content of water-repellent AAC was smaller than that of normal AAC. Whereas moisture content of the water-repellent AAC and the normal AAC was nearly the same at a relative humidity around 90%, which was the same relative humidity condition of the laboratory carbonation in this study. Therefore, the difference of water vapour sorption isotherms was not enough for explaining the higher carbonation resistance of the water-repellent AAC than the normal AAC.

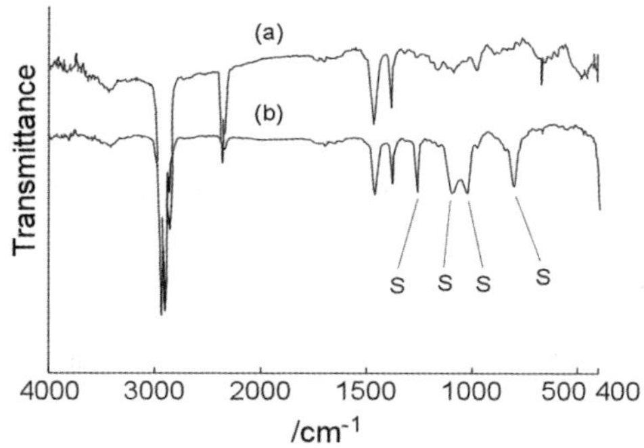

Figure 6. FT-IR spectra of normal AAC and water-repellent AAC (a: Normal AAC, b:Water-repellent AAC, S: Silicone-oil).

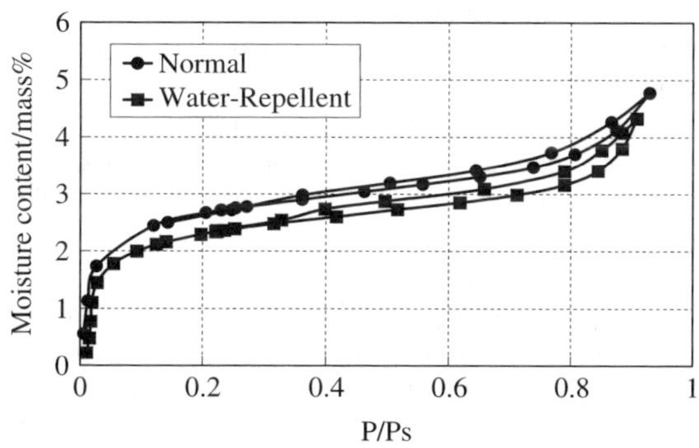

Figure 7. Water vapor sorption isotherms.

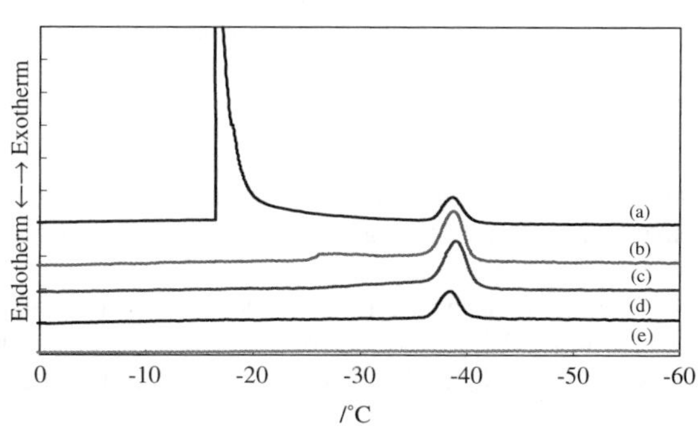

Figure 8. Cooling DSC patterns at various R.H. for normal AAC (a: R.H. = 98%, b: R.H. = 95%, c: R.H. = 85%, d: R.H. = 75%, e: R.H. = 65%).

3.3.2 *Cooling DSC*

A more detailed moisture property was investigated by cooling DSC. The results are shown in Figures 8 and 9 for normal and water-repellent AAC, respectively. In the cooling DSC data, an exothermic peak around −40°C corresponds to a statistical adsorbed water thickness of 4 to 6 molecular layers, while that around −20°C corresponds to a free adsorbed water (Tada 1996). In the normal AAC, no exothermic peak was detected at 65%R.H., while exothermic peaks around −40°C and around −20°C were detected at a relative humidity greater than 75 percent and than 95 percent, respectively. Whereas in water-repellent AAC, no peak was detected at a relative humidity smaller

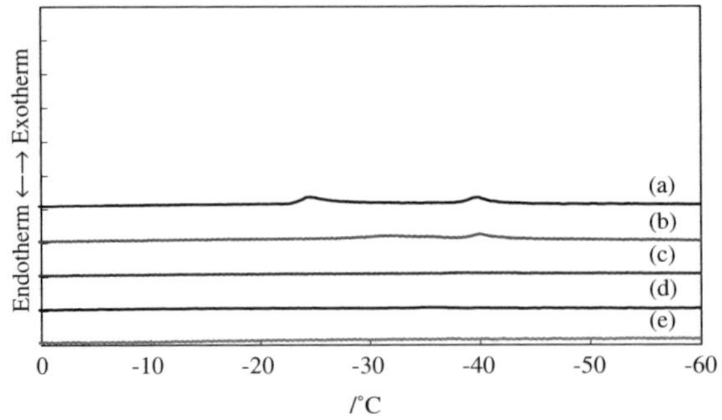

Figure 9. Cooling DSC patterns at various R.H. for water-repellent AAC (a: R.H. = 98%, b: R.H. = 95%, c: R.H. = 85%, d: R.H. = 75%, e: R.H. = 65%).

than 85 percent and exothermic peaks around −40°C and −20°C were detected at a relative humidity greater than 95 percent. The exothermic peak intensities of water-repellent AAC were remarkably smaller than those of the normal AAC. These differences in cooling DSC result between the water-repellent AAC and the normal AAC was considerable compared to that in water vapour sorption isotherms. The results suggested that the water-repellent AAC had distinct water vapour adsorption characteristics not only in the material surface but also in a comparatively thick adsorbed water film.

The phenomena may be explained by the hydrophobic hydration of adsorbed water. In the hydrophobic hydration theory, hydrophobic groups and water molecules may form a "frozen patch" structure in which each hydrophobic group is locked up in the water molecule polyhedron basket combined by the hydrogen bond or the "microscopic icebergs" structure (Hout & Jolicouer 1985). The significantly slight peak in the cooling DSC of water-repellent AAC irrespective of the amount of adsorbed water vapour could be due to the smaller degree of freedom of adsorbed water vapour due to the hydrophobic hydration. The degree of freedom of adsorbed water must have a strong effect on carbonation reaction that occur in the adsorbed water film itself. Therefore, the reason of the carbonation resistant characteristics of water-repellent AAC was attributed to the hydrophobic hydration of adsorbed water vapour by the hydrophobic groups of the silicone-oil.

4 CONCLUSION

(1) Water-repellent AAC showed twice as much carbonation resistance as that of the normal AAC under laboratory carbonation conditions; carbon dioxide concentration of 3% at a temperature of 20°C and an ambient relative humidity of 90%.
(2) No differences in the chemical composition, BET specific surface area, pore size distribution by mercury porosimeter and XRD patterns were observed between water-repellent AAC and the normal AAC. IR spectrum indicated the existence of silicone-oil for water-repellent AAC after autoclaving.
(3) Moisture content of the water-repellent AAC was lower than that of the normal AAC at a relative humidity less than 80% and was almost the same at relative humidities around 90%.
(4) The exothermic peak intensities of the water-repellent AAC were remarkably smaller than those of the normal AAC in cooling DSC. The reason of the carbonation resistant property of water-repellent AAC was attributed to the hydrophobic hydration of adsorbed water by the hydrophobic groups of silicone-oil.

ACKNOWLEDGMENTS

The authors are grateful to Prof. Dr. K. Okada and Dr. Y. Kameshima of Tokyo Institute of Technology, for their support in the measurement of water vapour sorption isotherms.

REFERENCES

Bonanno, A. S. & Griffiths, P. R. 1995. Characterization of silicones by direct-deposition GC/FT-IR, *J. High Resol. Chromatogr.*, 18: 289–294.

Fagerlund, G. 1978. The critical degree of saturation method assessing the freeze/thaw resistance of concrete, *Mater. Constr.*, 10: 217–229.

Goodier, A. & Matthews, S. 1997. In-service behavior of reinforced autoclaved aerated concrete planks, *Spec. Publ. Am. Concr. Inst.*, No.SP-170, 2: 1197–1214.

Hama, Y., Tabata, M., Watanabe, T. & Kamada, E. 1992. Frost resistant of increased density autoclaved aerated concrete, In Wittmann, F. H. (ed.), *Advances in Autoclaved Aerated Concrete*: 157–163, Rotterdam, Balkema.

Hout, J.-Y. & Jolicouer, C. 1985. Hydrophobic effects in ionic hydration and interactions, In Dogonadze, R. R. et al. (eds), *The Chemical Physics of Solvation*, Part A, Chap. 11, New York, Elsevier.

Japanese patent 1980, *Tokkai-shou*, 55–85452 (in Japanese).

Matsushita, F. & Shibata, S. 2000. Carbonation degree as durability criteria for autoclaved aerated concrete, *5th CANMET/ACI International Conference on Durability of Concrete*: 1123–1134.

Matsushita, F., Aono, Y. & Shibata, S. 2000. Carbonation degree of autoclaved aerated concrete, *Cement and Concrete Research*, 30: 1741–1745.

Nakamura, M., Kitano, T. & Arakawa, M. 1987. Surface modification of autoclaved light weight concrete, *Yougyou-kyoukai-shi*, 95(10): 1007–1011. (in Japanese).

Nakamura, M., Hata, Y. & Yamaoka, S. 1988. Inorganic surface modification of autoclaved light weight concrete, *Journal of the Ceramic Society of Japan*, 96(11): 1093–1097. (in Japanese).

Nakamura, M. & Ketani, K. 1990. Surface modification of ALC by low temperature plasma treatment, *Journal of the Ceramic Society of Japan*, 98(10) 1164–1168. (in Japanese).

Senbu, O. & Kamada, E. 1992. Mechanism of frost deterioration of AAC In Wittmann, F. H. (ed.), *Advances in Autoclaved Aerated Concrete*: 153–156, Rotterdam, Balkema.

Tada, S. 1992. Pore structure and moisture characteristics of porous inorganic building materials, In Wittmann, F. H. (ed.), *Advances in Autoclaved Aerated Concrete*: 53–63, Rotterdam, Balkema.

Tada, S. 1996. Microstructural approrach to freezing behavior of highly porous materials, *Proc. 7th International Conference on Durability of Building Materials and Components*: 299–308.

Autoclaved Aerated Concrete – Limbachiya and Roberts (eds)
© 2005 Taylor & Francis Group, London, ISBN 0 415 38356 0

Micro-mechanical analysis of AAC

I. Kadashevich & D. Stoyan
TU Bergakademie Freiberg, Institut für Stochastik, Freiberg, Germany

ABSTRACT: The paper describes simulations of the micromechanical properties of AAC, based on a stochastic-geometric structure model. The model is the "cherry-pit" model, which is a random system of partially interpenetrating spheres, which stand for the air voids. For the mechanical analysis the solid phase is approximated by an irregular spatial network obtained with the help of the so-called radical tessellation with respect to the hard spheres of the "cherry-pit" model. FE simulations are used for the analysis of the mechanical properties of this network, which are considered as approximations for that of AAC. They lead to information on the influence of the geometrical structure on the macroscopic Young modulus but also on the variability and spatial distribution of strain energy in the networks.

1 INTRODUCTION

The determination of the compressive strength of AAC is a difficult problem. While the paper Wolf et al. (2005) describes a statistical approach, a thorough physical understanding is only possible if the microstructure of AAC is studied, modelled and explicitly used in the mechanical calculations.

This paper presents such an approach. It is based on the stochastic-geometric structure model which is described in Stoyan & Kadashevich (2005). According to this model, the pores are spherical and partially interpenetrating. The matrix (the complement of the pore space) is then approximated by a lattice, a system of beams of spatially variable strengths and geometries. This lattice is used as a simplified mechanical model of the AAC structure. Its advantage is the possibility to carry out calculations with special and efficient FE software, see the discussion in Krajcinovic (1996), p. 105. In contrast to approaches where the matrix is approximated by voxels, very large structures can be analyzed which correspond to AAC samples of a size of some cm^3.

This model is used to determine the Young modulus E of various samples. Some of them model real AAC samples, using the results of Stoyan & Kadashevich (2005) . For these samples the true value of E is known from fracture tests and so the mechanical parameters of the beams can be calibrated to re-obtain the true values of E. Starting with these results, the relationship between microstructure and E is investigated. Additionally, the energy concentrations are studied to get some idea about the fracture processes in the lattices, which start with fractures of the most loaded beams.

2 CONSTRUCTION OF EQUIVALENT NETWORKS

In the approach of this paper the solid phase of AAC is replaced by a system of cylindrical beams, which form a spatial network or lattice. This network and the geometrical beam parameters are obtained by means of the so-called radical tessellation with respect to the pits or hard-cores of the "cherry-pit" model. According to this model the pores are partially interpenetrating spheres. These spheres consist of a hard non-interpenetrating spherical kernel (the "pit") and a soft spherical shell. Details are given in Stoyan & Kadashevich (2005). The following describes this tessellation and the construction of beams. Of course, it is the aim of the construction to obtain a realistic simplified substitute for the solid phase around the spheres which model the air voids.

If the spheres were identical, then it would be natural to construct the Voronoi tessellation (Okabe et al., 2000) with respect to the system of sphere centers. (A "tessellation" is a division of space into touching cells). Its cells are convex and polyhedral and the spheres are completely in the interior of the cells. The edges of the corresponding cells would then serve as axes of cylindrical beams and their lengths would be the cylinder heights. With well-chosen cylinder radii this lattice would be a realistic substitute for the solid phase.

But the spheres of the cherry-pit model used for AAC modelling are different, the pits have different radii r_i. If also in this situation the Voronoi tessellation would be used, sometimes cells could cut their corresponding spheres or touching spheres might not be neighbours in the tessellation. So another, more general and flexible tessellation construction is needed which is better adapted to the case of random radii, but yields nevertheless convex polyhedral cells with straight edges and planar faces. This tessellation should be constructed with respect to the sphere centers and the radii r_i of the hard cores or pits.

A tessellation that satisfies these demands is the *radical tessellation* (or Laguerre, weighted Voronoi, or power tessellation). It has been used until now for geometrical data analysis, namely for the statistical description of random packings of random spheres by statistical characterization of the distribution of cell parameters (Gervois et al., 2002; Richard et al., 2001; Medvedev, 2000), for modelling of grain growth (Telley et al., 1996), or for the definition of discrete-mechanical quantities for granular assemblies (Bagi, 1996; Satake, 2004). Another tessellation applicable for spheres with variable diameters, the S-Voronoi tessellation (Medvedev, 2000), is not useful for this paper since it generates cells with curved faces and edges.

The definition of the radical tessellation is not difficult, see Okabe et al. (2000) and Telley et al. (1996), but for making the paper self-contained it is repeated here. Equation 1 below gives the mathematical definition of the cell corresponding to the ith sphere with center P_i and radius r_i. It is the set of all points P of three-dimensional space satisfying

$$d^2(P,P_i) - r_i^2 \le d^2(P,P_j) - r_j^2 \tag{1}$$

for all other spheres with P_j, r_j, where d is the Euclidean distance.

All these cells are convex and polyhedral and have straight edges. The interior of different cells is of course disjoint. Sample cells of radical tessellations are presented in Figure 1.

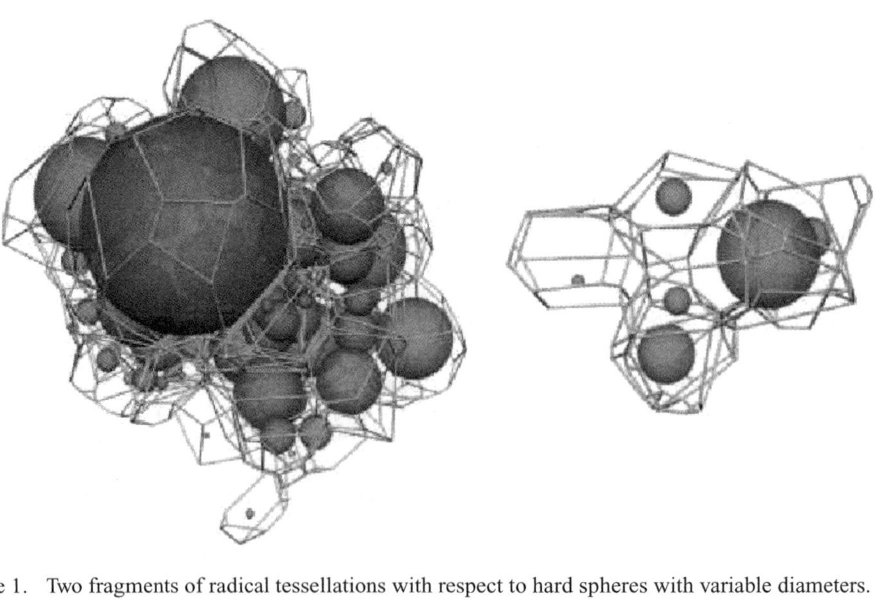

Figure 1. Two fragments of radical tessellations with respect to hard spheres with variable diameters.

By the way, the plane between two cells to which the joint face belongs is called "radical plane" of the corresponding spheres. All its points have the same tangency length or "power".

Algorithms for the construction of the radical tessellation are described in Telley et al. (1996a) and Gervois et al. (2002). For the present paper Luc Oger's program is used.

The authors had to learn that the tessellations obtained in this way are not always completely satisfactory from the standpoint of the cherry-pit model and the use of tessellation edges as beams of lattices. It can happen that edges go through cherry space between pits, what means in terms of AAC that they go through the air phase. Such edges are removed before beam construction and mechanical calculations. Furthermore, the network has to be trimmed in order to obtain the rectangular shape as in the samples investigated in the mechanical analysis.

The remaining edges are used to construct the cylindrical beams. The corresponding lines determine the cylinder axes, and their lengths are the cylinder heights. For the construction of the radius R_k of the cylinder around the kth edge there are several possibilities. One, which is used here, is simply to take the minimum distance of this edge to the spheres in the (usually three) cells that share this edge and to other nodes in the cells to which this edge belongs.

Figure 2 shows two slices through the lattices for two samples discussed below.

For the mechanical calculations, all beams are assigned linear elastic isotropic material properties with identical Young modulus E_{beam} and Poisson ratio ν in the whole network. (Because of different beam diameters and lengths the mechanical behaviour of the beams differs.) The numerical values of these parameters result from a pilot analysis described in Section 4 below, which ensures that macroscopic properties of test samples are reproduced.

The FE modelling is performed with the help of the commercial FE code ANSYS. According to its possibilities, the beams of the network are modelled as BEAM44 elements (3D beams with 6 degrees of freedom in each node). Indeed, a beam must have 6 degrees of freedom at each node, because not only the forces but also bending moments are responsible for the mechanical behaviour of the lattices.

At the end of this section, some remarks on alternative approaches may be useful. The classical (naive) FE approach consists in a discretisation of the solid phase with the help of many small polyhedral volume elements, which are usually much smaller than the voids, to obtain a precise geometrical reproduction of the solid phase. A successful method of this type is the digital-image-based FE method of Garboczi (see Garboczi, 1998; Garboczi & Day, 1995), where the porous medium is approximated by a system of cubic voxels. This way is suitable only for the analysis of small samples; according to Roberts & Garboczi (2002) a volume containing 100 pores is sufficient.

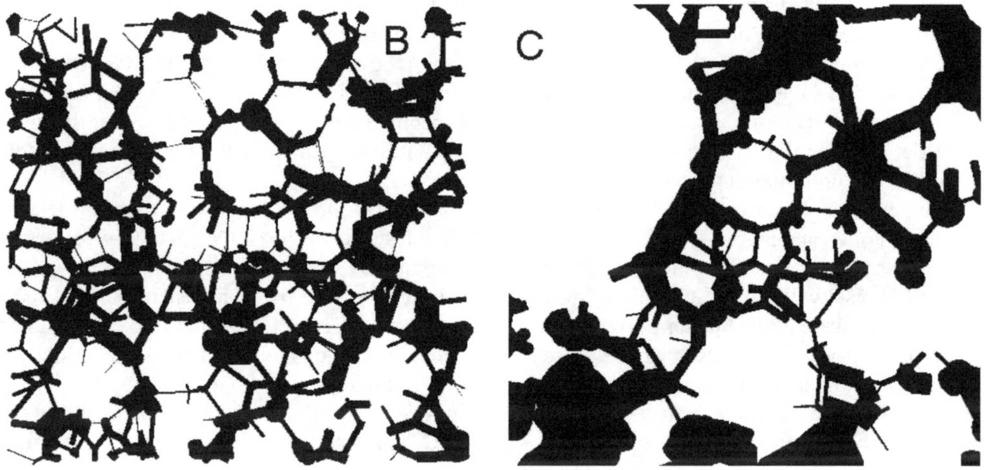

Figure 2. Thin slices of size $1 \times 10 \times 10$ mm from networks of samples B and C, which have different pore sizes.

Table 1. Structure parameters of the samples A–C.

Samples	Pure density ρ, g/cm^3	Bulk density ρ_R, g/cm^3	Total porosity P, %	Artificial porosity P_{Air}, %
A	2.632	0.814	69.07	16.56
B	2.632	0.485	81.57	52.99
C	2.632	0.535	79.67	43.97

Ryan & van Rietbergen (2005) approximated the solid phase of bones by systems of small quadratic columns and used then eight-noded brick elements. Ryan & van Rietbergen (2005), who speak about "micromechanical FEM", report that they used networks of around $2 \cdot 10^6$ nodes to model samples of a size of approximately 400 mm^3. Similar to the approach of the present paper is the skeletonization approach, see Pothuaud et al. (2000). It starts from voxel data coming from, for example, microtomography or magnetic resonance imaging. By some thinning procedure many voxels that represent the solid phase are omitted in such a way that the topological structure of the solid phase is retained. Networks quite similar to those in the present paper also appear in the study of elastic properties of three-dimensional open-cell solids (foams), see Roberts and Garboczi (2002) who, however, studied only Voronoi tessellations. In the literature of porous media beams appear already in Lin and Cohen (1982). They investigated sedimentary rocks and used networks whose vertices represent grains and whose branches correspond to or edges contacts between grains. Also Krajcinovic (1996) emphasizes the value of networks or (irregular) lattices in micromechanical calculations.

3 EXPERIMENTAL DETERMINATION OF MATERIAL PARAMETERS

For the beam elements of the lattice discussed in Section 2 mechanical parameters are needed, namely Young modulus E_{beam} and Poisson ratio ν. The Poisson ratio is simply taken as the constant standard value of 0.2 (for concrete). The Young modulus E_{beam} is chosen in such a way that empirically observed macroscopic values of E of AAC samples are reproduced. The way of calibrating the method, i.e. to come to the right values of E_{beam}, is described in Section 4 below, while here the experimental results are reported which yielded the macroscopic parameters.

A series of AAC samples of AAC types A–C as in Stoyan and Kadashevich (2005) (see Table 1 above) was tested by H.-J. Schneider from the material testing laboratory of the Institute of Ceramics, Glass and Construction Materials of TU Bergakademie Freiberg. The samples for the tests were cubes of side length 55 mm. Their orientation in the concrete block followed the direction of expansion, which is taken as the vertical direction. The cubes were cut out with a precision machine with a diamond cut-of wheel and then the specimens were wet-ground so that the sides were flat and parallel, with a tolerance of less than 0.1 mm on 100 mm. Finally, the cubes were dried at 50°C until mass constancy.

The samples were mechanically tested under uniaxial compression using a material testing machine. For each material type 30–50 stress-controlled tests according DIN EN 679 were carried out with constant loading rate 0.1 MPa/sec. For comparison and verification also two strain-controlled tests with constant compression rate of 1 mm/min were made for each material type. The samples were placed unsupported in the testing machine. The ultimate load was determined as the first maximum of the measured stress-displacement curve. The ultimate stress value was calculated as the ultimate load divided by the cross-sectional area of the sample. The results were recorded digitally with the help of the testing machine software (TIRAtest). Figure 3 shows the results for three samples each from the AAC types A–C. The macroscopic Young moduli obtained from the experiments are presented in Table 2.

The experimental results are in accordance with existing material-scientific theories: The values of E for the types A – C and some other samples not reported here follow the well-known relationship

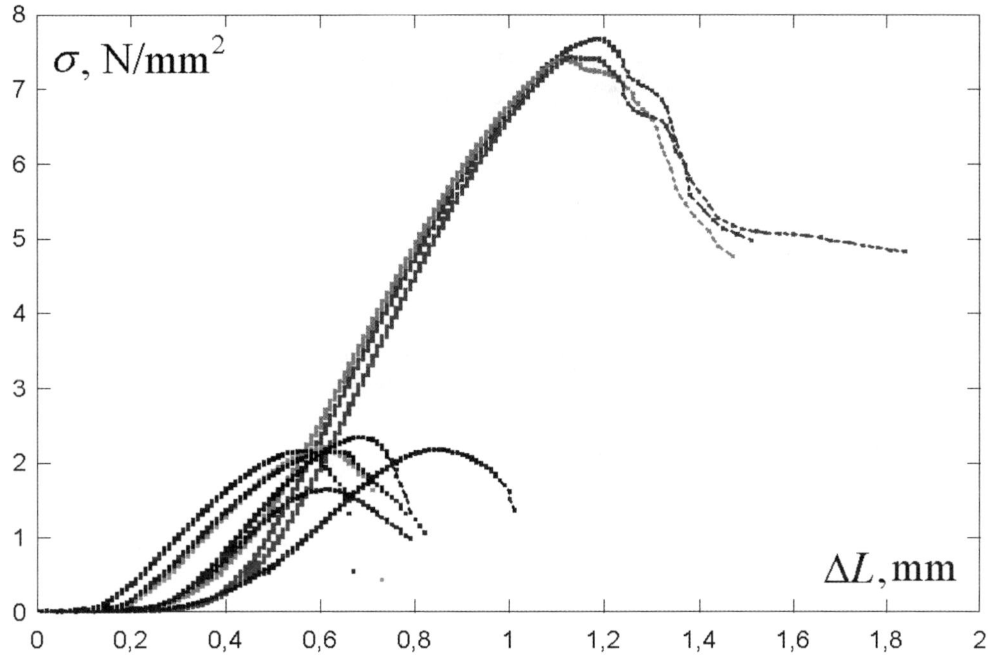

Figure 3. Stress-displacement curves for the samples A–C.

Table 2. Estimated Young modulus (macroscopic).

Samples	E_{min}, N/mm^2	E, N/mm^2	E_{max}, N/mm^2
A	643	695	753
B	309	346	399
C	303	389	471

between E and material density ρ for porous materials (Gibson & Ashby, 1997):

$$\frac{E}{E_{ref}} = \alpha \left(\frac{\rho}{\rho_{ref}} \right)^n \tag{2}$$

Here α is a constant, which is in the case discussed here equal to 1, since samples of the same material are compared. The parameter n takes the value 1.6 for our laboratory samples, which include A–C, while E_{ref} and ρ_{ref} are the parameters of some reference AAC.

However, the mechanical properties of AAC do not only depend on porosity. It is well known that important mechanical parameters depend also on other structure properties, in particular on the degree of inhomogeneity.

4 FE SIMULATION OF COMPRESSION TESTS

The authors understand under "simulation" the calculation of individual beam forces and displacements and macroscopic Young moduli of networks loaded analogously as in the fracture tests described in Section 3, with the help of FE software. These networks have free boundary conditions on the side surfaces if the samples are large enough (larger than the RVE) and symmetrical

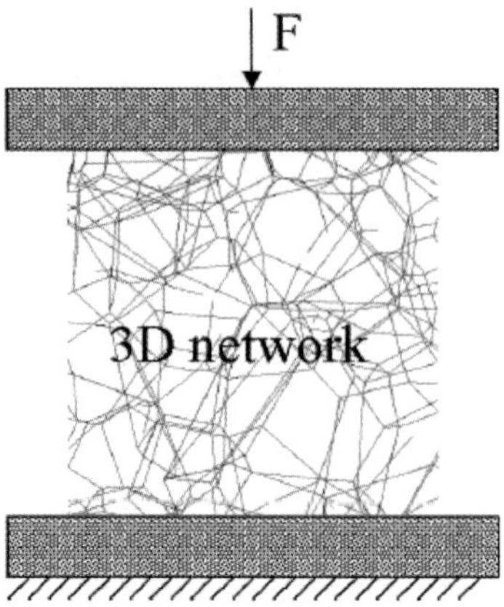

Figure 4. Sketch of the load situation for a lattice.

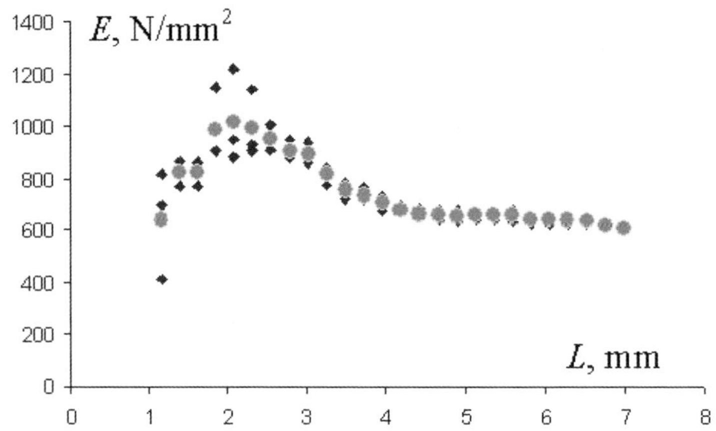

Figure 5. Bulk Young modulus E obtained for cubic samples with a structure as that of AAC type A in dependence of side length between approx. 1 mm and 7 mm. The RVE is a cube of side length 6.5 mm.

boundary conditions for smaller samples. They are fixed on the bottom and loaded with the force (displacements) on the top. The loading schema is shown in Figure 4.

The summary deformation in the simulations is less than 2% (similar to the experiments described in Section 3), therefore problem with contacts between edges in the network can be neglected.

For the determination of the macroscopic E it was necessary to explore the size of the representative volume element (RVE). Figure 5 shows a typical result, namely that for AAC type A. It turns out that the RVE for type A is a cube of side length 6.5 mm. (Note that the range of correlation of the point process of sphere centres is much shorter, about 0.3 mm.) For type B the side length is larger, 10 mm, corresponding to the greater degree of inhomogeneity of that material, and for C it is even 16 mm. It is perhaps better to characterize the RVE in terms of the number of beams of the sample. As Figure 6 shows, this RVE is for all samples nearly equal, 40,000 beams (or 2000 pores) seem to be sufficient. The value of the microscopic Young modulus $E_{beam} = 13540\,\text{N/mm}^2$ of the beams yields the macroscopic values of the Young modulus E shown in Table 3. They are close to the experimental values in Table 2. Fortunately, for all three types A–C the same E_{beam} can be used.

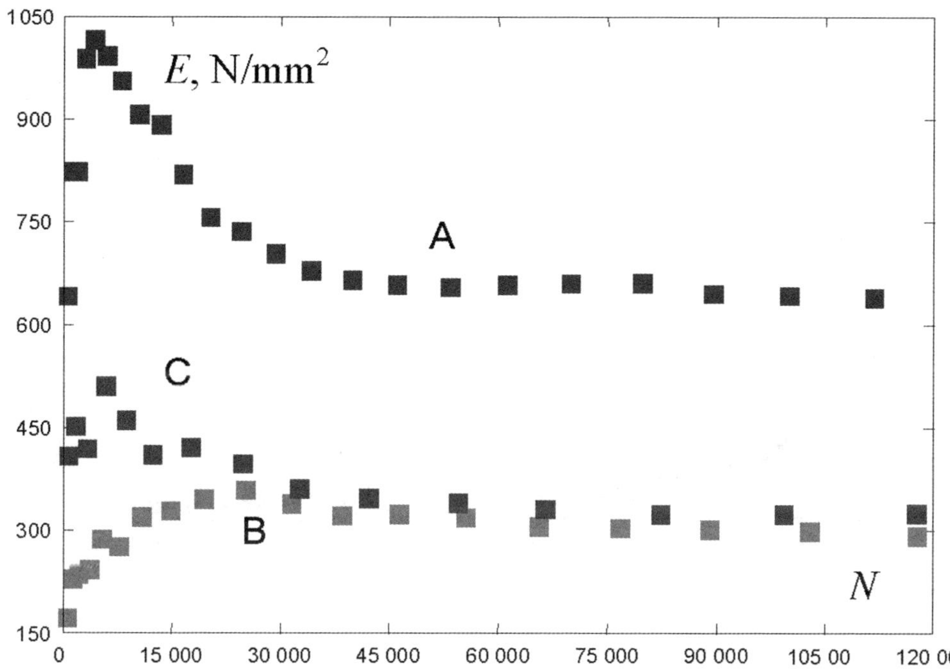

Figure 6. Relationship between macroscopic Young modulus E and sample size expressed in terms of beam numbers for samples A–C.

Table 3. Re-calculated macroscopic Young modulus.

Samples	E_{min}, N/mm^2	E, N/mm^2	E_{max}, N/mm^2
A	650	695	742
B	306	324	346
C	412	464	515

5 RESULTS

In order to study the influence of the geometrical structure of the air pore system on the macroscopic Young modulus, various simulated samples of porous material of fixed porosity of 53% (as AAC type B) with the same local mechanical parameters as the AAC samples were generated. We report here on the results for two cases, namely

– a material with identical pores (which never can be produced in reality), with pore radius of the hard cores of the cherry-pit model of 0.4 mm;
– a material with two different sphere radii of 0.4 mm and 0.8 mm with number fraction of smaller spheres of 75%.

The corresponding Young moduli are 430 N/mm^2 and 300 N/mm^2, i.e. the more regular material is stronger than AAC type B, while the bidisperse material is weaker.

Our calculations make it possible to explore also the distribution of the strain energies of the deformed beams in the network. Figure 7 shows the probability distribution of energy per beam

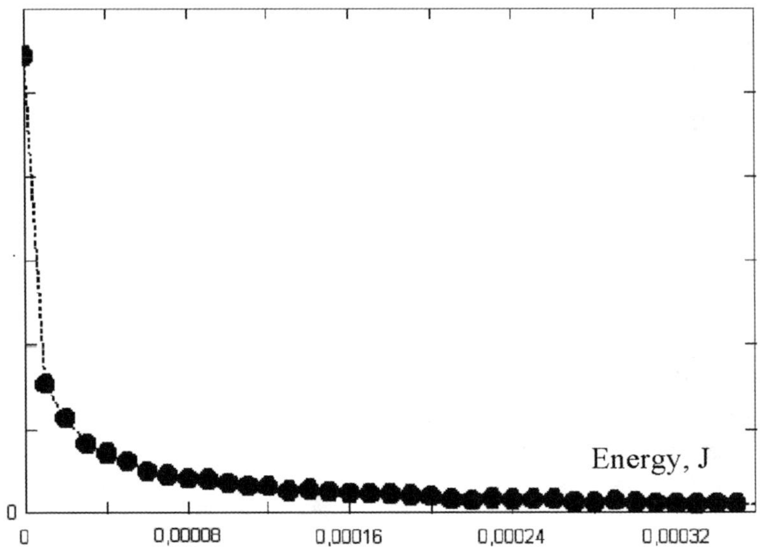

Figure 7. Strain energy distribution in the deformed sample A.

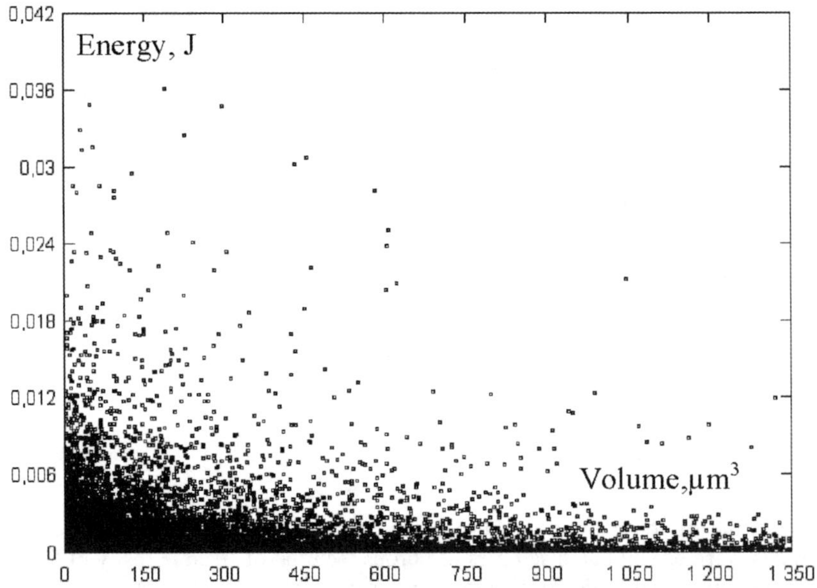

Figure 8. Strain energy distribution in the deformed sample A.

for sample A (under deformation 1.5%). The distribution looks like a power distribution with parameter $\beta = -1$

$$f(x) = \left(\frac{x}{b}\right)^{\beta} \tag{3}$$

In a further statistical analysis we investigated the relationship between strain energy and beam volume. Figure 8 shows the relationship for AAC type A.

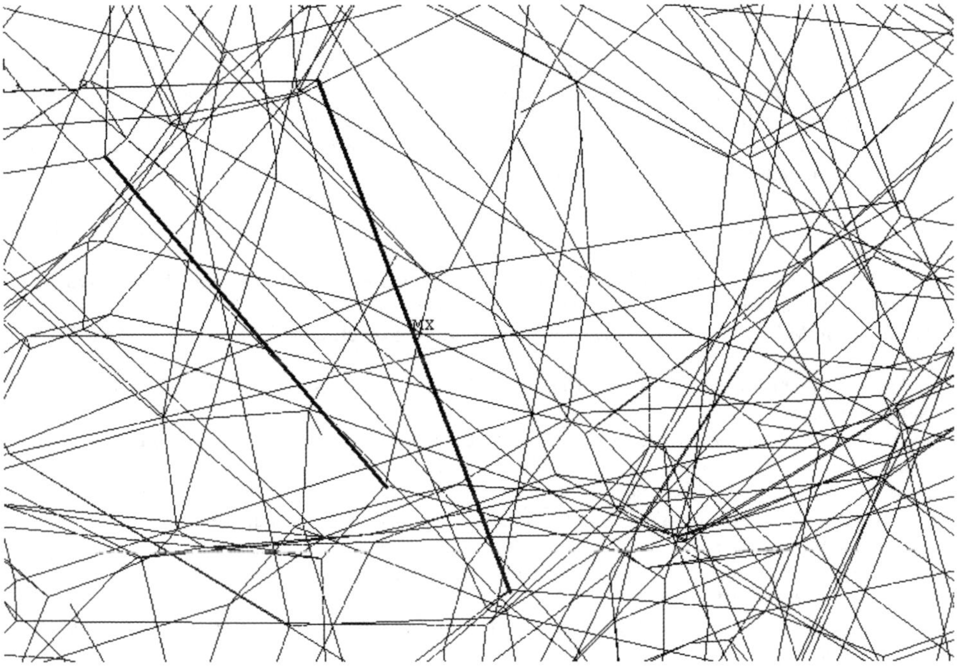

Figure 9. Cut-out of the deformed lattice of sample A with two strain energy concentrations.

Obviously, there is a tendency that beams of smaller volume have a higher strain energy. This shows also Figure 9 where two thin beams with the high energy are presented, which are nearly to the loading direction.

Finally, the spatial distribution of beams with high strain energy was analysed statistically. It turned out that these beams are randomly distributed in space; there are no clusters of high-energy beams. Figure 9 shows a small cut-out of the network for AAC type A with two isolated high-energy beams. Such a distribution is typical for the distribution of high-energy beams. The pair correlation function of the point process of beam centres is close to that of a Poisson process, the model for complete spatial randomness of point distribution. (Small deviations from Poisson process behaviour are natural since the beams are arranged around hard spheres.) This behaviour is different to the distribution of contact forces in loaded particle systems, where chains of highly loaded beams are observed, see Radjai et al. (1998).

6 CONCLUSIONS

The paper shows that porous materials such as AAC can be modelled by means of networks, which approximate the solid part of the material. The mechanical properties of these networks can be analysed by means of classical FE methods, what enables the analysis of large samples, which cannot be analysed by means of voxel-based FEM. These samples can model real AAC structures but also "mathematical" materials which result from simulations of stochastic models.

It was possible to find a unique Young modulus for the beams E_{beam} which leads to the macroscopic Young moduli for different AAC samples. The RVE in terms of beams is about 40,000. The spatial distribution of beams with high strain energy could be explored statistically and it turned out that these beams are randomly scattered in the sample. This supports assumptions in a statistical model of compression strength used in the paper Wolf et al. (2005).

Future investigations will concern the fracture process in networks, in order to contribute to the understanding of fracture behaviour of AAC.

REFERENCES

Bagi, K. 1995. Geometrical modeling of granular assemblies. *Acta Technica Acad. Sci. Hung.* 107: 1–16.

Garboczi, E.J. 1998. *Finite Element and Finite Difference Programs for Computing the Linear Electric and Elastic Properties of Digital Images of Random Materials*. NIST Internal Report 6269, Chapter 2.

Garboczi, E.J. & Day, A.R. 1995. An algorithm for computing the effective linear elastic properties of heterogeneous materials: Three-dimensional results for composites with equal Poisson radius. *J. Mech. Phys. Solids* 43: 1349–1362.

Gervois, A., Oger, L., Richard, P. & Troadec, J.P. 2002. *Computational Geometry and Applications* (CGA'02), Amsterdam.

Gibson, L.J. & Ashby, M.F. 1997. *Cellular Solids: Structure and Properties*.Cambridge: University Press.

Krajcinovic, D. 1996. *Damage Mechanics*. Amsterdam: North Holland.

Lin, C. & Cohen, M.H. 1982. Quantitative methods for microgeometric modeling. *J. Appl. Phys*. 53: 4152–4165.

Medvedev, N.N. 2000. *Voronoi-Delaunay Method for Non-crystalline Structures*. Novosibirsk: SB Russian Academy of Science.

Okabe, A., Boots, B., Sugihara, K. & Nok Chiu, S. 2000. *Spatial Tessellations, Concepts and Applications of Voronoi Diagrams*, 2nd edition. Chichester: John Wiley & Sons Ltd.

Pothuaud, L., Porion, P., Lespersailles, E. Benhamou, C.L. & Levitz, P. 2000. A new method for three-dimensional skeleton graph analysis of porous media: application to trabecular bone microarchitecture. *J. Microscopy* 199: 149–161.

Radjai, F., Wolf, D.E., Jean, M., et al. 1998. Bimodel character of stress transmission in granular packings. *Physical review letters* 80 (2) 61–64.

Richard, P., Oger, L., Troadec, J.P. & Gervois, A. 2001. A model of binary assemblies of spheres. *Eur. Phys. J. E* 6: 295–303.

Roberts, A.P. & Garboczi, E.J. 2002. Elastic properties of model random three-dimensional open-cell solids. *J. Mech. Phys. Solids* 50: 33–55.

Ryan, T.M. & van Rietbergen, B. 2005. Mechanical significance of femoral head trabecular bone struture in Loris and Galago evaluated using micromechanical finite element methods. *Amer. J. Phys. Anthropology* 126: 82–96.

Satake, M. 2004. Tensorial form definitions of discrete mechanical quantities for granular assemblies. *Int. J. Solids & Structures* 41: 5775–5791.

Stoyan, D. & Kadashevich, I. 2005. Stochastic modelling of the geometrical microstructure of AAC. This volume.

Telley, H., Liebling, T.M. & Mocellin, A. 1996. The Laguerre model of grain growth in two dimensions I. Cellular structures viewed as dynamical Laguerre tessellations. *Phil. Mag. B* 73: 395–408.

Wolf, S., Wiegand, S., Stoyan, D. & Walther, H., 2005. The compressive strength of AAC – a statistical investigation. This volume.

Autoclaved Aerated Concrete – Limbachiya and Roberts (eds)
© 2005 Taylor & Francis Group, London, ISBN 0 415 38356 0

Stochastic modelling of the geometrical microstructure of AAC

D. Stoyan & I. Kadashevich
Institut für Stochastik, TU Bergakademie Freiberg, Freiberg, Germany

ABSTRACT: The system of artificial air pores of AAC can be modeled by the penetrable-concentric-shell or "cherry-pit" model, which consists of partially penetrable spheres with soft shells randomly distributed in space, where the spheres stand for the pores. The corresponding hard cores form a random packing of hard spheres. The diameter distribution of the hard spheres and the thickness of the soft shells can be obtained from planar sections of AAC samples by means of stereological methods. For the diameter distribution a three-component distribution is suitable, which consists of two negative exponential and one Gaussian component.

1 INTRODUCTION

Between 60 and 90% of the volume of AAC consists of pores, which are classified as micro and macro capillaries and artificial air voids, and have diameters between some 100 nm and some mm. As for many materials, the microstructure of AAC has great influence on bulk properties such as strength, toughness, moisture transport, and heat transfer. In the case of AAC, it is well known that the most important material parameter is porosity or bulk density and that the "degree of inhomogeneity" of the geometrical pore structure plays a significant role, see Schlegel & Hums (2002). For theoretical studies of bulk properties of AAC, a precise geometrical model of its microstructure is needed. Such a model was introduced in Kadashevich et al. (2005). It starts from the usual assumption that single pores are spherical. The pore radii are variable and the corresponding radius distribution plays an important role. Based on the analysis of a series of AAC samples including image analysis of planar sections, a parameterized family of radius distributions is introduced by means of the Saltykov algorithm, a well-known stereological method. The corresponding parameters are then estimated by means of the maximum likelihood method. For the spatial arrangement of the pores the well-known penetrable-concentric-shell or "cherry-pit" model is used, i.e. it is assumed that the spheres, which stand for the pores, are randomly distributed in space and are partly (but not fully) penetrable. This model can be easily simulated and is therefore useful for future investigations of bulk properties of AAC. Comparison with samples of AAC shows that the model yields a good fit.

2 MATERIALS

Two series of AAC samples were investigated, one produced in the laboratory of the Institute of Ceramics, Glass and Construction Materials of TU Bergakademie Freiberg and the other in the Schrobenhausen plant of Xella Baustoffe GmbH, Germany.

The samples produced in Freiberg differ with respect to porosity, which ranges between 69 and 86%; but there are also samples of the same porosity but with different mean pore sizes. All these samples were produced under identical physical conditions; the structural differences result from quantity or type of powdered aluminium as pore agent. This paper reports on four typical AAC specimens, three from the laboratory and one from industry, which are considered as representative for the whole set of samples. They are denoted in the following as A, B, C, and D (the industry probe PP2-0,35 according to DIN 4165, with a porosity of 86.34%).

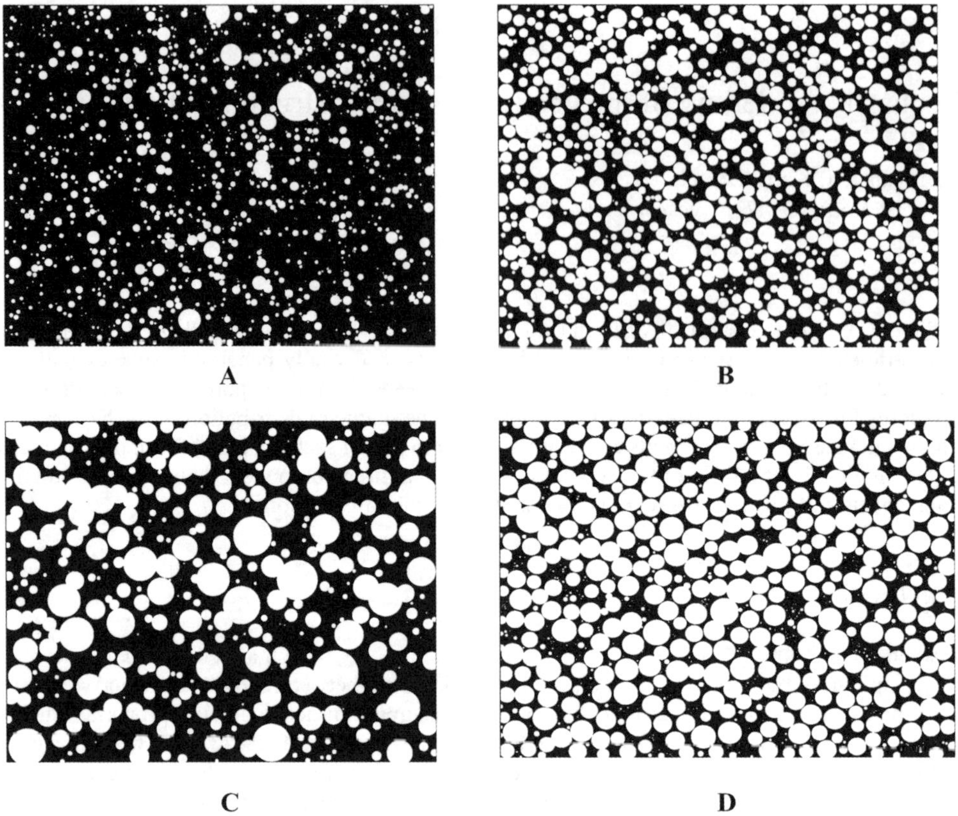

Figure 1. Results of disk approximation for the air voids in planar sections of the four samples A, B, C, and D.

3 EXPERIMENTAL

3.1 *Cutting of AAC samples*

From all AAC samples, three cubic subsamples were taken. Since the structure of AAC changes in the expansion direction (Grimer & Brewer (1965), Schwiete & Ludwig (1965)), the samples were cut out from the interior of the AAC bodies, and sample orientation followed the direction of expansion, which is taken as the vertical direction. The cubes were cut out with a precision machine with diamond cut-of wheel, the specimens were wet-ground and so the sides were flat and parallel and the final cube side was 55 mm. Finally, the cubes were dried at 50°C until mass constancy.

For image analysis the samples were carefully prepared by dry fine-grinding by means of abrasive paper, removal of debris, careful black coloring with water resistant alcohol based ink and filling of the pores with hydrated aluminum powder. The images were then taken with the macro objective, following the reflected light method. The prepared surfaces of sample cubes were processed using Image-C MATAM software of IMTRONIC and converted into bitmaps.

The first step of image analysis was determination of porosity on the surfaces, i.e. of the area fraction P_{Air} of the white phase of samples as in Figure 1. The value P_{Air} is very important for model fitting, because it serves as a target parameter.

In the second step the porous (white) phase is approximated by a system of overlapping disks as described in Kadashevich et al. (2005). The resulting radii are denoted as r_i. Figure 1 shows the resulting approximating disk systems for the four samples.

In the third step the pair correlation functions (pcf's) $g_A(r)$ (Hubalkova & Stoyan (2003)) for the planar point patterns of disk centers was determined. They describe in some way the spatial variability of the arrangement of voids and serve later for model test. Figure 3 in Section 4 below shows such functions.

4 METHODS

4.1 *Stereological determination of sphere radius distribution*

In the proposed three-dimensional model of AAC the voids are approximated as spheres. The corresponding radius distribution, which is given by the probability density function $\delta_V(r)$, is of course of great importance and must be carefully determined. For the AAC samples it was determined statistically, where the initial data are the disk radii of the constructed section profiles as in Figure 1.

Thus one is confronted with the classical Wicksell problem (Stoyan et al. (1995); Snyder et al. (2001); Ohser & Mücklich (2000)). This problem consists of the determination of the probability density function $\delta_V(r)$ of sphere radii, based on the probability density function $\delta_A(r)$ of radii of circles of intersection corresponding to a planar section of sphere system. In the given case , the surfaces of sample cubes are considered as (parts of) planes through the AAC bodies.

It is well known in stereology that the relationship between $\delta_A(r)$ and $\delta_V(r)$ is given by the integral equations

$$\delta_V(r) = -\frac{2rN_A}{\pi N_V}\int_r^{\infty}\frac{1}{\sqrt{x^2-r^2}}\frac{d}{dx}\left[\frac{\delta_A(x)}{x}\right]dx \tag{1}$$

and

$$\delta_A(r) = r\frac{N_V}{N_A}\int_r^{\infty}\frac{\delta_V(x)}{\sqrt{x^2-r^2}}dx. \tag{2}$$

Here N_V and N_A are the mean numbers of spheres and disks respectively, per unit volume and area, respectively. Equation 1 can be considered as the mathematical solution of the Wicksell problem. Unfortunately, $\delta_A(r)$ is usually given only numerically based on statistical data, the exact form is unknown and so the derivative of $\delta_A(r)$ appearing in Equation 1 cannot be calculated with sufficient precision. Therefore, numerical methods have to be used in order to determine $\delta_V(r)$ and N_V using planar information. A classical approach is the Saltykov method, which is based on a discrete approximation of Equation 2, which leads to a system of linear algebraic equations, which can be solved recursively. Details of this method are given in Stoyan et al. (1995) and Ohser & Mücklich (2000). The use of this stereological method is a very elegant approach since for AAC samples very good planar sections can be produced. Our experience shows that it is even better than the use of computerized microtomography, since also this method leads finally to numerical problems in the context of the determination of sphere diameters from voxel data.

Following Kopp-Schneider (2003), the Saltykov algorithm is used in this paper for exploring the general form of $\delta_V(r)$, i.e. to come to a family of probability distributions $\delta_V(r;\theta)$ which depend on some parameters, denoted here by a parameter vector θ. In a second step the parameter vector θ is estimated by means of the maximum likelihood method.

The likelihood function $L(r_1, \ldots, r_{n_j}; \theta)$ depends on the observed radii r_i of the constructed section disks and the parameter vector θ. It has the form:

$$L(r_1,\ldots,r_{n_j};\theta) = \prod_{i=1}^{n}\delta_A(r_i;\theta), \tag{3}$$

where $\delta_A(r;\theta)$ is given by Equation 2, which contains on its right-hand side the parameterized function $\delta_V(r;\theta)$. For every kind of AAC four planar sections (sample cube surfaces) offered the

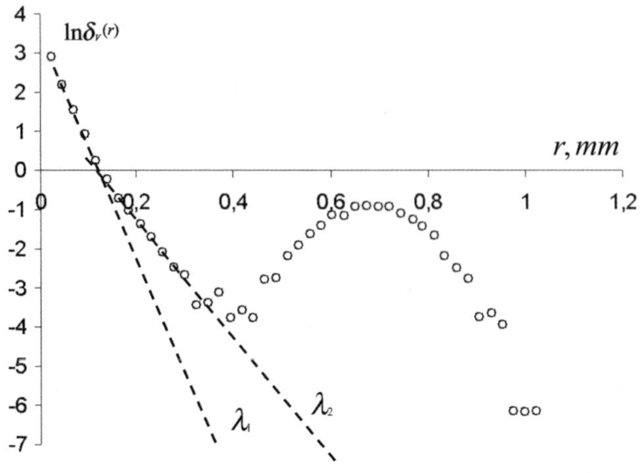

Figure 2. A rough approximation of $\delta_V(r)$ for sample D obtained by the Saltykov algorithm. The three-component structure is clearly recognizable.

data; n is the total number of radii (between 2000 and 5500). The maximum likelihood estimates were then obtained by maximization of $L(r_1, \ldots, r_n; \theta)$ for given r_i; in the numerical calculation the negative logarithm of $L(r_1, \ldots, r_n; \theta)$ was minimized by means of the method of steepest descent.

The stereological exploration based on the Saltykov algorithm led to the following parametric form for $\delta_V(r)$:

$$\delta_V(r) = p_1\delta_{V,1}(r) + p_2\delta_{V,2}(r) + p_3\delta_{V,3}(r) \tag{4}$$

with $p_1 + p_2 + p_3 = 1$ and

$$\delta_{V,i}(r) = \lambda_i e^{-\lambda_i(r-r_0)} \text{ for } r \geq r_0,\ i = 1,2 \text{ and } \delta_{V,3}(r) = \frac{1}{\sigma\sqrt{2\pi}} e^{-\frac{(r-\mu)^2}{2\sigma^2}}.$$

That means $\delta_V(r)$ is a mixture of three components with positive weights p_k. The first two components are shifted negative exponential distributions, which represent two subgroups of very small voids. The third component is Gaussian and corresponds to the larger voids (but not to the huge voids which belong to P_{Cracks}). Each AAC type is characterized by a parameter vector $\theta = (p_1, p_2, \sigma, \mu, \lambda_1, \lambda_2, r_0)$ of 7 independent variables.

Figure 2 shows an approximation of $\delta_V(r)$ for sample D obtained by the Saltykov algorithm. The three-component structure is obvious. In the semi-logarithmical scale there are two linear components and one parabolic, which are related to the exponential and Gaussian components of $\delta_V(r; \theta)$.

4.2 *Modelling the system of pores*

It seems to be natural to model the system of voids in AAC as a random system of spheres. As the planar sections of the samples show, there is frequent interpenetration of spheres. Thus a hard-sphere model is not realistic. On the other hand it is also not realistic to assume that the spheres are fully penetrable as in the Boolean model (Stoyan et al. (1995)), obviously the spheres are not fully penetrable: the pcf's of the section disk center systems in Figure 3 indicate some form of repulsion between the centers: $g_A(r)$ is smaller than 1 for small r, while for a structure with full overlaps a pcf is expected which is equal to unity for all r. Therefore the so-called "cherry-pit" model or concentric-shell model seems to be the right model. This model starts with a random system of hard (inpenetratable) spheres (the "pits") and then each of them is surrounded by a soft shell. This model is frequently used in materials science, see Torquato (2002), Sahimi (2003) and Bentz et al. (1999).

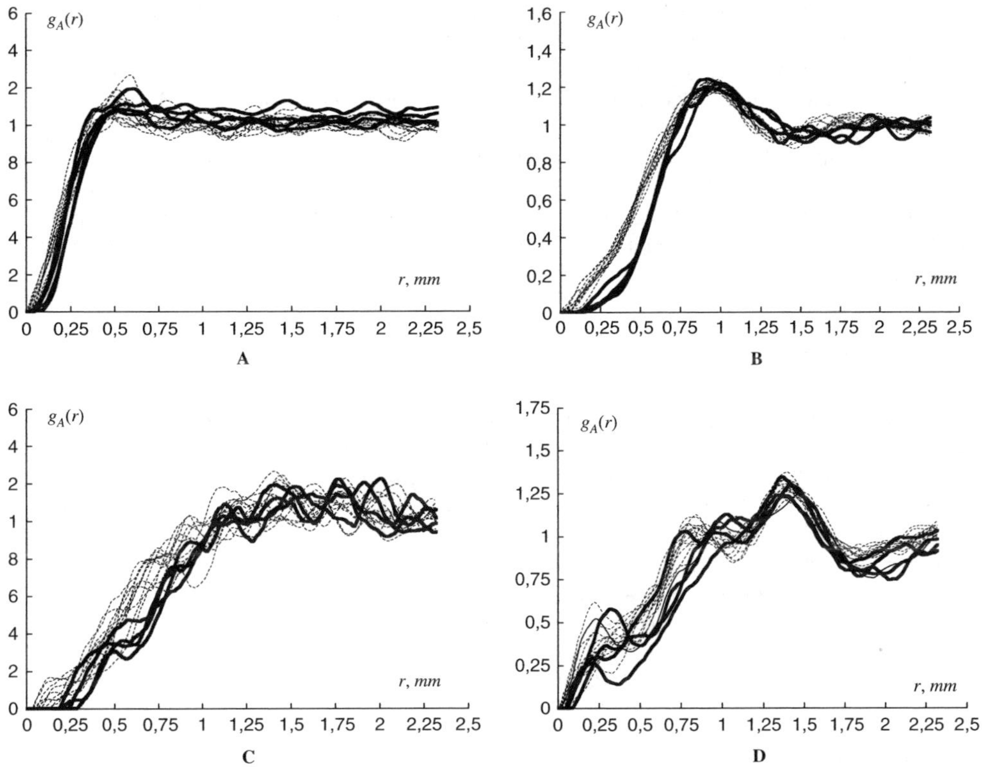

Figure 3. Empirical and theoretical pair correlation functions $g_A(r)$ for the four samples A–D. bold – empirical, from cube surface; normal – theoretical, from planar sections through simulated "cherry-pit" models.

While the verbal description of the "cherry-pit" model is simple, its mathematical treatment is extremely difficult. Simulation is the only way to obtain precise model characteristics. This is even the case for the volume fraction of the system of "cherries".

The "cherry-pit" model can be easily simulated if an algorithm for the simulation of the system of pits is available. When such a structure is given, then simply the radii have to be enlarged to the size of the "cherries" without respect to possible interpenetrations. The method for simulating systems of "pits" used in this paper is the force-based algorithm for the generation of random packings of hard spheres of random radii as described in Bezrukov et al. (2001), Bezrukov et al. (2002).

In the simulation of the "cherry-pit" model, the authors used the radius probability density function $\delta_V(r)$ as the probability density function of the "cherry" radii. The radii of the "pits" R_i^{Pit} are of course smaller than those of the "cherry" radii R_i and taken as

$$R_i^{Pit} = fR_i, \tag{5}$$

where f is a reduction factor. This f is the same for all spheres of a sample, but varies from sample to sample; it is a further model parameter. It is chosen so that the resulting "cherry-pit" model has the right porosity P_{Air}. Practically, f was obtained by iteration. With some start value of f, a random dense packing wit radii R_i^{Pit} and sphere density N_V was simulated. Then the spheres were extended to the true values R_i and the corresponding porosity was determined. If it is larger than P_{Air}, f was reduced, if smaller, f was increased, and so on.

Of course it is necessary to test the goodness-of-fit of the "cherry-pit" model. The natural way is to compare planar characteristics of this model with their counterparts in the concrete samples. Since the theoretical functions cannot be obtained analytically or numerically, the only way is to simulate the "cherry-pit" model and then to determine the functions statistically. As test characteristics we choosed the pcf's $g_A(r)$ corresponding to planar sections. We compared the experimental pcf's for the centers of constructed disks on the surfaces of sampling cubes as in Figure 1 and analogous functions for the section circle centers of simulated "cherry-pit" models. Since the stereological approach guarantees that there is the right relationship between the diameters of section disks and spheres in space, we did not consider them in the tests. Figure 3 shows both empirical pcf's and theoretical ones. The agreement is good, the differences between the samples and the corresponding cherry-pit models, which are characterized by different parameters, are reflected by different shapes of the pcf's.

5 CONCLUSIONS

This paper presents a stochastic model for the geometrical structure of the system of artificial air pores in AAC. This model has two ingredients, the pore radius distribution and the "cherry-pit" model. The form of the radius distribution is rather general; three components, two exponential, one Gaussian, offer a great variety of possible distributions. Different weights p_k can lead to qualitatively different distributions; a simple Gaussian distribution seems to be not appropriate. So the authors hope that the suggested radius distribution can be used for many types of AAC.

The "cherry-pit" model is a plausible stochastic structure model, which can be easily simulated. It is more realistic than the hard-sphere packing model of He (1988) and any form of a Boolean model with fully penetrable spheres. It will serve as a base for the investigation of relationships between microstructure and macroscopic material properties of AAC, see Kadashevich & Stoyan (2005).

REFERENCES

Bentz, D.P., Garboczi, E.J. & Snyder, K.A. 1999. A hard core/soft shell microstructural model for studying percolation and transport, In: *Three-Dimensional Composite Media*, NIST Internal Report 6265, Gaithersburg, Maryland.

Bezrukov, A., Bargiel, M. & Stoyan, D. 2001. Spatial statistics for simulated packings of spheres. *Image Anal. Stereol.* 20: 203–206.

Bezrukov, A., Bargiel, M. & Stoyan, D. 2002. Statistical analysis of simulated random packings of spheres. *Part. Part. Syst. Charact.* 19: 111–118.

Grimer, F.J. & Brewer, R.S. 1965. The within-cake variation of autoclaved aerated concrete. In: *Proc. Symposium Autoclaved Calcium Silicate Buildings Products, London.*

He, Tianchun, 1988. *Festigkeits- und Verformungsverhalten von Porenbeton.* PhD thesis, TU Berg-akademie Freiberg, Germany.

Hubalkova, J. & Stoyan, D. 2003. On a qualitative relationship between degree of inhomogeneity and cold crushing strength of refractory castables. *Cement and Concrete Research* 33: 747–753.

Kadashevich, I. & Stoyan, D. 2005. Micromechanical analysis of AAC. This volume.

Kopp-Schneider, A. 2003. Biostatistical evaluation of focal hepatic preneoplaria. *Toxicolog. Pathol.* 31: 121–125.

Ohser, J. & Mücklich, F. 2000. *Statistical Analysis of Microstructures in Materials Science.* Weinheim: Wiley.

Sahimi, M. 2003. *Heterogeneous Materials.* Heidelberg: Springer.

Schlegel, E. & Hums, D. 2002. Porenbeton – die Entwicklung von Gefüge und Eigenschaften eines Werkstoffes. *Mauerwerk* 6: 82–88.

Schwiete, E. & Ludwig, U. 1965. The porosity and the gas permeability of aerated concrete. In: *Proc. Symposium Autoclaved Calcium Silicate Buildings Products, London.*

Snyder, K.A., Natesaiyer, K. & Hover, K. 2001. Stereological and Statistical Properties of Entrained Voids in Concrete: A Mathematical Basis for Air Void System Characterization, In: S. Mindess, J. Skalny, (Eds.) *Proc. Materials Science of Concrete VI*, 129–214 American Ceramic Society.

Stoyan, D., Kendall, W.S. & Mecke, J. 1995. *Stochastic Geometry and its Applications.* Chichester: Wiley.

Torquato, S. 2002. *Random Heterogeneous Materials, Microstructure and Macroscopic Properties.* New York: Springer.

Autoclaved Aerated Concrete – Limbachiya and Roberts (eds)
© 2005 Taylor & Francis Group, London, ISBN 0 415 38356 0

Orientation effects of shear wall on RC frames

Santosh Kumar Rai & J. Prasad
Civil Engineering Department, IIT-Roorkee, Roorkee (U.A), India

ABSTRACT: Shear wall-frame system is the most commonly used building systems in concrete buildings. Locating the shear wall in the frame as well as using the right section of it can do further improvement in its performance by improving the rigidity and stability of entire structure. In the present study, an effort has been made to investigate, the effect of location and size of the shear wall in a building frame, forming part of a public building, subjected to gravity loading and lateral loads. A computer program based on stiffness matrix method has been used to analyse the frame. Various combinations of shear wall and frames have been compared for two parameters; drift and stresses in columns. Total drift and interstorey drift have been taken for drift analysis and comparison. To take into account the combined effect of axial-force and moment, steel requirement for the various cases in different columns has been compared.

1 INTRODUCTION

Shear wall systems are one of the most commonly used lateral load resisting systems in high-rise buildings [Smith et. al, 1984]. This paper presents a study of two-dimensional elastic behavior of a multistoreyed concrete building, having combination of rigid frame and shear wall. The building is subjected to lateral loads originating from wind or earthquake together with gravity load. In present study the emphasis is given on the drift pattern of the various shear wall frame combination and to find out the salient features of their behavior in this regard. As the height of the structures increases, effect of lateral load becomes more and more prominent and additional structural materials are required to resist it. In a shear wall-frame structure, the design can be done in a manner so that; the shear wall takes the entire lateral load and the frame can be designed for gravity loads only. However, the wall and frame together compliment each other synergistically and their performance in resisting the loads is better than the sum of their individual performances [Rosman, 1966]. Right placement and size of the shear wall in this system further improves its performance.

One simple parameter that offers an estimate of the lateral stiffness of a building is the drift index, defined as the ratio of the maximum deflection at the top of the building to the total height. In addition, the corresponding value of a single storey height, the inter-storey drift gives a measure of possible localized excessive deformation. Design drift index limits that have been used in different countries range from 0.001 to 0.005. As the height of the building increases, drift index should be decreased to the lower end of the range to keep the top storey deflection to a suitably lower level.

2 METHOD OF ANALYSIS

The purpose of the analysis is to check the adequacy of a design, when subjected to specified gravitational and lateral loads, from the safety and serviceability point of view. The performance of the structure must be appraised and checks made on whether the stress levels and lateral deflections fall within prescribed limits of established design criteria. In this analysis direct stiffness method has been used in computer program. The basic steps in this method are as follows.

i. Compilation and preparation of basic data of the idealized structure giving the position of the nodal points, elements orientation and properties, and support conditions.
ii. Identification of the local coordinates for the elements and calculation of stiffness matrix.
iii. Determination of transformation matrix for each element.
iv. Calculation of the element stiffness matrix in the global coordinate system $[K] = \{R\}^T [K]^* \{R\}$.
v. Evaluation of the stiffness matrix of the complete structure.
vi. Equivalent joint loads are obtained to form the nodal load vectors for the structure.
vii. Boundary conditions are incorporated and obtained the nodal deformations.
viii. Knowing the nodal deformations, stress resultants (internal forces) in various elements are obtained.

2.1 *Individual member stiffness*

The stiffness matrix of a beam/column element as shown in Figure 1, in local coordinate system, can be written as

$$\begin{bmatrix} S_1 \\ S_2 \\ S_3 \\ S_4 \\ S_5 \\ S_6 \end{bmatrix} = \frac{EI}{l^3} \begin{bmatrix} \frac{Al^2}{I} & 0 & 0 & -\frac{Al^2}{I} & 0 & 0 \\ 0 & 12 & 6l & 0 & -12 & 6l \\ 0 & 6l & 4l^2 & 0 & -6l & 2l^2 \\ -\frac{Al^2}{I} & 0 & 0 & \frac{Al^2}{I} & 0 & 0 \\ 0 & -12 & -6l & 0 & 12 & -6l \\ 0 & 6l & 2l^2 & 0 & -6l & 4l^2 \end{bmatrix} \begin{bmatrix} u_1 \\ u_2 \\ u_3 \\ u_4 \\ u_5 \\ u_6 \end{bmatrix}$$

2.2 *Stiffness matrix of shear wall element*

In the shear wall analysis, the shear deformations becomes very impartment therefore, effect of shear deformation is also included into the stiffness matrix of the shear wall, which is given below.

$$\begin{bmatrix} S_1 \\ S_2 \\ S_3 \\ S_4 \\ S_5 \\ S_6 \end{bmatrix} \begin{bmatrix} \frac{EA}{h} & 0 & 0 & -\frac{EA}{h} & 0 & 0 \\ 0 & \frac{12EI}{(1+\alpha)h^3} & \frac{6EI}{(1+\alpha)h^2} & 0 & -\frac{12EI}{(1+\alpha)h^3} & \frac{6EI}{(1+\alpha)h^2} \\ 0 & \frac{6EI}{(1+\alpha)h^2} & \frac{(4+\alpha)EI}{(1+\alpha)h} & 0 & -\frac{6EI}{(1+\alpha)h^2} & \frac{(2-\alpha)EI}{(1+\alpha)h} \\ -\frac{EA}{h} & 0 & 0 & \frac{EA}{h} & 0 & 0 \\ 0 & -\frac{12EI}{(1+\alpha)h^3} & -\frac{6EI}{(1+\alpha)h^2} & 0 & \frac{12EI}{(1+\alpha)h^3} & -\frac{6EI}{(1+\alpha)h^2} \\ 0 & \frac{6EI}{(1+\alpha)h^2} & \frac{(2-\alpha)EI}{(1+\alpha)h} & 0 & -\frac{6EI}{(1+\alpha)h^2} & \frac{(4+\alpha)EI}{(1+\alpha)h} \end{bmatrix} \begin{bmatrix} u_1 \\ u_2 \\ u_3 \\ u_4 \\ u_5 \\ u_6 \end{bmatrix} =$$

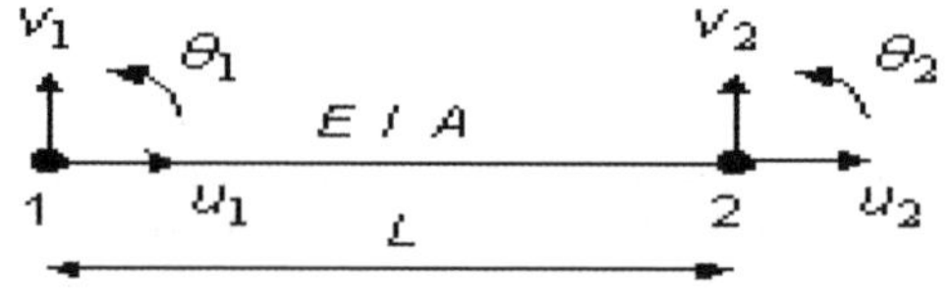

Figure 1. Beam/Column element.

where, $\alpha = 12EI/h^2GAr$ and the load-deflection equation in local coordinate system is written as

$$\{F\}^*=[K]^*\{\delta\}^* \tag{1}$$

where, $\{F\}^*$ is load vector and $\{\delta\}^*$ Displacement vector. It can be further written as for an element whose nodes are I and J.

$$\begin{bmatrix} S^I \\ F^I \\ M^I \\ S^J \\ F^J \\ M^J \end{bmatrix} = \frac{EI}{l^3} \begin{bmatrix} \frac{Al^2}{I} & 0 & 0 & -\frac{Al^2}{I} & 0 & 0 \\ 0 & 12 & 6l & 0 & -12 & 6l \\ 0 & 6l & 4l^2 & 0 & -6l & 2l^2 \\ -\frac{Al^2}{I} & 0 & 0 & -\frac{Al^2}{I} & 0 & 0 \\ 0 & -12 & -6l & 0 & 12 & -6l \\ 0 & 6l & 2l^2 & 0 & -6l & 4l^2 \end{bmatrix} \begin{bmatrix} \delta x^I \\ \delta y^I \\ \theta^I \\ \delta x^J \\ \delta y^J \\ \theta^J \end{bmatrix}$$

Transformation matrix relating member end deformations in local coordinate system to the deformation of frame in global co-ordinate system is

$$[R] = \begin{bmatrix} \lambda & \mu & 0 & 0 & 0 & 0 \\ -\mu & \lambda & 0 & 0 & 0 & 0 \\ 0 & 0 & 1 & 0 & 0 & 0 \\ 0 & 0 & 0 & \lambda & \mu & 0 \\ 0 & 0 & 0 & -\mu & \lambda & 0 \\ 0 & 0 & 0 & 0 & 0 & 1 \end{bmatrix}$$

where, $\lambda = \cos\alpha$ and $\mu = \sin\alpha$

Similarly, deformation transformation matrix can be written as

$$\{\delta\}^*=[R]\{\delta\} \tag{2}$$

And load transformation matrix as

$$\{F\}^*=[R]^*\{F\} \tag{3}$$

Substituting equation (2) and (3) into equation (1), one obtains

$$[T]\{F\}=[K]^*\{R\}\{\delta\} \tag{4}$$

Pre-multiplying equation (4) by $[R]^{-1}$, one obtains

$$\{F\}=[R]^{-1}[K]^*\{R\}\{\delta\} \tag{5}$$

Since $[R]^{-1} = [R]^T$
Therefore,

$$\{F\}=[R]^T[K]^*\{R\}\{\delta\} \tag{6}$$

Therefore, if the stiffness matrix in the local coordinate system is denoted by [K]* and stiffness matrix in the global coordinate system by [K], then [K] can be found by standard matrix equation as given below

$$[K]=[R]^T[K]^*\{R\} \tag{7}$$

Having known the stiffness matrix of the individual elements, all the members constituting the structure are assembled. Nodal stiffness matrix of complete discritised structure is thus obtained. After obtaining the nodal stiffness matrix and equivalent forces acting at the nodes of the discritised structure the governing equation is given as follows.

$$\{F\} = [K]\{\delta\} \tag{8}$$

The whole structure can be solved for nodal displacements subjected to support conditions.

3 ANALYTICAL INVESTIGATION

Here details about the structures considered, its properties, salient criterion for its selection and load cases used for the purpose of the analysis have been discussed. Analysis is done by considering each frame as independent substructure. An exact 3-D analysis is required for only a few special buildings. For the most of the buildings approximations can be introduced which will considerably reduce the computational efforts. Floors are assumed to be rigid in their own plane. The lateral loads are assumed to act at floor levels. Therefore, these loads are transferred to the columns and shear wall elements through the rigid floor diaphragms. This results in three displacement degree of freedom at each floor level-translations in X and Y directions and θ, the rotation about the vertical axis.

3.1 *Modeling of building*

For the purpose of preparing input, the building is separated in to a system of planer frames. The properties of a frame or shear wall are given with respect to its local stiffness direction. Floor level and column centerline are the basic reference lines used in the frame description. All lateral loads are applied at the floor levels and act on the complete structure. Plan of the building is an office building consisting of four bays as shown in Figure 2. Bay-1 and bay-4 is 4.0 m wide, bay-2 is 6.0 m and bay-3 is 5.0 m wide. Columns spacing in the lateral direction are kept 4.0 m. The building height is 25 storeys. The grade of concrete is M25 and steel is of grade Fe415. Other details about the building frames are given in Table 1.

To these initial structures, shear wall has been added in different bays and with varying sizes to find out the effect of the size and location of shear walls. Arrangements of shear wall in various bays of taken frame considered for analysis are shown in Figure 2, in the form of plan and Figure 3, in form of elevation arrangements is briefly described below.

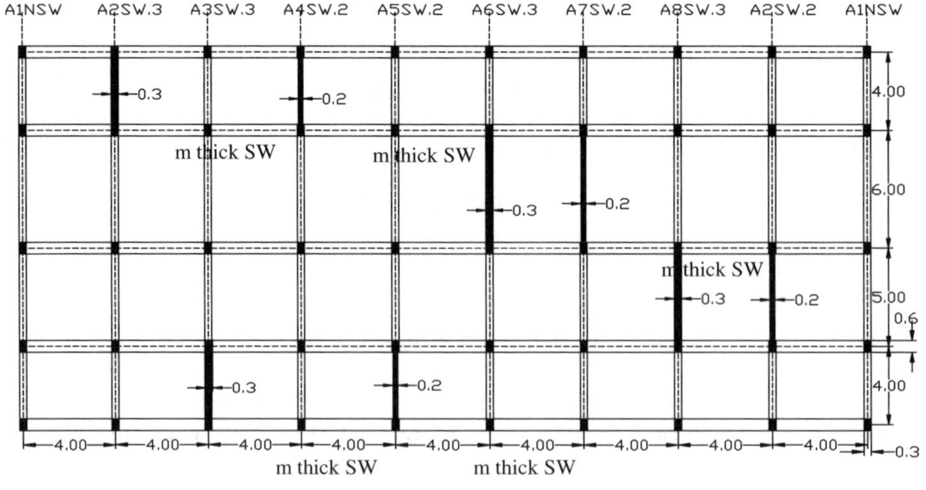

Figure 2. Arrangement of shear wall in the frame in X-Y direction.

Table 1. Description of building and frame.

Sl. No.	Particulars	Details/Values
1.	Number of storeys	25
2.	Number of bay in X-direction	4
3.	Spacing of frame in Z-direction	4.0 m
4.	Ground storey height	4.0 m
5.	Other storey height	3.5 m
6.	Size of columns	0.4 m × 0.6 m
7.	Slab thickness	150 mm
8.	Floor finishing	1.25 kN/mm^2
9.	Live Load	2 kN/mm^2
10.	Modulus of Elasticity of concrete	25000 MPa
11.	Poisson's Ratio	0.18
12.	Width of beam	300 mm

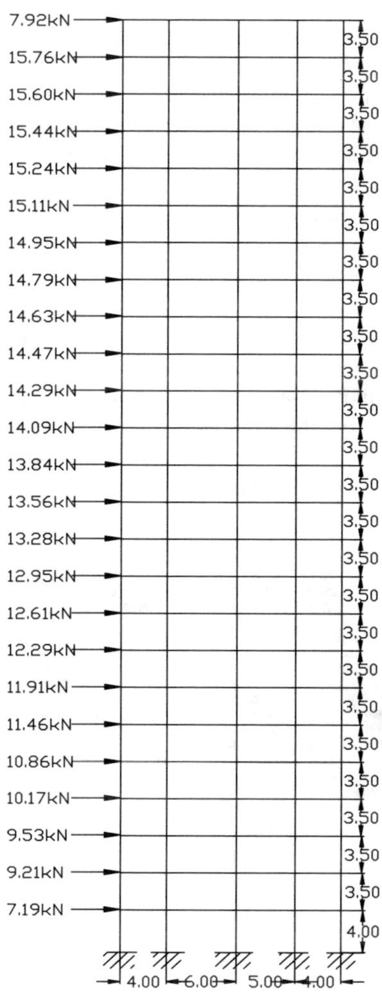

Figure 3. Front elevation of building in X-direction. *Note: All the dimensions given in the Figures 2 and 3 in meter.*

In arrangement A1NSW, only frame without shear wall has been considered for analysis. All the column sections are kept of 300 mm × 600 mm. The width of all the beams is kept 300 mm and the overall depth of the slab is kept 120 mm. The rib depth of bay-1 and bay-4 beam is kept 200 mm and bay-2 and bay-3 is kept 300 mm. To see the effect of location of shear wall, shear wall of thickness 300 mm is provided in arrangement A2SW.3 and A3SW.3. In case of A4SW.2 and A5SW.2 the thickness of shear wall is reduced to see the thickness effect of shear wall. In A6SW.3 and A8SW.3. the width of shear wall is changed and the thickness is kept at 300 mm. In A7SW.2 and A9SW.2 the width of shear wall is unchanged but the thickness is reduced to 200 mm.

3.2 *Identification of frame*

In all, nine cases, A1 to A9, are taken for analysis as shown in Figure 2. The wind load and gravity load are taken for analysis. SW represents the shear wall and the number following SW represents its thickness.

3.3 *Loads*

Analysis of the two-dimensional structures has been carried out for the combination of the gravity loads (dead load and live load) and lateral load (wind load). The lateral load is shown in Figure 3. The live load is taken as 3 kN/m^2. The wind load is calculated as per IS: 875 (3) -1987. The load combination is taken as per IS: 456 -2000. Based on the presumption of using limit state method of design, the load combination is formed for the analysis as per Table-18 of IS: 456-2000. The following load combinations were adopted.

Load: 1.2 (Self Weight +DL+LL+WL), 1.5 (Self Weight +DL+LL)

4 RESULT ANALYSIS

Computer program has option of providing the results of displacements, bending moments, axial-forces and shear-forces in columns, beams and shear wall has been used for analysis. Freedom of obtaining the results due to individual loadings or combination of more than one load case is incorporated in program. To find out the combined effect of axial load and bending moment in the columns, percentage steel requirement has been adopted as the criteria and the results for bottom columns has been presented for comparison.

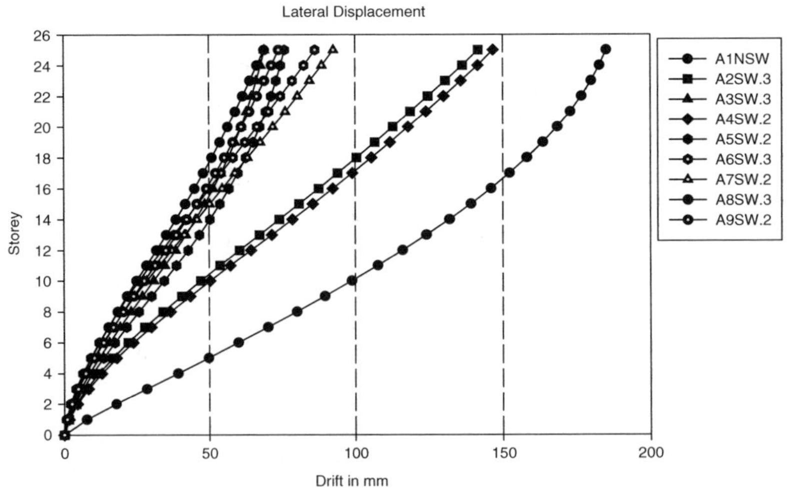

Figure 4. Lateral displacement of frames.

Lateral displacement of the columns has been presented in two ways; first is total drift, second is interstorey drift. For each case total drift and interstorey drift (ISD) is given in graphical form as shown in the Figure 4 and 5.

4.1 *Drift analysis*

The building frame is 25 storey in height with different arrangements of shear-wall and frame has been modeled and analysed as per the details of the analytical investigation and other particulars discussed above. The analytical results obtained with the help of computer program have been presented in the form of graphs, indicating the trends and patterns of the variables like total drift, interstorey drift and also in the form of percentage steel requirement in the bottom columns. Since all these columns are subjected to axial force as well as moments and the combined effect of axial force and the moments is of more of importance, therefore their combined effect is presented in the graphical form as the percentage steel requirement to resist them.

4.1.1 *Total drift*

The results are shown in Figure 4. Various cases are discussed here.

In this the 25-storeyed building frames subjected to wind load and gravity loads is taken for investigation. Maximum sway is there in the building frame A1NSW that is without the shear wall, having total drift 185.07 mm. In this the maximum total drift is for case A1NSW (total drift = 185.07 mm), and minimum total drift is for case A8SW.3 (total drift = 68.98 mm). However total drift for case A4SW.3 (total drift = 69.11 mm), A6SW.2 (total drift = 75.96 mm), A9SW.2 (total drift = 73.93 mm) show qualitative change from the corresponding cases in A-SW. From Figure 4 we can find that, as the thickness of the shear wall decreases, total drift increases, though very marginally. This is due to the fact that wind load is not dependent on the weight of the structure. Hence, loading remains same but there is decrease in the stiffness and increase in the drift.

Provision of shear wall of thickness 300 mm in case A2SW.3 (total drift = 141.57 mm) reduces the sway. However unlike the A4SW.2, when the thickness of the shear wall is reduced to 200 mm, (total drift = 146.71 mm), sway increases. For the case A3SW.3 (total drift = 69.11 mm), A5SW.3 (total drift = 75.96 mm) when the shear wall is provided in the leeward side, reduction of the sway is quite substantial as compared to A2SW.3 and A5SW.2 with the same cross section.

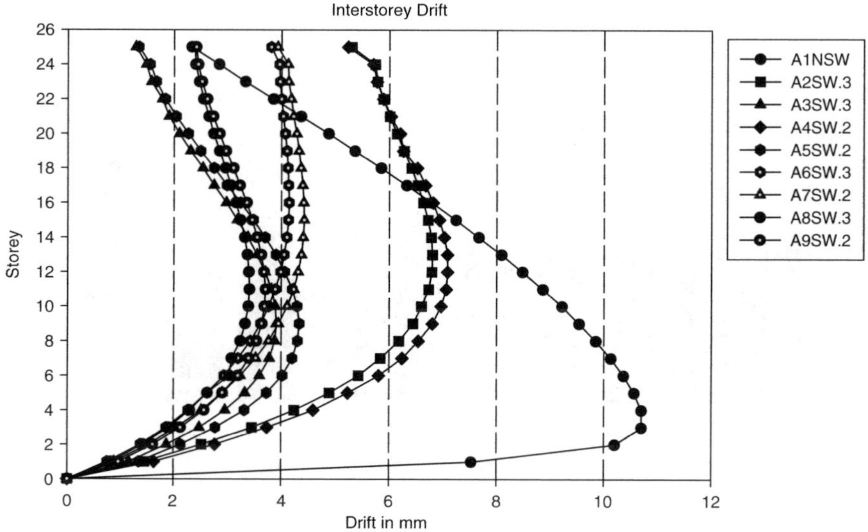

Figure 5. Interstorey drift of different arrangements.

In general the building frames taken for investigation, it is seen that best performance in respect to drift control is obtained for the cases in which the shear wall is placed away from the direction of the force. Reason for this is the shift in the neutral axis for the frame as a whole. Neutral axis in these cases is shifting towards the shear wall. Thus, area under tensile force in columns, due to lateral loads is increasing. Thus, in a larger portion of the structure, gravity loading and lateral loading is counteracting each other, which is resulting in the lesser drift.

4.1.2 *Interstorey drift analysis*

The interstorey drift is a very important parameter in analysis and design of buildings. If the interstorey drift (ISD) values at each floor level reach their maximum allowable limits, then the roof displacement will reach undesirable values. The storey drift in any storey as per IS: 1893-2002, due to minimum specified design lateral force shall not exceed 0.004 h. where h = storey height. Figure 5 shows the variation of the ISD with respect to height, details some interesting features. Maximum ISD for buildings without shear walls (A1NSW) is in the storey just above the ground level (between 2nd and 3rd storeys). It is because; the frame structure deflects in a shear configuration where rate of change of deflection goes on reducing with height.

For wall-frame structures maximum ISD is found in the range of 30 to 50% of the height of the structure. Deflection pattern is of flexure shape at lower heights in which rate of deflection increases and follows the shear configuration at higher heights. This means that the shear wall governs the deflection pattern at lower heights and at higher heights frame governs the deflected shape. Details of the various cases are discussed below.

In Figure 5, the maximum ISD index is for the case AlNSW (ISD = 0.31%) that is the structures without shear wall. A2SW.3 (ISD = 0.19%) show the ISD index for shear wall provided in bay-1 of thickness 0.3 m, but when thickness of wall reduced to 0.2 m for the same case A4SW.2 (ISD = 0.20%), the ISD index is increased. However, the best results are obtained when the shear wall thickness of 0.3 is provided in bay-4. In this case (A3SW.3) the ISD index is reduced to (0.11%).

4.2 *Forces in columns*

Figure 6 shows the percentage steel requirement in the bottom columns due to combined action of the bending moments and axial forces in the columns for various cases. Steel requirement is worked

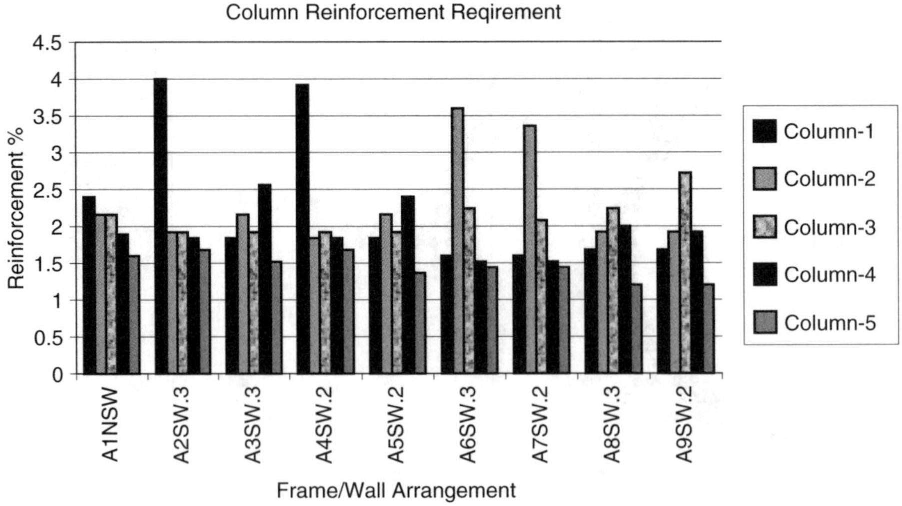

Figure 6. Column reinforcement percentages for various cases.

out to see whether the assumed sections are practically workable or not as well as to compare the cumulative effect of bending moments and axial forces on the columns.

Following salient features are observed.

(a) In general C1 is the most stressed column with maximum steel requirement of 4% in cases A2SW.3 and A4SW.2, which may be difficult to accommodate in the given section. However use of higher-grade materials such as M30 grade concrete and Fe500 grade steel, the percentage of steel can be brought within practicable range.
(b) Provision of shear wall is very useful for resisting the bending moments caused due to the lateral loads. Bending moments in other columns are being reduced very significantly due to the shear wall.
(c) Provision of shear wall in the bay-1 or bay-4 is shifting the neutral axis towards these bays. It is resulting in excessive increase or decrease in axial forces caused due to the lateral loads. In other cases due to the presence of shear wall, the reinforcement required in the column is increased, but reduced in all other columns.
(d) Placement of shear wall in the bay-3 and bay-4 giving the optimum steel requirements for the different cases. Hence, it seems to be the most suitable system for resisting the forces.

5 CONCLUSIONS

A 25 storeys building has been analysed. A computer program based on displacement matrix approach has been used for the purpose of the analysis. Various cases thus analysed show significant results. Performances of various systems with regard to drift control as well as to resist the forces have been evaluated. Inferences obtained from them are summarized below.

(a) Shear wall when placed in the bay away from the face where the lateral load is acting, is most effective in the controlling the drift. This is because, as major portion of the loading effects due to gravity and lateral force is counteracting each other. This can be effectively used in case of wind load where most serious prevailing wind direction is known e.g. in coastal areas.
(b) For wall frame structures interstorey drift is maximum between 30 to 50% height of the structure. Therefore to further control the drift this 10–20% should be further stiffened using the shear wall panels.
(c) Visco-elastic dampers are a mechanical method of controlling the drift. However these are very expensive. For wall frame structures, these dampers could be provided in that portion only where interstorey drift is more.
(d) Thickness of the shear wall has very little role to play whereas change in the width changes the performance appreciably.

REFERENCES

Smith, S.B. and Girgis, A. 1984. Simple analogous frames for shear wall analysis. Journal of Structural Division, A.S.C.E. Vol. 110, 2655–2666.

Rosman, R.1966. Laterally loaded systems consisting of wall and frames. Proc. of Sympo. on tall buildings. Southampton. University of Southampton. pp. 273–289.

Tall Building criteria and loading.1980. Vol.CL. Monograph on Planning and Design of Tall buildings. ASCE. pp 252–390.

Gere, J.M. and Weaver, W.J. 1965. Analysis of framed Structures, New York: Van Nostrand Rainhold.

Ghali, A. and Neville, A.M. 1997. Structural Analysis: A unified classical and matrix approach, London: E&FN SPON.

IS: 456-2000. Plain and Reinforced Concrete-Code of Practice, New Delhi: Bureau of Indian Standards.

IS: 875-1987 (Part 1,2,3). Design Loads (Other than earthquake) for Buildings and Structures-Code of Practice, New Delhi: Bureau of Indian Standards.

IS: 1893-2002. Criteria for Earthquake resistant design of structures (Part-1): General Provision and Buildings, New Delhi: Bureau of Indian Standards.

Stafford Smith, B., and Coull, A. 1991.Tall Building Structures: Analysis and Design, New York: John Willey and sons, Inc.

Przemieniecki, J.S. 1968. Theory of Matrix Structural Analysis, London: McGraw-Hill book Company.

Taranath, B. S. 1988. Structural analysis and design of tall buildings. McGraw-Hill Book company, United States of America.

Autoclaved Aerated Concrete – Limbachiya and Roberts (eds)
© 2005 Taylor & Francis Group, London, ISBN 0 415 38356 0

Moisture content, absorption and movement of low density Aircrete

A. Ahmed, A.N. Fried, M.C. Limbachiya & J.J. Roberts
Concrete and masonry Group, Kingston University, London

ABSTRACT: The moisture properties of low density Aircrete wallettes (compressive strength of 2.8 and 2 N/mm^2) are verified. This entails determining the moisture content after immediately removing from the pallet, measuring the water absorption rate and finally evaluating the shrinkage on drying. The tests were carried out in accordance with the relevant British and European Standards. Low density Aircrete has a low water absorption rate with minimal shrinkage on drying from a moistened state.

1 INTRODUCTION

Aircrete originated from Scandinavia in the 1950's (H+H Celcon Ltd 2003) and is used extensively in the United Kingdom. Aircrete is also known commercially as AAC (Autoclaved Aerated Concrete), Celcon, Durox, Thermalite and Topblock. Aircrete blocks provide both the thermal insulation and mechanical integrity expected from typical UK wall construction (Aircrete Bureau 2003a).

Aircrete is produced by mixing cementitious materials, cement and/or pulverised fuel ash (PFA), lime, sand, water and aluminium powder. The final process involves autoclaving for approximately 10 hours at high temperature and pressure (Mitsuda and Kiribayashi 1992, Isu and Mitsuda 1992, Pospisil 1992). Hence, the material is also known as Autoclaved Aerated Concrete. Aircrete is comprised of 60% to 85% of air by volume (70–85% for low density Aircrete). The solid material part is a crystalline binder, which is called Tobermorite by mineralogists. It is Tobermorite, which provides the high compressive strength of Aircrete in spite of the high proportion of pores in this construction material. This is why low density Aircrete is sufficiently strong for the construction of dwellings in spite of its considerably high air content.

The cellular (porous) structure (Tada 1992, Jacobs and Mayer 1992, Schober 1992) of the material ensures a lightweight construction. During installation, most Aircrete blocks can be lifted with one hand providing significant productivity advantages. The higher porosity of Low Density Aircrete ensures that the material is extremely lightweight (density $\leq 450\,kg/m^3$), thus, it is even easier to handle in comparison to medium and high density Aircrete. As a result transportation costs would be reduced and furthermore, houses would be built much more quickly.

1.1 *Sound/acoustic resistance*

The combination of the internal structure and the stiffness characteristics of Aircrete enable sound reduction performance for walls and partitions often superior to other types of masonry (Dubral 1992, Mitsuda and Kiribayashi 1992, Isu and Mitsuda 1992).

Sound absorption is a property relevant to particular applications. When exposed to sound, the aerated internal structure of Aircrete provides good sound absorption properties. Therefore, lower density Aircrete should impart superior sound insulation properties.

1.2 *Moisture movement and resistance to freezing in Aircrete*

Moisture movement through porous building materials is a very complex process (Hasegawa 1992, Wittmann 1993b) and for practical predictions simplifying assumptions have to be introduced.

There are at least three different origins of water in Aircrete. Immediately after autoclaving Aircrete contains typically about 30% water by weight of the dry material. This excess water is lost under normal conditions to the surrounding air after a few years (Dubral 1992). If the relative humidity of the surrounding air increases temporarily, Aircrete will take up water again by absorption and capillary condensation. If the surface of a structural element is in contact with liquid water the material absorbs water quickly by capillary suction (Mitsuda and Kiribayashi 1992, Hasegawa 1992, Wittmann 1993b). Although Low Density Aircrete has a very high proportion of air, the pores are fine and are not interconnected, therefore, the material offers good resistance to moisture penetration. Vapour resistivity of Aircrete is approximately 60 MNs/gm.

The resistance to freezing of a construction material is determined by its pore size distribution and, in particular, the percentage, size and shape of capillary pores and the mechanical strength of the inner pore walls (Wittmann 1993a, Senbu and Kamada 1992, Hama and Kamada 1992).

If the pores of the material become filled with water, which then freezes, the ice which has a volume 9% greater than water will cause pressure on the pore walls. When the tensile strength of the wall material is exceeded, cracking occurs. If the pores are filled with water to a critical degree, and if there is repeated freeze/thaw cycling, the whole structure may eventually be destroyed.

Aircrete possesses good resistance to freezing, which is proved by unrendered buildings, situated in areas where frequent freeze/thaw cycles occur, remaining undamaged. The reason for the good resistance is that the included spherical pores are almost all closed, meaning the material has comparatively low capillary suction and therefore the moisture content does not normally reach the critical degree. With Low Density Aircrete, the greater free volume of the material is better equipped for dealing with the pressures caused by freezing of water. Furthermore, as discussed earlier, as the pores are not interconnected, this radically reduces the possibility of water absorption. The high freeze thaw resistance in essence is due to the aerated internal structure of the material. The resistance to frost is superior to that of many stronger denser masonry materials although the degree of resistance is to some extent dependent on strength and density.

2 EXPERIMENTAL WORK

2.1 *Moisture content (By mass)*

Tests were carried out in accordance with BS EN 772-10 [British Standards a]. In essence, during this test, after drying to constant mass, the moisture content is calculated as the ratio of the loss of mass during drying to the mass after drying. According to BS EN 772-10 [British Standards a], a minimum of 6 representative portions from at least three units must be tested. The moisture content should also be measured immediately after removing the blocks from the pallets.

2.2 *Water absorption*

The method for determining the water absorption property is carried out in accordance with BS EN 772-11 [British Standards b]. After drying to constant mass a face of the Aircrete block is immersed in water for a specific period of time, as BS EN 771-4 [British Standards c] specifies that the coefficient of water absorption should be stated at 10, 30 and 90 minutes, tests were carried out at these immersion times. The water absorption of face of the unit exposed is measured. Tests were carried out on both 2 and 2.8 N/mm^2 standard Aircrete blocks. The test set – up is shown in Figure 1. Testing is undertaken on 6 specimens of each Aircrete material.

2.3 *Moisture movement*

The moisture movement of 2.0 and 2.8 N/mm^2 Aircrete is measured in accordance with BS EN 680 [British Standards d]. Prismatic test specimens are cut from prefabricated components (blocks) and, if necessary, moistened by underwater storage until their moisture content is at least 30% by mass.

Subsequently, the test specimens are stored in air under specified conditions until a moisture content of 4% by mass has been reached, and changes of length and mass are determined at appropriate intervals.

Finally, the test specimens are dried to constant mass to enable the calculation of the moisture content from the mass of the test specimen recorded at each measuring date. The samples and apparatus used are shown in Figure 2. In order to avoid faulty readings due to the presence of dirt, the DEMEC points were wiped carefully before each measurement. Points were attached to all four sides of prisms, which had a cross – section of 40×40 mm and a length of 160 mm. After the DEMEC points dried the first reading of the gauge length and the first measurement of mass, m_o, of the test specimens was made. The specimens were then moistened by underwater storage to ensure their moisture content was at least 30% by mass. Readings of the gauge length and accompanying measurements of the mass of the test specimens were executed at regular time intervals until the moisture content was less than 4% by mass; specimens were then dried to a constant mass.

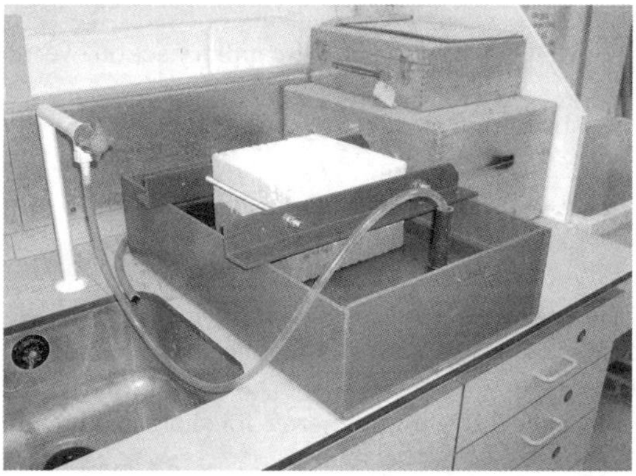

Figure 1. Test set-up for water absorption test. The block is immersed in water to a depth of 5 mm $\pm$ 1 mm. Continuous water flow ensures constant water level.

Figure 2. Test apparatus and specimens with DEMEC points attached to the surface used for the evaluation of moisture movement.

3 RESULTS

3.1 *Moisture content by mass*

Table 1 gives results of the moisture content of Aircrete specimens as delivered obtained for 2 and 2.8 N/mm^2 Aircrete blocks. The moisture content is calculated as follows:

$$W_s = \frac{M_{o,s} - M_{dry,s}}{M_{dry,s}} \times 100 \quad (1)$$

Where W_s = moisture content (%); $M_{o,s}$ = mass of specimen before drying, in grams; $M_{dry,s}$ = mass of specimen after drying, in grams.

The average moisture content therefore for both materials are:

(a) 2 N/mm^2 Aircrete = 13.7%
(b) 2.8 N/mm^2 Aircrete = 13.6%

3.2 *Water absorption*

Results of the coefficient of water absorption due to capillary action over short term exposure times are given in Tables 2 and 3.

The coefficient of water absorption is calculated as follows:

$$C_{w,s} = \frac{m_{so,s} - m_{dry,s} \times 10^6}{A_s\,(t_{so})^{0.5}} \quad [g/(m^2 \times s^{0.5})]$$

where $m_{dry,s}$ = mass of the specimen after drying (gm); $m_{so,s}$ = mass of the specimen in grams after soaking for time t, (gm); A_s = gross area of the face of specimen immersed in water (mm^2); t_{so} = time of soaking (sec).

The average $C_{w,s}$ for both 2 and 2.8 N/mm^2 Aircrete for short soaking times is summarised below on Table 4.

Table 1. Moisture content by mass for low density Aircrete.

Aircrete block strength (N/mm^2)	Mositure content, % by mass
2	11.7
	18.3
	7.4
	18.0
	11.3
	15.3
2.8	11.1
	11.2
	10.8
	13.3
	16.2
	18.9

In general, the coefficient of water absorption was found to be greater for 2.8 N/mm^2 Aircrete material. This is as expected as the rate of absorption is determined by the dry density of the material [AAC CEB 1978, Zhou 1992]. Also, the C_{ws} for both materials reduced with increasing soaking time as expected.

Table 2. Coefficient of water absorption of 2 N/mm^2 Aircrete blocks at short exposure times.

2 N/mm^2 Aircrete block sample	Mass (gm) Initial	Final	Time (Sec)[mins]	Area (mm^2)	$C_{w,s}$ [g/(m^2 × s$^{0.5}$)]
1	2202	2314	600 [10]	43460	105.2
		2356	1800 [30]	43460	83.5
		2398	5400 [90]	43460	61.4
2	2051	2141	600 [10]	42158	87.2
		2166	1800 [30]	42158	64.3
		2212	5400 [90]	42158	52.0
3	1884	1954	600 [10]	36575	78.1
		1986	1800 [30]	36575	65.7
		2026	5400 [90]	36575	52.8
4	2218	2459	600 [10]	44908	128.2
		2501	1800 [30]	44908	96.0
		2537	5400 [90]	44908	66.4
5	2145	2224	600 [10]	43263	74.5
		2261	1800 [30]	43263	63.2
		2305	5400 [90]	43263	50.3
6	2211	2376	600 [10]	4100	144.3
		2419	1800 [30]	4100	99.6
		2463	5400 [90]	4100	63.6

Table 3. Coefficient of water absorption of 2.8 N/mm^2 Aircrete blocks at short exposure times.

2 N/mm^2 Aircrete block sample	Mass (gm) Initial	Final	Time (Sec)[mins]	Area (mm^2)	$C_{w,s}$ [g/(m^2 × s$^{0.5}$)]
1	3486	3668	600 [10]	46216	160.8
		3744	1800 [30]	46216	131.6
		3870	5400 [90]	46216	113.1
2	3479	3668	600 [10]	47300	163.1
		3739	1800 [30]	47300	129.6
		3856	5400 [90]	47300	108.5
3	3478	3674	600 [10]	46866	170.7
		3747	1800 [30]	46866	135.3
		3871	5400 [90]	46866	114.1
4	3406	3588	600 [10]	46655	159.3
		3661	1800 [30]	46655	128.8
		3762	5400 [90]	46655	103.8
5	3488	3662	600 [10]	46434	153.0
		3731	1800 [30]	46434	123.4
		3848	5400 [90]	46434	105.6
6	3448	3631	600 [10]	46221	147.6
		3705	1800 [30]	46221	131.1
		3841	5400 [90]	46221	115.7

Table 4. Average short term soaking times for 2 and 2.8 N/mm^2 Aircrete.

Aircrete block strength (N/mm^2)	Soaking time (sec) [mins]	Average $C_{w,s}$ [g/(m^2 x s$^{0.5}$)]
2	600 [10]	102.9
	1800 [30]	78.7
	5400 [90]	57.8
2.8	600 [10]	159.1
	1800 [30]	129.0
	5400 [90]	110.1

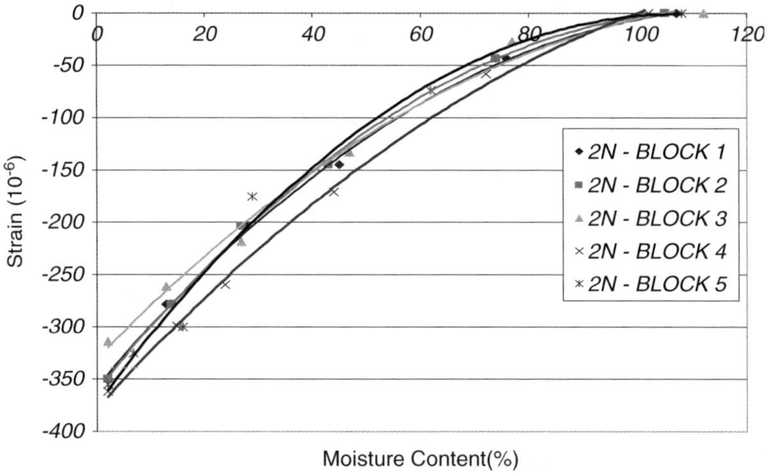

Figure 3. Graph of moisture movement for 2 N/mm^2 Aircrete using DEMEC points.

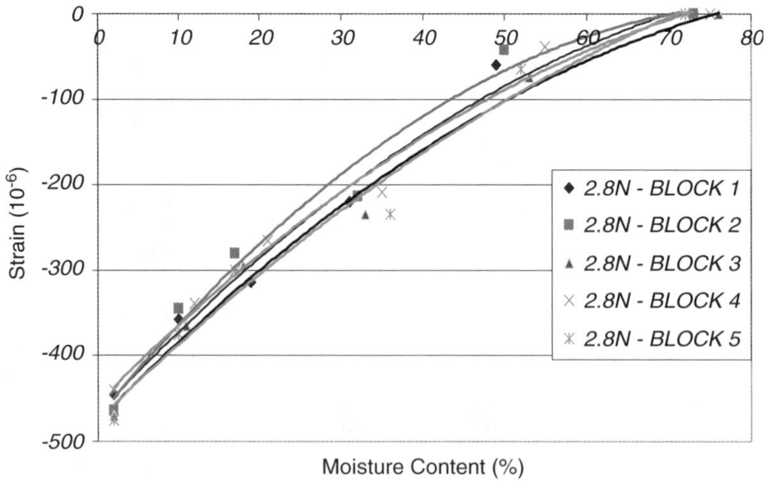

Figure 4. Graph of moisture movement for 2.8 N/mm^2 Aircrete using DEMEC points.

3.3 *Moisture movement*

Figures 3 and 4 show the graphs of moisture movement for both 2 and 2.8 N/mm^2 Aircrete blocks.

For comparative purpose, if the strain for both materials is assessed from 40% moisture content down to 2%, the shrinkage (or ε) is 200×10^{-6} and 310×10^{-6}, respectively for 2 and 2.8 N/mm^2 Aircrete. This is due to the greater density material having a higher shrinkage rate. These values, however, are very low in comparison to other construction materials. It should also be noted there is remarkably good consistency shown by both materials as depicted in Figures 3 and 4.

4 CONCLUSIONS

(a) Three different test methods have been used to assess and examine the moisture properties of Aircrete blocks.
(b) The average moisture content of 2 and 2.8 N/mm^2 blocks on immediate removal from the pallet was recorded to be 13.7% and 13.6% respectively.
(c) The average coefficient of water absorption ($C_{w,s}$) for 2 and 2.8 N/mm^2 blocks after 90 minutes soaking time was found to be 57.8 and 110.1 [g/(m$^2 \times$ s$^{0.5}$)], respectively. This indicates greater $C_{w,s}$ for the higher density material.
(d) Moisture movement tests revealed the higher density material imparting a larger shrinkage rate.

ACKNOWLEDGEMENTS

The authors would like to acknowledge the funding and support for the Low–Density Aircrete project given by the EPSRC and industrial collaborators, Autoclaved Aerated Concrete Products Association, H+H Celcon, Hanson-Thermalite, Quinn-Group and Tarmac Topblock, Building Research Establishment, Catnic-Corus UK, National House-Building Council and Office of The Deputy Prime Minister.

REFERENCES

AAC, CEB 1978, Manual of Design and Technology, The Construction Press, Lancaster, London, New York.
Aircrete Bureau, 2003a, Code of Best Practice for the Use of Aircrete Products, www.aircrete.co.uk.
British Standards a, BS EN 772, part 10, 1999 – Determination of moisture content of calcium silicate and autoclaved aerated concrete units.
British Standards b, BS EN 772, part 11, 2000 – Determination of water absorption of aggregate concrete, manufactured stone and natural stone masonry units due to capillary action and the initial rate of water absorption of clay masonry units.
British Standards c, BS EN 771, part 4, 2003 – Specification for masonry units, Autoclaved aerated concrete masonry units.
British Standards d, BS EN 680:1994 – Determination of the drying shrinkage of autoclaved aerated concrete.
Callister, W., Jr., 2000, Department of Metallurgical Engineering, The University of Utah, Materials Science and Engineering – An Introduction, Fifth Edition, John Wiley and Sons, Inc, ISBN 0 471 32013 7.
Daian, J. F., 1992, Laboratoire d'etude des Transferts en Hydrologie et Environment (IMG), Grenoble, France, Experimental determination of AAC moisture transport coefficients under temperature gradients, Advances in Autoclaved Aerated Concrete, Wittmann (ed.) © Balkema, Rotterdam. ISBN 90 5410 086 9.
Dubral, W., 1992, YTONG AG, Munich, Germany, Advances in Autoclaved Aerated Concrete, Wittmann (ed.) © Balkema, Rotterdam. ISBN 90 5410 086 9.
Frey, E., Hebel, AG. and Emmering, 1993, Germany, Recent results on thermal conductivity and hygroscopic moisture content of AAC.
Hama, Y. and Kamada, E., 1992, Hokkaido University, Japan, Frost resistance of increased density autoclaved aerated concrete, Advances in Autoclaved Aerated Concrete, Wittmann (ed.) © Balkema, Rotterdam. ISBN 90 5410 086 9.

Hasegawa, T., 1992, Department of Architecture, Faculty of Engineering, Hokkaido University, Sapporo, Japan, Investigation of moisture contents of autoclaved lightweight concrete walls in cold districts, Advances in Autoclaved Aerated Concrete, Wittmann (ed.) © Balkema, Rotterdam. ISBN 90 5410 0869.

H+H Celcon Ltd, 2003, Celcon House, Ightham, Sevenoaks, Kent, TN 15 9HZ. www.celcon.co.uk/downloads/RIBA%20CPD.pdf

Isu, N. and Mitsuda, T., 1992, Influence of quartz particle size on the chemical and mechanical properties of autoclaved lightweight concrete, Advances in Autoclaved Aerated Concrete, Wittmann (ed.) © Balkema, Rotterdam. ISBN 90 5410 086 9.

Jacobs, F. and Mayer, G., 1992, Institute for Building Materials, ETH Zurich, Switzerland, Porosity and Permeability of autoclaved aerated concrete, Advances in Autoclaved Aerated Concrete, Wittmann (ed.) © Balkema, Rotterdam. ISBN 90 5410 086 9.

Laurent, J. and Frendo-Rosso, C., 1992, Laboratoire d'etude des Transferts en Hydrologie et Environment (IMG), Grenoble, France, Application of image analysis to the estimation of AAC thermal conductivity, Advances in Autoclaved Aerated Concrete, Wittmann (ed.) © Balkema, Rotterdam. ISBN 90 5410 086 9.

Lippe, K., 1992, YTONG AG, R + D Centre, Schrobenhausen, Germany, The effect of moisture on the thermal conductivity of AAC, Advances in Autoclaved Aerated Concrete, Wittmann (ed.) © Balkema, Rotterdam. ISBN 90 5410 086 9.

Liu, C. and Wang, J., 1992, Suzhou Concrete and Cement Products Research Institute, China, An experimental study on thermal transmission properties of aerated concrete composite panels, Advances in Autoclaved Aerated Concrete, Wittmann (ed.) © Balkema, Rotterdam. ISBN 90 5410 086 9.

Millard, W., 1992, Marley Building Materials Limited, The thermal performance of European autoclaved aerated concrete, Advances in Autoclaved Aerated Concrete, Wittmann (ed.) © Balkema, Rotterdam. ISBN 90 5410 086 9.

Mitsuda, T. and Kiribayashi, T., 1992, Influence of hydrothermal processing on the properties of autoclaved aerated concrete, Advances in Autoclaved Aerated Concrete, Wittmann (ed.) © Balkema, Rotterdam. ISBN 90 5410 086 9.

Pospisil, F., 1992, Unit weight reduction of fly ash aerated concrete, Advances in Autoclaved Aerated Concrete, Wittmann (ed.) © Balkema, Rotterdam. ISBN 90 5410 086 9.

Senbu, O. and Kamada, E., 1992, Hokkaido University, Japan, Mechanism of frost deterioration of AAC, Advances in Autoclaved Aerated Concrete, Wittmann (ed.) © Balkema, Rotterdam. ISBN 90 5410 0869.

Schlegel, E. and Volec, J., 1992, Bergakademie, Freiberg, Germany, Application of autoclaved aerated concrete for high temperature insulation, Advances in Autoclaved Aerated Concrete, Wittmann (ed.) © Balkema, Rotterdam. ISBN 90 5410 086 9.

Schober, G., Hebel AG. and Emmering, 1992, Germany, Effect of size distribution of air pores in AAC on compressive strength, Advances in Autoclaved Aerated Concrete, Wittmann (ed.) © Balkema, Rotterdam. ISBN 90 5410 086 9.

Tada, S., 1992, Texte, Inc. and Nihon University, Japan. Pore structure and moisture characteristics of porous inorganic binding materials, Advances in Autoclaved Aerated Concrete, Wittmann (ed.) © Balkema, Rotterdam. ISBN 90 5410 086 9.

Thermalite, 2003, Marley Building Materials Ltd, Station Road, Coleshill, Birmingham B46 1HP. www.thermalite.co.uk.

Wittman, F.H., 1993a, Autoclaved Aerated Concrete: Properties, Testing and Design, RILEM Recommended Practice, RILEM Technical Committees 78 – MCA and 51 – ALC.

Wittman, F.H., 1993b, Autoclaved Aerated Concrete: Properties, Testing and Design, RILEM Recommended Practice, RILEM Technical Committees 78 – MCA and 51 – ALC.

Zhou, W., Feng, N. and Yan, G., 1992, Tsinghua University, Beijing, China, Fracture energy experiments of AAC and its fractal analysis, Advances in Autoclaved Aerated Concrete, Wittmann (ed.) © Balkema, Rotterdam. ISBN 90 5410 086 9.

Autoclaved Aerated Concrete – Limbachiya and Roberts (eds)
© 2005 Taylor & Francis Group, London, ISBN 0 415 38356 0

Structural fire protection with Autoclaved Aerated Concrete

G. Flassenberg
Bundesverband Porenbeton, Wiesbaden, Germany

ABSTRACT: Requirements relating to the fire protection of buildings arise from the Federal State building regulations and from the special building prescriptions and administrative regulations enacted on their basis. Depending on the ground plans and the height of the building, the building elements, such as the walls, ceilings and supports, as well as the building materials must comply with specific requirements. The technical classification of building materials and building elements with regard to fire protection takes place according to DIN 4102 or the equivalent corresponding European fire protection standards.

Hints for handling autoclaved aerated concrete building elements with respect to fire protection are given below with consideration of the determining regulations. These remain however restricted to the classification according to German standards.

1 REQUIREMENTS FOR AUTOCLAVED AERATED CONCRETE BUILDING ELEMENTS ACCORDING TO DIN 4102

According to DIN 4102-4 Section 2.2.1 d) "Mortar, concrete, reinforced concrete, autoclaved aerated concrete, light-weight concrete (...)" belong to the A1 class of building materials. They are considered as non-combustible. According to this data, all the autoclaved aerated concrete products listed in Table 1 belong to this category of building materials. This remains the same also if the surface is dressed with paints based on dispersion or alkyd resin or with normal paper wall coverings (wallpapers). Similarly, thin-layer mortar for masonry made of autoclaved aerated concrete units belongs to the A1 building material classification.

If resins, such as unsaturated polyester resin, or vinyl, acryloc, or a dispersion basis, are used as a bonding agent instead of cement, one is dealing with adhesives. Class 2 building materials belong to the adhesives category and are used for binding autoclaved aerated concrete wall components for non-loadbearing and loadbearing walls in the context of the technical approvals granted. In relation to autoclaved aerated concrete and under the condition that the thickness of the adhesive joints is not greater than 3 mm – in building practice, adhesive joints are 1.5 mm thick as a rule – the combustibility of the adhesive has no bearing in terms of the fire protection technology. The continuous layer of adhesive is rated like a barrier layer of inflammable building materials against ascending moisture and has no influence on the fire resistance of construction elements.

As follows from DIN 4102-4 Section 2.2.1 Item d), rendering finishes according to DIN 18550-2 – also with the usual parts of additives – belong likewise to Class A1 building materials. Should additives of organic composition be added beyond the framework established in DIN 18550-2, the "not combustible" property would be forfeited.

For the safety of a building in case of fire, however, not only the combustibility of the building materials is important, but mainly the fire resistance period of the building elements. On the basis of their fire resistance period, autoclaved aerated concrete building elements are classified according to the fire resistance grading period presented in Tables 28 to 42 of DIN 4102-4 (also please refer to Figures 1 and 2).

Table 1. Autoclaved aerated concrete building components for normal applications; dimensions and regulations.

Construction element	Dimensions in mm	Regulations
AAC masonry units with thin layer mortar	*Non-loadbearing*: Length: 332, 399, 499, 599, 624 Height: 249 Thickness: 50 to 150	DIN 1053 DIN 4166
	Loadbearing: Length: 332, 399, 499, 599, 624 Height: 249 Thickness: 115 to 365	DIN 1053 DIN V 4165
	Loadbearing: Length: 499, 599, 749, 999 Height: 498, 599, 623 Thickness: 115 to 365	DIN 1053, Technical approval
Reinforced roof and floor components of AAC	Length: $\leq$8000 mm Height: $\leq$750 mm Thickness: 100 to 300, Stagger 25	DIN 4223 Technical approval
Reinforced components of AAC – for non-loadbearing walls	Length: $\leq$8000 mm Height: $\leq$750 mm Thickness: 100 to 375, Stagger 25	DIN 4223 Technical approval
Reinforced components of AAC – for load bearing walls	Length: $\leq$3500 mm Height: 500 to 1500 Thickness: 100 to 375, Stagger 25	DIN 4223 Technical approval
Lintels of AAC for non-loadbearing walls	Length: 1190 to 1249 Height: 249 Thickness: 75 to 150	
Lintels of AAC for loadbearing walls	Length: 1229 to 2249 Height: 249 Thickness: 175 to 365	DIN 4223 Technical approval

2 FIRE WALLS MADE OF AUTOCLAVED AERATED CONCRETE

According to the Federal State building regulations, fire walls are walls for the separation or differentiation of fire compartments. They are designed to prevent the propagation of fire to other buildings or sections of buildings.

Fire walls made from autoclaved aerated concrete must fulfil the requirements made in Table 45 of DIN 4102-4 regarding slenderness and thickness and, in the case of reinforced components the axis-distance of the longitudinal reinforcement. The connections of non-loadbearing wall components of AAC to reinforced concrete or steel supports need to be carried out according to the figures 25, 26 and 28 to 30 of DIN 4102-4. Adding on masonry walls made of autoclaved aerated concrete to adjacent solid building elements needs to be achieved with full mortar joints according to DIN 1053-1, or implemented according to the data in figures 17 to 20 and 24 of the DIN 4102-4. Additional, further proven fire wall connections can be found in the Technical Report *Bericht 4*, Section 4.8, of the Federal Association for Autoclaved aerated concrete.

3 COMPLEX PARTITION WALLS MADE OF AUTOCLAVED AERATED CONCRETE

Complex partition walls are required by damage insurers for the separation of specific production areas or for the restriction of particular fire risks. In industrial construction complex partition

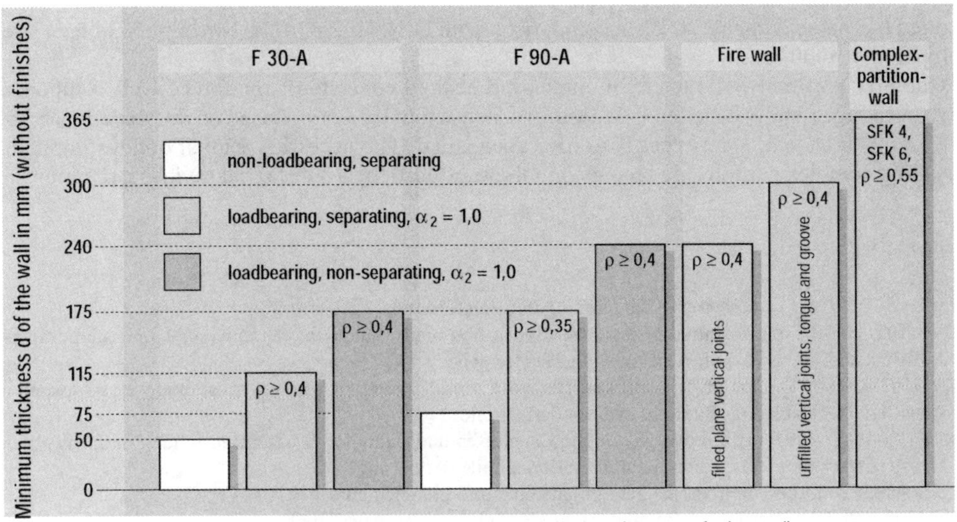

α_2 = Ratio of the applied design load to the design resistance of the wall
ρ = Class of density according to DIN V 4165 or Technical Approval
SFK = Class of compressive strength according to DIN V 4165 or Technical Approval

Figure 1. Minimum thickness of single-leaf walls made of autoclaved aerated concrete masonry according to DIN 4102-4 or test certificates.

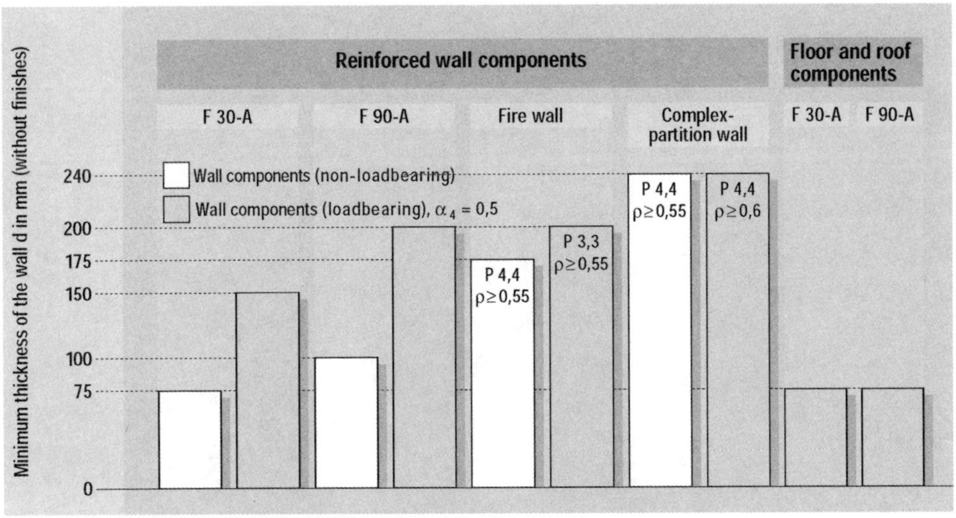

α_4 = Ratio of the applied design load to the design resistance of the wall
ρ = Class of density in accordance to DIN 4223 or Technical Approval
SFK = Class of compressive strength in accordance to DIN 4223 or Technical Approval

Figure 2. Minimum wall thickness of reinforced autoclaved aerated concrete according to DIN 4102-4 or test certificates.

walls made of autoclaved aerated concrete have proven to be resistant. Due to their significance in terms of fire protection and insurance engineering, greater demands are made on complex partition walls than on fire walls under DIN 4102-3. In the definition used in DIN 4102 complex partition walls are walls with the designation F 180-A, which possess an increased resistance to applied mechanical impact stress in comparison with DIN 4102-3 (4000 Nm applied impact stress compared to 3000 Nm).

Complex partition walls made of autoclaved aerated concrete or reinforced wall components have been successfully installed in industrial construction for years. Based on this experience, they offer an assurance of safety that is to date unequalled. The necessary capital outlays implied in installing complex partions are recouped in the short term by a substantial saving in premiums.

REFERENCES

DIN 1053-1: 1996-11 Mauerwerk; Berechnung und Ausführung

DIN 4102-3: 1977-09 Brandverhalten von Baustoffen und Bauteilen; Brandwände und nichttragende Außenwände; Begriffe Anforderungen und Prüfungen

DIN 4102-4: 1994-03 Brandverhalten von Baustoffen und Bauteilen; Zusammenstellung und Anwendung klassifizierter Baustoffe, Bauteile und Sonderbauteile

DIN 4102-4/A1: 2004-11 Brandverhalten von Baustoffen und Bauteilen; Zusammenstellung und Anwendung klassifizierter Baustoffe, Bauteile und Sonderbauteile; Änderung 1

DIN V 4165: 2003-06 Porenbetonsteine – Plansteine und Planelemente

DIN 4166: 1997-10 Porenbeton-Bauplatten und Porenbeton-Planbauplatten

Meyer-Ottens, C. Brandverhalten von Porenbetonbauteilen – Erläuterungen zu DIN 4102-4, Ausgabe März 1994; Berichtsheft 4 des Bundesverbandes Porenbeton, Wiesbaden 1997

Einbau von Feuerschutztüren und -toren, Berichtsheft 17 des Bundesverband Porenbeton, Wiesbaden 2002

Weber, H. Das Porenbeton-Handbuch – Planen und Bauen mit System; Bauverlag Wiesbaden, 5. Auflage 2002

Autoclaved Aerated Concrete – Limbachiya and Roberts (eds)
© 2005 Taylor & Francis Group, London, ISBN 0 415 38356 0

AAC based on fly ash in strategy of sustainable development

G. Zapotoczna-Sytek
R&D Center for Concrete Industry "CEBET", Warsaw, Poland

ABSTRACT: The paper presents 50 years of Polish experience in AAC production based on fly ash as well as the most important properties of AAC products. It has been demonstrated that production of AAC based on fly ash is particularly advantageous and meets the conditions of sustainable development due to both possibility of fly ash utilization and no-waste production process and harmless for environment. AAC products show low energy and binder consumption as well as high thermal insulation.

1 INTRODUCTION

One of the essential elements of sustainable development is to protect natural environment. In practice it means utilization of by-products and introduction of technological processes, which are harmless for environment. Utilization of fly ash – by-product from power stations – creates a serious problem in all countries, included Poland, where production of energy is based on solid fuels.

In the result of long-term investigations and productive experience in Poland we are able to indicate and realize an effective utilization of fly ash in the field of building material industry.

Our experience shows that the most rational and effective way for fly ash utilization is in AAC production (Bastian 1965).

Such utilization of fly ash has been used in Poland since early fifties. First time fly ash in production of cellular concrete was applied in Poland already in 1954 and as early as from the beginning of sixties fly ash technology in AAC production became preferred one in further development of this material, consistently realized on the base of Polish technologies, systems and equipment (Paprocki 1966, Jatymowicz 1984). 10 new factories have been built for production of AAC based on fly ash technology which utilized all together over 1 million tons of fly ash per year, producing about 2 millions cubic meters of AAC (what was about 40% of total AAC production in Poland at that time).

As the result of experience and achievements gained in this country there was considerable worldwide interest and profitable conditions for export of AAC factories. So far Poland exported 36 turnkey AAC factories, there of 12 based on fly ash technology (Jatymowicz 1980).

2 R & D ACTIVITY

Together with development of production of cellular concrete based on fly ash COBRPB Cebet carried out complex investigative works aiming at detailed recognition of properties of fly ashes, determination for them technical requirements in standards and instructions (Morawska 1987) as well as settlement of technological rules to be observed in dependence on quality of ashes.

The results of these works showed that compositions of fly ashes obtained from Polish power stations are significantly diversified in dependence on type of burned coal. In Table 1 average

Table 1. Chemical composition (average and limits) of fly ashes from hard and brown coals, %.

	Characteristic type of fly ash from		
		Brown coal	
Component	Hard coal (aluminosilicate type)	(alumino-silicate type) Turów	(calcium sulfate type) Konin
SiO_2	52,0 (43–57)	48,0 (41–50)	45,0 (30–50)
Al_2O_3	20,0 (18–23)	31,0 (30–35)	8,0 (5–9)
Fe_2O_3	13,0 (7–16)	7,0 (14–20)	5,0 (4–6)
CaO	6,0 (4–10)	3,0 (2–4)	32,0 (25–48)
MgO	3,0 (1–5)	1,5 (0,1–1,0)	3,0 (2–4)
SO_3	1,0 (0,5–3,5)	0,5 (0,5–0,7)	7,1 (5–10)
Na_2O+K_2O	1,8 (1–3)	1,5 (0,1–0,3)	0,2 (trace)

chemical compositions of fly ashes of aluminosilicate type from hard coal is presented as well as of calcium sulfate type and aluminosilicate type resulting from burning of brown coal.

The properties of fly ashes from brown coal originated from Konin Coalfield differ significantly from others because they contain large quantity of calcium oxide (CaO) and small quantity of silica (SiO_2).

In the result of long-term activity Polish original technologies for production of cellular concrete have been worked out both for use of fly ashes from hard coal from dry or wet removals as well as for use ashes from brown coal, taking into account specific properties of those ashes.

Economic analysis of all technologies showed that technology of AAC production based on fly ashes of calcium sulfate type from burning of brown coal is less effective economically than that one based on fly ash from hard coal. This results from a need of applying of a number of additional measures (such as preliminary slacking, enrichment into silica) due to properties of these ashes, which contain considerable quantity of free lime (CaO) and sulphur.

In the light of the above mentioned and because of the fact that in Poland there are very large quantities of fly ashes from hard coals which are not fully utilized, the utilization of fly ashes of calcium sulfate type in production AAC did not find so far practical usage. In production of cellular concrete in Poland there are applied exclusively fly ashes of aluminosilicate type from hard coal from dry removal. That's why the most research and development works is dedicated to these fly ashes in aspect of usage of cellular concrete.

Analysis of properties of fly ashes from hard coal as received by factories of cellular concretes showed that chemical composition of ashes from given power station does not indicate of essential variation, however, physical properties such as: fineness, water absorption, coal content as loss on ignition they change in considerably wider range. Changeability of the above mentioned physical properties influences directly onto process of cellular concrete production.

To minimize of an inhomogeneity of fly ashes there were carried out, in cooperation with power industry, investigation on influence of conditions of preparations and burning process of coal fuel on quality of received fly ash and next properties of AAC (Zapotoczna-Sytek 1973).

The investigations showed that inhomogeneity of fly ashes – concerning chemical and phase compositions as well as characteristic physical properties, such as water absorption and fineness – it could be possible to limit by suitable process of fuel preparation and mainly by proper process of its burning.

In the result of investigations parameters of these processes, assuring best quality of ashes, were qualified at simultaneous improvement of efficiency of work of boilers by elimination of phenomenon of slag deposit in chambers and protracted burning as well as forming of clear core of flame (where there are maximum temperatures) and forming suitable cooling zone.

Investigation of fly ashes from individual zones of fly ash disposal showed legitimacy of selective sending them from choose zones to cellular concrete factory in aim to select ashes of the bests quality.

Investigative works permitted to optimize production of AAC small size elements (blocks, plates). However after introducing to production AAC reinforced elements problem of cracking in big size reinforced elements appeared (Zapotoczna-Sytek 1978). In some AAC elements based on fly ashes of very similar chemical composition and similar physical properties occurred cracks while others were free of cracks. In elements of lengths above 4,80 m cracks and fractures occurred periodically (with different intensification) on the whole width of element. Investigation of chemical composition and physical properties, which indicated conformity with earlier fly ash specifications for small size elements, did not help in explanation of the problem. It was decided to undertake complex investigation together with executing of reinforced elements in scale 1:1 in experimental halls of COBRPB Cebet, using fly ashes from different power stations (Sytek 1977, 1978). During tests a number of technological measures were applied but we were not able to solve the problem and to eliminate cracking.

Tests on fly ashes included physical and chemical compositions as well as investigation of mineral compositions, determination of coefficients of thermal expansion Lt, heat of solution, electric conduction of water slurries, and measurement of fly ashes activity in autoclave conditions. In the result of so wide range of tests connections were affirmed among origin, properties of fly ashes and susceptibility to formation cracking in big size AAC reinforced elements (Stokłosa 1979, 1980). On this basis to the requirements for fly ash applied to production of reinforced elements an additional test was introduced so-called activity in autoclave condition.

The results and differences of coefficients of thermal expansion α_t for sand $11 \times 10^{-6}/°C$ and for fly ashes in range of $3{,}1$–$4{,}6 \times 10^{-6}/°C$ in comparison of these values with coefficient α_t for steel equal to $12 \times 10^{-6}/°C$ suggest that cracking occurs when large tensions in concrete during phase of temperature raise in autoclave exceed strength of AAC material gained to this time. More reactive fly ashes assure getting sufficiently larger strength of AAC material at the same period of autoclavization.

This interesting investigation on evaluation of fly ashes for reinforced AAC elements of lengths above 3 m should be continued.

In last years in Poland for protection of atmosphere there have been introduced different methods and systems of flue gases desulfurization from pulverized-fuel boilers as well as burning of coal in fluidized-bed boilers (Zapotoczna-Sytek 1994). Usually firms submitting tenders on processes of flue gases desulfurization do not propose any technology of utilization of desulfurization products which, depending on desulfurization processes, differ significantly from products received during traditional methods of burning of coal. To solve problems connected with utilization of by-products, received from different desulfurization methods, properties of these by-products have been tested and directions of theirs utilization were given for concrete industry. For part of these by-products (ashes and sulphates' materials) technologies of production of cellular concrete have been worked out (Zapotoczna-Sytek 1996). Both recognized properties of by-products and indicated directions of utilization do not exhaust indispensable investigative works in this range. There are necessary further investigations also of statistical character as well as checking in technical scale of the proposed directions of utilization (especially for technology with use of by-products from dry method of desulfurization and burning of coal in fluidized-bed boilers).

One of the essential elements of sustainable development concerns rational utilization of energy from renewable sources such as energy from biomasses. Their part in fuel balance of renewable energy in Poland grows from year in year out. Biomass is also introduced to burning in large industrial dust boilers (from this of type of boilers fly ashes applied to AAC production are taken). For this reason, there are new research works undertaken recently aiming at checking whether and to what degree properties of fly ashes from joint burning biomasses and coal are different as compare with fly ashes from burning only coal and whether possible change of properties would be accepted for application in AAC or in other building materials production (Zapotoczna-Sytek 2004).

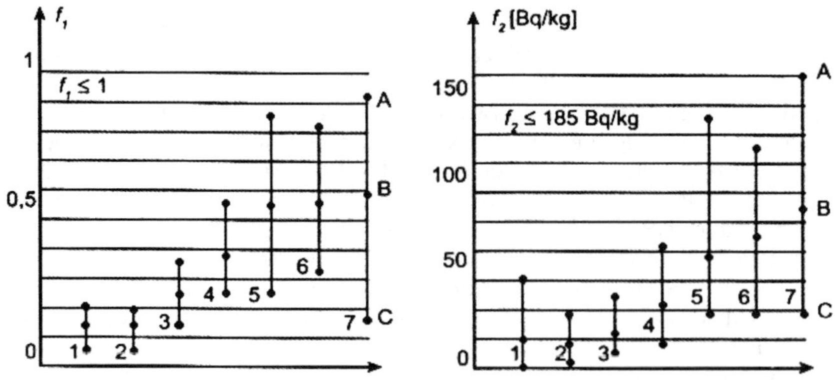

Figure 1. Variation of qualifying coefficients f_1 and f_2 in selected building materials: 1 – AAC based on sand, 2 – silicate brick, 3 – plain concrete, 4 – LWA concrete, 5 – ceramic products, 6 – slag concrete, 7 – AAC based on fly ash. Values: A – max, B – average, C – min concern all qualifying coefficients marked. On the drawing by dots.

3 LIMITATIONS IN FLY ASH APPLICATION

Since many years there have been in Poland some hindrances in developing application of fly ashes among others for AAC resulting from mass media activity, which are forming an opinion of society on harmfulness of fly ashes and products containing fly ashes.

All the time on columns of some mass medias such point of view is presented, and it happens also articles in which prefabricated elements from aggregate plain concrete (produced without any part of by-products) are attacked. However, prevailing opinion is that ceramic is the healthiest material. The tests carried out systematically since 1980 on mineral materials and materials of mineral origin from the point of view of existence in them concentrations of natural radioactive elements (defined by coefficients f_1 and f_2) they have showed that ceramic products are characterized by qualifying coefficients on the level of slag concrete and AAC produced on the base of fly ashes (Figure 1). During systematic investigations, over 25 years, on AAC based on fly ashes there was not single case of exceeding of admissible values of qualifying coefficients f_1 and f_2. It was affirmed also, that essential changes of this coefficients did not occurred in time.

The permissible values are:

$$f_1 = 0{,}00027\ S_K + 0{,}0027\ S_{Ra} + 0{,}0043\ S_{Th} \leq 1 \tag{1}$$

$$f_2 = S_{Ra} \leq 185\ \text{Bq/kg} \tag{2}$$

where: S_K, S_{Ra}, S_{Th} – are values of concentrations for potassium K-40, radium Ra-226 and thorium Th-228, respectively.

From 1 January 2003 there are in force in Poland new formulas and regulations concerning requirements on content of natural radioactive elements in raw materials and materials/products intended for use in buildings designed for stay of people and livestock. Tests carried out, on the base of new regulations, have indicated that AAC based on fly ash meets also new requirements.

Test on concentration of radon were also carried (for which it is responsible radium); in buildings made of AAC walls as well as for comparison in buildings from different building materials (10 in each material group). Buildings were inhabited, located in one region of Poland (within the radius of 30–40 km). Measurements showed (Table 2) that highest average level of radon (measured in Bq/m^3) it has occurred in wooden buildings. Taking into account that wood is not a source of radon the test results confirmed the results obtained in other countries that building material is not a decisive factor about increased concentration of radon but soil on which building have been founded as well as easiness of infiltration of radon from soil to interior of building (Mamont-Cieśla 1998).

Table 2. Average radon concentration on the ground floor in buildings constructed of different building materials [Bq/m^3].

			Cellular concrete based on		
Big size panel	Wood	Ceramic brick	Fly ash	Sand	Sand + Brick
47	92,3	86,2	56,7	61,2	68,2

Public opinion is informed about the presented above test results by research institutions and AAC manufacturers in order to change still existing, among part of society, opinions about harmfulness of fly ashes. Such opinions create conditions that some AAC manufacturers can give up from applying of fly ashes replacing them by sand. AAC manufacturers do not want to be perceived in opinion of their customers as users of by-products.

4 ASSORTMENT AND TECHNICAL PROPERTIES OF AAC

At present in Poland there are produced from AAC masonry elements of density classes in wide assortment from 300 to 700 (Soboń 2000–2005, Babiński 2005) as well as floor hollow blocks and reinforced elements, exclusively roof plates of lengths up to 3 m (classes 500 and 600), but only for individual orders.

In past years there were also produced reinforced wall elements: planks of lengths up to 6 m, and entirely factory finished big size wall units of lengths up to 6 m.

Significant limitation of industrialized building caused that at present there are mainly produced small size elements. In sure degree on above been mentioned Technological problems connected with quality of fly ashes also influenced to some extent (see: item.2) on the present range of AAC production.

In the result of long-term Polish experiences it can be stated that applying of fly ashes to AAC and keeping production process in accordance with established requirements for given technology assures getting small size elements and medium size reinforced (of length up to 3 m) of acceptable properties.

As confirmed in practice, usage of fly ashes facilitates getting AAC of lower density classes (300–500) and the compressive strength, for the given class, is higher 10–15% as compared with sand based AAC without fly ash.

Investigation of thermal properties (according to PN ISO 8302) they showed that advantage AAC is high thermal isolation among materials applied to masonry walls. The best isolation demonstrates AAC produced on the base of fly ashes because this concrete shows usually lower coefficient of thermal conductivity than AAC based on sand independently from density of concrete (Table 3) (Soboń 2005). This enables very advantageous solution for walls what in consequence results in sawing of thermal energy during exploitation of building.

In Table 4 the effect of coefficients of thermal conductivity onto the value of heat-transfer coefficients is presented for typical homogeneous external wall. These walls are particularly recommended for one-family houses because they enable the best utilization of insulation and structural advantages of AAC.

Investigations carried out, on walls from AAC in buildings after 30–40 years of exploitation, have confirmed that moisture content of partition from AAC based on sand amounted about 2,5% while for AAC based on fly ash – about 4,5% (Łaś 1999–2001). At such moisture content partitions from AAC show good thermo-insulating properties. Test results of moistures content of partitions deny propagated sometimes opinions about very large moisture content of walls from AAC.

Analyzing AAC properties one should take into consideration that AAC based on fly ashes shows often-larger shrinkage value than AAC based on sand. According to investigations (Hums 1992) as limit value of shrinkage, at which cracks do not occur in walls, it can be accepted value of 0,35 mm/m. From our experiences it results, that equally important, beside of value of shrinkage

Table 3. Declared and analytical values of thermal conductivity coefficients.

Lp.	AAC type	Dry density kg/m^3	Values of thermal conductivity coefficients W/(m × K) Declared λ_{D23}	λ_{D10}	λ_u analytical
1	Sand technology	400	0,095–0,115	0,091–0,110	0,110–0,130
		500	0,125–0,140	0,120–0,135	0,145–0,160
		600	0,140–0,170	0,130–0,165	0,160–0,195
		700	0,175–0,190	0,165–0,190	0,200–0,22
2	Fly ash technology	350*	0,070	0,068	0,095
		400	0,086–0,097	0,083–0,095	0,105–0,120
		500	0,110–0,120	0,105–0,115	0,130–0,140
		600	0,125–0,145	0,125–0,140	0,150–0,170
		700	0,150–0,160	0,145–0,155	0,175–0,190

* From one producer AAC (from 23 factories).

Table 4. Values of heat-transfer coefficients for typical homogeneous external wall.

Dry density of AAC [kg/m^3]	Type of technology	Analytical value of thermal conductivity coefficients [W/(m × K)]	Heat-transfer coefficients U[$W/m^2 \times K$] Wall thickness 36 cm	42 cm
Analytical values from producers for:				
350	Sand technology	0,095	0,25	0,22
	Fly ash technology	0,095	0,25	0,23*
400	Fly ash technology			
	– minimum value	0,105	0,28	0,24
	– maximum value	0,120	0,31	0,27
	Sand technology			
	– minimum value	0,110	0,29**	0,26
	– maximum value	0,130	0,34	0,29
500	Fly ash technology			
	– minimum value	0,130	0,34	0,29
	– maximum value	0,140	0,36	0,31
	Sand technology			
	– minimum value	0,145	0,37	0,32
	– maximum value	0,160	0,41	0,35

* Wall thickness 40 cm; ** Wall thickness 36.5 cm.

defined from tests, is moisture content of AAC at the moment of built in into wall as well as kind of mortar joining units in wall (Kruk 2003–2004).

5 EFFICIENCY OF FLY ASH UTILIZATION IN AAC PRODUCTION

Polish experience indicates the following effects:

- Lower costs of building and exploitation of dumping grounds for ashes as well as decrease of theirs surfaces. Economic analysis shows (in Polish conditions), that saving of expenditures born for storing of such quantity of ashes which would be used by average size AAC factory (of production capacity 200 000 m^3/year) in time of 8–10 years it permits to return at the same time building capital costs of new AAC factory (Jatymowicz 1984).

- Locating AAC factories in direct neighborhood of power station it enables to use of pneumatic transportation of ashes to factory, eliminating their burdensome transport as well as it permits utilization of dump steam from turbines in power station as technological steam and excludes a need of possession of own boiler room what essentially influences on decrease of investment and running costs of AAC factory.
- Decrease of energy and material consumption. Utilization of ashes for AAC production in relation to AAC based on sand allows to decrease:
  – Electric energy necessary for production of 1 m^3 of AAC about 40%,
  – Binder (cement + lime) for unit of production about 20–25%.

This is connected with natural fineness of ashes as well as with theirs puzzolanic properties i.e. bond ability of lime and aluminum oxide which are main components of ashes.

Cost of manufacture of 1 m^3 of AAC based on ashes is lower about 30% in relation with AAC based on sand.

Efficiency of utilization of fly ashes for AAC production compensates some shortcomings of ash resulting from its nature:

– it is dirtying and dusting, that's why it demands hermetic sealings in transport and at use as well as installing of dust extraction devices.
– it is mixture which composition is sensitive on quality of burned coal, its crumbling and conditions of burning and so it demands regular checking of properties.

6 CONCLUSIONS

– Necessity of endeavor to assurance of sustainable development of the world, for protection of climate of the Earth and rights of every man to breathing with clean air should make up motto of our activity. This activity should not be directed for satisfaction of needs not only today's generation but also future generations (Hums 1992). Often it is hard to find reference to such formulated thesis in present building's activity, engaging whole process of formations and existences of objects as well as minimization of effects on natural environment.
– Present-day technologies of AAC production, properties and characters of executed objects from this products show that both process of production as use of AAC enter into conditions of sustainable development (Zapotoczna-Sytek 2004). Application of AAC based on fly ashes is particularly advantageous, what has been underlined in this paper.
– Since 50 years production of AAC based on fly ashes has been consistently realized in Poland. Development of AAC production based on fly ashes is accompanied with complex investigation works connected with detailed recognition of properties of ashes, determination for them requirements as well as rules of technological procedures for elaborated production technology of AAC.
– From comparison of quantity of produced by-products and building materials it results, that on the world in coming decades utilization of by-products in building materials industry will become a civilization challenge.
– Utilization of by-products in many cases demands a change of traditional approach. It has to follow full understanding, that production technology cannot be considered only from the point of view of technical solutions, possibilities of sale of products but one should also consider ecological, economic and social questions. Real activity at present should be concentrated on propagating of actions being in favor of needs of sustainable development of the world.

REFERENCES

Babiński et al., 2005. Building with cellular concrete; *Poradnik-katalog*. Warszawa: Stowarzyszenie Producentów Betonów. [In Polish]

Bastian et al., 1965. Fly Ash. Warszawa: Instytut Techniki Budowlanej. Komisja wykorzystania popiołów lotnych. [In Polish]
Hums, D. 1992. Ecological aspects for the production and use of autoclaved aerated concrete. Wittmann (ed.), *Advances in Autoclaved Aerated Concrete*: 271–275. Rotterdam: Balkema.
Hums, D. 1999. Shrinkage of wall materials. *Materiały Budowlane* 98(4): 36–37. Warszawa. [In Polish]
Jatymowicz, H., Siejko, J. & Zapotoczna-Sytek, G. 1980. Technology of autoclaved cellular concrete. Warszawa: Arkady. [In Polish]
Jatymowicz, H. & Zapotoczna-Sytek, G. 1984. State of art and possible increase of fly ash for production of cellular concrete. *Materiały budowlane* 84(7): 17–19. Warszawa. [In Polish]
Kruk et al., 2003–2004. Testing on cracking susceptibility of masonry walls made of light classes of cellular concrete. Warszawa: COBRPB "Cebet". [In Polish]
Łaś et al., 1999–2001. Testing on thermal-moisture properties and durability of masonry walls made of cellular concrete units. Warszawa: COBRPB "Cebet". [In Polish]
Mamont-Cieśla, K. & Zapotoczna-Sytek, G. 1998. Radioactivity of cellular concrete – myth or reality? *Materiały Budowlane* 98(3): 22–23. Warszawa. [In Polish]
Morawska et al., 1987. Instruction for testing and evaluation of fly ash from hard coal used for production of cellular concrete. Warszawa: COBRPB "Cebet". [In Polish]
Paprocki, A. 1966. Betony komórkowe. Warszawa: Arkady. [In Polish]
Soboń et al., 2000–2005. Handbook for users of cellular concrete. *Czasopismo Stowarzyszenia Producentów Betonów.* Warszawa. [In Polish]
Stokłosa, J. et al., 1979–1980. Evaluation criteria for fly ashes applicable for production of full range of cellular concrete products. *Temat 3078*. Warszawa: COBRPB "Cebet". [In Polish]
Sytek et al., 1977, 1978. Comprehensive testing of fly ashes to enable production of full range of cellular concrete products. *Temat 3054.*. Warszawa: COBRPB "Cebet". [In Polish]
Zapotoczna-Sytek, G. & Siejko, J. Technological problems in industrial production of AAC based on fly ash. 81–87. Warszawa: COBRPB "Cebet". [In Polish]
Zapotoczna-Sytek, G. 1973. Usability of fly ashes from modern industrial boilers for cellular concrete production. *Praca doktorska*. Warszawa: Politechnika Warszawska. [In Polish]
Zapotoczna-Sytek, G. et al., 1994. Effect of flue desulphurization technology and burning process of coal in fluidized boilers on chemical composition and physical-chemical properties of furnace by-products. *Sprawozdanie z realizacji projektu badawczego nr 602879101*. Warszawa: COBRPB "Cebet". [In Polish]
Zapotoczna-Sytek, G. & Jarosz-Jaroszewski, P. 1996. Evaluation of properties and usability for production of building materials by-products from various desulphurization methods . *Materiały Konferencji Naukowo-Technicznej; Zagadnienia Materiałowe Inżynierii Lądowej, Kraków, 19–22 czerwca 1996:* 416–423. [In Polish]
Zapotoczna-Sytek, G. 2004. Sustainable development and process of production of building products. *III Miedzynarodowa Konferencja Naukowa; Energia i Środowisko w technologiach materiałów budowlanych, Szczyrk-Opole:* 259–269. [In Polish]
Zapotoczna-Sytek, G. 2004. Present-day technologies for cellular concrete – prognosis in the light of principles of sustainable development. *Materiały XIX Konferencji Naukowo-Technicznej; Beton i Prefabrykacja, Jadwisin:* 137–149. Warszawa. [In Polish]
Zapotoczna-Sytek, G. et al., Górska, B. et al., Romanowski, J. 2004. Application of ashes from joint burning of biomass for production of building materials. *Konferencja; Współspalanie biomasy i paliw wtórnych w kotłach energetycznych, Zakopane, 12–14 maja 2004.* [In Polish]

Autoclaved Aerated Concrete – Limbachiya and Roberts (eds)
© 2005 Taylor & Francis Group, London, ISBN 0 415 38356 0

Carbonation of AAC

M. Haas
Xella Technologie und Forschungsgesellschaft mbH, Germany

ABSTRACT: Carbonation of AAC is a natural process which takes places over a long period of time and which improves the properties of AAC under normal conditions. This improvement requires well developed CSH phases i.e. tobermorite and not amorphous phases like CSH-I and CSH-II. If not, the properties of the AAC material can be debased.

Long-term investigations about the behaviour of well mineralized AAC as well as of insufficient autoclaved material over a period of more than ten years under normal climate conditions (i.e. no increased CO_2-content) have been made. The development of the characteristics under these conditions and contrary under increased CO_2 concentrations will be shown in this paper.

1 INTRODUCTION

Carbonation of tobermorite is a complex, multistage process. Humidity in the material and in air plays an important role besides CO_2 concentration (Siauciunas 2004). Fundamental investigations showed "changes in micro-level structures … were more significant under accelerated conditions than under working conditions" (Matsushita 2000 & 2004).

Investigations by ourselves done in the past showed the same improvement of material properties like compressive strength under normal working conditions (Figure 1). No influence on the thermal conductivity could be noticed, even if an increase in raw density can be measured (Figure 2).

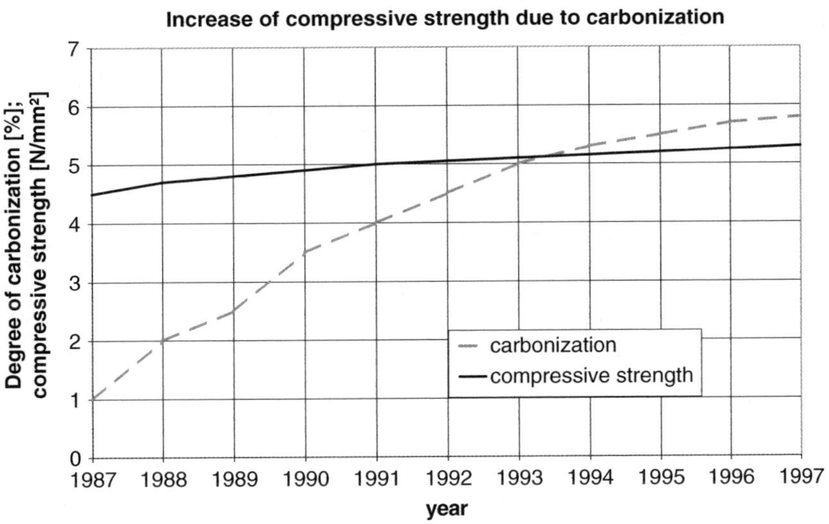

Figure 1. Development of carbonation and compressive strength under normal atmospheric conditions.

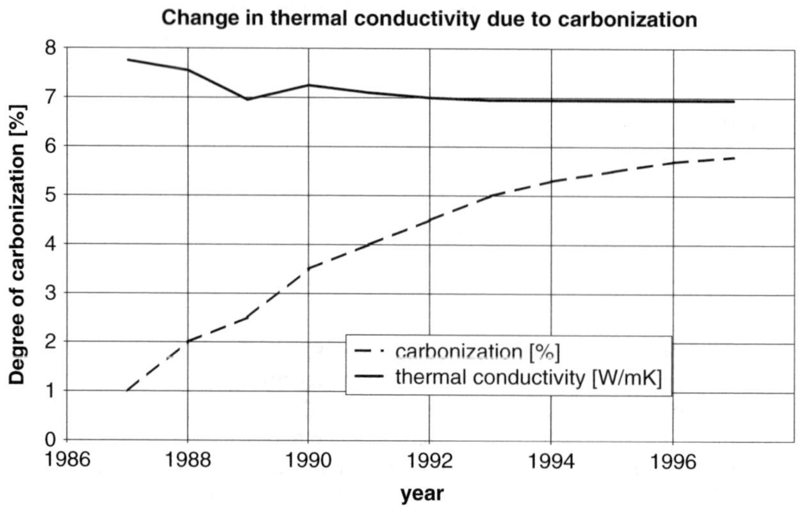

Figure 2. Development of carbonation and thermal conductivity under normal atmospheric conditions.

Table 1. CO_2 concentrations and their impact on human beings.

CO_2 Concentration	Impact
• 0.03%	Normal CO_2 concentration in air
• 0.08%	CO_2 amount in bad ventilated rooms
• 0.1%	Appearance of tiredness and lack of concentration
• 3–5 Vol-%	Irritation of breathing
• 7–10 Vol-%	Unconsciousness by lack of oxygen

2 CARBONATION OF AAC UNDER ATMOSPHERIC CONDITIONS

Normally, properties of AAC are improving while the material is ageing under normal atmospheric conditions. This is valid if:

- The initial phase formation of tobermorite in terms of crystal formation was sufficient. This can be easily achieved by correct and secure autoclaving. e.g. having a sufficient long constant pressure phase in the correct temperature range between 180°C and 200°C.
- The material suffers only normal CO_2 concentrations of about 0.03 to 0.08 Vol.-%.

3 ARTIFICIAL CARBONATION OF AAC

It is also known that AAC can be damaged by CO_2 chemisorption. In building applications this problem can easily be avoided by a sealed coating if necessary. However, damaging of AAC can occur if:

- AAC has a insufficient initial phase formation of tobermorite e.g. forced by failure during autoclaving.
- The CO_2 concentration the material suffers is too high (Table 1).

In Xella AAC plants the quality of the phase formation of tobermorite is examined and regulary checked by X-ray diffraction. Nevertheless this is only an internal measurement method used by our own. Up to now, no official standards for testing methods of the durability of AAC exist.

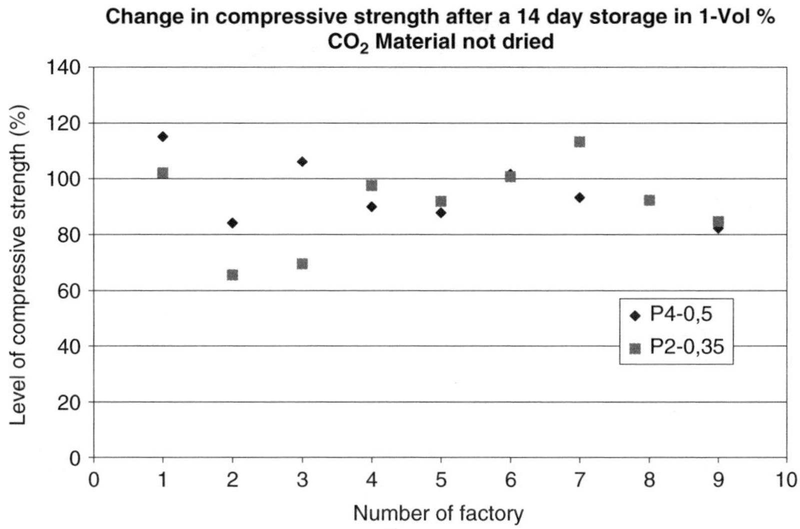

Figure 3. Change in compressive strength after 14 day storage in 1 Vol.-% CO_2 (material with production humidity).

Gaining for an easy to use method to distinguish between AAC being sensitive to carbonation or not, the effect on AAC properties like compressive strength after storing the material in different climate conditions have been investigated in several laboratories and institutes in Germany. However, these interesting results reflect the testing conditions but not the natural ageing of AAC. Only the most important results are shown in Figure 3.

Material of different raw densities was taken from different AAC factories and stored under different climate conditions:

- Series 1: Cubes with dimensions of $100 \times 100 \times 100\,mm^3$, production humidity, put into reference climate (20°C, 65% humidity) immediately after production stored in 14 days in 1 Vol.-% CO_2.
- Series 2: Cubes with same dimensions like before, production humidity, put into reference climate (20°C, 65% humidity) immediately after production stored in 28 days in 1 Vol.-% CO_2.
- Series 3: Cubes with dimensions like before, dried to 6–8%, put into reference climate (20°C, 65% humidity) immediately after drying stored in 14 days in 1 Vol.-% CO_2.
- Series 4: Cubes with dimensions like before, dried to 6–8%, put into reference climate (20°C, 65% humidity) immediately after drying stored in 28 days in 1 Vol.-% CO_2.

Afterwards the material was dried in an oven at a temperature of 60°C to a humidity of $6 \pm 2\%$ before the tests of compressive strength were carried out. This rest – humidity plays an important role on the results as we will see later on.

The results of the 14 day-storage are shown in Figures 3 and 4. Obviously, a great part of the material with production humidity shows a decrease in compressive strength after the 14 day carbonization. For the dried material instead, in most cases an increase in compressive strength was noticed (Figure 4).

Another series of humid and dried material was stored in a climate of 1-Vol.-% CO_2 for 28 days. During this time a failure at the CO_2-control happened and the material had to suffer a concentration of 5 Vol.-% CO_2 and more. The results are not shown here in detail, but the humid material showed a massive decrease in compressive strength after this experiment. The dried material instead showed the same behaviour like in Figure 4.

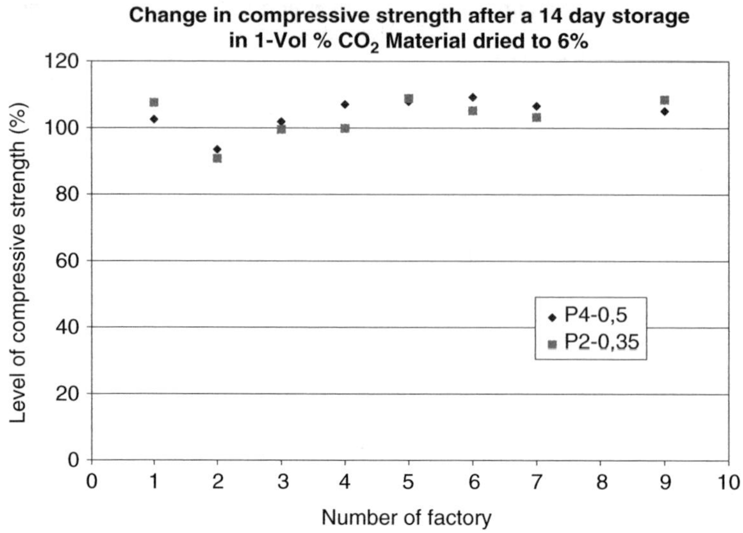

Figure 4. Change in compressive strength after a 14 day storage in 1 Vol.-% CO_2 (material dried to 6–8%).

The failure in the CO_2 control showed that the results of artificial carbonation are very sensitive to ambient conditions e.g. CO_2 concentration and humidity of material and air.

This influence will be discussed below.

4 ANALYSIS OF CARBONATION DATA USING ARTIFICIAL NEURAL NETWORKS

Due to the fact, that a lot of additional information was collected during the above described experiments, we were able to analyze the experimental data by artificial neural networks.

The main advantage of using this analysis method is, that all measured values can be sorted, analyzed and correlated at the same time. The software tool we used was the so called "Profiler" from the Austrian manufacturer "eudaptics" (www.eudaptics.com).

The software clusters all datasets by their most similar attributes and shows the results in so called "self organizing maps" (SOM). Each attribute (e.g. compressive strength, storage climate, rest – humidity before testing the compressive strength, etc.) is visualized in a separate window (Figure 5).

Afterwards a profile of selected values (e.g. those with massive decrease in compressive strength can be investigated in detail (Figure 6). Using this method, it is very easy to visualize multi – relationships between the data.

As show in Figure 6, not only the climate in which the material was stored (1 Vol.-% of CO_2 for 28 days.), but also the rest – humidity of the material while testing the compressive strength had an influence on the results of the compressive strength.

This was done not only for material with a decrease but also with an increase of compressive strength. However, the statistical analysis to explain the interrelation of the data are the same.

The most important result of this investigation of the experimental data is, that the test – humidity of the material has the same order of influence than the storage conditions.

Therefore it is not possible to distinguish between test – humidity and storage conditions as the reason for the change in compressive strength.

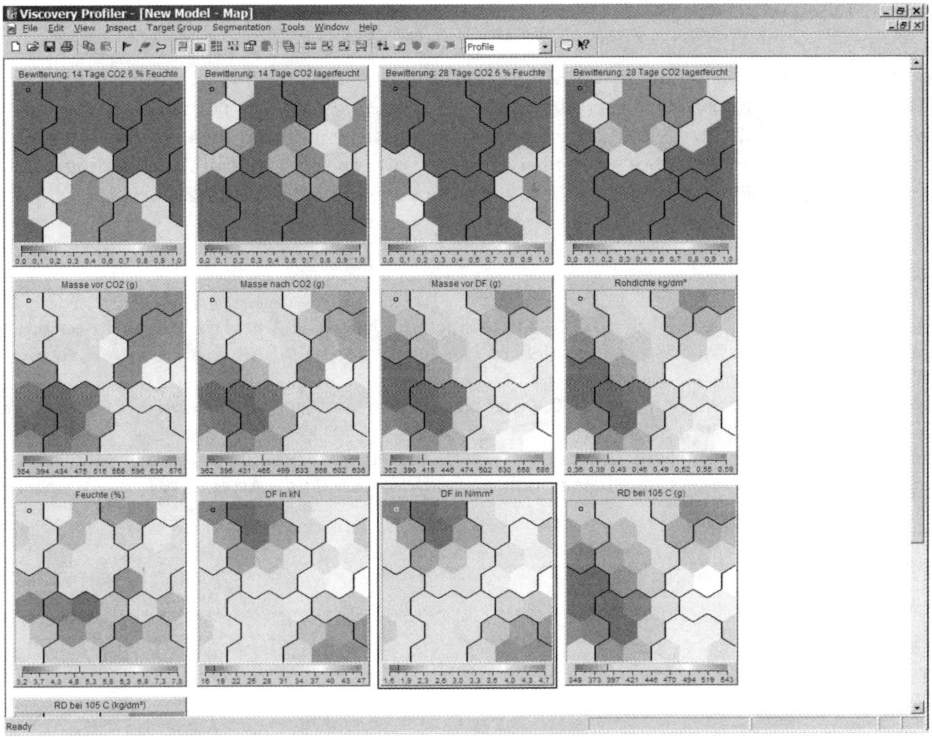

Figure 5. SOM Ward clusters of carbonation data

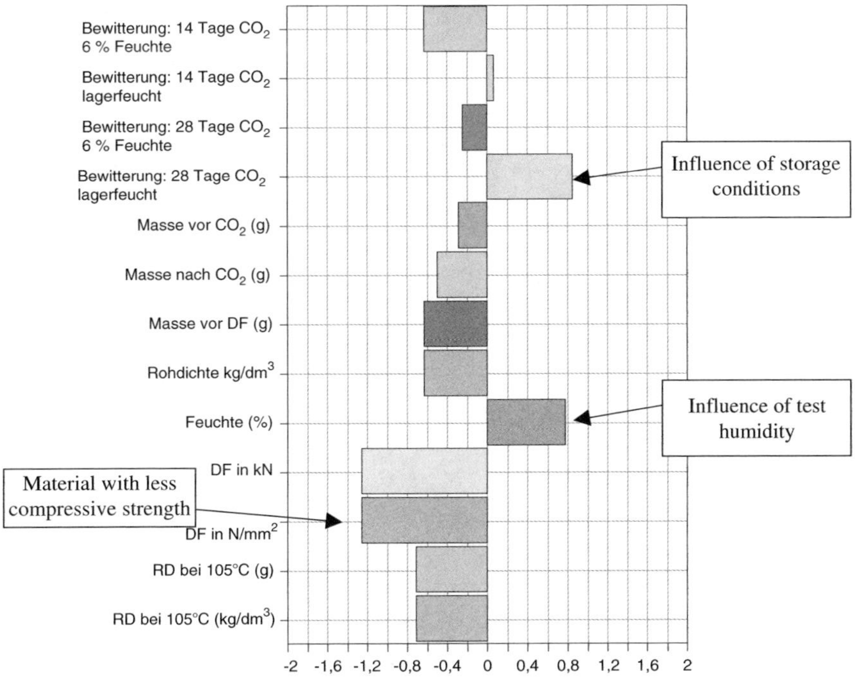

Figure 6. Profile of experimental results with strong decrease in compressive strength.

5 CONCLUSIONS

Artificial carbonation using increased CO_2 concentration in the test climate is not an appropriate method for the determination of the long time behaviour of AAC. It is not only that the changes in micro-level structure are quite different under accelerated conditions than normal, but also that the ambient conditions of the test setup play a decisive role on the results of the experimental data, especially the humidity of the material before testing the compressive strength.

REFERENCES

Matsushita F., Yoshimichi A., Shibata S. 2004 Microstructure Changes in Autoclaved Aerated Concrete during Carbonation under Working and Accelerated Conditions, Journal of Advanced Concrete Technology Vol. 2, No. 1, 121–129.

Matsushita F., Yoshimichi A., Shibata S. 2000. Carbonation degree of autoclaved aerated concrete, Cement and Concrete Research 30: 1741–1745.

Siauciunas R., Rupsyte E., Kitrys S., Galeckas V. 2004. Influence of tobermorite texture and specific surface area on CO_2 chemisorption. Colloids and Surfaces A. Physicochem. Eng. Aspects. 197–204. Elsevier B.V.

Theme Three: Codes, standards and regulations

Autoclaved Aerated Concrete – Limbachiya and Roberts (eds)
© 2005 Taylor & Francis Group, London, ISBN 0 415 38356 0

Aircrete blockwork subjected to concentrated loading

N. Bright
Consulting Engineer, Visiting Professor, Kingston University, Concrete and Masonry Research Group-Kingston upon Thames, Surrey, UK

ABSTRACT: National Design Codes, such as BS5628 Code of practice for use of masonry, which are used for the structural design of Aircrete blockwork, will be replaced by Eurocode 6: Design of Masonry Structures – Part 1-1: Common rules for reinforced and unreinforced masonry structures in European countries which are members of CEN. This will occur after a period of co-existence.

One aspect of Eurocode 6 which differs from existing UK practice in BS5628 is the treatment of Aircrete blockwork subjected to concentrated loads.

Both codes allow enhancement of the maximum permitted stress in the Aircrete blockwork under a concentrated load. These enhancement factors for stress immediately below the bearing vary as a result of different calculation procedures.In addition, Eurocode 6 and BS5628 assume different load spreads from concentrated loads to a level where the stress is assessed for the design of the wall.

These differences result in changes in the maximum concentrated load which may be applied to Aircrete blockwork. Some simple examples are compared and consequences considered.

1 INTRODUCTION

The structural design for concentrated loads on Aircrete blockwork in Eurocode 6, Masonry Structures – Part 1-1 Common rules for reinforced and unreinforced masonry structures (Final version 2004) and in BS5628: Part 1: Structural use of unreinforced masonry (1992) are compared.

Each code has a different approach to the enhancement factor permitted immediately under the bearing and to the total spread of the concentrated load when calculating the stress in the masonry at a specified level below the concentrated load. Both codes limit the maximum magnitude of a concentrated load to the lower of the enhanced bearing stress immediately under the concentrated load or the stress spread from the concentrated load at a prescribed distance below the bearing.

The simple example of a single leaf wall built of Aircrete blockwork laid in a full bed of mortar with a concentrically loaded bearing over the full thickness of the wall (a) at a point within the length of the wall and (b) at a point at the end of the wall is considered. To keep the comparison simple, the wall is of sufficient length to enable the concentrated loads to fully spread along the length of the wall to the level in the masonry where the stress is considered according to the requirement of the particular code.

The concentrated load considered is applied concentrically through a stiff 150 mm by 100 mm plate bearing across the full 150 mm thickness of a 2.5 m high wall.

In the first case, the concentrated load is applied within a length of the wall so that the load may spread in both directions along the wall to the maximum extent of spread allowed by the relevant code. In the second case, the concentrated load is applied at the end of a wall where the load may only spread in one direction. To fully compare the results of using the two codes it will be necessary to also compare the treatment of concentrated loads which do not bear over the full thickness of the wall and which are applied concentrically and eccentrically. In addition to this paper being limited to a simple comparison of the different treatments of concentrated loads, the differences in the treatment of uniformly distributed loading are not considered. These factors are the subject of further research being undertaken at Kingston University.

Figure 1. Detail from Aircrete concentrated load test panel.

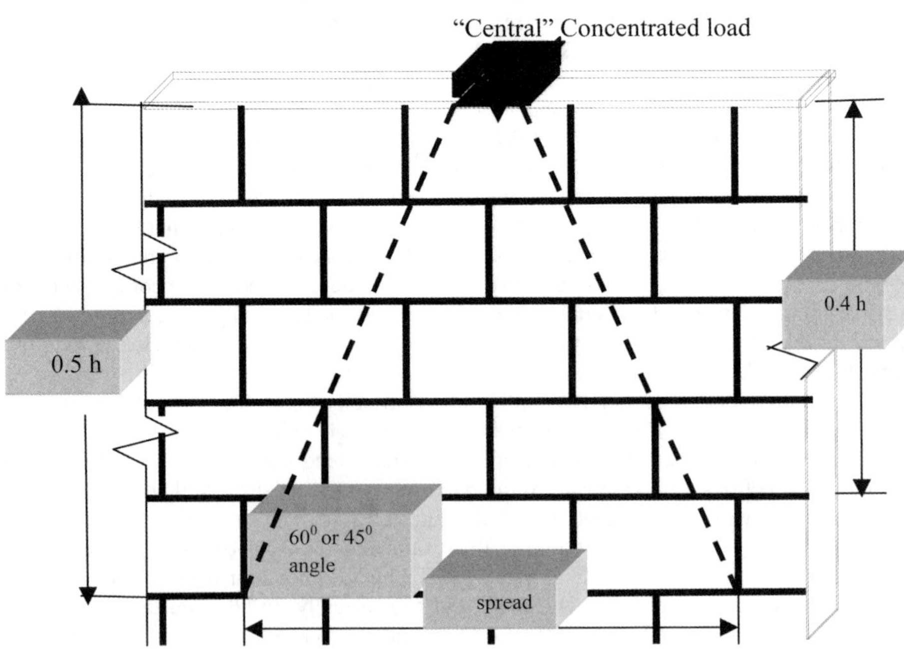

Figure 2. "Central" load.

End Concentrated load

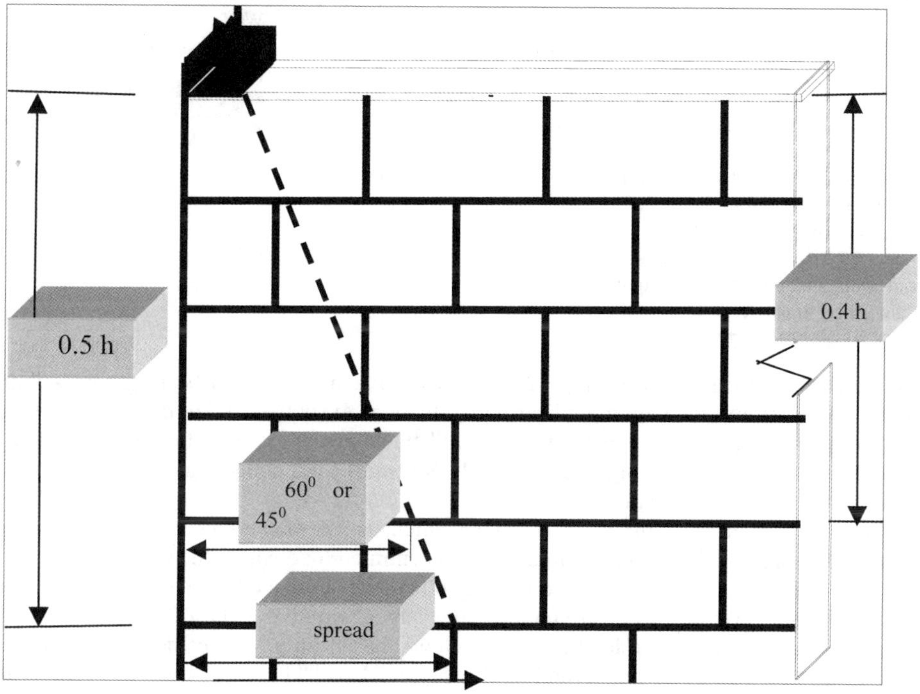

Figure 3. End load.

2 BRITISH STANDARD BS5628

Increased local stresses may be permitted beneath the bearing of a concentrated load of a purely local nature according to BS5268. The element applying the load should be sensibly rigid such as a beam, column or lintel, or a suitable spreader. The code illustrates in diagrammatic form the limits of what constitutes a purely local nature for different types of concentrated loads and whether a 1.25 enhancement or a 1.5 enhancement applies. The limits are a function of "t", the thickness of the wall and "x", the length of bearing perpendicular to one face of the wall. Or in the case of a beam spanning in the plane of the wall, the length of bearing perpendicular to the end of the wall. The case of a concentrated load remote from both faces of the wall does not appear to be explicitly illustrated.

The concentrated load may be assumed to be uniformly distributed over the area of the bearing and dispersed in two planes within a zone contained by lines extending downwards at 45° from the edges of the loaded area. The effect of the local load combined with stresses due to other loads should be checked:

a) at the bearing, assuming a local design bearing strength of $1.5fk/\gamma m$ for a bearing across the full thickness of the wall where fk is the characteristic strength of the masonry and γm is the partial safety factor for the material. For blocks with a ratio of height to least horizontal dimension between 2.0 and 4.0 with a compressive strength up to 5.0 N/mm^2, the characteristic compressive strength is equal to the block strength.
b) at a distance of 0.4 h below the bearing where the stress due to the design loads should be not greater the design strength. The stress from the concentrated load should be combined with that from the general loading above. The overlapping stress from more than one concentrated load may also need to be considered.

Table 1. BS 5628 calculations.

	Distance from wall end, m	h (m)	0,4 h (m)	Wall thickness (t, m)	Bearing width l (m)	Max. spread (m)
Concentrated load remote from the ends of the wall (bearing type 2)	Not less than 1.00	2.5	1.0	0.15	0.10	2.10
Concentrated load at end of the wall from beam spanning in the plane of the wall (bearing type 2)	0	2.5	1.0	0.15	0.10	1.10
Concentrated load at end of the wall from beam spanning perpendicular to the wall (bearing type 1)	0	2.5	1.0	0.15	0.10	1.10

For the first load case (bearing type 2 with an enhancement of 1.5) bearing across the full thickness "t" (150 mm), the maximum bearing width "l" is 4t (600 mm). For the second load case (also bearing type 2), again bearing across the full thickness t (150 mm) but applied at the end of the wall, the maximum bearing width "l" is 2t (300 mm) for a concentrated load from a beam spanning in the plane of the wall.

For both these load cases the value of enhancement is 1.5 of the characteristic compressive strength of the masonry immediately under the concentrated load where there is no reduction for slenderness. (In the second case, if the concentrated load was from a beam spanning perpendicular to the wall it would be bearing type 1 and the enhancement would only be 1.25.)

The calculated spread of the load (at 0.4 h below), based on a 45° spread, is 2.10 m for the first load case and 1.10 m for the second load case i.e. 21 and 11 times the width of the bearing respectively. See Table 1.

3 FINAL DRAFT EUROCODE 6 PART 1-1

According to EC6, when a wall with thickness "t", built with Aircrete units laid on a full bed of mortar is subjected to a concentrated load without any eccentricity in the plane of the wall, the design value of the vertical load resistance of the wall N_{Rdc} is given by:

The product of an enhancement factor β (calculated according to Equation 1), loaded area A_b and design strength of the wall f_d.

$$\text{Where } \beta = \left[1 + 0{,}3\frac{\alpha_1}{h_c}\right]\left[1{,}5 - 1{,}1\frac{A_b}{A_{ef}}\right] \quad (1)$$

and

α_1 is the distance from the end of the wall to the nearer edge of the loaded area,

h_c is the height of the wall to the level of the load,

A_b is the loaded area.

The loaded area is the product of the dimension of the bearing in the plane of the wall "l" and the thickness of the wall "t" where the bearing is across the whole thickness of the wall.

A_{ef} is the effective area of bearing.

and

$\frac{A_b}{A_{ef}}$ is not to be taken greater than 0.45.

Limitations for the maximum values of β are 1.5 or $1.25 + \alpha_1/2h_c$ whichever is the lesser.

(This implies that the load spread is assumed to be 45° although it is only allowed to be taken as 60°.)

Table 2. EC6 calculations.

	α, m	h_c, m	A, m^2	A_{ef}, m^2	t, m	l, m	β	$1.25 + \alpha_1/2h_c$	Max spread, m
Concentrated load remote from the ends of the wall	Not less than 1.25	2.5	0.015	0.232	0.15	0.10	1.55	1.5	1.54
Concentrated load at end of the wall	0	2.5	0.015	0.123	0.15	0.10	1.37	1.25	0.82

The effective length of the bearing is determined at the mid height based on a load spread of not more than 60° to the vertical from the edges of the concentrated load plus the width of the bearing "l".

The effective area of bearing is the product of the effective length of the bearing and the thickness of the wall "t" where the bearing is across the whole thickness of the wall.

For the first load case, the calculated value of enhancement β is 1.55 but the calculated cut off $(1.25 + \alpha_1/2h_c)$ for the maximum enhancement is 1.5 and therefore the limit of 1.5 of the design strength of the wall (f_d) applies in this case. See Table 1.

For the second load case, the calculated value of enhancement β is 1.37 but the calculated cut off $(1.25 + \alpha_1/2h_c)$ for the maximum enhancement is 1.25 and therefore limit of 1.25 of the design strength of the wall (f_d) applies in this case. See Table 1.

EC6 does not appear to distinguish between an end bearing from a beam spanning in the plane of the wall and one from a beam spanning perpendicular to the wall.

The calculated spread (effective length of bearing at mid height) is 1.54 m for the first load case and 0.82 m for the second load case based on a 60° spread i.e.15.4 and 8.2 times the width of the bearing. See Table 2.

4 COMPARISON OF THE RESULTS FROM CALCULATIONS ACCORDING TO THE TWO CODES

4.1 *Comparison of maximum dimensions of a concentrated load*

In BS5628 what is a "purely local nature" is shown in diagrammatic form for different types of concentrated loads. For a load bearing across the full thickness "t" at a distance greater or equal to 0.5x from the end of the wall, the maximum bearing area is $4t^2$. For a similar load at the end of the wall the maximum bearing area is $2t^2$.

In EC6, "$\dfrac{A_b}{A_{ef}}$ is not to be taken greater than 0.45" which appears rather generous.

4.2 *Comparison of assumed spreads in the codes*

The maximum spread for calculating the stress in the masonry is different in each code. EC6 only allows a maximum spread of 60° compared with the maximum of 45° in BS5628. This results in a reduction of approximately 40% in the maximum load which can be applied when designing to EC6 compared with BS5628. However, while EC6 calculates the area of dispersion at the mid height of the wall below the concentrated load, BS5628 calculates it "at a distance of 0.4 h below the bearing". This has the effect of a reduction of 20% for BS5628 relative to EC6. The precise difference depends on the width of the bearing plate.

In the examples chosen for this paper the spreads, for walls where there is no limitation from ends of walls or positions of movement joints etc on the assumed spreads at the level in the masonry where the design stress is calculated, range from:

For loads remote from the ends, 1.54 m in EC6 based on a 60 deg spread and 2.10 m at 0.4 of the storey height below the load for BS5628 based on a 45 deg spread.

For loads at an end, 0.82 m in EC6 based on a 60 deg spread and 1.1 m at 0.4 of the storey height below the load for BS5628 based on a 45 deg spread.

When the stress at mid wall height or 0.4 blow the concentrated load, as the case may be, is the limiting factor, the stress according to EC6 is 36% to 34% higher. Therefore, a given wall will support a concentrated load which is that much higher when designed to BS5628. EC6 and BS5628 both consider the resulting stress including that from overlapping spreads from adjacent concentrated loads.

4.3 *Comparison of enhancement factors*

The maximum enhancement factor for stress permitted immediately under the type of bearing being considered varies between 1.25 and 1.5 depending on the position of the load and differences between each code.

In EC6, the expression $1.25 + \alpha_1/2h_c$ has an odd effect for "central" concentrated loads. The term α_1 (the distance from the edge of the bearing to the end on the wall) only allows the maximum enhancement of 1.5 to be used when α_1 is not less than $h_c/2$ which implies a 45° spread but 60° is all that is allowed in the spread calculation. The expression also limits the maximum enhancement for an end load to 1.25.

In BS5628 the enhancement factor is either 1.5 or 1.25 generally, depending on what proportion of the wall thickness is supporting the bearing, when the load can spread to the maximum allowed.

An Aircrete block with a characteristic compressive strength of 2.8 N/mm^2 can therefore be subjected to a maximum compressive stress of 4.2 N/mm^2 or 3.5 N/mm^2 immediately beneath the concentrated load according to BS5628. The same block has a compressive strength of 2.9 N/mm^2 according to BS EN 772-1 Methods of test for masonry units – Part 1: Determination of compressive strength and can therefore be subjected to a maximum compressive stress of 4.35 N/mm^2 or 3.63 N/mm^2 immediately beneath the concentrated load according to EC6.

4.4 *Comparison of stress distribution immediately under the bearing*

Both codes consider concentrated loads to exert uniform pressure under the bearing plate.

In BS5628, the assumption of uniform distribution does not apply in the special case of a spreader located at the end of a wall and spanning in its plane. In this case the distribution of stress "should be based on an acceptable elastic theory" and the maximum stress should not exceed twice the characteristic stress over the partial safety factor for the material. In both codes, considering that generally the assumption is that there is a uniform bearing under the load, there is a remarkable lack on emphasis on the care needed to achieve a uniform bearing.

The results of the calculations from each of the codes are shown in Table 3.

Table 3. Enhancement factor under load and design length in Aircrete blockwork.

	Central		End	
Code	Enhancement factor under load	Design length in masonry (spread), m	Enhancement factor under load	Design length in masonry (spread), m
EC 6	1.5	1.54	1.25	0.82
BS5628	1.5	2.1	1.5 (for concentrated load from beam spanning in the plane of the wall (bearing type 2)	
			1.25 (for concentrated load at end of the wall from beam spanning perpendicular to the wall (bearing type 1)	1.1

5 CONCLUSIONS AND RECOMMENDATIONS

1. As shown in Table 3, the enhancements and spreads of concentrated loads on Aircrete blockwork in these simple examples are different when using each code. Although the enhancement factors are not always different, they are arrived at by different calculation procedures. The spreads in the Aircrete blockwork are significantly different.
2. In the case of end concentrated loads, EC6 should distinguish between those from beams spanning in the plane of the wall and those from beams spanning perpendicular to the wall.
3. BS5628 requires simpler calculations than EC6 to derive the enhancement factor. Consideration should be given to whether the increased complexity is justified taking account of the full range of variables including eccentric in the plane of the wall bearings.
4. A simplified approach is recommended when designing to EC6 to minimize calculations as in BS5628 e.g.
   - "For a "central" concentrated load applied concentrically and able to spread both sides at 45° to mid height of the wall below the concentrated load on an Aircrete wall, the enhancement factor is 1.5 where the bearing is across the whole thickness of the wall. The spread equals1.14h + bearing width."
   - "For an end concentrated load applied concentrically and able to spread to one side at 45° to mid height of the wall below the concentrated load on an Aircrete wall, the enhancement factor is 1.25 where the bearing is across the whole thickness of the wall. The spread equals 0.57h + bearing width."
5. The foregoing comments and conclusions regarding the differences between BS5628 and EC6 apply to all types of solid masonry units with the same compressive strengths and dimensions.

REFERENCES

BRITISH STANDARD BS 5628-1:1992 Code of practice for use of masonry. Part 1: Structural use of unreinforced masonry, BSI 1992

EUROPEAN STANDARD Final Vote Draft EN 1996-1-1 Eurocode 6. Design of masonry structures. Part 1-1. Common rules for reinforced and unreinforced masonry structures, CEN 2004

EUROPEAN STANDARD BS EN 772-1 Methods of test for masonry units – Part 1: Determination of compressive strength CEN 2000

Autoclaved Aerated Concrete – Limbachiya and Roberts (eds)
© 2005 Taylor & Francis Group, London, ISBN 0 415 38356 0

Development of ASTM standards for AAC

F.H. Fouad & A.C. Vichare
Dept. of Civil Engineering, University of Alabama at Birmingham, Birmingham, Alabama, USA

R.E. Barnett
Technical Director, Aercon Florida LLC, Haines City, Florida, USA

ABSTRACT: This paper provides an overview of the American Society for Testing and Materials (ASTM) standards that were developed in the U.S for AAC materials and products. These standards include ASTM C 1386 "Standard Specifications for Precast Autoclaved Aerated Concrete Wall Construction Units," ASTM C 1452 "Standard Specification for Reinforced Autoclaved Aerated Concrete Elements," ASTM C 1555 "Standard Practice for AAC Masonry," and most recently ASTM C 1591 "Standard Test Method for Determination of the Modulus of Elasticity of AAC." These standards cover a wide variety of technical information related to AAC including specifications, test methods, and construction practices. The first three standards were developed based on an extensive research program performed at the University of Alabama at Birmingham on domestically produced AAC. In the United States, the use of AAC has been somewhat limited. Only a few manufacturers are currently in existence, the material is relatively new, and technical information in the form of standards and specifications has not been readily available. It is believed that the development of ASTM standards on AAC is a step forward in the process to familiarize and promote the overall usage of the material in the U.S.

1 INTRODUCTION

ASTM (American Society for Testing and Materials) International is one of the largest voluntary standards development organizations in the world. It is a trusted source for technical standards for materials, products, systems, and services. ASTM develops standard test methods, specifications, practices, guides, classifications, and terminology in 130 areas covering subjects such as metals, paints, plastics, textiles, petroleum, construction, energy, the environment, consumer products, medical services and devices, computerized systems, electronics, and many others. Known for their high technical quality and market relevancy, ASTM International standards have an important role in the information infrastructure that guides design, manufacturing and trade in the global economy.

AAC has been used extensively through Europe, but only recently have major production plants been constructed in the United States. The first major production plant was not constructed in the U.S. until 1996. Two large European conglomerates have led the introduction of AAC in the U.S. The German Hebel Group through Hebel USA and Ytong Inc. subsidiary of a German company Ytong Holding began production of AAC in the United States in 1996 and 1997 respectively. Initially houses were constructed from imported AAC blocks and reinforced panels from Europe.

A building system composed of floor and roof panels with external block walls was marketed in the U.S. and successfully used for residential construction in resorts and housing developments in Florida. Today, AAC is gaining rapid acceptance as a new building product in the U.S. as a result of the increasing importance placed on conservation of energy (both the energy savings produced by thermal insulation and the amount of energy required for the mass production of the product). The rising cost of lumber and increasing environmental concerns have also played a role in the increasing interest in AAC. The availability of technical information and ASTM standards on AAC will serve as a tool to further promote the use of the material.

This paper provides a review of the existing literature associated with sustainability of AAC construction in comparison to other construction materials, including comparisons of embodied energy with the intent to show the advantages of AAC as an environmentally friendly material.

2 DEVELOPMENT OF ASTM STANDARDS FOR AAC

Research activities on AAC in the U.S. began in the late 1980s on imported AAC blocks and panels. However, once manufacturing plants were established in the late 1990s, it was evident that additional research on U.S.-manufactured AAC was needed. Research was sponsored by the AAC manufacturers and focused on the material properties of AAC and the structural behavior of steel reinforced AAC elements. The research studies, which were generally led by the University of Alabama at Birmingham and the University of Texas at Austin, formed the basis for the ASTM standards on AAC.

The process of developing ASTM standards on AAC started in 1992 through the efforts of ASTM Subcommittee C27.20 "Architectural and Structural Products." As the work on AAC standards increased to a level beyond the subcommittee scope, it was decided to organize a separate ASTM subcommittee that would specifically focus on AAC. In 1999 Subcommittee C27.60 "Precast Autoclaved Aerated Concrete" was formed, and around the same time Subcommittee C15.10 "Autoclaved Aerated Concrete Masonry" was also organized to address issues that are specifically related to construction of AAC masonry. Since that time, the members of both subcommittees have been working collaboratively in an effort to produce ASTM standards that would most benefit the AAC industry.

Through the efforts of ASTM Committees C27 and C15, four AAC standards were developed and published by ASTM over the past ten years. These standards are

- ASTM C 1386-98 "Standard Specifications for Precast Autoclaved Aerated Concrete Wall Construction Units"
- ASTM C 1452-00 "Standard Specification for Reinforced Autoclaved Aerated Concrete Elements"
- ASTM C 1555-03a "Standard Practice for Autoclaved Aerated Concrete Masonry"
- ASTM C 1591-04 "Standard Test Method for Determination of the Modulus of Elasticity of AAC".

3 HIGHLIGHTS OF CURRENT AAC STANDARDS

3.1 *ASTM C 1386-98*

The ASTM C 1386-98 "Standard Specifications for Precast Autoclaved Aerated Concrete Wall Construction Units" covers AAC block units used in the construction of solid nonload-bearing and load-bearing walls. The standard provides test procedures for determining the compressive strength, the moisture content, and the bulk density of AAC.

Details of a drying shrinkage test are also provided. The physical requirements of AAC, including the compressive strength, nominal dry bulk density, and average drying shrinkage for each of three strength classes of AAC are specified in the standard (Table 1). Dimensional tolerances for AAC standard units are also prescribed.

The compression test prescribes cube specimens of 10 mm edge length tested in an air dried condition at about 5 to 15% by mass moisture content. A minimum of three specimens are to be tested and whenever possible one specimen shall be obtained from the upper third of the block unit, one from the middle, and one from the lower third, determined in the direction of the rising of the mass during manufacture. The direction of rise shall be noted on all specimens.

Table 1. Physical requirements (Source:-ASTM C 1386-98).

Strength class	Compressive strength (MPa) Average	Min	Nominal dry bulk density (kg/m^3)	Density limits[3] (kg/m^3)	Average drying shrinkage (%)
PAAC-2	2.5	2.0	400	350–450	
			500	450–550	
PAAC-4	5.0	4.0	500	450–550	
			600	550–650	
			700	650–750	$\leq$0.02
			800	750–850	
PAAC-6	7.5	6.0	600	550–650	
			700	650–750	
			800	750–850	

The moisture content of AAC is determined by drying the samples in a ventilated oven at 100 to 110°C until two successive determinations of mass at 2-hour intervals show an increment of loss not greater than 0.2% of the last previously determined mass of the specimen. The exact dimensions are measured with a caliper gage and the volume of the specimens is determined. The dry bulk density of each specimen is determined by dividing the dry mass by its volume.

To determine the drying shrinkage the specimens shall be packaged in plastic and stored for a minimum of 24 hrs at 18 to 22 degree Celsius for uniform moisture distribution. It is then stored on grid to allow sufficient movement of air around it at a temperature of 20 ± 2 degree Celsius and a relative humidity of 45%. The specimens mass and length is determined regularly until the moisture content has decreased to below 4%. The moisture content of these samples shall then be determined and the average values for the relative change in length and the moisture content for each reading shall be shown graphically and connected by a curve.

3.2 *ASTM C 1452-00*

The ASTM C 1452 "Standard Specification for Reinforced Autoclaved Aerated Concrete Elements," covers load-bearing and nonload-bearing steel reinforced autoclaved aerated concrete floor, roof, wall, and stair elements used as components for building construction. For the wire reinforcement, the standard specifies minimum yield strength of 485 MPa and minimum tensile strength of 550 MPa. A minimum concrete cover over the steel reinforcement of 10 mm is specified.

The test for corrosion protection of the steel reinforcement in AAC is performed by immersing the reinforced AAC specimens in an aqueous sodium chloride solution, 3% NaCl by mass, for 2-hour periods at intervals of three days. This is repeated for a total of 10 test cycles. After completion, the specimens are allowed to air dry for 4 hrs and the AAC around the steel reinforcing is removed from both the reference and the test specimens. The area of rust covering the steel is determined by visual inspection and is expressed as a percentage of the total area of specimen.

The standard also provides a procedure to determine the weld-point shear strength of the prefabricated steel wire cages. The specimen is taken at random from the welded reinforcement cage before coating with the corrosion protection compound. The wire with the largest diameter shall be selected as the test specimen. The shear specimen shall be gripped in the test fixture such that the tension bar is centrally located and rotation of the anchoring bar is prevented. The loading rate shall not exceed 0.5 kN/s. A typical device for gripping the test specimen is illustrated in the standard.

As part of this standard, a full-scale test procedure to determine the lateral deflection and load bearing capacity of reinforced AAC floor and roof panels in bending is outlined. A symmetrical four-point loading scheme is used (Figure 1). The bearing length (S_b) is adjustable such that the

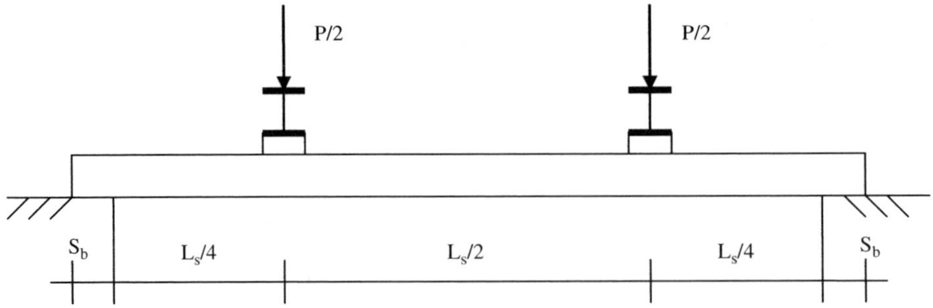

Figure 1. Transverse loading (flexural) test.

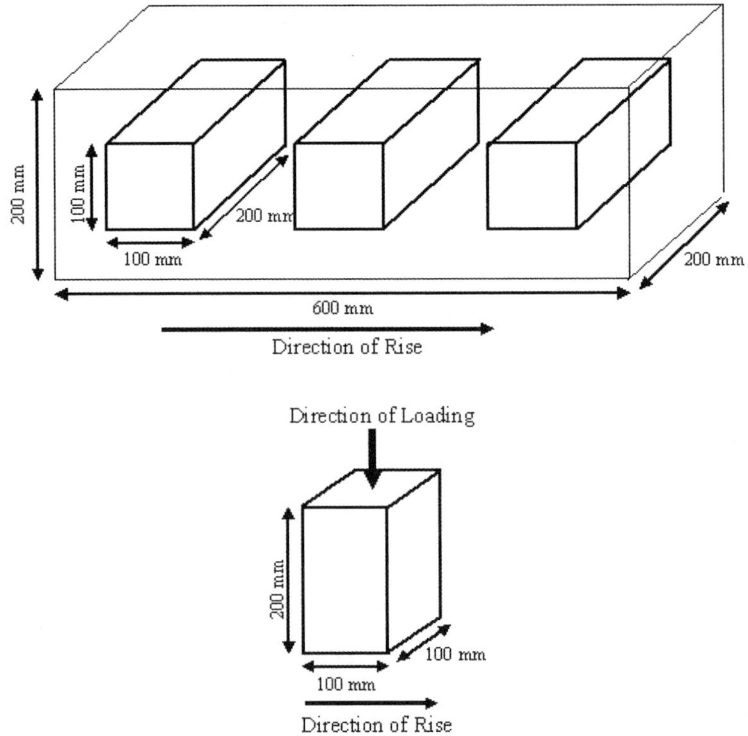

Figure 2. Number and orientation of compression prism specimens.

minimum bearing length is equivalent to the least bearing length supplied by the manufacturer or a minimum of 40 mm.

The two point loads are transmitted to the surface of the AAC element by steel profiles having sufficient bearing area to produce an even bearing pressure not exceeding 50% of the compressive strength of the AAC material. The ultimate load shall be recorded and observations made as to the mode of failure of the specimen.

3.3 *ASTM C 1591-04*

ASTM C 1591 "Standard Test Method for Determination of the Modulus of Elasticity of AAC" is the most recent ASTM test standard on AAC, which appeared in print in 2005. It covers the determination of the modulus of elasticity of AAC in compression as derived from stress-strain test data.

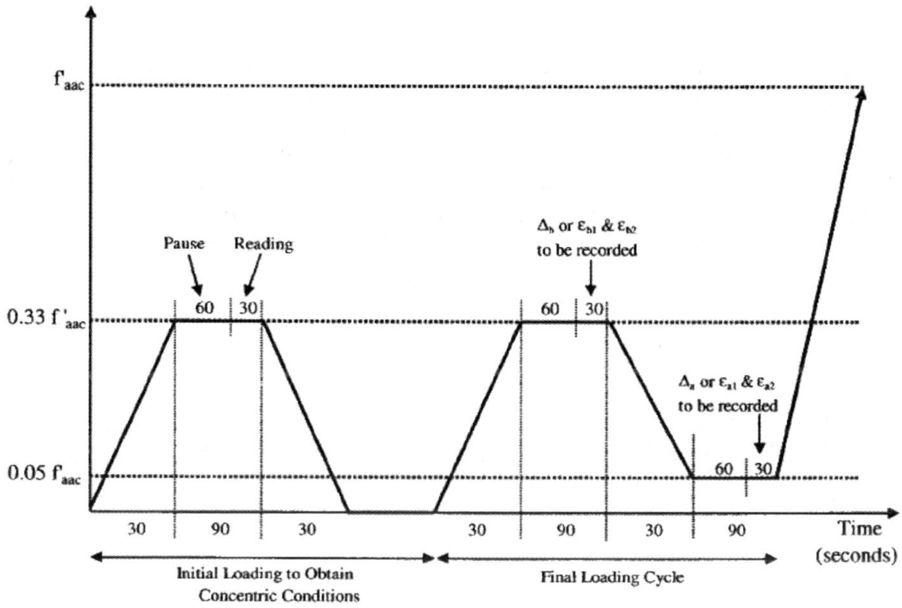

Figure 3. Modulus of elasticity loading cycle diagram.

The test specimens are prisms having the dimensions of 100 mm × 100 mm × 200 mm. The specimens are prepared such that the loading is applied to the 100 mm by 100 mm surface and is perpendicular to the direction of rising during manufacturing (Figure 2).

Electric strain gages, or a mechanical compressometer, for determining the strain of the specimen during loading are attached to two opposite longitudinal surfaces of the specimen in the vicinity of the middle portion. A base load is applied, which is equal to 0.33 times the ultimate expected compressive strength, f'_{aac}, for the material, and maintained for 90 seconds. The corresponding strains, ε_{b1} and ε_{b2}, are measured during the last 30 seconds of the applied load. If ε_{b1} and ε_{b2} deviate by more than 20%, the applied loading is considered eccentric. The specimen should then be unloaded, realigned, reloaded to 0.33 f'_{aac}, and the corresponding strains measured. When the strain readings at 0.33 f'_{aac} are within 20%, decrease the load gradually until a value of 0.05 f'_{aac} is reached (this should take approximately 30 s). This load is maintained for 90 s, and the corresponding strains, ε_{a1} and ε_{a2}, are measured during the last 30 seconds of the applied load. If the difference in readings from the two gages ($\varepsilon_{b1} - \varepsilon_{a1}, \varepsilon_{b2} - \varepsilon_{a2}$) corresponding to the applied loads of 0.33 f'_{aac} and 0.05 f'_{aac} deviate by more than 20%, the applied loading is considered eccentric. The specimen should then be unloaded, realigned, and the test repeated. If the difference in readings from the two gages ($\varepsilon_{b1} - \varepsilon_{a1}, \varepsilon_{b2} - \varepsilon_{a2}$) corresponding to the applied loads of 0.33 f'_{aac} and 0.05 f'_{aac} are within 20% of each other, the loading cycle above is repeated. The loading is taken to 0.33 f'_{aac}; read ε_{b1} and ε_{b2} and calculate the average ε_b; decrease the load to 0.05 f'_{aac}; read ε_{a1} and ε_{a2} and calculate the average ε_a. These values will be used to calculate the modulus of elasticity, E_{aac}. After completion of this second loading cycle the compressometer shall be removed, and the specimen loaded to failure. The complete loading cycle is illustrated in Figure 3.

The modulus of elasticity, E_c, is determined by:

$$E_c = \frac{f_b - f_a}{\varepsilon_b - \varepsilon_a}$$

where f'_a = stress recorded at 0.05 f'_{aac}; f_b = stress recorded at 0.33 f'_{aac}; ε_a = average strain calculated at 0.05 f'_{aac}, and ε_b = average strain calculated at 0.33 f'_{aac}.

3.4 *ASTM C 1555-03a*

ASTM C1555-03a "Standard Practice for Autoclaved Aerated Concrete Masonry" was developed to specifically address materials, tests, and workmanship of masonry made of AAC units. Direct reference is made to ASTM C 1386 for material specifications and testing. A note in the standard recommends using the modulus of elasticity test for AAC once it is developed by Subcommittee C27.60 (currently ASTM C 1591). The workmanship requirements of the ACI/ASCE Specification for Masonry Structures are adopted and supplemented by additional requirements that are specific to AAC masonry. A liquid permeability test for exterior surface treatments, which are used to protect AAC masonry exposed to weather, is described in detail.

4 FUTURE STANDARDS

Additional ASTM standards for AAC have been suggested and are currently under development. Test methods for flexural strength, shear strength, and freeze-thaw durability are examples of such standards. These tests are needed as current test methods for conventional concrete or masonry may not be directly applicable to AAC.

5 CONCLUSION

AAC is a fairly new construction material in the U.S. and as such technical information is needed to provide guidance on the use as well as testing of the material. The development of AAC standards through ASTM is a forward step to accomplish this goal. As the use of the AAC in the U.S. increases, it is expected that additional standards will be needed to address various performance aspects of the material.

REFERENCES

ASTM C1386-98: Standard Specification for Precast Autoclaved Aerated Concrete (PAAC) Wall Construction Units, Annual Book of ASTM Standards, American Society of Testing and Materials, West Conshohocken, PA, 1999.

ASTM C1452-00: Standard Specification for Reinforced Autoclave Aerated Concrete Elements, Annual Book of ASTM standards, American Society for Testing and materials, West Conshohocken, PA, 2001.

ASTM C1591-04: Standard Test Method for Determination of the Modulus of Elasticity of AAC, Annual Book of ASTM standards, American Society for Testing and materials, West Conshohocken, PA, 2005.

ASTM C1555-03a: Standard Practice for Autoclaved Aerated Concrete Masonry, Annual Book of ASTM standards, American Society for Testing and materials, West Conshohocken, PA, 2003.

ASTM WK56: Practice for Autoclaved Aerated Concrete Masonry, Annual Book of ASTM standards, American Society for Testing and materials, West Conshohocken, PA, 2005.

Dembowski, J., *A Study of the Material Properties and Structural Behavior of Plain and Reinforced AAC Components*, Thesis, University of Alabama at Birmingham, Birmingham, Alabama, 2001.

Fouad, F.H., "Physical and Mechanical Properties of AAC Produced in the United States," Report to the Autoclaved Aerated Concrete Products Association, 2002.

Snow, C., *A comprehensive study of the material properties and structural behavior of AAC products*, Thesis, University of Alabama at Birmingham, Birmingham, Alabama, 1999.

Autoclaved Aerated Concrete – Limbachiya and Roberts (eds)
© 2005 Taylor & Francis Group, London, ISBN 0 415 38356 0

The compressive strength of AAC – a statistical investigation

S. Wolf
TU Bergakademie Freiberg, Institut für Stochastik, Freiberg, Germany

S. Wiegand
Siltronic AG, Freiberg, Germany

D. Stoyan
TU Bergakademie Freiberg, Institut für Stochastik, Freiberg, Germany

H.B. Walther
Xella Technologie- und Forschungsgesellschaft mbH, Kloster Lehnin, Germany

ABSTRACT: The compressive strength of AAC is usually determined using original blocks or test cubes of 100 mm edge length. Conversion factors are used in order to transfer the results from cubes to blocks. These factors base on old studies with only a small number of tested blocks and should be statistically assured. New systematic tests were carried out in order to improve this situation, which have yielded the following results: (1) the compressive strength follows the Weibull distribution, (2) there are significant differences between different moulds, and (3) the strengths of blocks in the same mould can be considered as stochastically independent, even if close together. The question of the dependence of strength on block size remains open. The paper shows that there are plausible stochastic models which lead to different size effects.

1 INTRODUCTION

The intention of this paper is to contribute to the understanding of the behaviour of compressive strength of AAC, in particular to study the spatial variation of compressive strength in the mould and to explore the size effect of AAC, i.e. the dependence of strength on block size. The compressive strength of AAC is an important quality parameter and takes an decisive role in the quality assurance and control. It is practically determined by testing whole blocks as well as cubes (100 mm × 100 mm × 100 mm). If the very elegant method of testing cubes is used, the question arises how to transform the results to larger blocks. Here conversion factors are used, which were obtained by earlier experiments; they are closely related to the size effect. Until now there is no mathematical-statistical or physical theory for the size effect of AAC. Empirical results dominate, but such a theory is necessary for a scientifically based compressive strength test system. Furthermore, for such a system also knowledge of the probability distribution of compressive strength of cubes and blocks is important.

This paper reports on investigations carried out in 2003. In its first part the distribution pattern of compressive strength in two moulds is investigated. In the second part the authors present a statistical theory for the size effect of AAC in terms of stochastic models. Unfortunately, the experimental results are unclear and so there is a need to make further experiments to explain the size effect of AAC satisfactorily.

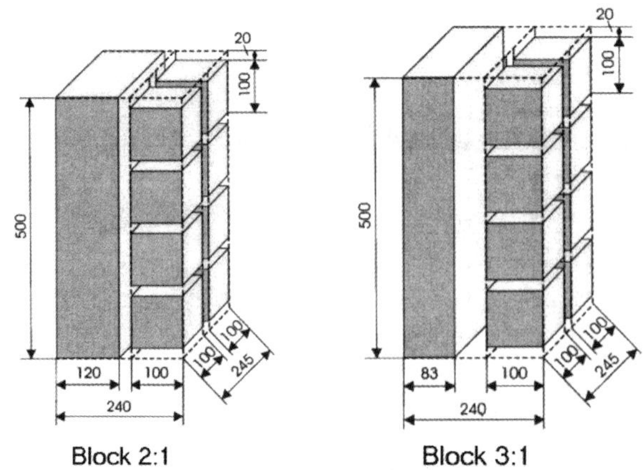

Figure 1. Test specimen view from above to top of the test faces. The cubes appear in four levels, where level 4 is the highest.

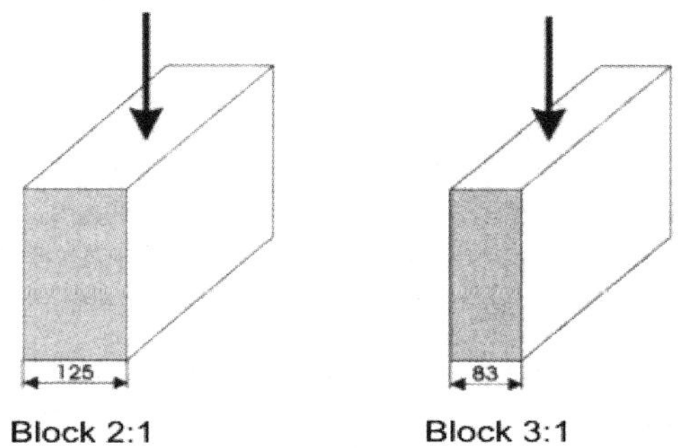

Figure 2. Load direction for the blocks.

2 EXPERIMENTAL DESIGN

In 2003 the R&D centre of Xella carried out extensive investigations dealing with the compressive strength of AAC to obtain the data required for a thorough statistical analysis. The material investigated was a PP4/0.5 produced in Schrobenhausen. Two subsequently produced moulds chosen arbitrarily from the current production were investigated. Each mould yielded 120 blocks of the size 240 mm × 250 mm × 500 mm. Each second row of blocks was cutted into smaller blocks and test cubes as shown in Figure 1. The values of compressive strength of 117 blocks of two different sizes and of 941 cubes were determined according to EN 771-4. The dimension of the cubes is 100 mm × 100 mm × 100 mm, and there are two block formats: block 2:1 (500 mm × 125 mm × 245 mm) and block 3:1 (500 mm × 83 mm × 245 mm). Also blocks 1:1 of size 500 mm × 240 mm × 245 mm were analysed, but the results are so unclear that they cannot be reported here. The original height of 250 mm of the blocks was decreased to 245 mm during the preparation of the smooth and parallel test faces required for the tests. The test blocks and cubes were carefully prepared. After grinding and drying the blocks and cubes at 60°C the compressive strength was determined using equipment from Toni Technik. All blocks and cubes were loaded contrary to the rising direction. Figure 2 shows the load direction for the two block sizes. The test specimen were mixed several times during preparation and drying. Therefore, the examination of the specimen took place in random order.

The following table shows the design of the compressive strength experiment.

Table 1. Basic parameters of the experiment.

	Block size n		Volume V (dm^3)	
	Mould 1	Mould 2	Mould 1	Mould 2
Block 2:1	30	28	15.24	15.21
Block 3:1	29	30	10.12	10.10
Cube	469	472	1.00	1.01

Table 2. Statistical characteristics of the strength (in N/mm^2) for blocks 1:1, 2:1, 3:1 and cubes.

Mould 1	Block 2:1	Block 3:1	Cube
n	30	29	469
$\bar{\sigma}$	5.11	5.02	5.11
s^2	0.02	0.03	0.09
s	0.13	0.18	0.30
Min	4.83	4.68	3.77
25%-quantile	5.02	4.85	4.93
Median	5.12	5.02	5.15
75%-quantile	5.20	5.12	5.31
Max	5.35	5.31	5.76
Mould 2	**Block 2:1**	**Block 3:1**	**Cube**
n	28	30	472
$\bar{\sigma}$	4.94	4.82	4.95
s^2	0.02	0.02	0.06
s	0.15	0.15	0.25
Min	4.63	4.48	4.25
25%-quantile	4.85	4.76	4.79
Median	4.99	4.81	4.99
75%-quantile	5.04	4.92	5.14
Max	5.16	5.04	5.52

3 STATISTICS OF THE COMPRESSIVE STRENGTH

3.1 *Basic statistical results*

Table 2 presents the basic statistical results for both moulds. Obviously, there exist considerable differences: The mean strengths $\bar{\sigma}$ (in the following we use only the term 'strength', omitting 'compressive') for mould 1 are greater than those for mould 2. A statistical test (the Welch test, see Montgomery and Runger, 1994, p. 413) showed clearly that the differences of the means for all blocks and cubes for mould 1 and mould 2 are significant; the p-value is usually smaller than 0.05.

Fortunately, the qualitative behaviour of the strength values for both moulds is similar, so that the same statistical theory can be applied; only some parameters differ. Therefore, in the following mainly mould 1 is considered.

The most interesting result in Table 2 is perhaps the behaviour of the mean strengths $\bar{\sigma}$ given in the second line. Obviously, the $\bar{\sigma}$-values are so close together that the hypothesis makes sense that for the mean strength there is no size effect. Indeed, the Welch test shows that there is no significant difference of the mean strengths between cubes and blocks 2:1 for both moulds. In the other cases the differences must be considered as significant. However, the Welch test assumes Gaussian distributions, which are not given in the case of AAC strength. A refined approach is the bootstrap method (see Chernick, 1999; Davison and Hinkley, 1997), which uses the empirical distribution (and its great variability) and is so more realistic. Using this method, all differences of mean strengths for mould 1 can be considered as not significant. Obviously the variance of strength decreases with increasing block size.

Unfortunately, the interpretation of these results is not easy. Perhaps they simply show random fluctuations around a constant mean value for cubes and smaller blocks without a size effect; or there may be a size effect of that type that increasing blocks have increasing strength. The reader should note that the cube values are avarages over cubes from all levels 1–4.

3.2 *Spatial variability of compressive strength in a mould*

The strength of cube samples clearly follows the rising direction. Consequently, the mean strengths depend on the levels, they decrease from below to above, see Table 3.

The Welch test can be used to show that the mean strength of the lower two cube levels 1 and 2 can be regarded as equal, while the other observed differences must be considered as significant.

In order to explore the spatial variability of strength in the horizontal direction, statistical methods of geostatistics (see e.g. Isaaks and Srivasta, 1989) were used. They showed clearly that in the two moulds there is practical no spatial correlation: The range of correlation is around 250 mm. Consequently the strength values of cubes and blocks can be considered as stochastically independent, they fluctuate randomly around the mean strength of the whole mould.

3.3 *The Weibull distribution*

The Weibull distribution is a classical strength distribution, which is based on the weakest link assumption, see Section 4 below. It depends on two parameters, a scale parameter σ_0 and a shape parameter m, which is called Weibull modulus. The Weibull distribution function is given as

$$F(\sigma) = 1 - \exp\left\{ -\left(\frac{\sigma}{\sigma_0} \right)^m \right\} \quad , \ \sigma \geq 0 \tag{1}$$

Under homogeneity assumptions, the parameter σ_0 can be expressed in the form

$$\sigma_0 = \left(\frac{m}{aV} \right)^{\frac{1}{m}}, \tag{2}$$

where a is a further material parameter.
The mean of the Weibull distribution is

$$\bar{\sigma} = \sigma_0 \Gamma\left(1 + \frac{1}{m}\right) \tag{3}$$

and the variance

$$Var\sigma = \sigma_0{}^2 \left(\Gamma\left(1 + \frac{2}{m}\right) - \left(\Gamma\left(1 + \frac{1}{m}\right) \right)^2 \right), \tag{4}$$

where $\Gamma(x)$ denotes the gamma function.

Table 3. Mean strengths for the different levels for mould 1 and mould 2.

	Mould 1				Mould 2			
	Level 1	Level 2	Level 3	Level 4	Level 1	Level 2	Level 3	Level 4
$\bar{\sigma}$	5.27	5.29	5.09	4.79	5.09	5.12	4.91	4.69

We could show that the strength of all AAC blocks follows the Weibull distribution in good approximation, however with different parameters. As an example, Figure 3 shows the histogram for the cubes of mould 1 in comparision with the Weibull probability density function. Here the parameters are $\hat{m} = 20.38$ and $\hat{\sigma}_0 = 5.24$. Figure 4 shows the corresponding distribution in the traditional doubly logarithmic plot.

According to the Weibull theory (Bazant and Planas, 1998; Krajcinovic, 1996) the parameters m and a should be the same for all block sizes, and block size should enter the strength distribution via volume V and Formula 2. In contrast, we found that a and m depend on V, see Table 4.

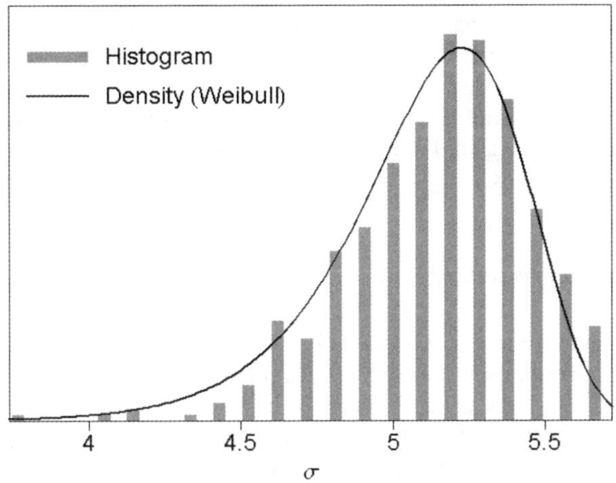

Figure 3. Histogram and Weibull density function for the cubes of mould 1.

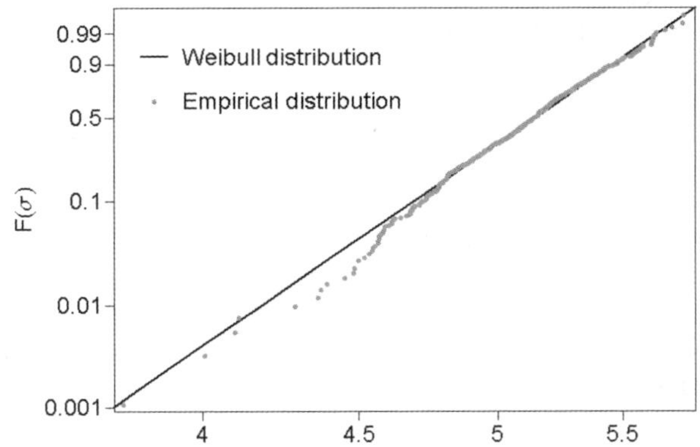

Figure 4. Theoretical and empirical distribution function for the cubes of mould 1.

Table 4. Estimates of a and m for the blocks of mould 1.

	V	$\hat{a}$	$\hat{m}$
Cube	1.00	$4.40 \cdot 10^{-14}$	20.38
Block 3:1	10.12	$2.01 \cdot 10^{-22}$	31.36
Block 2:1	15.24	$2.66 \cdot 10^{-33}$	46.33

Table 5. Means and standard deviations for strength according to the Weibull theory.

	V	$\bar{\sigma}$	$s(\sigma)$
Cube	1.00	5.10	0.3023
Block 3:1	10.12	4.56	0.2704
Block 2:1	15.24	4.47	0.2650
Block 1:1	29.40	4.32	0.2561

The Weibull theory also implies a decrease of $\bar{\sigma}$ increasing V. In order to demonstrate this effect, we fix the parameters $\hat{a} = 4.40 \cdot 10^{-14}$ and $\hat{m} = 20.38$ found for the cubes of mould 1 for all observed volumes. Formula 3 then yields the values in Table 5 for mean strength $\bar{\sigma}$ and standard derivation $s(\sigma)$. Since theoretical calculations are carried out, also the case of 1:1 blocks can be considered.

We observe clear deviations from the true values of $\bar{\sigma}$ in Table 2 and conclude that the Weibull theory cannot be applied to AAC.

4 A NEW STATISTICAL THEORY

More sophisticated models can explain other forms of size effect, for example nearly constant strength for increasing size or even increasing strength, as it was sometimes observed for AAC.

We present here such models which belong to the large class of stochastic models with size effects, which are well-known for brittle or quasi-brittle materials (Bazant and Planas, 1998; Krajcinovic, 1996). Note that it can happen that Weibull distributions appear but the Weibull theory is nevertheless not true.

The following stochastic model tries to explain the observations. It depends on three material parameters a, α and m, which are the same for all blocks of the same mould, while all distributional differences for different block sizes result simply from variation of block volume V. The model is closely related to the Weibull failure model. It will be explained so that both models, Weibull and the new model, can be jointly understood.

It is assumed that there are randomly distributed flaws in the material. These flaws are passive first. With increasing load σ more and more flaws are activated. The mean number of activated faults for load σ is

$$\Lambda(\sigma) = a\sigma^m V/m\,, \tag{5}$$

where a is a material constant, m another material constant (which could be called Weibull modulus) and V block volume. The random values of load $\sigma_1, \sigma_2, \ldots$, at which flaws are activated, follow a non-homogeneous Poisson process with intensity function

$$\lambda(\sigma) = a\sigma^{m-1}V\,, \tag{6}$$

$$\text{where } \Lambda(\sigma) = \int_0^\sigma \lambda(\tau)d\tau\,.$$

This model leads to the Weibull model if the weakest link assumption is made: The first activated flaw leads to fracture. The corresponding random variable σ_1 has then the Weibull distribution function

$$F(\sigma) = 1 - \exp\left\{-\frac{aV}{m}\sigma^m\right\}, \ \sigma \geq 0\,. \tag{7}$$

Table 6. Means and standard derivations of the sample model.

	Model				Experiment	
	Exp		Const			
	$\bar{\sigma}$	$s(\sigma)$	$\bar{\sigma}$	$s(\sigma)$	$\bar{\sigma}$	$s(\sigma)$
Cube	5.05	0.48	4.92	0.40	5.11	0.30
Block 3:1	5.07	0.23	5.07	0.16	5.02	0.18
Block 2:1	5.08	0.18	5.08	0.13	5.11	0.13
Block 1:1	5.09	0.13	5.09	0.09		

Clearly, both elementary understanding of the mechanical properties of AAC and the experimental results show that the weakest link model cannot be true. There is obviously some form of failure accumulation. This leads to the following assumption:

(a) Each flaw has some random size; the size of the i-th flaw, which is activated at load σ_i, is a random variable S_i. The S_i are independent and identically distributed. (Below it is assumed that the S_i are either exponentially distributed or constant.)
(b) The block suffers fracture when the cumulative sum of the S_i, linearly interpolated between the jumps, exceeds some level. More precisely, the block breaks for σ given as

$$\sigma = \sigma_{k-1} + \frac{\alpha V - \sum_{i=1}^{k-1} S_i}{S_k}(\sigma_k - \sigma_{k-1}), \tag{8}$$

where

$$\sum_{i=1}^{k-1} S_i < \alpha V \quad \text{and} \quad \sum_{i=1}^{k} S_i > \alpha V .$$

That means we assume that the critical accumulated failure load is proportional to block volume. The random strength of the block is then equal to σ.

It may look that assumption (b) ensures that there is no size effect. However, this is not the case. For example, for $S_i \equiv 1$ (all flaws are of the same size 1), cubes are significantly weaker than blocks. In contrast, the assumption that the S_i follow an exponential distribution with mean 1 makes that the mean $\bar{\sigma}$ is practically independent of V. Formulas from the applied probability literature (Rolski et al., 1999) can be used to derive the theoretical strength distribution.

For the following parameters we obtained values comparable with those for mould 1:

$\hat{m} = 15.0,$
$\hat{\alpha} = 0.5,$
$\hat{a} = 1.84 \cdot 10^{-10}.$

These parameters were chosen so that the mean strength of 5.08 is obtained, the average of $\bar{\sigma}$ for the three smaller block sizes. The corresponding means and standard derivations are given in Table 6. Also here results for 1:1 blocks can be given since theoretical results are reported.

Figure 5 shows the theoretical strength density functions for the model with the parameters given. They are quite similar to the empirical distributions, e.g. exponential that in Figure 3 for cubes. Figure 6 shows empirical distribution functions of simulated samples of random strengths in the Weibull diagram (log-log-plot). The curves are nearly straight lines, which confirms the fact of being close to Weibull distributions.

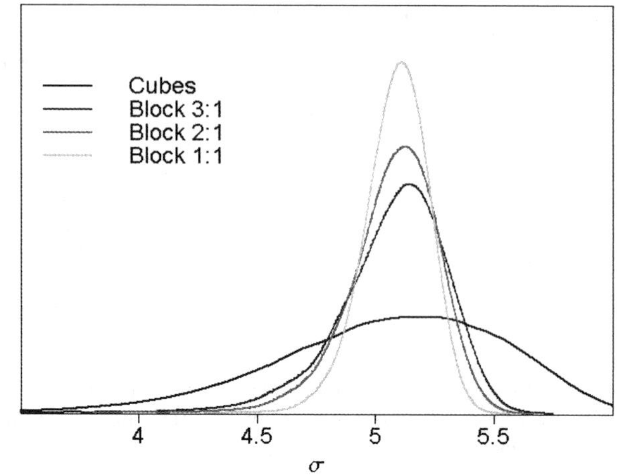

Figure 5. Probability density functions for the failure accumulation model.

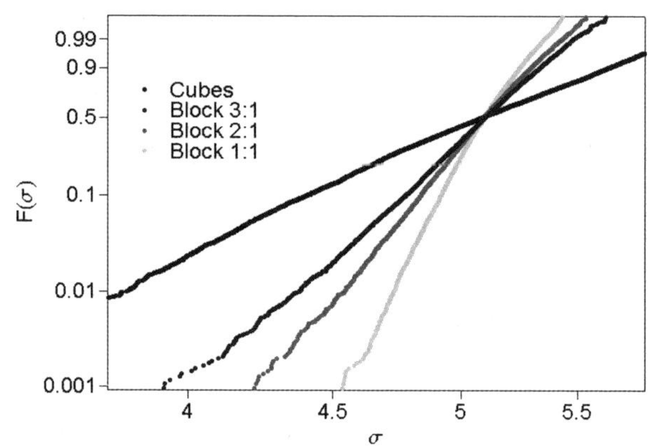

Figure 6. Simulated strength distribution functions for the failure accumulation model for blocks and cubes.

5 DISCUSSION AND CONCLUSIONS

The statistical analyses have clearly shown that there is a considerable vertical trend of compressive strength in the mould; test samples taken from the bottom are significantly stronger than those from the top. Furthermore, in horizontal direction there is no spatial correlation. It can be assumed that there is statistical independence, the strengths of cubes and blocks of the same mould fluctuate randomly around the corresponding mean strength for the whole mould.

The size effect of AAC is to be still an open question. There are experimental data which say that there is not a size effect, while others say that strength increases with increasing block size. As the theoretical results of the paper show, statistical fracture theories cannot help to solve the problem since plausible models can be created which have arbitrary size effects, increasing, constant or decreasing strength with increasing block size. Consequently, new experiments with larger number of samples of large blocks are necessary. Perhaps a further development of the investigations described in Kadashevich & Stoyan (2005) may support such experiments theoretically.

REFERENCES

Bazant, Z.P. & Planas, J. 1998. *Fracture and Size Effect in Concrete and other Quasibrittle Materials*. Boca Raton, Boston, London, New York, Washington, D.C.: CRC Press.

Chernick, M.R. 1999. *Bootstrap Methods: A Practitioner's Guide.* New York: Wiley.

Davison, A.C. & Hinkley, D.V. 1997. *Bootstrap Methods and their Application.* Cambridge: Cambridge University Press.

Isaaks, H.E. & Srivasta, R.M. 1989. *Applied Geostatistics.* New York, Oxford: Oxford University Press.

Kadashevich, I. & Stoyan, D. 2005. *Micromechanical analysis of AAC.* This volume.

Krajcinovic, D. 1996. *Damage Mechanics, Applied Mathematics and Mechanic.* Amsterdam, Lausanne, New York, Oxford, Shannon, Tokyo: Elsevier. *Volume 41.*

Montgomery, D.G. & Runger, G.C. 1994. *Applied Statistics and Probability for Engineers.* New York: Wiley.

Rolski, T., Schmidli, H., Schmidt, H. & Teugels, J. 1999. *Stochastic Processes for Insurance and Finance.* New York: Wiley.

Autoclaved Aerated Concrete – Limbachiya and Roberts (eds)
© 2005 Taylor & Francis Group, London, ISBN 0 415 38356 0

The proposed ACI guide for the use of AAC panels

F.H. Fouad
The University of Alabama at Birmingham, Birmingham, Alabama, USA

R.E. Klingner
The University of Texas at Austin, Austin, Texas, USA

R.E. Barnett
AERCON Florida, LLC, Haines City, Florida, USA

J.E. Tanner
The University of Wyoming, Laramie, Wyoming, USA

ABSTRACT: The paper provides an overview of the American Concrete Institute (ACI) proposed recommended practice for the use of reinforced AAC panels (ACI 523.5R). The recommended practice has not been officially published but is in its final stages of development. It was conceived in an attempt to provide the U.S. engineer with provisions for the analysis and design of AAC factory-produced reinforced panels. The practice is based on various research studies and experiences on the use of AAC in the USA.

The ACI recommended practice addresses the materials, manufacture, and structural design of reinforced AAC panels. Additional design considerations such as the erection and construction of reinforced AAC panels are also covered. Typical construction details are illustrated to ensure that the engineer can incorporate the use of AAC products in conventional construction. Sample design calculations are provided to illustrate the use of the various provisions of the recommended practice. The design provisions are not intended for use with unreinforced, masonry-type AAC units. Design of those units is covered by provisions currently under development by the Masonry Standards Joint Committee.

1 BACKGROUND

1.1 *AAC development in the U.S.*

Autoclaved Aerated Concrete (AAC) has been in use for over 60 years. It was first produced in Sweden in 1929 based on a patented process by Johan Eriksson, and since that time the material has gained wide acceptance in Europe as well as across the world. More than 300 production facilities are now in operation in Europe and many parts of the world including Japan, Australia, China, the Middle East, Central and South America, and the USA.

Although AAC was used on a small scale in the U.S. during the sixties, the first major production plant was not constructed until 1996. The plant was built by a German company (Hebel) and is located in Adel, Georgia. In 1997 a second production plant was in full operation in Haines City, Florida; this plant was built and operated by another German company (Ytong). Both plants were state-of-the-art that produced a wide range of AAC units for commercial and residential construction. The Georgia plant stopped production in 2002 due to business reasons and currently the Florida plant is operating under the ownership of Aercon. The constant rising cost of lumber, increase in environmental awareness, and the need for a more durable construction material should pave the way for the entry of AAC into the U.S. market. The material has been gaining rapid acceptance and numerous construction projects are currently underway in the southeastern and south central regions of the U.S. Additional production plants are now in the planning or construction stage by investor companies, and it is expected that the material will soon be available to all parts of the country (Fouad 1998).

1.2 *The need for design standards*

AAC may be used in unreinforced or reinforced construction. Unreinforced construction is basically construction using AAC blocks. Traditionally, the design has been based on elastic theory using service loads and permissible unit stresses. Although some of the design rules in ACI 530 for masonry structures, and ACI 318 (Chapter 22-Structural Plain Concrete), may be used, the information in these codes must be modified to account for the AAC material properties. Leadership in the design of AAC masonry has been assumed by the Masonry Standards Joint Committee (MSJC), which is sponsored by the American Concrete Institute (ACI), the American Society of Civil Engineers (ASCE), and The Masonry Society (TMS). Very recently, the MSJC produced the 2005 Building Code Requirement for Masonry Structures (ACI 530-05/ASCE 5-05/TMS 402-05), which includes an appendix on the design of AAC masonry.

In the case of reinforced elements, the reinforcement adds significantly to the complexity of the design. ACI 318 does not address AAC; but European standards and codes such as DIN, RILEM, etc. are available and do provide detailed coverage of the material. These standards and codes, however, differ in their treatment of AAC design. Some use an ultimate strength approach, with load factors assigned to service loads and capacity reduction factors to account for understrength, and others utilize a simpler allowable stress design approach.

In the United States, the use of AAC has been somewhat limited due to marketing conditions, lack of applicable building codes, lack of familiarity with design methodologies, and limited understanding of the advantages of AAC by the design and construction industry. The availability of the technical information will encourage the engineer to specify and design AAC and hence promote the overall use of the material. To that end, ACI Committee 523 has taken the lead in developing a document on the design of AAC reinforced panels.

2 ACI GUIDE FOR THE USE OF AAC PANELS

2.1 *Formation of the ACI subcommittee on AAC*

ACI Committee 523 "Cellular Concrete" was created in 1958 with the mission of developing and reporting information on cellular concrete materials and products. The committee addressed both cast-in-place and precast cellular concrete and produced the following three committee reports:

- Guide for Cast-in-Place Low-Density Cellular Concrete (ACI 523.1R)
- Guide for Precast Cellular Concrete Floor, Roof, and Wall Units (ACI 523.2R)
- Guide for Cellular Concrete Above 50 pcf (ACI 523.3R).

The focus of ACI Committee 523, however, was on cellular concretes produced in the field using preformed foam or foam concentrates blended in high speed mixers.

It was not until late 1997 when the membership of the committee expressed interest in forming a separate committee to address AAC. Since AAC was just an emerging product in the U.S. at that time, it was decided to organize a subcommittee under the jurisdiction of ACI 523 that would focus on AAC. As such, Subcommittee 523A "Autoclaved Aerated Concrete" was formed, and its first official meeting took place in the spring of 1998. The mission statement of ACI 523A stated to "develop and report information on the manufacture, design, construction, and handling of autoclaved aerated concrete products."

The first order of business for ACI Subcommittee 523A was the development of a Guide for the use of AAC panels. The scope of the report was agreed upon, assignments were made, and the work on the document began in 1998. It was evident that such a document will be of benefit to the industry and will provide the design engineer and user with the necessary technical information for the proper design and construction of AAC.

2.2 *Scope of ACI Guide*

The ACI Guide for the Use of AAC Panels, hereinafter referred to as the ***Guide***, is written primarily for the structural designer. It addresses the design of factory-reinforced AAC panels with an AAC density of 50 lb/ft^3 (800 kg/m^3) or less.

The Guide consists of six chapters and three appendices. Chapter 1 provides a general introduction including a brief history of AAC, typical physical and mechanical properties of the material, and some of the more common applications of AAC panels. The manufacture of AAC, including processing and curing of the raw materials, is addressed in Chapter 2. Dimensional tolerances of AAC panels as well as block units are also given in the chapter. Chapter 3 outlines the basic design assumptions and the proposed design provisions for AAC subjected to different straining actions including flexure, shear, axial, and bearing. Handling and construction requirements specific to AAC are addressed in Chapter 4. Chapter 5 illustrates typical construction details. Chapter 6 provides a list of references that were used primarily in the development of the Guide. Details of the design provisions are given in Appendices A and B, and Appendix C provides sample design examples.

The remainder of the paper will provide an overview of the structural design philosophy and requirements for reinforced AAC panels since this topic represents new information that was developed specifically for AAC based on research activities in the U.S.

3 STRUCTURAL DESIGN OF REINFORCED AAC PANELS

3.1 *Overview*

The design provisions of ACI 318 "Building Code Requirements for Structural Concrete" address the strength design of a wide variety of conventional reinforced concrete elements similar to AAC applications, including prefabricated wall panels. Based on the combination of test results from The University of Texas at Austin, the University of Alabama at Birmingham, and elsewhere, a proposed strength design approach, which is consistent with ACI 318-05 philosophy, was developed for factory reinforced AAC panels (Argudo 2003, Dembowski 2001, Fouad 2002, Snow 1999, Tanner 2003). This approach includes design provisions for flexure, shear, and anchorage that are generally similar to current ACI 318 strength design provisions for reinforced concrete elements.

3.2 *Basic design assumptions*

The design provisions ACI recommends for reinforced AAC panels are based on the same principles used for the strength design of conventional reinforced concrete elements: strain compatibility between AAC and reinforcement, stress-strain behavior of AAC and reinforcement, and equilibrium. AAC's design strength in compression is based on a specified design compressive strength, f'_{AAC}, which is verified by compressive strength testing using ASTM C1386 once the AAC panels are fabricated. AAC's design strength in tension is proposed as a function of the specified compressive strength. The design strength of reinforcement in tension is proposed as the specified yield strength (ACI 523.5R Draft Committee Report 2005).

3.3 *Combinations of flexure and axial load*

Using principles similar to those for conventional reinforced concrete, AAC panels are designed for combinations of flexural and axial load. Nominal capacity is computed assuming plane sections;

strain in tensile reinforcement depends on whether the section is tension- or compression-controlled; the stress in compressive reinforcement is computed based on its strain and its stress-strain behavior; and the distribution of compressive stress in the AAC is approximated by an equivalent rectangular stress block. For the complete flexure and axial load behavior, an interaction diagram can be constructed using the principles described above for various levels of axial load (ACI 523.5R Draft Committee Report 2005).

The factor β_1, which is used in defining the equivalent rectangular stress block, is taken as 0.67 for AAC and the minimum reinforcement of flexural members ($A_{S\,min}$) is calculated by Equation (1).

$$A_{S\,min} = \frac{0.33\sqrt{f'_{AAC}}}{f_y} bd \tag{1}$$

where b = width of cross section, mm; d = distance from extreme compression fiber to centroid of tension reinforcement, mm; f'_{AAC} = specified compressive strength of AAC, MPa; and f_y = yield strength, MPa.

3.4 *Bond and development of reinforcement*

AAC panels are reinforced with prefabricated steel wire cages that must be coated prior to casting the slurry to provide additional protection against steel corrosion. Deformed reinforcement installed in 3- to 4-in. grouted cores is also used as the panels are erected. The bond and development requirements for deformed reinforcement in grout are identical to those used for concrete or masonry construction. The anchorage mechanism for the prefabricated steel wire cages embedded in AAC is quite different from that of conventional concrete because the bond of the steel wire cage to AAC is negligible. Hence, the development of the flexural longitudinal reinforcement is primarily due to the cross wires bearing against the AAC. For typical cross-wire spacings, local crushing of the AAC under the cross wires can be assumed to redistribute the bearing stresses under the cross wires, leading to a uniform bearing strength of f'_{AAC} under the cross-wire. Multiplying this stress by the number of cross wires and by the bearing area of each cross-wire gives the maximum force that can be developed in the steel wire cage (Figure 1).

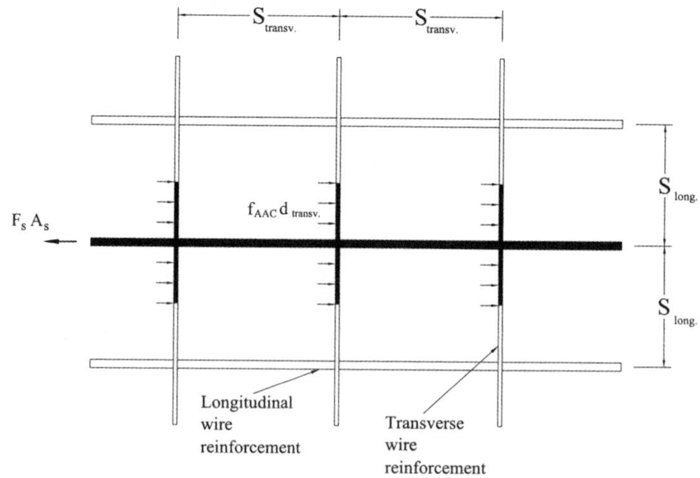

Figure 1. Anchorage mechanism of steel wire cages in AAC.

This maximum force in the steel wire cage can limit the flexural capacity of a reinforced AAC panel. The number of cross wires on both sides of the point of maximum moment of the panel shall equal or exceed the value required to satisfy Equation (2).

$$n_{cross,\min} = \frac{M_u/\phi}{(d - a/2) \cdot d_{cross} \cdot l_{cross} \cdot f'_{AAC}} \tag{2}$$

where a = depth of the equivalent rectangular stress block, mm; d_{cross} = diameter of cross-wires, mm; l_{cross} = length of reinforcement bearing on AAC, mm; M_u = factored moment at section, N-mm; d = distance from extreme compression fiber to centroid of tension reinforcement, mm; f'_{AAC} = specified compressive strength of AAC, MPa; and ϕ = strength reduction factor.

The number of cross-wires within a distance of 2h (h = panel thickness), measured in each direction from each end of the panel, shall equal or exceed one-half the value calculated in the equation, but not less than 2. A minimum of one cross wire shall be located in each end of the panel directly above the bearing area. Additional details regarding the spacing and size of the cross-wires are recommended in the Guide (ACI 523.5R Draft Committee Report 2005).

3.5 *Shear*

The nominal shear resistance of the AAC panel (V_n) is computed as the sum of the shear resistance of AAC (V_{AAC}) and the shear resistance provided by shear reinforcement oriented perpendicular to the axis of the member (V_s). For an applied factored shear force (V_u), the design of the cross section is based on Equation (3).

$$V_u \leq \phi V_n \leq \phi(V_{AAC} + V_S) \tag{3}$$

For members subjected to shear and flexure only, the shear resistance provided by AAC is given by Equation (4).

$$V_{AAC} = 0.07\sqrt{f'_{AAC}}\, b_w d \tag{4}$$

where b_w = the width of the compression face, mm.

The Guide also provides equations for the shear strength of AAC incorporating the effect of axial loads and for the shear strength of AAC walls subjected to in-plane loading.

The capacity of the shear reinforcement perpendicular to the axis of the member is calculated by the following Equation (5), which is limited to the bearing capacity of the AAC on the longitudinal reinforcement.

$$V_S = \frac{A_v f_y d}{s} \leq d_{long} s f'_{AAC} \tag{5}$$

where A_v = area of shear reinforcement within a distance s, in^2 (mm^2); d_{long} = diameter of longitudinal reinforcement, in. (mm); and s = spacing of shear reinforcement, mm (ACI 523.5R Draft Committee Report 2005).

For the shear resistance due to transverse reinforcement embedded in grout, the Guide recommends computing the resistance based on the cross-sectional area of the transverse reinforcement crossing a hypothetical 45-degree crack in the AAC. As with conventional reinforced concrete, this shear resistance may also be limited by bond and development of the reinforcement (ACI 523.5R Draft Committee Report 2005).

3.6 *Bearing*

To prevent local crushing of AAC, the design bearing strength on AAC is limited to $\phi(0.85\ f'_{AAC})$ multiplied by the loaded area. Special precaution should be taken into consideration when AAC floor or roof panels bear on AAC walls as shear failure of the edge of the wall may occur (ACI 523.5R Draft Committee Report).

4 CONCLUSION

Technical information on the design of AAC factory-reinforced panels is needed to familiarize the engineer with the design of AAC and hence promote the overall use of the material in the U.S. The proposed ACI Guide for the Use of AAC Panels" is a forward step to accomplish this goal. To facilitate the use of the Guide, its provisions are presented in a form very similar to that of ACI 318 Building Code. The Guide is in its final stages of development, and it is expected to be published in the very near future.

REFERENCES

ACI 523.5R Draft Committee Report, "Guide for Using Autoclaved Aerated Concrete Panels.," 2005.

Argudo, J.F., "Evaluation and Synthesis of Experimental Data for Autoclaved Aerated Concrete," MS Thesis, The University of Texas at Austin, 2003.

Autoclaved Aerated Concrete: Properties, Testing and Design, RILEM Recommended Practice, RILEM Technical Committees 78-MCA and 51-ALC, E & FN Spon, London, 1993.

Dembowski, J., *A Study of the Material Properties and Structural Behavior of Plain and Reinforced AAC Components*, Thesis, University of Alabama at Birmingham, Birmingham, Alabama, 2001.

Fouad, F.H., "PAAC – A New Precast Product in the U.S.," *Manufactured Concrete Magazine*, Winter 1998.

Fouad, F.H., "Physical and Mechanical Properties of AAC Produced in the United States," Report to the Autoclaved Aerated Concrete Products Association, 2002.

Snow, C., *A comprehensive study of the material properties and structural behavior of AAC products,* Thesis, University of Alabama at Birmingham, Birmingham, Alabama, 1999.

Tanner, J.E., *"Design Provisions for Autoclaved Aerated Concrete (AAC) Structural Systems"* Ph.D. Dissertation, The University of Texas at Austin, 2003.

Theme Four: Design and performance

Autoclaved Aerated Concrete – Limbachiya and Roberts (eds)
© 2005 Taylor & Francis Group, London, ISBN 0 415 38356 0

Design of low-rise housing in the UK using AAC masonry

C.A. Fudge
Autoclaved Aerated Concrete Products Association Limited, Sevenoaks, Kent, UK

G. Harding
Office of the Deputy Prime Minister, Buildings Division, London, UK

ABSTRACT: This paper describes the structural use of aac masonry in the UK to design low-rise dwellings, based on the guidance to the Building Regulations for England and Wales. The majority of houses and low-rise buildings are sized and wall thicknesses determined using guidance based on simple rules. The guidance has been based on a combination of testing and experience, as well as traditional tried and tested practices over decades. Background information is given for the testing programme. The paper outlines the basic requirements for aac masonry units, both in cavity and solid wall construction, and reports on typical constructions used in the UK today. The guidance was amended in 2004 to include references to new European Standards for masonry units and this paper reports on the compressive strength requirements of the changes.

1 INTRODUCTION

Most housing in England and Wales is built in accordance with the guidance given in the so-called 'Approved Documents' to the Building Regulations. These are in the form of prescriptive rules for sizing masonry elements, although the overall stability of the structure is also considered. The origins of the rules go back many centuries. In the main, guidance in the documents, especially in relation to wall thickness, materials strengths and position of openings is based upon what has traditionally been shown to work in practice. Although the rules for sizing certain walls have not changed significantly during the last few decades, a number of trends have developed in construction methods in both the way loads are supported and the materials and methods of construction. The widespread adoption of trussed rafter timber roofs in place of the heavier traditional timber rafter and purlin construction significantly changed the distribution of loading to be carried by the walls. Similarly, the common construction of walls changed from 225 mm thick solid brickwork to a cavity wall with a 100 mm brick outer leaf and an inner leaf of aac blockwork. These two developments alone made it reasonable to question the applicability of the rules, which limited the sizes of openings and returns permitted under the schedule of design.

Other changes in design, for example, the omission of chimneybreasts and the more widespread use of picture windows, led to a general reduction in the robustness of house construction. The rapid increase in the use of metal straps and joist hangers instead of building roof and floor members into the external walls appeared to have made significant reductions in the mutual restraint of horizontal and vertical elements. Consequently, in the early 1980's the Building Research Station (as it was at the time) considered that a programme of experimental work should be carried out to assess the strength enhancement that would exist in housing constructed by virtue of the interaction between elements.

2 THE ROBUST HOUSE TEST

The shell of a full-scale semi-detached domestic house was constructed in a laboratory based on a typical house plan chosen from a feedback study at the time. The construction used, wherever possible, the minimum requirements of the rules, but took into account practical considerations. Thus, the external walls were built of a 102 mm clay brickwork outer leaf and 100 mm aac blockwork (compressive strength of 2.8 N/mm^2 measured using British Standard test methods) as the inner leaf of the cavity wall. Openings at the front and rear elevations of this semi-detached house occupied two-thirds of the overall width to simulate a limiting condition of the guidance rules. At the point where the timber floor was supported by the inner leaf of the cavity wall, steel joist hangers were used. It was considered that this would provide a more severe loading case for the wall than had the joists been built-in. This is also more common practice in modern construction, since it is deemed to provide a more airtight external envelope.

A test programme was carried out looking at compression loading (Edgell, 1983) of the front and rear walls, overall positive pressure on the gable wall (Edgell, 1985) (Templeton 1985) and accidental damage (Templeton, 1985) to the front and rear walls by removal of masonry sections The following conclusions were drawn from the work:

- The floor, whether spanning parallel to the wall and connected to it by straps or spanning perpendicular to it and supported by joist hangers, effectively restrained the wall when it was subjected to vertical load.
- A considerable proportion of the load applied to either front or rear wall was transmitted into the side gable and party walls.
- Wire butterfly wall-ties were effective in transmitting a small amount of vertical load from the inner leaf to the outer leaf.
- The house successfully sustained the loads that exceed those generally applied to domestic constructions.
- It appeared that the recommended connections between the floor and walls given in the regulations were adequate and conservative.
- In a test in which an overall positive pressure was applied to the whole of the gable wall, an adequate factor of safety against failure was achieved despite there being a 3 m opening in the floor adjacent to the wall.
- In the overall positive pressure test on the gable wall, the walls containing openings acted as good buttresses; the returns used in the project were only 525 mm.

In summary, the robust house test confirmed that simple rules that had been developed for slender external walls in low-rise buildings.

3 TYPICAL HOUSE CONSTRUCTION USING AAC MASONRY

The vast majority of houses built in England and Wales are of masonry construction and of these the most common form of construction is to have an external cavity wall with clay brickwork on the outside and concrete blockwork on the inner leaf. It is estimated that over half of the inner skin market is constructed using aac masonry. The thickness of blockwork in this position is usually in the range of 100 mm to 140 mm. Cavities are typically between 50 mm to 100 mm and for new building works these are filled with a full width of thermal insulant or partially filled with a residual 50 mm air cavity. The overall structural wall thickness is generally therefore less than 300 mm. Whilst these walls may be seen to be more slender than other countries throughout Europe, this form of construction does make it competitive when compared to lightweight framed construction. In addition, consideration has to be given to the weight of individual masonry units and with aac units of 100 mm or so, the bricklayer is able to construct at a favourable rate compare to heavier weight counter-parts thus giving a cost effective overall solution. A typical storey height would be of the order of 2.5 m clear height and intermediate floors are generally of timber construction. Ground floors vary, but a popular form of construction, particularly in poor ground conditions, is to use a suspended concrete floor of inverted 'T' beams of pre-cast pre-stressed concrete with concrete block, often aac, infill units. There is growing tendency to have a 'room-in-the-roof' thus reducing the ground floor footprint but maximising on the overall available living area. Virtually all roofs are of timber construction, typically of trussed rafters.

There are some areas of the UK where a different external finish is preferred. This could be where a local stone material dominates, although in some areas of the country it is quite acceptable to use rendered blocks as the outside finish. It is usually the choice of the local authorities to decide what is or is not acceptable in any location. Solid walls of aac do have a market in the UK and have proved to be acceptable in what can be a harsh climate in terms of wind-driven rain. However, this form of construction is most popular for the construction of rear extensions or smaller types of buildings.

4 GUIDANCE RULES FOR THE DESIGN OF HOUSES

The generally accepted method to design houses is to use the guidance documents to the Building Regulations in this Approved Document A. In addition there is a design code, BS8103-2, (British Standards Institution, 2005) which follows the same general principles. The code is written for those with expertise in building construction but not necessarily in structural engineering design. Low-rise buildings constructed within the limitations stated in the relevant clauses do not require additional specialist advice. For any conditions outside the limitations of this code, appropriate specialist advice is needed. The code states that it ought not be expected that the recommendations made can be proved by calculation as they are based on traditional prescriptive guidance substantiated by long experience.

In formulating the guidance of the code, the worst combination of circumstances likely to arise was taken into account. It might be appropriate to consider a minor departure from the recommendations of this part and show adequacy by calculation. When using BS8103-2 it is important to assess that the overall stability of the building is achieved, and that the work of any specialist engaged is properly co-ordinated. BS8103-1 covers the stability aspect of low-rise housing. The Approved Document, however, contains the necessary information on how to achieving stability, for example, by tying in the floors to walls for example.

The recommendations of the guidance are intended to provide economic safe designs without the need for calculations of loading and strength criteria.

The guidance generally gives recommendations for walls above ground level damp-proof course (DPC) and walls between ground floor level and top of foundation level in the following range of buildings of traditional construction:

- low-rise housing comprising detached, semi-detached and terraced houses and flats (with not more than four self-contained dwelling units per floor accessible from one staircase) of not more than three storeys above ground, intended for domestic occupation and as governed by stated limiting conditions given in this code;
- certain small single storey non-residential buildings, e.g. domestic garages and annexes to residential buildings not exceeding 36 m^2 in floor.

For the purposes of the guidance, additional habitable accommodation in the roof space constitutes a storey of the house. This does not include the design of basements, but providing the basement is of one level only and is designed to provide a firm platform at ground level, the provisions of this code may apply to the superstructure above the basement. In addition, the structural design recommendations described are not intended to cover other considerations such as fire resistance, thermal and sound insulation, resistance to damp penetration and durability for which reference should be made to other documentation.

The guidance provides certain limitations, which are generally not found to be too restrictive in practice. Rules, for example, are given for the difference in ground levels either side of a wall and maximum wind loading.

The recommendations given in the guidance documents are governed by the following limiting dimensions;

(a) The maximum length of loadbearing wall between vertical lateral supports, i.e. between centre-lines of buttressing walls, piers, return walls etc., should be 9 m.
(b) No part of a wall or roof should be higher than 15 m above the lowest adjacent ground level.

(c) The maximum external wall height should be 12 m.
(d) The maximum clear span of any roof should be 12 m.
(e) The maximum clear span of any floor should be 6 m.
(f) No opening in a loadbearing wall should have a clear span greater than 3 m.
(g) There is a maximum floor area given for stability.
(h) The maximum height from top of foundation to top of ground floor should be:
    (1) 2.7 m where the floor provides lateral restraint to the wall; and
    (2) 1 m for all other cases.
(i) A maximum storey height is given.

In terms of the minimum wall thickness, internal non-loadbearing aac masonry walls should have a minimum thickness of 75 mm (including plaster finish, but excluding dry lining) and should be adequately restrained at the head. The minimum thickness of internal loadbearing walls is given as 90 mm. This is increased to 140 mm in certain conditions. In external cavity walls, the general requirement is to have a 90 mm minimum thickness for each leaf. In solid wall construction a minimum of 190 mm is required. There is an over-riding rule that the thickness should also not be less than the restrained height divided by sixteen.

As far as AAC masonry units are concerned, they should conform to current British Standards, which includes the BSEN 771-4, Specification for aac masonry units, have a declared compressive strength not less than the values given in a table for the guidance. In summary, this requires the use of units tested whole to have a minimum of 2.9 N/mm^2 for two-storey work or the upper two storeys of a three-storey building. In three storey buildings, the lower storey is required to have a compressive strength of 7 N/mm^2. If the external wall of a two-storey construction is of solid construction of at least 300 mm thickness, the AAC masonry units can have a compressive strength of 2.0 N/mm^2.

Minimum requirements for lintel bearings are given for various configurations, but this is normally a minimum of 150 mm. The leaves of a cavity wall have to be tied together by wall ties with a maximum spacing of 900 mm horizontally and 450 mm vertically (density of 2.5 ties/m^2) and normally in a staggered pattern. Guidance is given for cavity widths of 50 mm up to 300 mm. Additional ties have to be provided within 225 mm of the vertical edge of all openings so that there is at least one tie for each 300 mm of height of the opening. For the purposes of the guidance the following apply;

(a) Vertical unreturned edges are tied as for openings.
(b) Sloping unreturned edges, such as the roof verge, are tied as openings, i.e. they are provided with at least one tie for each 300 mm of vertical height and within 225 mm of the edges.

The masonry adjacent to vertical movement joints in the outer leaf are tied as for openings using ties tolerant of some horizontal movement.

To provide vertical restraint, the ends of every loadbearing wall are bonded or otherwise securely tied throughout the full height of the wall to a buttressing wall, pier or chimney. Long walls may be provided with intermediate buttressing support, effectively dividing the wall into distinct lengths. Any buttressing wall, pier or chimney usually has to provide support from the base to the full height of the supported wall. For internal buttressing, the actual wall has to be bonded or securely tied to the wall, or loadbearing leaf of a cavity wall, which it is supporting. The other end of the buttressing wall is then bonded or securely tied to a buttressing wall, pier or chimney. In the case of a connection between a loadbearing wall on foundations and a non-loadbearing wall supported on a ground-bearing slab, it is preferable to tie, not bond the walls. This reduces the risk of cracking due to differential vertical movement.

In dwelling houses, a non-loadbearing masonry wall can be used as a buttressing wall to provide support to other walls. The thickness of such a wall should not be less than 75 mm if the supported wall does not, as a whole, exceed six metres in height or nine metres in length. In all other cases, the thickness of a buttressing wall should be not less than 90 mm.

The number, size and position of openings and recesses are designed so that it does not impair the stability of a wall in which they are included or the lateral support afforded by a buttressing wall to a supported wall. Also, construction over openings and recesses should be adequately supported. The limitations of the position of openings and recesses above the ground floor is given in the guidance. Openings should be not greater than three metres in width. No openings should be provided in walls below ground floor except for small holes for services and ventilation etc., which should be limited to a maximum area of 0.1 m^2 at not less than two metre centres. The clear height of an individual opening is limited and should be not greater than 0.9 times the clear floor to ceiling height, and the depth of the lintel including any masonry over the opening should be not less than 150 mm.

5 SUMMARY

The majority of low-rise dwellings in the UK are designed using 'simple rules', which being based on experience and testing, have proved to be satisfactory in practice. Millions of dwellings have been designed in this manner and have withstood some extreme climatic conditions as well as excessive loading in practice. The 'robust house' programme has also given designers additional confidence to use the rules to their limits to provide cost effective houses of AAC masonry that are not only structurally sound, but also meet the necessary performance requirements relating to energy conservation, acoustic and fire regulations.

REFERENCES

British Standards Institution: 2005; *BS8103 Structural design of low rise buildings, Part 2: Code of practice for masonry walls for housing*, BSI, London

Edgell, G.J and de Vekey, R.C. 1983. The robustness of domestic house, Part 1; Compressive load tests in walls', *B Ceram RA, TN 350*

Edgell, G.J and de Vekey, R.C. 1985. The robustness of domestic house, Part 2; Wind suction tests on gable walls', *B Ceram RA, TN 364*

Templeton, W., Edgell, G.J and de Vekey, R.C. 1985. The robustness of domestic house, Part 3; Positive wind pressure tests on gable walls', *B Ceram RA, TN 370*

Templeton, W., Edgell, G.J and de Vekey, R.C. 1985. The robustness of domestic house, Part 4; Accidental damage', *B Ceram RA, TN 371.*

Autoclaved Aerated Concrete – Limbachiya and Roberts (eds)
© 2005 Taylor & Francis Group, London, ISBN 0 415 38356 0

Development of seismic design provisions for Autoclaved Aerated Concrete: an overall strategy for the United States of America

Prof. Richard E. Klingner
Dept. of Civil Engineering, The University of Texas, Austin, Texas, USA

Prof. Jennifer E. Tanner
Dept. of Civil Engineering, University of Wyoming, Laramie, Wyoming, USA

Prof. Jorge L. Varela
Dept. of Civil Engineering, Autonomous University of Yucatán, Mérida, Yucatán, Mexico

Ronald E. Barnett
Technical Director, Aercon Florida LLC, Haines City, Florida, USA

ABSTRACT: Autoclaved Aerated Concrete (AAC), a lightweight cementitious material originally developed in Europe more than 70 years ago and now widely used around the world, has recently been introduced into the US construction market. AAC products include unreinforced masonry-type units, factory-reinforced panels, and floor and roof planks. AAC elements can contain conventional reinforcement in grouted cores, either alone or with factory-installed reinforcement. To facilitate the use of AAC in the US market, an integrated seismic-qualification program has been carried out, involving general seismic design provisions, specific element design provisions, and material specifications. This paper is an overview of that program. The general approach is summarized; the organization of the program is described; the research required to produce each set of provisions is reviewed; and sample results of each type of research are given. This paper is intended to provide an overview of the entire seismic-qualification strategy for AAC for the US. Two other papers submitted for review and possible publication in the *Proceedings* of the 4th International Conference on Autoclaved Aerated Concrete are intended to present more details about key phases of the qualification effort.

1 INTRODUCTION TO AUTOCLAVED AERATED CONCRETE (AAC) MASONRY AS USED IN THE US

Autoclaved Aerated Concrete (AAC) is a low-density cementitious product of calcium silicate hydrates in which the low density is obtained by the formation of macroscopic air bubbles, mainly by chemical reactions within the mass during the liquid or plastic phase (RILEM 1993). The air bubbles are uniformly distributed and are retained in the matrix on setting, hardening, and subsequent curing with high-pressure steam in an autoclave, to produce a cellular structure (Figure 1). Material specifications for this product in the US are prescribed in ASTM C 1386.

In Table 1, typical mechanical and thermal characteristics of AAC are compared with those of conventional concrete, including conventional concrete made with lightweight aggregates. AAC typically has one-sixth to one-third the density of conventional concrete, and about the same ratio of compressive strength, making it potentially suitable for cladding and infills, and for bearing-wall components of low- to medium-rise structures. Its thermal conductivity is one-sixth or less that of concrete, making it potentially energy-efficient. Its fire rating is slightly longer than that of conventional concrete of the same thickness, making it potentially useful in applications where fire resistance is important. Because of its internal porosity, AAC has very low sound transmission, making it potentially useful acoustically.

AAC was first produced commercially in Sweden, in 1923. Since that time, its production and use have spread to more than 40 countries on all continents, including North America, Central and South America, Europe, the Middle East, the Far East, and Australia. This wide experience has produced many case studies of use in different climates, and under different building codes.

In the US, modern uses of AAC began in 1990, for residential and commercial projects in the southeastern states. US production of plain and reinforced AAC started in 1995 in the southeast, and has since spread to other parts of the country. AAC products include masonry-type units, reinforced panels, and specialty elements such as lintels, floor or roof planks, and stairs (Figure 2).

Masonry units of Autoclaved Aerated Concrete (AAC) can be produced with dimensional tolerances as small as 1.5 mm. As a consequence, AAC masonry units can be laid with mortar

Figure 1. Cellular structure of AAC.

Table 1. Typical mechanical and thermal characteristics of AAC.

Characteristic	AAC	Conventional concrete
Density, kg/m^3	400–800	1280–2400
Compressive strength, f_{AAC}, MPa	2.5–7.5	6.9–69
Thermal conductivity, kjoule-mm/m^2-hr-C	400–600	3000–5000
Fire rating, hours (200 mm thick)	≤ 8	≤ 6

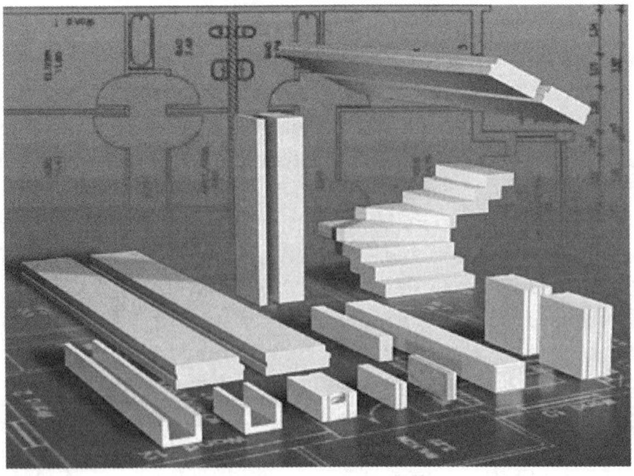

Figure 2. Sample AAC elements produced in the US.

joints about 10 mm thick, and also with thinner joints about 1.5 mm thick. The latter are more common in the US. The exterior face of the resulting AAC masonry wall is then protected from the elements using an exterior wythe of masonry, a cladding system, or a breathable coating resistant to penetration by liquid water. The interior face can be plastered, covered with gypsum board, or painted.

In the US, AAC masonry units are laid a top a leveling bed of ASTM C 270 mortar, and are shimmed if necessary to achieve a bed course that is plumb, level and true (Figure 3). Most AAC masonry in the US is "thin-bed masonry," laid with joints about 1.5 mm thick. Subsequent courses of thin-bed masonry are laid using special thin-bed mortar, using a special notched trowel available from the AAC manufacturer (Figure 4). Units are laid in alignment with either the inside or the outside plane of the wall. Minor adjustments can be made by sanding the exposed faces of the units with a sanding board.

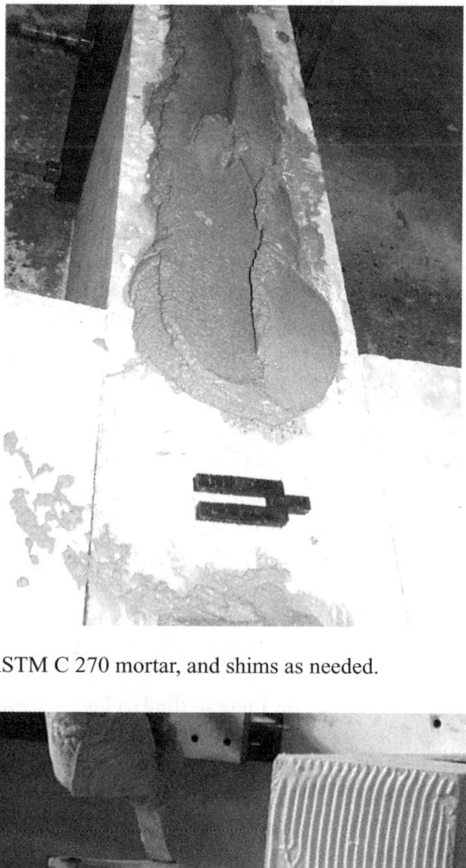

Figure 3. Leveling bed of ASTM C 270 mortar, and shims as needed.

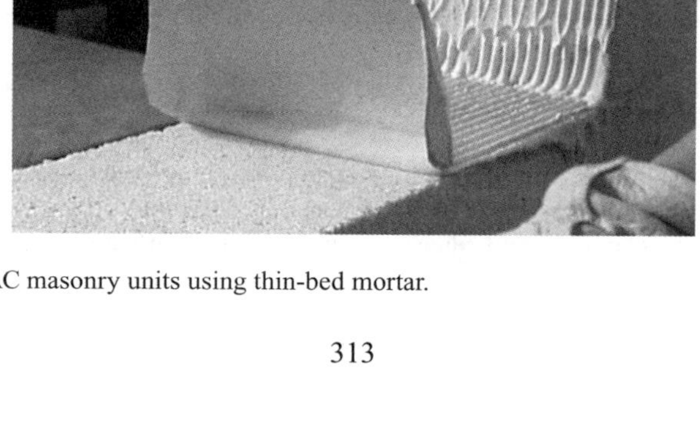

Figure 4. Laying AAC masonry units using thin-bed mortar.

AAC masonry units themselves are unreinforced. Field reinforcement can be installed horizontally in lintel units, or vertically in grouted cells 75 to 100 mm in diameter. The cells to be grouted are wetted thoroughly; a small-diameter vibrator is inserted into the empty cell; the cell is filled with high-slump grout; and the vibrator is withdrawn, consolidating the grout.

Mortar for thin-bed AAC masonry is a polymer-fortified mixture of portland cement and fine sand, produced by mortar manufacturers to meet performance standards approved by AAC manufacturers. In general, AAC manufacturers have approved lists of AAC mortar suppliers, based on internal performance criteria for AAC mortar. Those criteria generally address minimum dry compressive strength, minimum wet compressive strength, minimum bond strength, minimum open time, and minimum working time. They also require that the mortar provide sufficient bond to the AAC masonry unit so that flexural tensile strength is controlled by the flexural tensile strength of the units rather than by the bond between units and mortar.

2 US RESEARCH ON AAC MASONRY

Because it has been used extensively in Europe for more than 70 years, AAC has been extensively researched there. Extensive manufacturer testing is available world-wide. In the US, basic evaluation-service testing has been carried out on structural performance, fire resistance, and thermal and acoustical properties. Prior to the work described here, structural testing had been conducted at the University of Texas at Arlington (by Ytong), at Clemson University (by Ytong), and at the University of Alabama at Birmingham (by Hebel, subsequently known as Matrix, and then known as Babb). That prior testing verified compressive strength, tensile strength, diagonal tensile strength, modulus of rupture, and modulus of elasticity. Results were circulated internally to groups developing design provisions, but were not published in the US. More recently, extensive structural testing was conducted at The University of Texas at Austin, and the results have been published (Tanner et al. 2005a,b and Varela et al. 2004).

Seismic qualification of AAC masonry is based in general on experience in the Middle East and Japan, and on an extensive analytical and experimental research program conducted at The University of Texas at Austin. That seismic qualification program was intended to develop design models, draft code provisions, analytical models, and R and C_d factors. It involved confirmatory material tests, more than 19 reversed cyclic load tests on different configurations of AAC shear walls, and finally proof tests of a full-scale, two-story AAC assemblage under quasi-static, reversed cyclic loading. That seismic research is the topic of two other papers at this conference, and is not discussed further here.

3 APPROVAL APPROACHES FOR AAC MASONRY IN THE US

Until late 2003, proposed AAC masonry buildings in the US had to be approved on a case-by-case basis. After late 2003, product approvals could be obtained through evaluation services of model code agencies; designs are carried out in accordance with industry guidelines; and project approvals are obtained through local building officials.

This approval approach, while feasible on a limited scale, was cumbersome and expensive. It was judged better to have AAC masonry buildings approved on a national basis. AAC material would then be addressed by ASTM material specifications; design provisions, developed under ANSI consensus procedures, would be referenced by legally adopted building codes; and project approvals would be obtained automatically.

To achieve the goal of a national basis for approval of AAC masonry buildings, the following steps were required:

- Evaluate the existing and probable future code framework for masonry and for concrete, and determine where within that framework should design provisions be developed for AAC masonry and for reinforced AAC panels, and where within that framework should material specifications be developed.

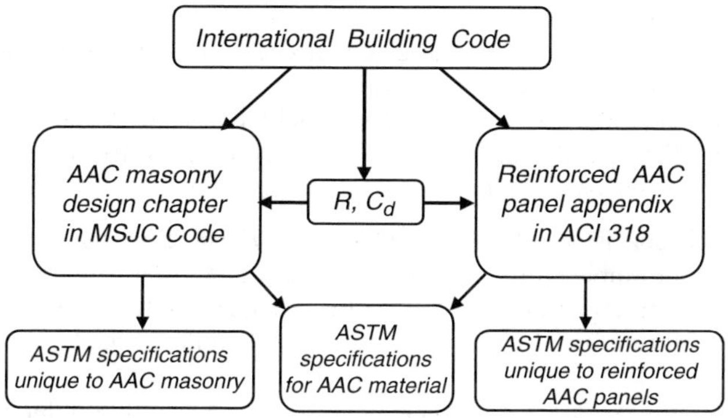

Figure 5. Overall framework for national design basis for AAC structures.

- Using a comprehensive, integrated research program, synthesizing existing research and conducting new research when necessary, develop the performance data necessary to justify a proposed design approach for AAC masonry, including seismic design. For seismic design, carry out the experimental and analytical research necessary to propose and justify R and C_d factors. The research program would be specifically set up to deliver draft code provisions, commentary, and technical justification.
- Prepare a draft set of code provisions, code commentary, and code "super-commentary" (extensive technical justification for proposed provisions and commentary), for design of AAC masonry.
- Prepare a draft set of code provisions, code commentary, and code "super-commentary" (extensive technical justification for proposed provisions and commentary), for design using reinforced AAC panels. The design provisions for AAC masonry and reinforced AAC panels, while possibly different in format, should produce similar results.
- Prepare a draft set of ASTM specifications for AAC masonry, and for other aspects of AAC construction.
- Propose R and C_d factors for AAC masonry and reinforced AAC panel construction.

That overall code framework is shown schematically in Figure 5, and is discussed in detail in the rest of this paper.

4 INTEGRATED US DESIGN FRAMEWORK FOR AAC STRUCTURES

4.1 *Review of Code Development Process in the US*

In the US, structural design provisions are developed under ANSI consensus procedures, and the resulting documents are referenced by model codes. Over the past 10 years, model-code harmonization has resulted in the development of a predominant harmonized model code, the International Building Code (IBC), whose first issue was the 2000 IBC, and which is updated at 3-year intervals. Another model code is published by the National Fire Protection Association (NFPA). Those model codes acquire legal standing when they are adopted by local governmental jurisdictions. Loads for structural design are developed by ASCE 7, whose latest load standard is ASCE 7-05, which will be updated at 3-year intervals. That load standard is referenced by model codes. Material specifications are usually developed by the American Society for Testing and Materials (ASTM).

4.2 *Design provisions for AAC masonry*

In the US, development of masonry design provisions by an ANSI consensus process is the responsibility of the Masonry Standards Joint Committee (MSJC), sponsored by the American Concrete

Institute (ACI), the American Society of Civil Engineers (ASCE), and The Masonry Society (TMS). Beginning with its 2003 edition, the IBC has essentially referenced the MSJC provisions.

The MSJC design provisions cover a wide variety of design approaches (strength, allowable-stress, empirical) and materials (clay, concrete, glass block). Based on the combination of test results from The University of Texas at Austin, the University of Alabama at Birmingham, and elsewhere, a proposed design approach was developed for AAC masonry, with the following characteristics:

- A strength approach, consistent with Chapter 3 of the 2002 MSJC provisions. Strength provisions are preferred over allowable-stress provisions because they offer a more consistent factor of safety against structural failure; because they offer a better opportunity for harmonization with design provisions for reinforced panels; and connote reliability at the level of the material and the structural system.
- Design provisions for flexure, shear and anchorage that are generally similar to current strength-design provisions for other types of masonry, and that produce final designs similar to those produced by the proposed ACI provisions for reinforced AAC elements.

The first set of proposed design provisions, commentary, and "super-commentary" was introduced to the AAC Masonry Subcommittee of the MSJC early in 2002. It was refined in response to MSJC Main Committee ballot comments and additional research results, and was approved for the 2005 MSJC Code.

4.3 *Proposed design provisions for reinforced AAC panels*

In the US, development of design provisions for reinforced concrete under the ANSI consensus process is the responsibility ACI Committee 318. The 2000 IBC essentially references ACI 318-99; the 2003 IBC essentially references ACI 318-02; and this is expected to continue for future IBC cycles.

The design provisions of ACI 318 address the strength design of a wide variety of conventional reinforced concrete elements similar to AAC applications, including prefabricated wall panels. Based on the combination of test results from The University of Texas at Austin, the University of Alabama at Birmingham, and elsewhere, a proposed design approach has been developed for reinforced AAC elements, with the following characteristics:

- A strength approach, consistent with ACI 318-02.
- Design provisions for flexure, shear and anchorage that are generally similar to current ACI 318 strength-design provisions for reinforced concrete elements, and that produce final designs similar to those produced by the proposed MSJC provisions for AAC masonry.

The first set of proposed design provisions, commentary, and "super-commentary" was introduced to ACI Subcommittee 523A (Autoclaved Aerated Cellular Concrete) in the fall of 2002. Because ACI 523A is a relatively new subcommittee, the design provisions, commentary, and "super-commentary" were introduced as appendices to a non-mandatory design guide on AAC. It is anticipated that the guide will be refined in response to balloting within Subcommittee 523A and Committee 523. After it has been approved by Committee 523, it will be offered to ACI 318 as a basis for mandatory-language design provisions and commentary, as an appendix to ACI 318. Because these provisions must be discussed and refined within ACI Committee 318 as well as ACI 523, their timetable for approval will probably be extended longer than for their counterpart provisions for AAC masonry. The authors believe that a reasonable goal would be approval within the 2011 ACI 318 cycle.

4.4 *Proposed R and C_d factors for AAC structures*

Because seismic design is an important aspect of the proposed design provisions for AAC structures, whether of masonry units or reinforced panels, it is necessary to develop R and C_d factors for use with ASCE 7, the seismic load document referenced by model codes such as the 2003 IBC.

The seismic force-reduction factor (R) specified in seismic design codes is intended to account for energy dissipation through inelastic structural deformation (ductility), and structural overstrength. The factor (R) is based on observation of the performance of different structural systems in previous strong earthquakes, on technical justification, and on tradition.

For structures of Autoclaved Aerated Concrete (AAC), the force-reduction factor (R) and the corresponding displacement-amplification factor (C_d) must be based on laboratory test results and numerical simulation of the response of AAC structures subjected to earthquake ground motions. The proposed factors must then be verified against the observed response of AAC structures in strong earthquakes.

Values of R and C_d for AAC structures are being proposed in three code-development arenas.

- In the fall of 2002, values were proposed to ICBO ES (a model-code evaluation service), as part of a proposed ICBO ES listing for AAC structural components and systems. That listing was approved in October 2003 (ICC ES 1371, 2004). It listing is intended to facilitate the use of such systems on a nationwide basis, until the consensus design provisions proposed above are incorporated in MSJC and ACI documents, and referenced by model codes.
- After approval by ICBO ES, the same values were proposed for adoption by the Building Seismic Safety Council, a code resource group supported by the federal government of the US (NEHRP 2003). When adopted by NEHRP, they will form a part of the design package for AAC structural elements.
- After approval by ICBO ES, the same values were proposed for adoption by ASCE 7. When adopted by ASCE 7, they will form a part of the design package for AAC structural elements.

4.5 *Proposed ASTM specifications for AAC construction*

ASTM traditionally deals with specifications for materials and methods of test. For the past several years, standards-development work regarding AAC has been going on in two committees:

- In 1998, ASTM Subcommittee C-27.60 (Precast Concrete Elements of AAC) developed a material standard for AAC: ASTM C 1386-98 (Standard Specification for Precast Autoclaved Aerated Concrete Wall Units). Subcommittee C27-60 has also developed a standard for reinforced AAC panels: ASTM C 1452-00 (Standard Specification for Reinforced Autoclaved Aerated Concrete Units). That subcommittee has also developed a standard method of test for determining the modulus of AAC (ASTM C 1591-04).
- In 2002, ASTM Subcommittee C-15.10 (Autoclaved Aerated Concrete Masonry) developed a standard practice for Autoclaved Aerated Concrete Masonry, now ASTM C1555-03a (Standard Practice for Autoclaved Aerated Concrete Masonry). It references the AAC material provisions of ASTM C 1386-98, and also contains construction provisions. Many of the construction provisions of that standard have been incorporated into the 2005 MSJC Specification.

5 SUMMARY

Autoclaved Aerated Concrete (AAC), a lightweight cementitious material originally developed in Europe more than 70 years ago and now widely used around the world, has recently been introduced into the US construction market. In this paper, code-development efforts in the US related to AAC are summarized, with emphasis on AAC masonry.

Using coordinated research at The University of Texas at Austin, the University of Alabama at Birmingham, and elsewhere, integrated design provisions, commentary, and "super-commentary" (extensive technical justification) have been developed for AAC masonry (including field reinforcement), and for factory-reinforced AAC panels. Design provisions for AAC masonry have been incorporated into the 2005 MSJC Code and Specification; design provisions for reinforced AAC panels are being studied and refined within ACI Subcommittee 523A; values for the seismic-design factors R and C_d have been accepted by ICBO ES and are being evaluated by ASCE 7;

and material specifications and standard practices have been developed by ASTM Subcommittee C-15.10 and C-27.60.

An innovative aspect of this code-development strategy for AAC in the US is the planning of research deliverables to include specific draft design provisions, commentary, and supporting "super-commentary." That aspect could usefully be applied to other materials as well.

ACKNOWLEDGMENTS

Much of the information reported in this paper was developed in the course of a research project on the Seismic Behavior of Autoclaved Aerated Concrete, carried out at The University of Texas at Austin with the support of the Autoclaved Aerated Concrete Products Association.

REFERENCES

ACI 318-02: Building Code Requirements for Reinforced Concrete, ACI International, Farmington Hills, Michigan, 2002.

ASCE 7-05: Minimum Design Loads for Buildings and Other Structures, American Society of Civil Engineers, Reston, VA, 2005.

ASTM C 270: Standard Specification for Mortar for Unit Masonry, American Society for Testing and Materials, West Conshohocken, PA, 2004.

ASTM C 1386: Standard Specification for Precast Autoclaved Aerated Concrete High-Precision Wall Construction Units, American Society for Testing and Materials, West Conshohocken, PA, 1998.

ASTM C 1452: Standard Specification for Reinforced Autoclaved Aerated Concrete Units, American Society for Testing and Materials, West Conshohocken, PA, 2000.

ASTM C 1555: Standard Practice for Autoclaved Aerated Concrete Masonry, American Society for Testing and Materials, West Conshohocken, PA, 2003.

ASTM C 1591: Standard Method of Test for Determining the Modulus of AAC, American Society for Testing and Materials, West Conshohocken, PA, 2004.

IBC 2003: International Building Code, 2003 Edition, International Code Council, Falls Church, Virginia, 2003.

ICC ES 1371: Autoclaved Aerated Concrete (AAC) Block Masonry Units, Report ESR-1371, ICC Evaluation Service, Whittier, California, October 2004.

MSJC 2005a: Building Code Requirements for Masonry Structures (ACI 530-05 / ASCE 5-05 / TMS 402-05), American Concrete Institute, Farmington Hills, Michigan, American Society of Civil Engineers, Reston, Virginia, and The Masonry Society, Boulder, Colorado, 2005.

MSJC 2005b: Specifications for Masonry Structures (ACI 530.1-05 / ASCE 6-05 / TMS 602-05), American Concrete Institute, Farmington Hills, Michigan, American Society of Civil Engineers, Reston, Virginia, and The Masonry Society, Boulder, Colorado, 2005.

NEHRP 2003: Recommended Provisions for Seismic Regulations for New Buildings and Other Structures, 2003 Edition, Part 2: Commentary. Building Seismic Safety Council, Washington, D.C.

RILEM 1993: Autoclaved Aerated Concrete: Properties, Testing and Design, RILEM Recommended Practice, RILEM Technical Committees 78-MCA and 51-ALC, E & FN Spon, London, 1993.

Tanner et al. 2005a: Tanner, J.E., Varela, J.L., Klingner, R.E., "Design and Seismic Testing of a Two-story Full-scale Autoclaved Aerated Concrete (AAC) Assemblage Specimen," *Structures Journal*, American Concrete Institute, Farmington Hills, Michigan, vol. 102, no. 1, January - February 2005, pp. 114-119.

Tanner et al. 2005b: Tanner, J.E., Varela, J.L., Klingner, R.E., Brightman M. J. and Cancino, U., "Seismic Testing of Autoclaved Aerated Concrete (AAC) Shear Walls: A Comprehensive Review," *Structures Journal*, American Concrete Institute, Farmington Hills, Michigan, vol. 102, no. 3, May - June 2005, pp. 374-382.

Varela et al. 2004: Varela, J. L., Tanner, J. E. and Klingner, R. E., "Development of Seismic Force and Displacement Modification Factors for Design of AAC Structures," *Proceedings*, 13th World Conference in Earthquake Engineering, Vancouver, Canada, August 1-6, 2004.

Autoclaved Aerated Concrete – Limbachiya and Roberts (eds)
© 2005 Taylor & Francis Group, London, ISBN 0 415 38356 0

The influence of material moisture on heat conductivity of AAC

Torsten Schoch
Xella Technologie – und Forschungsgesellschaft mbH, Germany

ABSTRACT: As regards the discussion about the influence of moisture on heat technical characters of building materials – especially on their heat conductivity – our competitors increasingly try to publish particularly negative reports on the building material AAC in terms of the aggravation of its characters with cumulative moisture. As a matter of course this is done in the context of cited would-be top class characters of the building material brick. These specifications are authorized by publication of diverse authors but also by naming European standards such as DIN EN 1745 and DIN EN ISO 10456 which rank as harmonized standards. If you take calculational instructions of these standards as a basis this would result in immense additional values in the production moisture field which could finally lead to increased heat losses at the building site and may cause damages.

1 INTRODUCTION

It is also known, that measurings, which form a relation between heat conductivity and moisture were ranging almost exclusively in the hygroscopic moisture are a (i.e. close to balance moisture). Consequentially this was reflected to the hyperhygroscopic area. The correctness of this assumption is not or incompletely validated.

Within the framework of a measuring project of department construction engineering the connection based on former trials has been examined once again.

Up to now there is no calculation procedure which is able to include the influence of greater moisture contents on the heat conductivity of AAC. Available standardized procedures are – as already stated – only valid for the balance moisture area (adsorptive bound moisture). Measurings that have been carried out recently reconfirm the results of former Ytong investigations, i.e. the influence of greater moisture contents can be calculated as follows:

$$\lambda_{1\,moisture} = \lambda_R / 1{,}05 + 0{,}0070 \times u \times \rho$$

with $\lambda_{1moisture}$ – heat conductivity with uM.-% moisture in W/(mK); λ_R – calculation value of heat conductivity in W/(mK); u – mass related humidity content in M.-%; ρ – apparent density in kg/dm^3.

The increase of heat conductivity of AAC caused by moisture augmentation is considerably lower than fixed in the European standards, even with currently typical production moisture of 35–45 M.-% the λ_R value is at most doubled but not sextupled as postulated in the European standards. If you take typical drying-out procedures in the construction area as a basis, the heat conductivity value should after a short time be close to the declared rated value. A considerable increase of heat energy consumption of AAC buildings or even a inclination of mould spores at the building material AAC cannot be associated with the above mentioned interrelation.

Details: There have been cumulative questions in many places, lately, on how high material moisture could effect the heat conductivity of AAC, to which extent the energy use of a building would thereby change and if there is a danger of mould formation with thermal bridges. The existing standards and scientific publications do not give any precise support since they are only valid for the balance moisture area (reference moisture). This leads to incorrect results when using in connection with AAC. With balance moisture the moisture is located adsorptive at the inner surface and therefore physically very fix whereas in case of greater moisture contents the humidity in terms of water is mainly bound by capillary force in pores.

According to the current European standards the influence of moisture is calculated with the following formula

$$\lambda_u = \lambda_{tr}\ x\ F_m$$

With λ_{tr} – heat conductivity in dry condition in W/(mk); F_m – is the moisture conversion coefficient according to EN ISO 1045, which is calculated as follow

$$F_m = e^{fu\ (u2-u1)}$$

f_u – is mass related according to DIN EN 1745 table A.10 for AAC 4 (kg/kg); u_1 – is mass related moisture content of first clause of boundary conditions; u_2 – is mass related moisture content of second clause of boundary conditions.

The appliance of this formula leads to extreme values due to the exponential function with high moisture contents.

According to DIN V 4108-4 annex A, one can provide evidence for an own moisture conversion factor as regards the balance moisture of AAC. This fact led to the formula

$$\lambda_u = \lambda_{tr}\ x\ F_m \quad \text{with } F_m = 1{,}05$$

In case this F_m value is used for greater moisture rates, however, this led to undervalued aggregate values which were clearly too low.

Many measurement results were existent from the time when heat conductivity of AAC was fixed by measurements at walls in dry and moist condition. There are no measurements in the 30 to 45 M.-% area. By application of this measurement procedure the values at lower moisture contents have certainly been falsified.

Each of the measurements took at least 3 days and at the same time as a matter of fact transport activities have been carried out. In the book Krischer/Kasten "Drying Technics", 1978, volume 1, page 269 ff interesting theoretical considerations about the influence of moisture and temperature on heat conductivity of porous materials have been developed. In the fifties measurements at AAC have been executed with a specifically developed probe in the frame of a dynamical measurement procedure. The quality of this procedure can be hardly evaluated. The accuracy can certainly not be compared to standardized stastical measurement procedures. The measurement results as well as the results of theoretical conductance can be taken from the diagram on page 6.

Already in recent years Ytong has tried to measure with increased moisture contents. Formerly, there were only experiences with measurements of moisture contents of approximately 10 V.-%. The decisive development was established with measurements at plates of 40 mm thickness during a measuring period of 1 to 1½ hours.

2 RESULTS

Based on existing experiences new measurements have been carried out starting with the production moisture (40 and 45 M.-% respectively) up to the dry stage. The measurements have been executed three stones such as P2 0,35 from Schrobenhausen (one sample each) and at two stones P4 0,60 from Alzenau (two samples each).

At the plants freshly produced stones were unloaded and directly packed in moisture sealed foil. In the laboratory the stones were dryly cut into plates of 220 × 220 × 40 mm thickness and subsequently ground coplanar. The plates were stored at 10°C packed in foil and following to that the heat conductivity was measured. The measurement period took 1½ hours. Afterwards the plates were dried up at 60°C by 5–10 M.-%, stored at 10°C and measured. This procedure was retained up to the drying process at 105°C.

The measurement results for P2 0,35 plates (real apparent density 0,350 kg/m^3) taken from three stones are shown in Table 1 as well as in the correspondent diagram.

The measurements show a very slight dispersion. With the P2 stone R^2 is resulting in 0,9436 with the P4 in 0,9971. Under "allgemein" (in general) you can find the conversion formula which is valid for all different apparent densities, whose derivative will be founded in the next paragraph.

In Table 2 you can see the measured data referring to measurements at two plates taken from one stone P4 0,60 (real apparent density 0,525 kg/dm^3) including diagram.

Table 1. Measurement at P2 0,35 from Schrobenhausen.

M.-%	W/(mK)
0,00	0,094
0,00	0,099
0,00	0,094
10,71	0,127
12,20	0,141
13,18	0,135
15,41	0,142
16,59	0,154
17,83	0,150
20,34	0,149
21,74	0,167
23,37	0,164
24,92	0,158
27,30	0,184
27,96	0,177
30,76	0,162
31,85	0,182
32,38	0,197
38,59	0,194
39,08	0,199
39,21	0,216

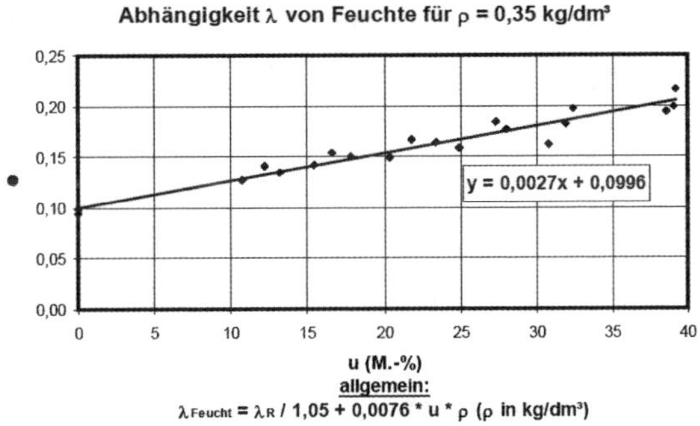

Table 2. Measurements at P4 from Alzenau.

M.-%	W/(mK)
0,00	0,127
0,00	0,131
5,60	0,149
5,66	0,147
10,40	0,170
11,09	0,170
17,40	0,193
18,67	0,196
20,36	0,204
23,83	0,213
26,33	0,220
27,50	0,226
45,56	0,283
45,57	0,282

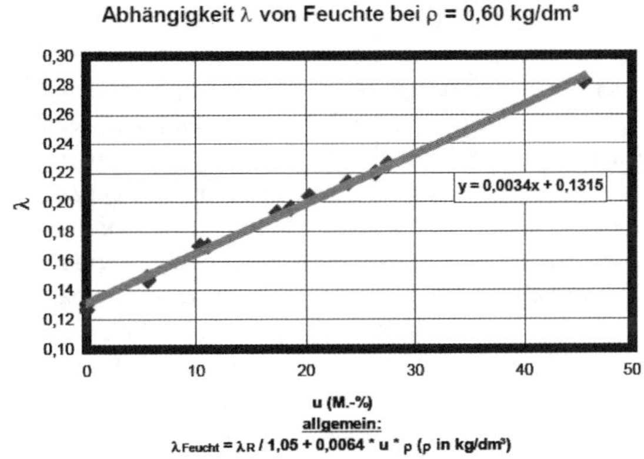

3 CONCLUSION

The slight dispersion of measuring data testifies, that the boundary conditions of measurements have been strictly observed. This is also a hint on the non-existence of systematic measurement errors, since otherwise there would have been an increased dispersion with augmented moisture contents to be expected. Some of the prior investigations in the balance moisture area at 23°C and 80% relative air moisture have shown, that with a measuring period of <3 hours there are practically no visible moisture transportation procedures, with longer measuring periods, however, these procedures are significantly dependent on the quality of mineral phase formation.

The following diagram compares the result and calculations according to the European standards for the apparent density 400 kg/m^3 and λ0,10 W/(mK).

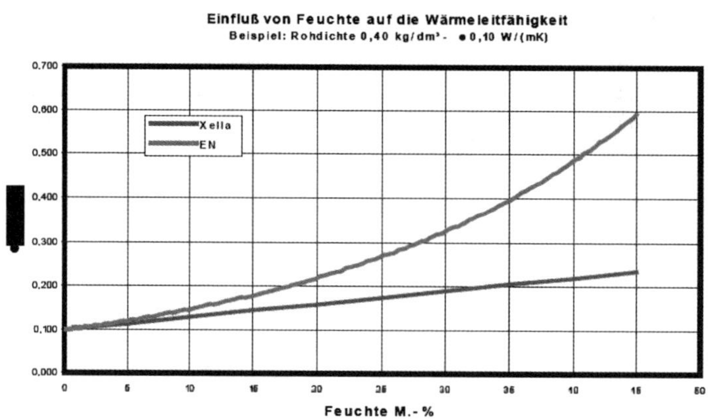

The measurements show a considerably lower additive allowance and no exponential increase in dependence of moisture. There is no logical explanation for the fact that the moisture content has no linear influence the more so as in the small pores of AAC there are virtually heat conductance processes only.

The additive allowance on heat conductivity is also dependant on the apparent density, since it is about moisture in pores, i.e. about the volume of compound but not about the adsorptive bound water on the crystal surface of balance moisture. That is the reason why you have to calculate on the basis of volume factor or as just recommended with M.-% and the factor apparent density.

The moisture allowance is also dependant on the "quality" of AAC, as prior investigations up to 10 V.-% could have been shown. A direct connection could not have been deduced. For the internal conversion a median allowance $F_z = 0{,}0070$ was established and has been used for all measurement up till now. Surprisingly, in recent measurements this value could have also been confirmed for greater moisture contents.

The above mentioned diagram for AAC with an apparent density of $p = 649$ kg/m^3 and $\lambda_{tr} = 0{,}146$ W/(mK) has been taken from the book "drying technics", Krischer/Kasten (measurements were made before 1956, propably deriving from slag or flue ash). The diagram shows the heat conductivity λ in dependence on relation between moisture content and porosity ψ_w/ψ.

Our own measurement results are included in that picture, they are diagrammed as a blue curve and can be calculated if new measurements of an apparent density of 640 kg/m^3 and λ_{tr} of 0,146 W/(mK) will be converted. Should you need to compare different values you have to take into consideration that Krischer/Kasten's measurements and calculations can be only executed at temperatures of >20°C. For measurements at 10°C as carried out by Xella the curve has to range below the 24°C line and should be closer to a straight line. The difference of 0,05 W/(mK) between Xella measurements and Krischer/Kasten details therefore was to be expected. The specifications of Krischer/Kasten therewith confirm the results of our own investigations.

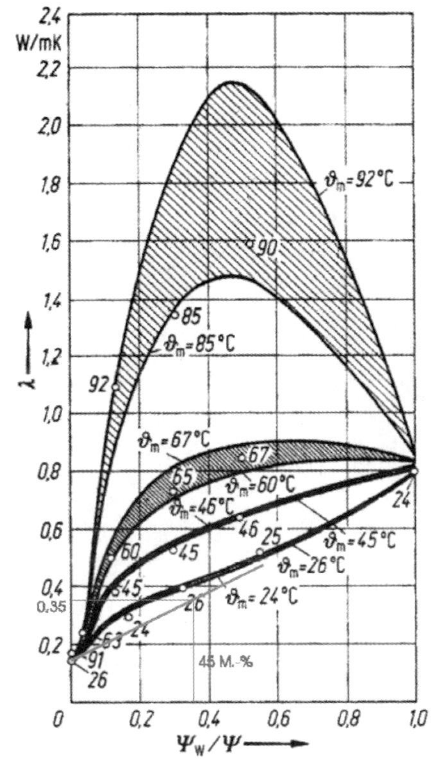

For the calculation of AAC's heat conductivity at different moisture stages we therefore suggest to use the following formula:

$$\lambda_{1_{moisture}} = \lambda_R / 1{,}05 + 0{,}0070 \times u \times \rho$$

with $\lambda_{1moisture}$ – heat conductivity with u M.-% moisture in W/(mK); λ_R – calculation value of heat conductivity in W/(mK); F_m – moisture conversion factor in W/(mK); F_z – moisture allowance for V.-% 0,0070; u – mass related humidity content in M.-%; ρ – apparent density in kg/dm^3

Autoclaved Aerated Concrete – Limbachiya and Roberts (eds)
© 2005 Taylor & Francis Group, London, ISBN 0 415 38356 0

Technical basis for US design provisions for Autoclaved Aerated Concrete masonry

Prof. Jennifer E. Tanner
Dept. of Civil Engineering, University of Wyoming, Laramie, Wyoming, USA

Prof. Jorge L. Varela
Dept. of Civil Engineering, Autonomous University of Yucatán, Mérida, Yucatán, Mexico

Prof. Richard E. Klingner
Dept. of Civil Engineering, The University of Texas, Austin, Texas, USA

Prof. Fouad H. Fouad
Dept. of Civil Engineering, The University of Alabama, Birmingham, Alabama, USA

Ronald E. Barnett
Technical Director, Aercon Florida LLC, Haines City, Florida, USA

ABSTRACT: In this second of three related papers in the Proceedings of the 4th International Conference on Autoclaved Aerated Concrete, the technical basis for US design provisions for AAC masonry is presented. That basis consists of extensive tests on AAC shear walls and confirmatory tests on a two-story, full-scale AAC assemblage. The shear-wall tests involved 19 specimens, with aspect ratios from 0.6 to 3, tested under in-plane reversed cyclic loads at the University of Texas at Austin. The results of these tests were used to develop predictive models for the behavior of AAC shear walls with respect to flexural cracking, web-shear cracking, flexure-shear cracking, nominal flexural capacity, crushing of the diagonal strut, and sliding shear. These predictive models permitted the development of reliable design equations for AAC shear walls, the primary lateral force-resisting element of AAC structural systems. The models were confirmed by testing of a two-story, full-scale AAC shear wall specimen, under reversed quasi-static loads representative of those experienced in a strong earthquake. The specimen withstood repeated reversed cycles to story drifts of about 0.3% & displacement ductility ratios of about 3. The tests confirmed the validity of the predictive equations for wall behavior, and also the validity of the overall seismic design approach.

1 INTRODUCTION AND BACKGROUND

Autoclaved Aerated Concrete (AAC), a lightweight cementitious material originally developed in Europe more than 70 years ago and now widely used world-wide, has recently reached the US construction market. Because AAC is a new material in the United States, technical information is required to produce proposed design provisions for AAC structural systems. Although significant material exists on material and thermal properties of AAC produced in Europe (Tanner et al. 2005a, b), little information exists on the performance of AAC shear walls subject to in-plane loads. The Autoclaved Aerated Concrete CEB Manual of Design and Technology (1978) and the CEB-FIB Model Code (1990) are silent on the design of AAC shear walls.

A research program at The University of Texas at Austin was developed to test 19 shear-wall specimens made from a variety of AAC elements, including masonry-type units and reinforced panels, laid either horizontally or vertically. Following this, tests were conducted on a two-story

Table 1. AAC shear-wall specimens tested at The University of Texas at Austin.

Specimen	Length (m)	Height (m)	Thickness (m)	Axial load (kN)	Exterior reinforcement	Interior reinforcement
1	6.10	3.91	0.203	694	2-B7 25 mm dia. at 0.61 m from ends	None
2	6.10	3.91	0.203	694	2-B7 25 mm dia. at 0.61 m from ends	None
3	6.10	3.86	0.203	534	2-B7 25 mm dia. at 0.61 m from ends	None
4	6.10	3.91	0.203	534	2-B7 25 mm dia. at 0.61 m from ends	16 mm @ 1.22 m
5	6.10	3.86	0.203	267	2-B7 25 mm dia. at 0.61 m from ends	None
7	3.66	3.86	0.203	356	2-B7 25 mm dia. at 0.61 m from ends	None
9	2.44	3.91	0.203	267	2-B7 25 mm at wall ends	None
11	1.22	3.86	0.203	111	2-B7 25 mm at wall ends	None
13	1.83	3.91	0.203	111	None	16 mm at 0.30 m from ends
14a	1.42	3.91	0.254	22	None	16 mm at 0.10 m from ends
14b	1.42	3.91	0.254	22	None	16 mm at 0.10 m from ends
15a	2.84	3.91	0.254	111	None	16 mm at 0.20 m from ends
15b	2.84	3.91	0.254	111	None	16 mm at 0.20 m from ends
16	2.84	3.91	0.254	111	None	16 mm at 0.20 m from ends
17	2.84	3.91	0.254	111	None	16 mm at 0.10 m from ends
18	3.66	3.86	0.203	200	2-B7 25 mm dia. at 0.30 m from ends	16 mm at 0.30 m from ends
19	5.49	3.86	0.203	258	2-B7 25 mm dia. at 0.61 m from ends	16 mm @ 1.22 m
Assemblage	6.10 each wall	5.08 each story	0.203	133 each wall	None	16 mm at 0.20 m from ends

Note: "B7" refers to ASTM A 193 Grade B7 threaded rods, specified ultimate strength 860 MPa.

AAC assemblage. AAC masonry units are typically 200 mm by 200 mm by 610 mm, and AAC panels are 200 mm by 610 mm by 6.1 m, although the thickness is variable. The aspect ratio of the specimens (ratio of height to base length) varied from 0.6 to 3. The geometry and axial load of each shear-wall specimen are presented in Table 1 (Tanner et al. 2005a, b). This paper discusses the behavior of AAC shear walls made either of masonry units or of factory-reinforced AAC panels, both with field-installed reinforcement. It then discusses the assemblage.

2 SETUP AND PROCEDURES FOR AAC SHEAR-WALL TESTS

The shear-wall specimens tested at UT Austin were constructed on a concrete foundation post-tensioned to the strong floor of the Ferguson Structural Engineering Laboratory. Each shear wall

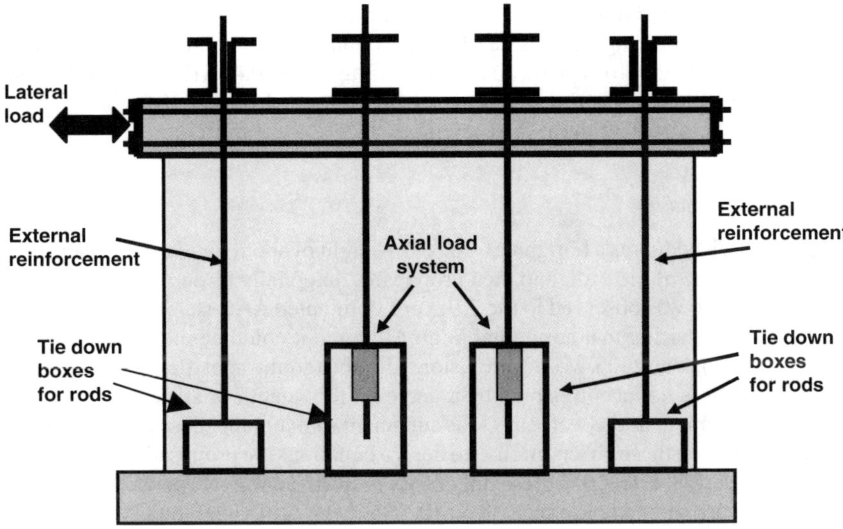

Figure 1. Test setup for AAC shear-wall specimens at UT Austin.

was constructed on a leveling bed of ASTM C 270 Type S cement-lime masonry mortar placed on the concrete foundation. A very stiff longitudinally post-tensioned loading beam was placed on top of the specimen (Figure 1). Reversed cyclic lateral load was applied to each specimen using either a single hydraulic ram or dual hydraulic rams mounted to a strong wall and supplied through a manually controlled hydraulic pump (Tanner et al. 2005b).

Axial load was applied through hydraulic rams, post-tensioned rods or a combination of the two. Net axial force was maintained approximately constant under reversed cyclic loading by using a mechanical "load maintainer" and by ensuring that some post-tensioning force remained in the external rods. The test setup and results are presented in Tanner et al. (2005b).

The geometry, flexural reinforcement, and axial force for each specimen were selected to induce either shear- or flexure-dominated behavior, as desired. The shear-dominated specimens were over-reinforced in flexure using external threaded rods as flexural reinforcement. In most specimens, these supplemental rods were also pre-tensioned to increase the axial load in the wall. The flexure-dominated specimens were lightly reinforced in flexure (Table 1).

3 MSJC DESIGN EQUATIONS BASED ON RESULTS OF AAC SHEAR-WALL TESTS

One or more of the following behavioral modes were observed in each of the AAC shear-wall specimens: flexural cracking; flexure-shear cracking; web-shear cracking; crushing of the diagonal strut; and nominal flexural capacity. Each behavior described below was initially predicted based on reinforced concrete theory, expressed in terms similar to the current US masonry design provisions, and modified based on experimental results. Finally, the corresponding design equations from the 2005 MSJC Code (MSJC 2005a) are given.

3.1 *Flexural cracking*

In flexural cracking of AAC shear walls, a flexural crack forms at the base of the tensile side of the wall, between the AAC and the leveling bed. Flexural cracking is governed by the lower of modulus of rupture of the AAC, or by the tensile bond between the AAC material and the leveling bed if such a joint is present in the shear wall under consideration. Flexural cracking was observed in 15 shear-wall specimens tested at UT Austin, always in the masonry leveling bed. Using the lateral loads at which flexural cracking was observed, tensile bond strengths were back-calculated for each shear wall (Tanner et al. 2005b). The mean modulus of rupture is 0.46 MPa, and the corresponding 20% fractile is 0.34 MPa. The proposed design value (Section A.1.8.3 of MSJC 2005a) is 0.34 MPa if a leveling-bed joint is present in the AAC element.

3.2 *Flexure-shear cracking*

A flexural shear crack begins as a horizontal crack at a height of about one-half the plan length of the wall (l_w) above the base of the wall, and then propagates diagonally through the center of the wall. Flexure-shear cracking was observed in the 7 flexure-dominated AAC shear-wall specimens tested at UT Austin, always starting in a horizontal joint. Although it could be quite accurately predicted (Tanner et al. 2005b), the 2005 MSJC provisions do not require that flexure-shear cracking be checked, because it was not accompanied by a decrease in strength or stiffness of the specimens, and vertical reinforcement at the wall ends was sufficient to maintain shear capacity after flexure-shear cracking. Based on these observations, no design equations are proposed to limit flexure-shear cracking. To consider the effect of horizontal cracks in horizontal AAC joints the proposed value for modulus of rupture should not exceed 80 psi (0.55 MPa) if horizontal joints containing thin-bed mortar are used in an AAC shear wall.

3.3 *Web-shear cracking*

Web-shear cracks formed in the shear-dominated specimens tested at UT Austin and elsewhere (Tanner et al. 2005b). The 2005 MSJC design equation for capacity as governed by web-shear cracking began with an equation giving the base shear at which the maximum principal tensile stress reaches the splitting tensile strength of the AAC material (Equation (1)). When the splitting tensile strength is expressed in terms of the compressive strength of the AAC material, Equation (2) results.

$$V_{AAC} = \frac{2 l_w t}{3} f_{tAAC} \cdot \left[1 + \left(\frac{P}{f_{tAAC} l_w t}\right)\right]^{0.5} \tag{1}$$

$$V_{AAC} = 1.6\, l_w\, t \sqrt{f_{AAC}} \sqrt{1 + \frac{P}{2.4\sqrt{f_{AAC}}\; l_w\, t}} \tag{2}$$

f_{AAC} *in psi, P in lbs,* l_w *and t in in.*

The lateral loads at which web-shear cracking was observed and predicted using Equation (2) are presented in Table 2. The mean of the ratios of observed to predicted strength is 0.72 with a COV of 13.8%. So that it would correspond to the 10% lower fractile of the ratios of observed to predicted capacities, the leading coefficient of 1.6 in Equation (2) was multiplied by 0.59 to give Equation A-12a of Section A.3.4.1.2.1 of the 2005 MSJC.

3.4 *Sliding shear capacity*

An AAC shear wall of horizontal panels or masonry-type blocks exhibits a bed-joint crack when the shear stress on the bed joints exceeds the interface shear capacity. Sliding shear capacity is the product of the coefficient of friction across the potential failure interface, and the force acting normal to that interface. This mechanism commonly referred to as "shear friction," includes normal forces from reinforcement crossing the interface, and from externally applied load normal to it.

In traditional shear friction, sliding over a rough interface causes widening of the interface, stressing vertical reinforcement crossing it and creating additional clamping force. Under reversed cyclic loading of AAC, this resistance can degrade, however, leaving primarily resistance from dowel action of reinforcement crossing the interface. This contribution is limited by the product of

Table 2. Results of predicted and observed lateral loads for web-shear cracking of AAC shear-dominated specimens.

Specimen	Axial load, P kips (kN)	Observed V_{AAC} kips (kN)	Predicted V_{AAC} kips (kN)	Observed V_{AAC}/ Predicted V_{AAC}
1 (UT)	156.0 (694)	164.2 (730)	127.7 (568)	1.29*
3 (UT)	120.0 (534)	81.3 (362)	111.4 (495)	0.73
4 (UT)	120.0 (534)	110.5 (492)	132.5 (589)	0.83
5 (UT)	60.0 (267)	62.2 (277)	117.4 (522)	0.53
7 (UT)	80.0 (356)	57.4 (255)	68.7 (305)	0.84
9 (UT)	60.0 (267)	37.7 (168)	55.9 (249)	0.67
11 (UT)	25.0 (111)	15.6 (69)	26.9 (120)	0.58
18 (UT)	45 (200)	39 (172)	50.1 (223)	0.77
19 (UT)	58 (258)	56 (249)	72.8 (324)	0.77
Assemblage (UT)	25.0 (111)	59.5 (268)	96.7 (430)	0.54
3.6 (Hebel)	36.8 (164)	27.7 (123)	39.3 (175)	0.71
4.7 (Hebel)	62.4 (277)	46.7 (208)	57.7 (256)	0.81
4.8 (Hebel)	178.2 (792)	61.5 (273)	80.3 (357)	0.77
			Mean	0.72
			COV (%)	13.8

* Data not included in average because it is anomalously high with respect to the remaining tests.

the area of the reinforcement perpendicular to the potential failure interface and the yield strength of that reinforcement in shear, about $0.6f_y$.

Sliding shear was observed in Shear Wall Specimen 4 and the Two-story Assemblage Specimen. The latter is discussed later in this paper. The effectiveness of the dowel action was reduced in both specimens where sliding occurred. As the damage increases, the resistance provided by bearing on the dowel is reduced. For this reason, the corresponding design equation (Equation A-13b of Section A.3.4.1.2.3 of the 2005 MSJC) conservatively neglects the contribution of the dowels.

3.5 *Crushing of a diagonal strut*

As axial load increases, the shear capacity of an AAC shear wall as governed by web-shear cracking also increases. If these axial loads are accompanied by significant shear, as in a low-rise wall, the wall's shear capacity may be limited by crushing along the compression diagonal. Diagonal crushing can be predicted using a strut-and-tie model consisting of two elements: a diagonal compression strut (F_{strut}); and a tension tie-down force (T). The compressive force in the diagonal strut is the resultant of the vertical tie-down forces and the applied horizontal load. Because the system is statically determinate, the vertical component is the summation of the force in the tie-down rods, and the horizontal component is the applied lateral load (V).

Crushing of the diagonal strut was observed in Shear Wall Specimen 1, over a length of one-quarter of the plan length of the wall. Based on the applied load at crushing and the geometry of the wall, Equation (3) was developed to give the horizontal force (V) at crushing based on equilibrium of horizontal forces at the base of the diagonal strut (Tanner et al. 2005b).

Based on this, Equation A13-a of Section A.3.4.1.2.2 of the 2005 MSJC was developed, was used in designing subsequent shear-wall specimens, and was validated by the fact that crushing of the diagonal strut was not observed in subsequent specimens. Although this equation is far from perfect, it was a reliable tool for avoiding crushing of the diagonal strut in the AAC shear walls tested in this study.

$$V_{ds} = f_{AAC} \cdot t \cdot w_{strut} \left[\frac{h \cdot (l_w - w_{strut})}{h^2 + (l_w - w_{strut})^2} \right] \tag{3}$$

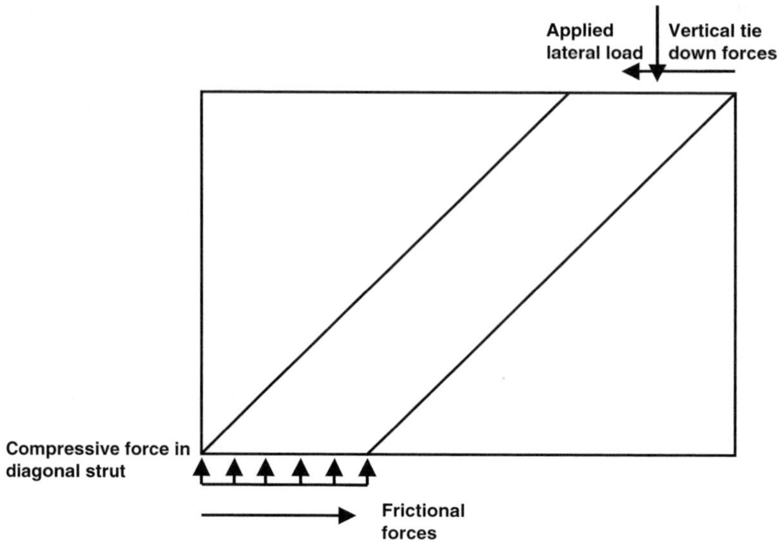

Figure 2. Diagonal compressive strut in an AAC shear wall.

3.6 *Nominal flexural capacity*

The nominal flexural capacity of AAC shear walls can be determined based on equilibrium of a cross-section. General assumptions for flexural behavior were verified in the flexure-dominated shear-wall specimens. The compressive zone is determined based on a linear strain relationship, using a maximum useful compressive strain in the AAC of 0.003, and an equivalent rectangular stress block whose height is $0.85f_{AAC}$ (or $0.85f_m$), and whose depth is $\beta_1 c$, where $\beta_1 = 0.67$.

Observed versus predicted nominal flexural capacities can be compared for flexure-dominated Shear Wall Specimen 14a, 14b, 15a, 15b and 17. During the testing of Shear Wall Specimens 13 and 16, the actuators used to apply the supposedly constant axial load inadvertently reached the end of their travel. As increasing lateral drifts were applied, axial load on the wall inadvertently increased. To successfully interpret those test results, the probable axial load applied to the walls was back-calculated from the predicted flexural capacity, and those two tests were excluded in verifying observed versus predicted flexural capacity (Tanner et al. 2005b).

Nominal flexural capacity was calculated using a steel yield strength of 490 MPa based on mill reports. The ratios of observed to predicted nominal flexural capacity range from 1.11 to 1.34, with an average of 1.22 and a COV of 6.8%. Using a refined analysis considering strain-hardening, the ratio of observed to predicted nominal flexural capacity has a mean of 1.06 and a COV of 7.0%.

4 TEST SETUP AND PROCEDURES FOR TWO-STORY ASSEMBLAGE

The Two-story Assemblage Specimen consisted of two flanged walls connected by AAC floor slabs (Figure 3) (Tanner et al. 2005a). The walls were constructed of vertically oriented Autoclaved Aerated Concrete (AAC) panels with internal reinforcement and additional field-placed longitudinal reinforcement, and the floor slabs were constructed of internally reinforced AAC panels. On the upper level, the floor panels were oriented longitudinally; on the lower level, transversely.

The reinforcement in the shear walls of the Two-story Assemblage Specimen consisted of flexural (longitudinal) reinforcement and foundation dowels (Figure 4). The flexural reinforcement continued up the height of the specimen with a splice just above the first elevated slab. The dowels extended 24 in. (61 mm) above the foundation, and were included to increase the sliding shear

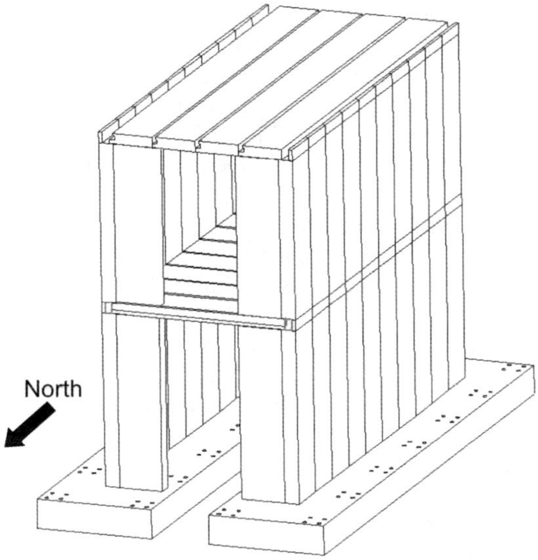

Figure 3. Isometric view of Two-story AAC Assemblage Specimen.

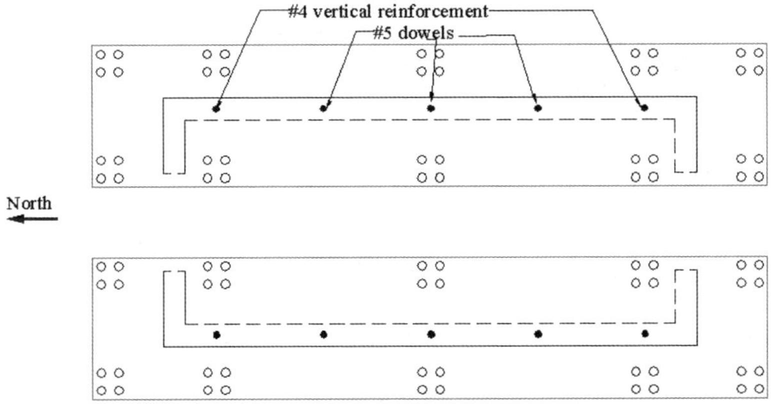

Figure 4. Plan view of horizontal section at base, showing flexural reinforcement and dowels in Two-story AAC Assemblage Specimen.

capacity of the specimen. They were also placed at the level of the first elevated slab to prevent sliding at the bed of leveling mortar placed between the vertical panels and the slab.

Instrumentation was used to measure global and local behavior of the Two-story Assemblage Specimen. Global behavior is characterized by the in-plane force-displacement relationship. The horizontal displacement of each wall was measured at each story using string potentiometers. The applied load was measured through pin load cells which had been successfully used in previous tests of shear wall specimens at UT Austin. Applied loads were verified using pressure transducers.

The local behavior of individual components was measured and used to characterize the deformation patterns in the specimen. The following measurements were recorded for both stories of the east and west shear walls:

- Vertical deformations of wall;
- Diagonal deformation of each story in each wall;
- Profile of floor-slab displacements with respect to the loading beam; and
- Slip between loading equipment and specimen.

The vertical displacements were measured to assess flexural deformations. The diagonal displacements were used to calculate the shearing deformation in each wall. Profiles of floor-slab displacements were measured to determine abnormal deformation patterns or indicate slip between floor panels on the second level slab. Slip between elements was measured using linear potentiometers mounted between the adjacent elements. Slip values were used to detect relative movement between the loading beams and the floor slabs, between the walls and the foundation, and between the floor slabs and the walls. Slip at other interfaces was determined visually using relative movement of the ends of lines drawn across the interfaces.

Tests were performed to predict the splitting tensile strength and the compressive strength of the AAC. ASTM C1006 tests performed on AAC modular blocks from the same shipment of AAC indicated a splitting tensile strength of 45 psi (0.31 MPa) with a COV of 15% (moisture content of 16%). Compression tests performed on 4 in. by 8 in. cylinders cored from the same units and air-dried to a moisture content of 6% indicated an average strength is 495 psi (3.4 MPa) with a COV of 6%. A yield strength of 75 ksi (517 MPa) was used; this is consistent with the mill reports for the longitudinal reinforcement.

Lateral load was applied parallel to the webs of the shear walls, through reinforced concrete loading beams attached to the slabs. Equal load was applied at each story level, based on nonlinear dynamic analyses. The lateral loading history consisted of a series of reversed cycles to monotonically increasing maximum load or displacement. The actuators were loaded manually to target load values, based on predicted loads of events that would cause a significant change in the behavior of the specimen. In this case the predicted major events were flexural cracking and the nominal flexural capacity. After the maximum load was reached, the specimen was loaded based on target displacements for the tip of the wall. These were 13 mm, 20 mm and 38 mm. A constant axial load was applied through the self-weight of the specimen and the loading equipment. The total axial load on the specimen is 60 kips (267 kN).

5 BEHAVIOR OF TWO-STORY ASSEMBLAGE

Overall behavior of the Two-Story Assemblage Specimen is shown in Figure 5. Total base shear is the summation of the equal shears applied to each floor level. Positive displacements are to the south; negative, to the north. The following behaviors were observed in the assemblage:

- flexural cracking at the base;
- minor vertical cracking on the north end of the east wall;
- web-shear cracking in the webs of both walls in the lower story;
- yielding of the flexural reinforcement at all four corners of the assemblage;
- separation of the flanges from the webs in the lower story; and
- separation of the vertical joint at the location of reinforcement at the northeast corner.

Flexural cracking was observed at the base of the west wall at a total base shear of 215 kN, corresponding to a calculated tensile bond stress of 0.2 MPa. Flexural cracking was observed at the base of the east wall at a total base shear of 364 kN, corresponding to a calculated tensile bond stress of 0.4 MPa. Both values fall below the average value of the 14 shear wall specimens previously tested at UT Austin, which was 0.50 MPa with a COV of 29% (Tanner et al. 2005b). This could be attributed to shrinkage cracking at the leveling bed, or to a reduced area of leveling bed mortar.

At the location of flexural reinforcement at the north end of the east wall, a vertical crack divided the wall into two individual walls. If the in-plane lateral stiffness of the specimen were dominated by flexure, this crack would significantly decrease the stiffness and strength of the specimen. For the specimen, the ratio of shear stiffness to flexural stiffness is 1.35. Since this value is close to one, the shearing and flexural deformations contribute about equally to the flexibility of the specimen. Because the in-plane lateral stiffness of the specimen was influenced about equally by shear and by flexure, the stiffness did not change much as a result of the vertical crack. Also, since the crack

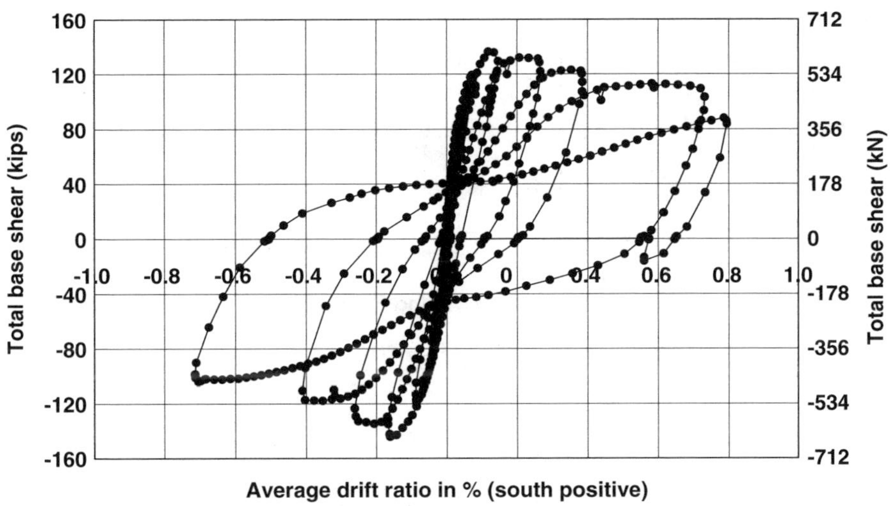

Figure 5. Overall hysteretic behavior of Two-story AAC Assemblage Specimen.

occurred two feet from the flange, the loss of flexural stiffness was less than it would have been had the crack occurred near the middle of the web.

Web-shear cracking was first observed in the east wall while loading to the north at a total base shear of 464 kN, and while loading to the south, at a total base shear of 520 kN. The corresponding values for the west wall are 530 kN and 520 kN. Tests conducted to confirm the splitting tensile strength gave an average value of 0.3 MPa with a COV of 15%, corresponding well with the predicted total base shear capacity of 528 MPa at a total axial load of 267 kN.

As the second set of web-shear cracks formed, the flexural reinforcement began to yield. Based on strain gage readings, the flexural reinforcement in the east wall yielded at a total base shear of 118 kips (525 kN) while loading to the north and to the south. The flexural reinforcement in the west wall yielded at a base shear of about 130 kips (580 kN) in each direction. The initially predicted total base shear at flexural yielding at the base of each wall, excluding the effect of dowels, under the applied axial load of 267 kN, was 344 kN. The base shear is determined by dividing the flexural capacity by the effective height or M/V ratio (Figure 6). The increase in observed capacity is due to the dowels. The base shear at yielding can be predicted based on the flexural capacity at two critical sections:

- at the base, considering the contribution of the dowels to the flexural capacity; and
- at a critical section just above the ends of the dowels.

Including the contribution of the dowels would significantly increase the flexural capacity at the base. Assuming the critical section at the base, the base shear capacity at yield would be determined by calculating the flexural capacity at yield and converting it to an equivalent base shear yield capacity by dividing by the M/V ratio presented in Figure 6. Assuming the critical section at the point where the dowels end, the flexural capacity would be calculated without the dowels, and would be converted to a base shear yield capacity by dividing by the reduced lever arms between each load and the critical section. The results of these calculations are presented in Table 3. The base shear at yielding of the flexural reinforcement falls between the limiting cases defined by the two critical sections noted above.

After the flexural reinforcement yielded, both walls exhibited flexural behavior, consisting largely of rigid-body rocking. Vertical displacements were observed at the wall bases on the tension side, due to yielding and bond deterioration of the tensile reinforcement. Crushing of the compression toe was avoided, due to lateral support by the flanges. The maximum loads observed in each wall

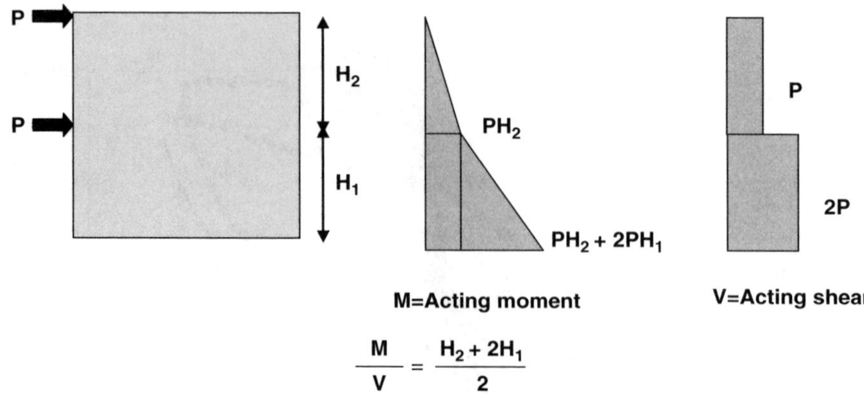

Figure 6. Shear span used to calculate the base shear capacity of a wall corresponding to a given flexural capacity.

Table 3. Base shear capacity for the Two-story Assemblage Specimen for yielding of the flexural reinforcement and nominal flexural capacity.

	Critical section	
	Base of wall	Top of dowels
Base shear at flexural yield, kips (kN)	145 (644)	107 (477)
Base shear at nominal flexural capacity, kips (kN)	197 (877)	117 (521)

was 605 kN loading to the south and 641 kN loading to the north. The tested flexural capacity of each wall falls between the bounds presented in Table 3. After flexural yielding, distributed web-shear cracks continued to form in the walls at the lower level.

Diagonal cracks formed around the dowels, separating those dowels from the webs of the AAC shear walls. This reduced the effectiveness of the dowel action, which in turn reduced the sliding-shear capacity of each wall to 48 kips (214 kN) at an axial load of 30 kips (134 kN). Degradation of dowel action is also identified by spalling of AAC around the diagonal cracks. Slip between the AAC shear walls and the foundation exceeded 0.5 in. (13 mm) after Load Point (LP) 677, corresponding to a total base shear of 142.6 kips (634 kN). The wall displacement, corrected for this slip, is shown in Figure 7.

After 3 cycles of flexural rocking, to displacement drift ratios of 0.32% (loading south) and 0.24% (loading north), vertical cracks began to form at the interface between the web and the flanges. As the displacements increased, the flange panel did not slide with the web in the direction of loading, resulting in local damage to the flange and finally instability of the flange at both the north and south ends of the specimen. At the north end, the flange damage was accompanied by a large vertical crack in the east web. Testing of the Two-story Assemblage Specimen was halted due to this damage.

6 SUMMARY OF RESPONSE OF ASSEMBLAGE

Including base slip, the Two-story AAC Assemblage Specimen reached drift ratios between 0.7% and 0.85 %. Final displacement ductilities (final displacement divided by the displacement at yielding of the flexural reinforcement), ranged from 8.3 to 11.7 (Table 4).

After removing the base slip, the Two-story Assemblage Specimen reached drift ratios between 0.24% and 0.42%, and final displacement ductilities ranged from 2.8 to 5.8 (Table 5).

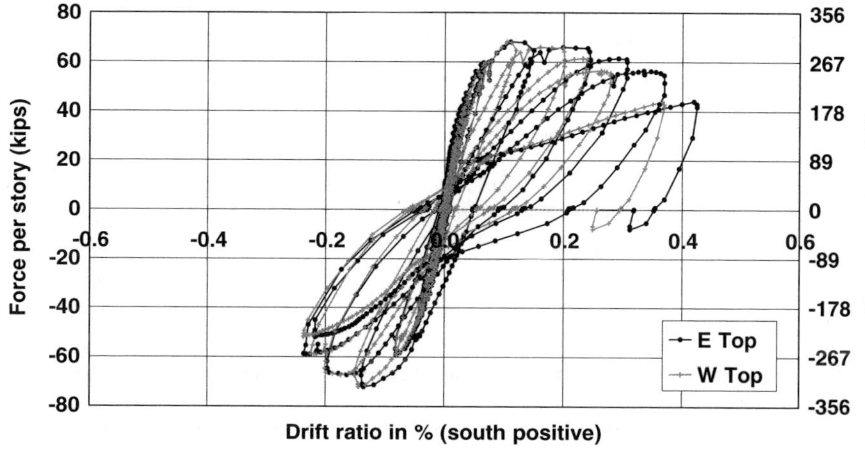

Figure 7. Hysteretic behavior of Two-story AAC Assemblage Specimen with slip removed.

Table 4. Drift ratios and displacement ductilities for each wall, including base slip.

Shear wall/Direction of loading	Displacement ductility	Drift ratio (%)
East wall/south	11.7	0.85
East wall/north	8.3	0.70
West wall/south	8.7	0.74
West wall/north	8.6	0.73

Table 5. Drift ratios and displacement ductilities for each wall, with base slip removed.

Shear wall/Direction of loading	Displacement ductility	Drift ratio (%)
East wall/south	5.8	0.42
East wall/north	2.8	0.24
West wall/south	4.4	0.37
West wall/north	2.8	0.24

For design purposes, these results justify an assumption of an available flexural ductility of at least 3.0, reasonably consistent with that observed in previous tests of flexure-dominated AAC shear walls at UT Austin (Tanner et al. 2005b), and consistent with the proposed R and C_d values of 3.0 (Varela et al. 2004).

7 CONCLUSIONS

Based on test results from The University of Texas at Austin and elsewhere, procedures and corresponding provisions were developed for the design of AAC shear walls made of masonry-type units or horizontally oriented reinforced panels. The provisions, now incorporated into the 2005 edition of the MSJC Code and Specification, address flexural cracking; flexure-shear cracking; web-shear cracking; sliding shear; and crushing of a diagonal strut. The proposed provisions are reliable, with low coefficients of variation (generally below 15%) and values close to or exceeding 1.0 for the ratios of observed capacity to the capacity predicted using equations based on the tested strength. The proposed provisions were validated in tests of a full-scale, two-story AAC shear wall assemblage specimen under quasi-static, reversed cyclic loading representing strong earthquake. Results from assemblage tests show an available flexural ductility of at least 3.0 for flexure-dominated specimens, consistent with proposed values for seismic design coefficients.

ACKNOWLEDGMENTS

Much of the information reported in this paper was developed in the course of a research project on the Seismic Behavior of Autoclaved Aerated Concrete, carried out at The University of Texas at Austin with the support of the Autoclaved Aerated Concrete Products Association.

REFERENCES

ASTM C 270: Standard Specification for Mortar for Unit Masonry, American Society for Testing and Materials, West Conshohocken, PA, 2004.

Boutros and Saverimutto, 1997: Boutros, M. and Saverimutto, L., "Anchorage Capacity of Reinforcing Bars in Autoclaved Aerated Concrete Lintels" Materials and Structures, November 1997, pp. 552–557.

CEB 1978: "Autoclaved Aerated Concrete CEB Manual of Design and Technology," Comite Euro-International du Beton, The Construction Press, Lancaster, England, 1978, 89 pp.

CEB 1992: "CEB-FIB Model Code," CEB, Comite Euro-International du Beton, Thomas Telford Services Ltd., Lausanne, Switzerland, 1992, 437 pp.

MSJC 2005a: Building Code Requirements for Masonry Structures (ACI 530-05/ASCE 5-05/TMS 402-05), American Concrete Institute, Farmington Hills, Michigan, American Society of Civil Engineers, Reston, Virginia, and The Masonry Society, Boulder, Colorado, 2005.

MSJC 2005b: Specifications for Masonry Structures (ACI 530.1-05/ASCE 6-05/TMS 602-05), American Concrete Institute, Farmington Hills, Michigan, American Society of Civil Engineers, Reston, Virginia, and The Masonry Society, Boulder, Colorado, 2005.

Tanner et al. 2005a: Tanner, J.E., Varela, J.L. and Klingner, R.E., "Design and Seismic Testing of a Two-story Full-scale Autoclaved Aerated Concrete (AAC) Assemblage Specimen," Structures Journal, American Concrete Institute, Farmington Hills, Michigan, vol. 102, no. 1, January–February 2005, pp. 114–119.

Tanner et al. 2005b: Tanner, J.E., Varela, J.L., Klingner, R.E., Brightman, M.J. and Cancino, U., "Seismic Testing of Autoclaved Aerated Concrete (AAC) Shear Walls: A Comprehensive Review," Structures Journal, American Concrete Institute, Farmington Hills, Michigan, vol. 102, no. 3, May–June 2005, pp. 374–382.

Varela et al. 2004: Varela, J.L., Tanner, J.E. and Klingner, R.E., "Development of Seismic Force and Displacement Modification Factors for Design of AAC Structures," Proceedings, 13th World Conference in Earthquake Engineering, Vancouver, Canada, August 1–6, 2004.

Autoclaved Aerated Concrete – Limbachiya and Roberts (eds)
© 2005 Taylor & Francis Group, London, ISBN 0 415 38356 0

The emergence of solid AAC external walls for UK construction

C.R. Marshall
Tarmac Topblock Limited, Wolverhampton, United Kingdom

ABSTRACT: Although the use of solid AAC external wall constructions are firmly established in continental Europe, traditional and construction related factors have to date limited its appeal in the UK. The current construction market is encouraging builders to consider more efficient methods of building. Solid AAC external walls offer several advantages over traditional cavity construction including significantly improved build times and simplified construction details.

Aesthetically solid AAC can be finished in a number of ways to satisfy planning and customer requirements. The use of externally applied insulation not only permits a variety of finishes but also helps with the energy efficiency of the insulated fabric by utilising the thermal mass of the blockwork to help minimise the fluctuations on heating demand.

Because cavity construction is a tried and tested solution in providing adequate resistance to rain penetration in all geographical areas of the UK, the use of properly designed solid AAC walls has to be based on experience in similar climates as well as a high degree of reliability in the design and in the component materials that form part of the overall specification.

1 INTRODUCTION

This Paper highlights the advantages and performance of solid AAC external walls. Although specifiers and builders in the UK may not be as familiar with solid AAC external walls as they will be for cavity construction, solid walls are relatively straightforward to design and construct. They can also contribute towards simplifying the construction process and therefore improve the speed of building.

In the UK, there has previously been much discussion about solid walls versus cavity constructions. It is acknowledged that cavity walls do, and will continue to fulfil a major role in the construction of all buildings. Nevertheless, there have been situations in recent times where solid AAC walls have been used successfully. When thermal insulation standards were improved in the 1970's and 1980's, AAC was often specified as the primary insulation material. Where it was used as an outer leaf or as a solid wall, it was recommended that they should be finished with render or an alternative cladding treatment such as tile hanging or weatherboarding.

The limiting factors to the growth of AAC walls at the time were mainly the predominance of brickwork as an aesthetically acceptable material. In addition restrictions on the use of traditionally rendered solid walls due to exposure limitations coupled with conservatism in situations where such construction could be technically acceptable meant that there was a general reluctance to accept something different.

Today there is a strong desire to consider alternative methods of construction as a means of meeting the shortfall in the required number of new homes. This ambitious building programme is set to take us well beyond the next decade. It is apparent that many non-masonry build systems being evaluated do not necessarily incorporate a facing brick veneer. This is simply because it may not suit the method of construction and it is a relatively expensive cosmetic treatment that does not usually contribute anything towards the structural stability of the building. This observation equally applies to solid AAC external walls, a material that is widely known and understood. However, the benefit of solid AAC is that it can be readily combined with other fast build methods such as thin jointed masonry and pre-cast flooring to provide a low risk building method. Although such construction is commonplace in many parts of continental Europe, experience has shown that transferring techniques is by no means a straightforward exercise. There are examples of technically proven building practice that can be combined with current AAC products to deliver cost effective construction methods and yet at the same time provide specifiers and building users with all the inherent advantages of traditional masonry.

2 ADVANTAGES OF SOLID AAC EXTERNAL WALLS

2.1 *General advantages*

It is widely acknowledged that single leaf walls are quicker to build and more cost effective than comparable size cavity walls. The benefits arising from AAC units accrue from the use of fewer units and less mortar if one or more of the cavity leaves comprise of brickwork, as well as the elimination of wall ties and ancillary cavity components. Productivity will be significantly increased. Scaffolding for AAC walls only needs to be erected on one side of the wall as opposed to two sides for cavity walls. The installation of lintels and like components is simplified. The inherent stability of single leaf construction of minimum 190 mm thickness ensures that delays during construction due to the immaturity of the blockwork are kept to a minimum. Furthermore, if thin layer mortars they are probably eliminated all together.

Solid AAC walls can be built using conventional or thin layer mortar. The latter is associated with installations requiring acceleration in build time and improvement in overall build quality. For example the flatness of the wall will be superior to those built with conventional mortar making them particularly suitable to receive sprayed surface finishes of minimal thickness.

Additional thermal insulation material is straightforward to install and to supervise. Insulation could be applied externally before rendering or cladding takes place, or fixed internally onto the wall in the form of a plasterboard laminate or similar. Again, the flatness of the wall arising from the use of thin layer mortars is of particular benefit when applying external insulation material.

No special skills are needed to construct single leaf AAC external walls other than those associated with any type of masonry construction.

2.2 *Programming*

The specification of a single masonry material for the external wall simplifies the procurement process. AAC units are normally readily available from stock and, depending on market conditions, can be supplied at relatively short notice. For large building projects, it is recommended that the delivery programme be agreed with the AAC manufacturer to ensure continuity of supply. Avoidance of excessive lead times contributes significantly towards predicting the construction time and cost of external envelope.

It is recommended that the use of AAC blockwork external walls are combined with internal walls of the same material thus eliminating the problems normally associated with combining building materials with different characteristics.

AAC can also be incorporated into floors, for example as infill blocks to beam and block floors, which effectively results in the construction team handling one masonry material for the entire structure.

2.3 *Supervision*

Compared to cavity walls, single leaf AAC blockwork is comparatively easy to inspect and supervise. During construction it will normally be possible to view the work from both sides of the wall as well as avoiding the uncertainties of inspecting cavities.

Table 1. Properties of contemporary production from manufacturers web sites.

Grade of AAC	Density (kg/m^3)	Compressive strength* (N/mm^2)	Thermal conductivity (W/mK)
Low density	420–480	2.8–3.5	0.10–0.12
Medium density	600–630	3.5–5.0	0.15–0.17
High strength	715–770	7.0–8.4	0.19–0.20

*Measured to BS6073

2.4 *Technical performance*

The performance requirements for most projects can be realised through the technical properties of AAC units incorporated into a suitably designed external wall. Properties of contemporary production from UK manufacturers web sites are shown in Table 1. Relevant performance criteria are discussed elsewhere in this Paper.

3 STRUCTURAL PERFORMANCE

3.1 *Design methods*

The principles of the structural design of solid AAC blockwork are the same as for any type of masonry. The design of dwellings may be undertaken using one of the following design methods:

- BS5628: Part 1: Structural use of unreinforced masonry
- BS8103: Part 2: Code of Practice for masonry walls for housing
- The Building Regulations Approved Document A

BS8103 and Approved Document A both use a 'simple rules' approach. The walls of buildings or those parts of buildings that are outside the scope of the rules should be designed in accordance with BS5628: Part 1. Whichever method is adopted there is likely to be considerable advantages over cavity walls in terms of structural performance including vertical loadbearing capacity. This is due to enhanced slenderness ratios and the availability of a considerably increased bearing width. A small but important distinction is that the effective thickness is the actual thickness. The 'simple rules' approach will yield a minimum wall thickness of 190 mm and unit strengths of 2.8 and 7.0 N/mm^2 for two and three storey dwellings respectively. However given the considerable reserves of strength in solid walls it is conceivable that lower strength units, perhaps as low as 2 N/mm^2, can be used in the majority of dwellings regardless of storey height. Although at this time it is acknowledged that strengths below 2.8 N/mm^2 are not yet recognized in UK design codes or methods.

3.2 *In General purpose mortar*

The structural design of AAC blockwork should use general purpose mortars that are compatible with AAC blocks. Over strong general purpose mortars should not be specified as their use may lead to cracking. Lower strength general purpose mortars are more compatible with AAC.

The following mortars are generally suitable above ground:

- 1:1:5-6 (Cement:lime:sand)
- 1:2:8-9 (Cement:lime:sand)

Plasticised mortars, which are no stronger, are also suitable. Below ground, the mortar should be stronger (e.g. 1:½:4–4½) and sulfate-resisting cement may be required, depending on the soil and ground water conditions.

3.3 *In thin layer mortar*

Thin joint AAC blockwork is normally constructed using a special formulated thin layer mortar. Current practice is to use this above ground but to revert to a conventional cement and sand mortar below ground. The structural design of thin-joint walls should be in accordance with BS5628: Part 1, assuming the characteristic compressive strength is that given in Table 2 of BS5628, for designation (i) mortar.

The characteristic flexural strength should normally be taken as given for blocks laid in a designation (i) mortar (BS5628: Part 1).

If bed joint reinforcement is used, for example to control drying shrinkage stresses in the masonry, it should be of a type intended specifically for use with thin-joints.

Large format blocks that are convenient to handle may be used to increase site productivity.

4 DURABILITY AND RESISTANCE TO RAIN

4.1 *Durability*

The inherent durability of AAC material makes it eminently suitable for single leaf external walls. There is a track record stretching back several decades where AAC single leaf construction have been used in this way. Generally, such walls were finished with render – subject to exposure category – or an impervious cladding such as tile hanging and weatherboarding. In such situations AAC blockwork would be expected to have a service life equivalent to other types of masonry. There is no practical reason why lower density AAC blocks cannot be considered. The advantages include improved thermal performance; lower self-weight and structural economy. Currently the lowest strength blocks that can be considered for walls are specified in Building Regulations at $2.8N/mm^2$.

4.2 *Resistance to rain*

In the context of encouraging fast-build construction with low technical risk, external walls can be finished with an insulated render system or an alternative cladding treatment. This could include brick slips, as well as tile or timber cladding. In such circumstances the use of AAC walls can be considered for use in 'severe' and 'very severe' exposure zones as described in BS8104 .

It will be appropriate to verify the suitability of proprietary render and cladding systems to suit the conditions of exposure with the manufacturer.

Modern insulated render systems are compatible with the requirements of fast-track applications. These rely on machine application avoiding costly labour in mixing, hoisting and placing render on the wall. Hidden savings, for example, in the form of early scaffold release, contribute to all round shorter build programmes.

5 THERMAL INSULATION

5.1 *Design*

It is relatively straightforward to design single leaf AAC walls to meet the requirements of the Building Regulations. Due to the good insulation properties of AAC the amount of added insulation – whether applied externally or internally – can be kept to a minimum.

Unlike cavity walls, added insulation can be applied after the AAC wall is erected which simplifies the construction process. Thin-layer mortar joints reduce thermal transmittance compared with conventional mortar joints and large format AAC blocks reduce thermal transmittance and improve air tightness due to the reduced number and area of mortar joints.

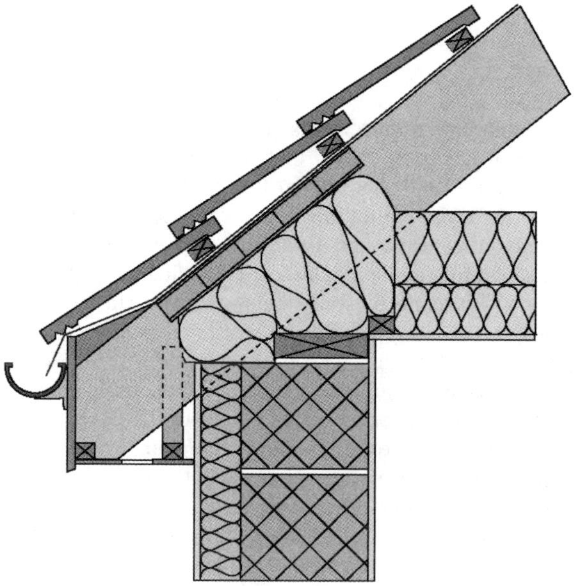

Figure 1. Example Robust Detail at eaves level to limit thermal bridging.

5.2 *Thermal bridging, air leakage and condensation*

Sustained improvement in the standards of thermal insulation in UK Building Regulations has focussed attention on steps to minimise thermal bridging. The current Regulations, which are soon to be upgraded, have addressed these issues in greater detail. The concept of "Robust Details" (Limiting thermal bridging and air leakage, 2001) was introduced for dwellings and similar buildings to avoid unnecessarily complex calculations (Figure 1).

Following these details should minimise any inherent or potential risks associated with such aspects as interstitial and surface condensation as well as undue air leakage, both inward and outward. A number of Robust Details have been developed and include details for single leaf walls. The details cover U-values in the range 0.35 to 0.47 W/m^2K. It is envisaged that the Robust Details will be upgraded to take account of future changes in Building Regulations.

Placing insulation on the exterior of the AAC walls overcomes many of the difficulties in designing for thermal insulation. The supporting wall construction falls within the warm zone and is thus kept warm and dry, and will act as a heat store. Thermal bridging and interstitial condensation are all but eliminated.

5.3 *Thermal inertia*

With rising standards of thermal insulation, acknowledgement of the benefits of the thermal inertia of building materials is growing. Building Regulation compliance tends to rely on calculations based on steady-state conditions that do not occur in practice. The thermal mass of a building influences how fast its temperature rises and cools on a daily cycle and over seasons. Higher thermal mass will result in a longer delay in the rise of the internal temperature. This delay is beneficial both in terms of thermal comfort as well as avoiding heat losses due to rapid dissipation. AAC provides a useful thermal inertia contribution and when combined with good levels of thermal insulation means that AAC structures can avoid the extremes of internal temperature compared with buildings made of lighter materials. Studies have shown that designing structures to make them future proof for climatic change will rely heavily on ensuring sufficient thermal inertia is present to help prevent summer overheating and the need for cooling.

6 SOUND INSULATION

6.1 *Requirements*

Currently there are no regulatory requirements for the acoustic performance of solid external walls in the UK. However, it has been shown that AAC performs relatively well in this context. Laboratory work (Luckin et al. 1986) has shown that AAC follows a mass law that exceeds the performance predicted by the more commonly accepted mass law for denser materials. External walls associated with separating walls and floors need to be able to limit the flanking sound transmission around the separating element. Current UK guidance (Approved Document E. 2003) is to limit this transmission by using masonry having an as laid weight of at least 120 kg/m^2. For standard density UK AAC (630 kg/m^3) this means that a minimum external wall thickness of 175 mm would be required. In practice, for reasons discussed earlier, solid external walls tend to be at least 190 mm thick and so have no problems in achieving the mass guidelines. However, thinner AAC walls, such as 100 mm thick inner leaves of cavity walls, have been shown to consistently perform satisfactorily in practice. In order to show compliance with the current UK Building Regulations, builders are required to provide evidence that their constructions will actually achieve the minimum levels of sound insulation given in the Regulations. Evidence can be provided either by on site acoustic testing (a minimum of 10% of the new walls and floors are required to be tested) or alternatively the builder can subscribe to a third party quality assured scheme known as Robust Details. The drawback with having to provide site test evidence is that the tests can only be performed at a very late stage in the construction programme. Consequently the financial risks associated with failure are a great burden for the builder to bear.

6.2 *Robust details*

The Robust Details (Edition 2. 2005) contain material specifications and construction details that have been proven, by an extensive series of on site tests, to out perform the Regulatory requirements by a significant and comfortable margin. Importantly for the builder, although the details are by their very nature conservative, the risks associated with site testing are alleviated.

Significantly for AAC, the Robust Details include provisions for the use of 100 mm thick blocks (450-800 kg/m^3) for the flanking walls to both separating walls and separating floors. The results of the site tests bear out the previously mentioned findings that AAC performs relatively well when compared to denser materials.

Although solid AAC external walls will provide a substantial mass and, therefore, provide adequate resistance to flanking sound transmission, the dilemma for builders is that details for solid external walls are not given in the Robust Details. The reason for this was not a performance issue but simply that the required number of site tests were not available during the development of the Robust Details.

Several projects have been completed and tested since the introduction the 2003 Regulations. However, the details of these schemes have been developed on an ad-hoc basis. Work is currently ongoing to develop standard junction details for solid external walls, which will be assessed under laboratory conditions before being site tested.

7 FIRE RESISTANCE

7.1 *Requirements*

The fire resistance requirements for AAC walls will be in accordance with Building Regulations (Approved Document B) plus any additional requirements imposed by agencies such as insurance companies. The minimum periods of fire resistance for residential buildings specified in Building Regulations range from 30 minutes to 120 minutes. This range reflects the heights of buildings-more than 30 metres at the extreme end-and where the external wall is located with distinctions made between basements and walls above ground. Given that external walls are likely to be constructed from AAC units of minimum 190 mm thickness and the fire resistance of such a construction in a loadbearing wall is 3 to 4 hours depending on the AAC density, it is apparent that the performance of the AAC wall alone is sufficient to meet all likely fire resistance requirements. In reality it is the performance of applied finishes and linings that need to be evaluated for their ability to limit fire spread. Externally attention should be given to the location and amount of combustible materials for cladding framework, or of combustible thermal insulation as an over cladding or in ventilated cavities. Internally, surface linings must meet the classification of linings appropriate to their location as specified in building regulations.

8 FINISHES

8.1 *External*

AAC walls exposed to the external environment normally require some form of additional finish to prevent moisture penetration and to protect the surface of the blockwork from the effects of the elements. Where walls are exposed to extreme conditions, then it is advisable to clad the wall with an impermeable cladding, such as timber boarding or tile hanging. There has always been a strong preference in the UK for a traditional brick appearance, particularly for housing. This has, to a certain extent, hampered the progress of AAC solid walls, which are normally finished with some form of render or a cladding finish. However, recent developments in external cladding systems have seen the growth of brick slip systems, which give a more traditional appearance.

8.1.1 *Render*

AAC provides a good background for the application of renders. One of the basic principals of the application of any render is that applied coats should not be stronger than the background on to which they are applied. This is normally achieved either by reducing the thickness of subsequent coats or by ensuring that weaker mixes are used. In the UK, the strongest render mix recommended for the use on AAC is a 1:1:6 cement:lime:sand mix, which equates to a strength of 3.5-4.0 N/mm^2. In situations where a directly bonded render is required preference should be given to pre-mixed renders to help avoid inconsistency in site batching. The manufacturer should be consulted for the most suitable AAC product to be used in such situations.

Generally, a 20 mm thick render (consisting two coats) on a 215 mm thick block, is suitable to resist the passage of wind driven rain for exposure conditions up to and including severe (BS5628 Pt3). However, proprietary renders are available which can be applied at thicknesses of 6–8 mm. These are particularly suitable for use on thin jointed AAC walls as these types of walls tend to be built to better tolerances than traditional walls.

8.1.2 *Externally insulated systems*

Modern systems can produce any type of appearance required; from render to an ashlar stone look facade. As we have already discussed, additional insulation is often required when building solid AAC walls. The opportunity is taken to introduce external insulation, which serves two purposes. Firstly, it provides the necessary insulation required, and secondly it provides the support and acts as backing to the finish.

Although insulated renders have been around for a number of years in the UK, they have predominantly been used in the refurbishment sector for over cladding of existing structures. The introduction of brick slip faced systems is fairly recent. The main attraction of this type of system for the UK market is the opportunity to maintain a traditional facing brick appearance.

External insulation offers a distinct thermal advantage over internally applied insulation in that it enables the thermal mass of the AAC blocks to be utilized and therefore minimize heating costs.

8.1.3 *Cladding*

For very severe conditions, a cladding system is generally required. Such a system may consist of proprietary steel sheets, or more commonly for housing, timber boarding or tile hanging.

Both the latter systems are normally fixed onto vertical or horizontal timber battens, which in turn are mechanically fixed to the blockwork. AAC solid walls offer an advantage, when compared to denser materials, as nailed anchors can fix the battens. This eliminates the costly alternative of having to use screws and plastic or metal plugs, as well as pre-drilling of the battens to receive the screws.

8.2 *Internal*

Although AAC can be left unfinished, or simply painted, in areas of limited access (such as garages or plant rooms), an additional finish is often applied for decorative purposes.

8.2.1 *Plaster finishes*

AAC provides a good background for the application of a wet plaster finish. As with external renders, it is important that strong cement based mixes are avoided and the 1:1:6 mix mentioned previously is the strongest that should be used.

Most proprietary plasters can be directly applied without the need for further surface treatments. However, a major benefit of AAC is that it can be manufactured to high dimensional tolerances and is therefore an ideal background for the use of thin coat projection plasters, particularly when built using thin joint mortars. Sprayed thin coat plasters offer a distinct advantage to the developer in that drying times are significantly reduced when compared to those normally associated with traditional wet plasters. These types of plasters can also be supplied pigmented so that further savings can be made on decoration costs.

8.2.2 *Plasterboard systems*

Plasterboard systems have become a very popular finish onto masonry walls over recent years. The big advantage they offer is that the installation procedure is effectively a dry process, so effectively eliminating the need for lengthy drying out periods before decorations can be applied. Plasterboards may be fixed onto battens, which have been previously fixed to the background blockwork, or, more commonly, fixed onto walls directly using adhesive dabs.

An advantage of dry lining solid external walls is that it gives the opportunity to add insulation internally if required. This may be placed between battens fixed to the walls or, alternatively, thermal laminate plasterboards can be used. The latter consist of plasterboards bonded with insulation and offer convenience and ease of installation.

9 CONSTRUCTION ASPECTS

As is common for all masonry, solid walls should be constructed in accordance with the recommendations of BS8000:Part 3 and BS5628:Part 3. The construction process is simplified as only one masonry leaf is considered thus eliminating wall ties, cavity trays, cavity filling and similar cavity related components and associated workmanship. The AAC units will normally be laid in half bond with all joints fully filled with mortar. Due to the relative lightness of AAC it should be possible for conventional mortared walls to be constructed to the maximum recommended height in a day of 1.5 m. However BS5628:Part 3 does acknowledge that it may be possible to safely carry lightweight units higher this. No limitation is placed on walls built using thin layer mortar. The blockwork is normally laid from the inside of the building using the ground and intermediate floors as the working platform. An independent external scaffold is required to provide access and allow the intermediate floors and roof to be installed as well as providing access for the application of any externally applied insulation and finishes.

Figure 2. Machine application of render onto a solid external wall.

10 OPPORTUNITIES FOR IMPROVING THE SPEED OF CONSTRUCTION

The adoption of solid AAC walls will considerably speed up the construction process. Maximum productivity will result by taking advantage of the rapid setting characteristics of thin layer mortar in conjunction with large format blocks that are convenient to handle. It is a realistic prospect that storey-heights could be completed in a day. The potential for rapid building can be extended to include other elements that are quick to install together with the productivity benefits of using machine applied finishes.

Pre-cast floors, whether wide slab or beam and block incorporating AAC infill units, can be considered and offer the advantage of an immediate working platform after erection. Roofs can be quickly assembled using pre-formed which can be erected soon after the wall plate is firm. These units are pre-insulated and are suited to the current trend in housing of utilising the roof void as part of the habitable area.

Machine applied finishes have been used on an ad-hoc basis for some years. Machine application avoids costly labour in mixing, hoisting and placing render on the wall. Hidden savings in the form of early scaffold release contribute to all round shorter build programmes.

Their use can be considered for proprietary rendering as well as sprayed plasterwork. Walls built with thin layer mortar are particularly suitable for this application as they are highly accurate in construction leaving a flat internal surface. Although perceived to be more costly than traditional finishes it is their unfamiliarity that is the biggest obstacle to overcome. Advantages also accrue by a reduced thickness of material, 3–8 mm is typical, and high productivity if the work is planned to give the plasterer sufficient work ahead. Compared to site mixed materials the quality of machine-applied finishes is based on factory prepared materials.

11 CONCLUSIONS

There is a track record of using solid AAC external walls going back several decades. There are considerable advantages in considering solid wall construction as a rapid construction method to meet today's housing requirements. AAC can be combined with other fast build methods such as thin joint masonry as well as prefabricated floor and roof units to provide an efficient and low risk building solution. The emergence of proprietary wall plasters and renders, including insulated render systems, provide an almost limitless opportunity to finish AAC walls internally and externally with durable and aesthetically pleasing finishes. The design of solid walls can be readily undertaken using established design methods and consequently there should be no difficulty in meeting the requirements of building regulations. There are of course some areas where more performance data would be beneficial, particularly in the area of sound insulation, but given that pre-completion testing of separating walls is now a regulatory requirement current data suggests that there should be no difficulty in obtaining this as the market for solid walls becomes established.

REFERENCES

BS5628 Code of practice for use of masonry, Part 1: Structural use of unreinforced masonry 1992. BSI London

BS5628 Code of practice for use of masonry, Part 3: Material and components, Design and workmanship 2001. BSI London

BS6073 Precast concrete masonry units, Part 1: Specification for precast concrete masonry units 1981. BSI London

BS8000 Workmanship on building sites, Part 3: Code of Practice for masonry. 2001. BSI London

BS8103 Structural design of low-rise buildings, Part 2: Code of practice for masonry walls for housing. 1996. BSI London

BS8104 Code of Practice for Assessing exposure of walls to wind-driven rain. 1992. BSI London

Limiting thermal bridging and air leakage: Robust construction details for dwellings and similar buildings. 2001. HMSO. London

Luckin, K R., Jones, A J & Engledow, G. 1986. Sound Insulation Performance of Autoclaved Aerated Concrete, BMS Proceedings N°2

Robust Details Part E Resistance to the Passage of Sound. 2005. Robust Details Ltd, Milton Keynes

The Building Regulations 2000. Approved Document A. Structure. 2004. HMSO. London

The Building Regulations 2000. Approved Document B. Fire safety. 2002. HMSO. London

The Building Regulations 2000. Approved Document E. Resistance to the Passage of Sound 2003. HMSO. London

Autoclaved Aerated Concrete – Limbachiya and Roberts (eds)
 ISBN 0 415 38356 0

Design examples for AAC masonry structures using US provisions

Michael J. Robinson
TruStone America, Providence, Rhode Island, USA

Ronald E. Barnett
Aercon Florida LLC, Haines City, Florida, USA

Prof. Jennifer E. Tanner
Dept. of Civil Engineering, University of Wyoming, Laramie, Wyoming, USA

Prof. Jorge L. Varela
Dept. of Civil Engineering, Autonomous University of Yucatán, Mérida, Yucatán, Mexico

Prof. Richard E. Klingner
Dept. of Civil Engineering, The University of Texas, Austin, Texas, USA

ABSTRACT: In this third of three related papers proposed for publication in the *Proceedings* of the 4th International Conference on Autoclaved Aerated Concrete, current US design provisions for AAC masonry are applied to prototypical elements of reinforced AAC masonry, including a wall loaded out of plane, a lintel, and a shear wall. AAC masonry is seen to be a practical material for each application.

1 INTRODUCTION TO AUTOCLAVED AERATED CONCRETE (AAC) MASONRY AS USED IN THE US

Autoclaved aerated concrete (AAC) is produced in the US in a variety of specified compressive strengths, according to ASTM C 1386. It is described in terms of "strength classes", each of which corresponds to the specified compressive strength in MPa.

Table 1. Typical mechanical characteristics of different strength classes of AAC.

Strength class of AAC	Specified compressive strength (MPa)	Nominal dry density (kg/m^3)
AAC 2.0	2.0	400 ± 50 500 ± 50
AAC 4.0	4.0	500 ± 50 600 ± 50 700 ± 50 800 ± 50
AAC 6.0	6.0	700 ± 50 800 ± 50

2 BASIC STRUCTURAL BEHAVIOR OF LOW-RISE, BEARING-WALL BUILDINGS

Low-rise, bearing-wall buildings resist lateral loads as shown in Figure 1. This resistance mechanism involves three steps:

- Walls oriented perpendicular to the direction of lateral load transfer those loads to the level of the foundation and the levels of the horizontal diaphragms. The walls are idealized and designed as vertically oriented strips.
- The roof and floors act as horizontal diaphragms, transferring their forces to walls oriented parallel to the direction of lateral load.
- Walls oriented parallel to the direction of applied load must transfer loads from the horizontal diaphragms to the foundation. In other words, they act as shear walls.

This overall mechanism demands that the horizontal roof diaphragm have sufficient strength and stiffness to transfer the required loads.

Because the walls are reinforced, design is straightforward and reasonably familiar even to those with little or no experience in masonry design.

- Vertical strips resisting combinations of gravity loads and out-of-plane loads act as reinforced beam-columns. Behavior of reinforced masonry is quite similar to that of reinforced concrete, and can be described by a moment–axial force interaction diagram for out-of-plane bending.
- Shear walls act in flexure as cantilever beam-columns. Behavior of reinforced masonry is quite similar to that of reinforced concrete, and can be described by a moment–axial force interaction diagram for in-plane bending.

The overall starting point for reinforcement for AAC masonry structures is shown in Figure 2.

Structural design is carried using the strength provisions of the 2005 MSJC Code and Specification, because this document is referenced by most model codes, and its strength provisions can be easily learned by designers familiar with strength provisions for reinforced concrete (MSJC 2005a,b).

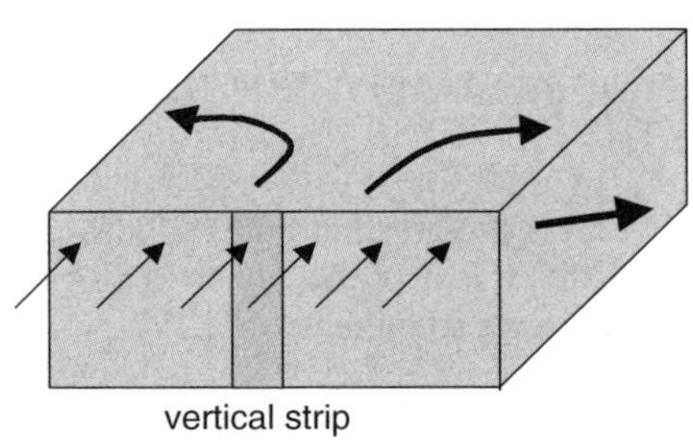

Figure 1. Basic structural behavior of low-rise, bearing-wall buildings.

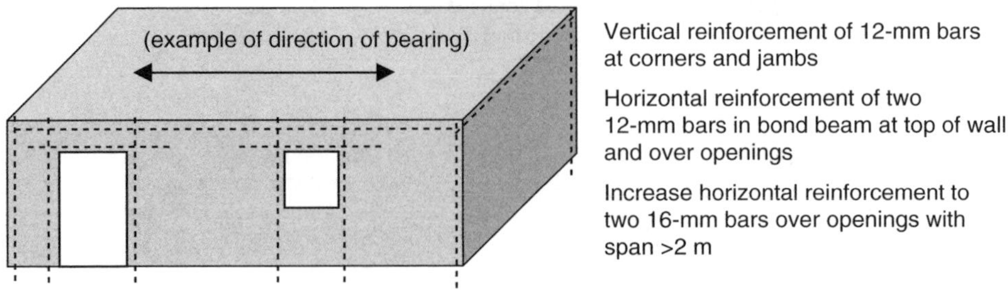

Figure 2. Overall starting point for reinforcement of AAC masonry structures.

Reinforced masonry walls loaded out-of-plane are essentially reinforced masonry beam-columns. As in reinforced concrete, they are designed using moment–axial force interaction diagrams. Combinations of axial force and moment lying inside the diagram represent permitted designs; combinations lying outside, prohibited ones.

Unlike reinforced concrete, however, reinforced masonry beam-columns rarely take the form of isolated rectangular elements with four longitudinal bars and transverse ties. The most common form for a reinforced masonry beam-column is a wall, loaded out-of-plane by eccentric gravity load, alone or in combination with wind.

For example, Figure 3 shows a portion of a wall, with a total effective width of $6t$ prescribed by the strength-design provisions of the 2005 MSJC Code (Section 3.3.5.2).

Using the strength approach, interaction diagrams that represent combinations of axial and flexural capacity are constructed. This is most easily done with the help of a computer spreadsheet. The position of the neutral axis is varied across the depth of the cross-section; compressive forces are taken as positive; and compressive stresses in reinforcement are neglected, because the reinforcement cannot be laterally supported. Net axial force on the cross-section is the difference between compressive and tensile forces; and moment is obtained by summing the product of each force by its distance from the plastic centroid of the cross-section.

3.1 *Example: moment–axial force interaction diagram by the strength approach (spreadsheet calculation)*

Construct the moment-axial force interaction diagram by the strength approach for a nominal 200-mm AAC wall, with $f'_{AAC} = 4$ MPa and reinforcement consisting of 16-mm bars at 1.2 m, placed in the center of the wall. The results are shown in Figure 4.

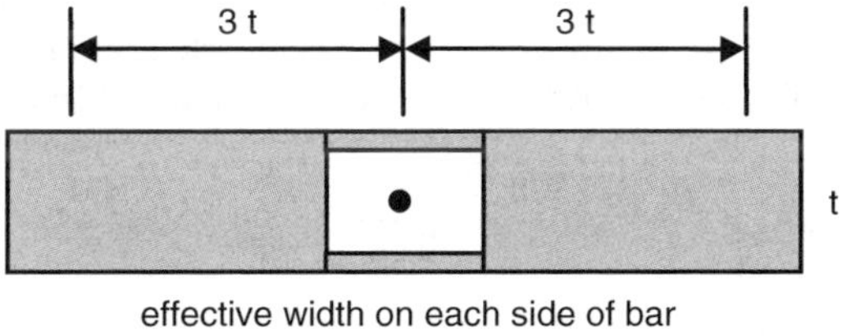

Figure 3. Effective width for AAC walls loaded out of plane.

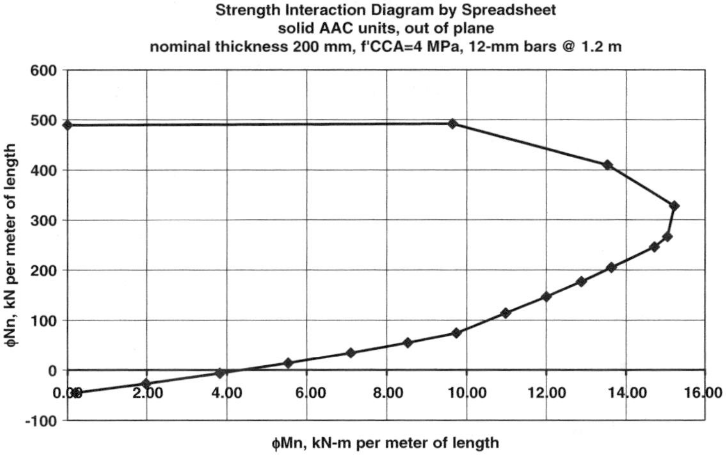

Figure 4. Strength interaction diagram by spreadsheet for AAC masonry wall loaded out of plane.

3.2 *Example #2: design of masonry walls loaded out-of-plane, strength approach*

Once we have developed the moment–axial force interaction diagram by the strength approach, the actual design simply consists of verifying that the combination of factored design axial force and moment lies within the diagram of nominal axial and flexural capacity, reduced by strength-reduction factors. Consider the bearing wall shown below. It has an eccentric axial load plus out-of-plane wind load of 1500 N/m^2.

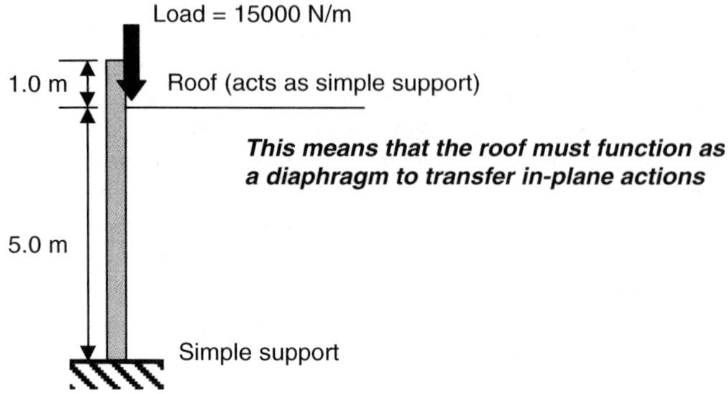

At each horizontal plane through the wall, combinations of factored axial load and moment must lie within the moment–axial force interaction diagram, reduced by strength-reduction factors. Because flexural capacity increases with increasing axial load, the most critical loading combination according to ASCE 7-05 is probably 0.9D + 1.6 W. The critical point on the wall is at the midspan of the lower portion. The maximum moment is close to that occurring at mid-height. The moment from wind load is the superposition of one-half moment at the upper support due to wind load on the parapet only, plus the midspan moment in a simply supported beam with that same wind load. Due to wind alone, the unfactored moment at the foot of the parapet (roof level) is:

$$M = \frac{qL^2}{2} = \frac{1500\ N/m \times 1^2\ m^2}{2} = 750\ N-m$$

At midheight of the wall, the unfactored moment is:

$$M_{midheight} = \frac{-750}{2} + \frac{qL^2}{8} = \frac{-750\ N-m}{2} + \frac{1500\ N/m \times 5^2\ m^2}{8} = 4313\ N-m$$

Unfactored moment diagrams due to eccentric axial load and wind are as shown below:

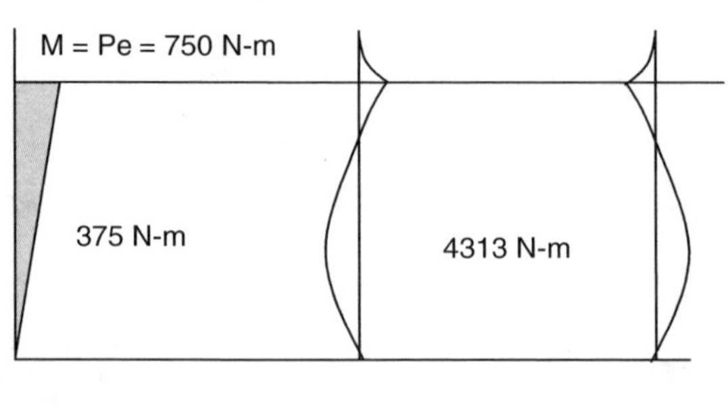

At the midheight of the wall, the axial force due to 0.9D is:

$$P_u = 0.9\,(10000\ N) + 0.9\,(1\ m + 2.5\ m) \times 1000\ N/m = 12{,}150\ N$$

At the midheight of the wall, the moment due to the combination of dead load and wind is:

$$M_u = P_u \frac{e}{2} + M_{u\,wind}$$

$$M_u = \left(\frac{1}{2}\right) 0.9 \times 10000\ N \times 0.05\ m + 1.6 \times 4313\ N-m$$

$$M_u = 225 + 4313\ N-m = 4{,}540\ N-m$$

On each meter of wall, design actions are $P_u = 12.15$ kN and $M_u = 4.54$ kN-m. This combination lies within the design interaction, and the design is satisfactory. The wall also needs to be checked for slenderness and out-of-plane shear, but those won't affect the final design.

4 STRENGTH DESIGN OF REINFORCED AAC MASONRY LINTELS

The most common AAC masonry beam is a lintel, whose design is addressed in Section A.4.2.4 of the 2005 MSJC Code and Specification, and involves the following steps:

(1) Shear design: compute the necessary depth so that shear reinforcement is not required; and
(2) Using that depth, compute the required flexural reinforcement.

The depth of the lintel is often imposed by architectural considerations.

4.1 *Example: Strength design of an AAC masonry lintel*

Design the lintel below, assuming a distributed load of 15,000 kN/m, plus the self-weight of the lintel, and an average weighted load factor of dead load and live load of 1.5. The AAC masonry has $f'_{AAC} = 4$ MPa, and a thickness of 200 mm. The parapet has a height of 1.0 m, plus a distance between the roof level and the top of the door opening of another 1.0 m, giving a total available depth of 2.0 m, or 10 courses (Figure 5). The span of the lintel is the dimension of the opening, plus half the bearing length on each side. The bearing length is half the length of a unit, or 200 mm. The span is therefore 3.00 + 0.10 + 0.10 m, or 3.20 m.

$$M_u = \frac{w_u \ell^2}{8} = \frac{(15000 \cdot 1.5 + 2.0 \cdot 1200 \cdot 1.4) \cdot 3.20^2}{8} N-m = 33.1\ kN-m$$

$$V_u = \frac{w_u \ell}{2} = \frac{(15000 \cdot 1.5 + 2.0 \cdot 1200 \cdot 1.4) \cdot 3.20}{2} N = 41{,}380\ N = 41.4\ kN$$

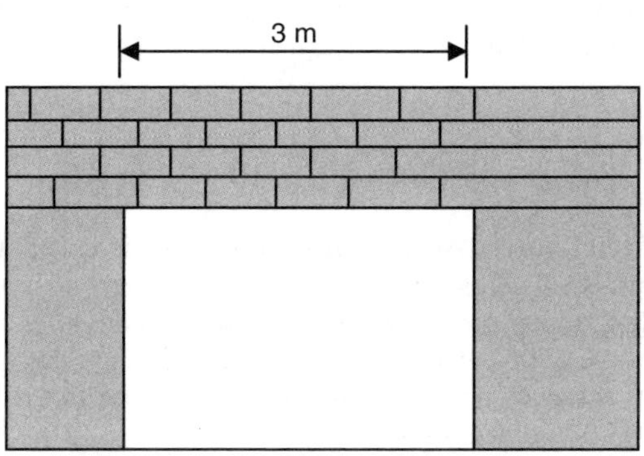

Figure 5. Lintel to be designed.

Because this is a reinforced element, its shear capacity is calculated using Section A.3.4.1.2.1 of the 2005 MSJC Code.

$$V_m = 0.08\, \ell_w\, t \sqrt{f'_{AAC}} \sqrt{1 + \frac{P_u}{0.2\sqrt{f'_{AAC}}\; \ell_w\, t}}$$

Using the entire effective depth of 2.0 m, the thickness of 200 mm, and neglecting any axial force, the capacity is:

$$V_m = 0.08 \times 2000\ mm \times 200\ mm \sqrt{4.0} \sqrt{1 + \frac{0}{0.2\sqrt{f'_{AAC}}\; \ell_w\, t}} = 64{,}000\ N = 64.0\ kN$$

This nominal capacity, reduced for the capacity-reduction factor of 0.8 for shear, gives a design capacity of 51.0 kN, greater than the factored design shear of 41.4 kN. The effective depth of the lintel will be approximately the total depth, less half the height of one course, or

$$d = 2.0 - \frac{0.20}{2} = 1.90\ m$$

Now check the required flexural reinforcement:

$$M_n = A_s f_y \left(int\,ernal\ lever\ arm)\right)$$
$$M_n \approx A_s f_y \cdot 0.9d$$

In our case,

$$M_n^{required} = \frac{M_u}{\phi} = \frac{M_u}{0.9} = \frac{33.1\ kN - m}{0.9} = 36.8\ kN - m$$

$$A_s^{required} \approx \frac{M_n^{required}}{0.9 d f_y} = \frac{36.8 \times 10^3\ N - m}{0.9 \cdot 1.90\ m \cdot 420\ N/mm^2} = 51.24\ mm^2$$

This can easily be satisfied with a 12-mm bar. Other 12-mm bars will be placed at the roof level and at the top of the parapet. Section A.3.4.2.2.2 of the 2005 MSJC Code also requires that the nominal capacity not be less than 1.3 times the cracking capacity. Because the reinforced placed is greater than that required, this requirement does not govern. Nor do the maximum reinforcement requirements, discussed next in the context of shear walls.

5 DESIGN OF REINFORCED AAC MASONRY SHEAR WALLS

The flexural capacity of reinforced AAC masonry shear walls is calculated using moment–axial force interaction diagrams. Such walls usually have multiple layers of flexural reinforcement, and the calculation of such diagrams is facilitated considerably by computer spreadsheets.

In Section A.3.4.1.2 of the 2005 MSJC Code, nominal shearing resistance is the summation of the resistance of the AAC masonry itself, plus the resistance from shear reinforcement. The latter includes only deformed reinforcement in bond beams.

$$V_n = V_m + V_s$$

The shearing resistance from AAC is given in Section A.3.4.1.2.1 of the 2005 MSJC Code as:

$$V_m = 0.08\,\ell_w\,t\,\sqrt{f'_{AAC}}\,\sqrt{1+\frac{P_u}{0.2\sqrt{f'_{AAC}}\,\ell_w t}}$$

For AAC masonry, the 2005 MSJC Code assumes that shear resistance from reinforcement embedded in grout is due to reinforcement in tension crossing a hypothetical failure surface oriented at 45 degrees to the direction of the applied shear:

$$V_s = A_v f_y n$$

$$V_s = A_v f_y \left(\frac{d}{s}\right)$$

Finally, an upper bound is placed on total shear resistance. This is consistent with a strut-and-tie mechanism in which shear resistance could be limited by diagonal crushing of the AAC.

5.1 *Example: design of a 4-story reinforced AAC shear wall*

Consider the design of the reinforced AAC masonry shear wall shown in Figure 6.

The lateral loads at each story level are due to earthquake, and are shown below, along with the corresponding shear and moment diagrams (Figure 7).

The load factor for seismic loads is 1.0 (ASCE 7-05). Assume AAC of Strength Class 4.0, a nominal thickness of 200 mm, and an effective depth of 7.85 m.

Unfactored axial loads are given in Table 2.

The critical loading combination is $0.9D + 1.0E$ (ASCE 7-05). At the base of the wall, $P_u = 0.9(600) = 540$ kN.

Check shear resistance for the assumed thickness. According to Section A.3.4.1.2 of the 2005 MSJC Code,

$$V_n = V_m + V_s$$

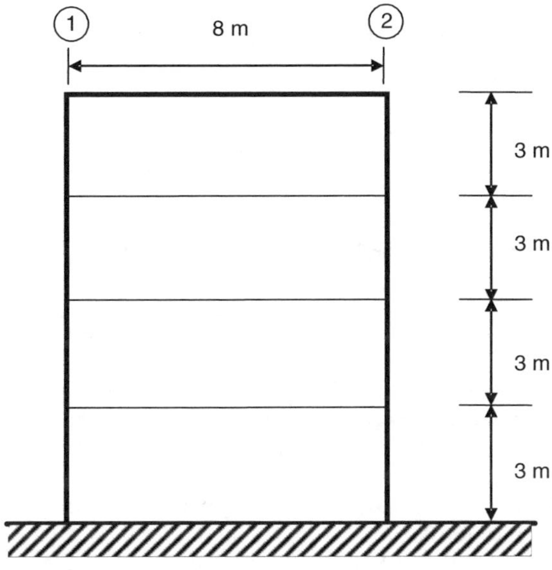

Figure 6. AAC masonry shear wall.

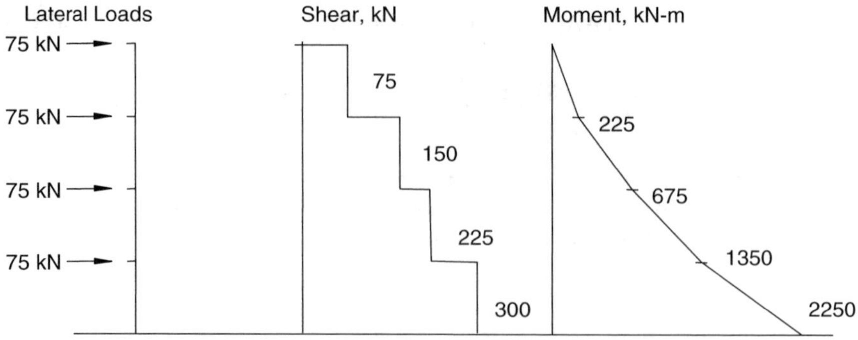

Figure 7. Shear and moment diagrams for AAC masonry shear wall.

Table 2. Unfactored axial loads on AAC shear wall.

Level on wall	Dead load, kN	Live load, kN
4	150	75
3	300	175
2	450	275
1	600	375

$$V_m = 0.08\, \ell_w\, t \sqrt{f'_{AAC}} \sqrt{1 + \frac{P_u}{0.2\sqrt{f'_{AAC}}\; \ell_w\, t}}$$

$$V_m = 0.08 \times 8000\, mm \times 200\, mm \sqrt{4} \sqrt{1 + \frac{540{,}000}{0.2\sqrt{4} \times 8000 \times 200}}$$

$$V_m = 256{,}000\sqrt{(1 + 0.84)}\, N = 256{,}000\,(1.36)\, N = 347{,}609\, N = 347.6\, kN$$

$$\phi V_n > V_u \quad \phi = 0.80 \quad V_n = V_m = 347.6\, kN$$

$$0.80(347.6\, kN) = 278\, kN < V_u = 600\, kN$$

Shear reinforcement is necessary:

$$V_s = A_v f_y n$$

$$V_s = A_v f_y \left(\frac{d}{s}\right)$$

With two 12-mm bars in bond beams and intermediate bond beams,

$$V_s = A_v f_y \left(\frac{d}{s}\right) = 2 \times 113\, mm^2 \times 420\, N/mm^2 \times \frac{7.85\, m}{3.0\, m} = 248.6\, kN$$

Shear design is satisfactory with a little shear reinforcement. Section A.1.3 of the 2005 MSJC Code will be checked a little later. Now check flexural capacity, using a spreadsheet-generated interaction diagram, with 9-mm bars every 2.0 m. This is equivalent to 64 mm² every 2.0 m. The resulting diagram is shown in Figure 8.

At a factored design axial load of 0.9D, or 540 kN, the design capacity in flexure is 2300 kN-m, and the design is satisfactory for flexure.

Now check Section 3.1.3 of the 2005 MSJC Code (capacity design). At an axial load of 540 kN, the nominal flexural capacity of the wall is 2300 kN-m, divided by the capacity reduction factor

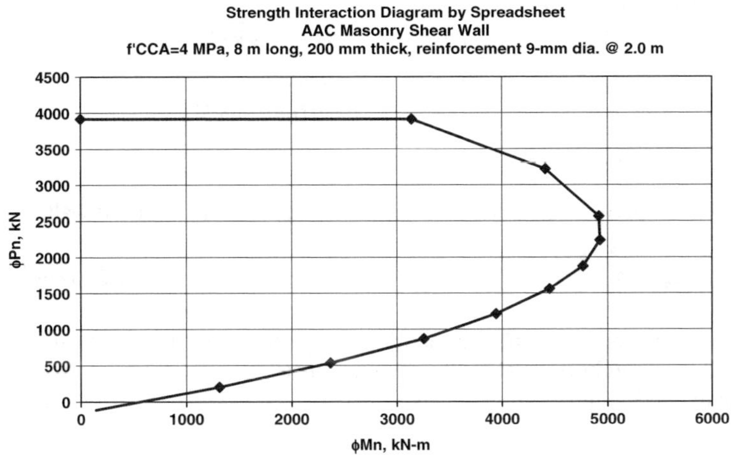

Figure 8. Moment–axial force interaction diagram for AAC shear wall.

of 0.9, or 2556 kN-m. The ratio of this nominal capacity in flexure divided by the factored design moment is 2556 divided by 2250, or 1.14. Including the additional factor of 1.25, the ratio is 1.42.

$$\phi V_n \geq 1.42\, V_u$$

$$V_n \geq \frac{1.42}{\phi} V_u = \frac{1.42}{0.8} V_u = 1.77\, V_u = 1.77 \times 300 = 532.4\ kN$$

The wall is satisfactory:

$$V_n = V_m + V_s = 348\ kN + 249\ kN = 597\ kN$$

Finally, check the 2005 MSJC Code requirements for maximum reinforcement, assuming that the wall is "intermediate," with a factor $\alpha = 3.0$. These requirements are intended to prevent crushing of the compression toe of the wall under a critical strain gradient, under which the AAC is at its maximum useful strain, and the extreme tensile reinforcement is at some multiple α of yield strain:

$$\rho_{max} = \frac{0.85 \times 0.67 \left(\frac{f'_m}{f_y} \right) \left(\frac{\varepsilon_{mu}}{\alpha\, \varepsilon_y + \varepsilon_{mu}} \right) - \frac{N_u}{\phi \cdot bd \cdot f_y}}{\left[1 - 2 \left(\frac{\varepsilon_{mu}}{\alpha\, \varepsilon_y + \varepsilon_{mu}} \right) \right]}$$

According to the 2005 MSJC Code, Section A.3.3.5.1(d), the governing axial loading combination N_u is $D + 0.75\,L + 0.525\,Q_E$, and the axial load is (600 + 0.75 × 375 kN), or 881.3 kN.

$$\rho_{max} = \frac{0.85 \times 0.67 \left(\frac{f'_m}{f_y} \right) \left(\frac{\varepsilon_{mu}}{\alpha\, \varepsilon_y + \varepsilon_{mu}} \right) - \frac{N_u}{\phi \cdot bd \cdot f_y}}{\left[1 - 2 \left(\frac{\varepsilon_{mu}}{\alpha\, \varepsilon_y + \varepsilon_{mu}} \right) \right]}$$

$$\rho_{max} = \frac{\left(\frac{0.85 \times 0.67 \times 4\ MPa}{420\ MPa}\right)\left(\frac{0.003}{3(0.00207)+0.003}\right) - \frac{881.3 \times 10^3\ N}{0.9 \times (200\ mm)\,7850\ mm \times 420\ MPa}}{\left[1 - 2\left(\frac{0.003}{3(0.00207)+0.003}\right)\right]}$$

$$\rho_{max} = 8.08 \times 10^{-4}$$

Check the maximum permitted of flexural reinforcement for every 2.0 m of wall:

$$A_{s\,max} = \rho_{max} b \times 2000\ mm = \left(8.08 \times 10^{-4}\right)(200\ mm)\ 2000\ mm = 323.3\ mm^2$$

We have only 64 mm^2 every 2.0 m, and the design is satisfactory. Use vertical reinforcement of 9-mm bars @ 2.0 m, and horizontal reinforcement consisting of two 12-mm bars in bond beams and intermediate bond beams.

6 SUMMARY

Autoclaved aerated concrete (AAC), a lightweight cementitious material originally developed in Europe more than 70 years ago and now widely used around the world, has recently been introduced into the US construction market, and AAC masonry design provisions have been introduced into the latest (2005) edition of the MSJC Code and Commentary, used for masonry design throughout the US. Using the AAC masonry design provisions of the 2005 MSJC Code and Commentary, designs are carried for a typical reinforced AAC masonry wall loaded out of plane; a typical reinforced AAC masonry lintel; and a typical reinforced AAC masonry shear wall. In each case, AAC masonry is seen to be a feasible construction material, and its design by the 2005 MSJC Code is quite similar to the design of conventional masonry in the US.

REFERENCES

ASCE 7-05: *Minimum Design Loads for Buildings and Other Structures*, American Society of Civil Engineers and Structural Engineering Institute, Reston , Virginia, 2005.

MSJC 2005a: *Building Code Requirements for Masonry Structures (ACI 530-05/ASCE 5-05 / TMS 402-05)*, American Concrete Institute, Farmington Hills, Michigan, American Society of Civil Engineers, Reston, Virginia, and The Masonry Society, Boulder, Colorado, 2005.

MSJC 2005b: *Specifications for Masonry Structures (ACI 530.1-05/ASCE 6-05/TMS 602-05)*, American Concrete Institute, Farmington Hills, Michigan, American Society of Civil Engineers, Reston, Virginia, and The Masonry Society, Boulder, Colorado, 2005.

Autoclaved Aerated Concrete – Limbachiya and Roberts (eds)
© 2005 Taylor & Francis Group, London, ISBN 0 415 38356 0

Strength-size relationships for aircrete under compression

A.T. Vermeltfoort
Technische Universiteit Eindhoven, Eindhoven, The Netherlands

ABSTRACT: The paper discusses the effects of size, volume and moisture content on the compressive strength of aircrete. One hundred and eight compression tests on specimens of various sizes were performed. Specimens with three different dimensions of the loaded areas in combination with five slendernesses were used. Each 'size' was tested in sextuple. To minimize the effects of production, specimens were taken at the bottom level of one mould. The fracture processes were analyzed and Young's moduli established. Slender specimens showed vertical splitting cracks, squat specimens showed the well known hour glass fracture model. Multiple regression analysis with linear models, in which thickness, length, slenderness, moisture content and position were incorporated showed that only size and position were significant, but with a contribution of at most nine percent of averaged strength. It is reasonable to use a shape factor equal to one as described in the Dutch code.

1 INTRODUCTION

When specimens are loaded in compression the lateral expansion is usually confined by friction between the loading platens and the surfaces of the specimen. The higher the specimen, the smaller this effect is. This confining also affects the fracture pattern.

In order to have an appropriate design parameter, i.e. the compressive strength, a value without effects of confinement at the loaded ends of the specimen, the slenderness of the specimen and load eccentricity, should be obtained, [NEN 6790:2005], [ENV 2002-1-1:2002].

Therefore, to allow for comparison of the results when the dimensions of the test specimens vary, a size correction factor is prescribed. The size correction factor depends on the material that is tested. In contradiction with [EC6], for Aircrete, a factor equal to one is proposed in the Dutch code [NEN 6790:2005]. This would mean that the load platen confinement nor the volume have an effect on strength.

Aircrete is a product that is made from a lime – sand slurry which rises like a cake in a mould. Consequently, there is some variance of the properties over the height of the mould. To minimize this effect, only material from the bottom 2/3 of the mould is used for testing. All specimens were cut from eleven, 150 mm thick slices. These eleven slices were cut from one, $650 \times 500 \times 1850\,mm^3$ block.

The moisture content of a material also affects the compressive test result, [NEN 6790:2005], [ENV 2002-1-1:2002] and [Bijen]. Therefore, in this project, the specimens were stored for at least seven days to allow for drying. Then, the moisture content of the specimens was established.

2 TEST PROGRAM AND SET UP

2.1 *Specimen sizes and cutting position*

As stated above, the purpose is to establish a slenderness correction factor for Aircrete when loaded in compression. Specimens with fifteen different dimensions were used. Three different sizes were used for the loaded sections, i.e. 150 mm by 150 mm, 70 mm by 70 mm and 150 mm by 70 mm, each with five slendernesses (h/t) ranging between h/t = 5 and h/t = ½. Each 'size' was tested in sextuple. The sizes of the specimens are presented in Table 1.

Xella, an aircrete manufacturer in the Netherlands, provided eleven slices of 650 × 750 × 150 mm. These slices were cut as shown in Figure 1 from one large block in the 'green' situation, i.e. in the factory before putting them in an autoclave.

The specimens were cut from the bottom 2/3 of this block, in four layers numbered from 1 below until 4 as the top row, Figure 1. The 'rise' effect is minimal in this part of the block. Specimens to establish the dry specific mass were cut separately from the same block. All specimens were cut with a water cooled diamond circular sawing machine.

The surfaces to be loaded were smoothened before testing by grinding them with grinding paper. This was possible because the cutting process only left some cutting traces. Further, the surfaces were level and plan parallel.

2.2 *Load control*

The specimens were positioned in the centre of the loading platens of a 250 kN Schenk testing machine. The specimens with the largest loaded area were loaded via a steel plate since the standard load platen of the machine was too small.

The displacement speed of the load platens was controlled in order to reach failure within three to six minutes. A specimen with a height of 60 mm failed after approximately six minutes with a load platen displacement speed of 0.1 mm/min. The code [EN 772-1:2000] prescribes a rate so that the maximum load is reached in not less than approximately one minute. However, in this test program a duration between 3.5 and 7.5 minutes was aimed for. Therefore, the speed was adjusted in proportion to the height.

2.3 *Deformation measurements*

The deformation of six 150 × 150 mm^2, 600 mm high specimens was measured with four LVDTs per specimen. A reading was taken every second. After testing, the fracture pattern of ninety specimens was sketched.

Table 1. Specimen sizes. Each size was tested in sextuple.

		h/t ratio	½	1	2	3	4	5
Loaded	A	150 × 150 mm^2	75	150	300	450	600	
area	B	70 × 70 mm^2		70	140	210	280	350
	C	150 × 70 mm^2		70	140	210	280	350

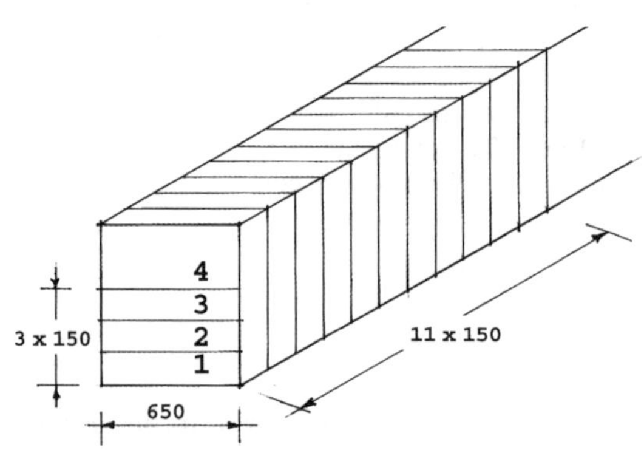

Figure 1. Cutting position of the specimens.

3 RESULTS

3.1 *Moisture content*

The moisture content of each specimen was established to allow for the estimation of its effects. The dry specific mass (γ_{dry}) was established by drying specimens in an oven, until a negligible weight change over a 24 hour period was obtained [EN 772-10:2000]. This resulted in an averaged dry specific mass of 579 kg/m^3.

After cutting, the specimens were left to dry for at least 7 days under laboratory conditions, i.e. at room temperature between 15 and 20°C and a relative humidity of approximately 50%. To establish the moisture content, each specimen was weighed and the wet specific mass (γ_{wet}) calculated. The weight decrease in the last few days before testing was 5.6 kg/m^3 in average.

Assuming that the specific weight of the dry material was equal for all specimens ($\gamma_{dry} = 579$ kg/m^3) the moisture content (mc) was calculated with:

$$\text{m.c.} = \frac{\gamma_{wet} - \gamma_{dry}}{\gamma_{dry}} 100\% \tag{1}$$

The averaged moisture contents per series are given in Table 2. In the three series that were performed first, the moisture content of the specimens was relatively high and the boundary conditions were not optimal. Therefore these series were repeated.

The moisture distribution over the volume remains unknown. It is assumed that the outer surfaces were more moist because they were wetted during cutting.

3.2 *Strength*

The averaged compressive strengths per series of six specimens are given in Table 2. The averaged strength for all one hundred and eight specimens was 4.87 N/mm^2. Omitting the noted firstly

Table 2. Averaged strength per series. Each size was tested in sextuple.

Series number		Dimensions l × t × h mm × mm × mm	Strength N/mm^2	Spec. mass kg/m^3	l/t –	h/t –	Moisture content % m/m
A	1	150 × 150 × 600	5.11	671	1	4	13.7
	2	150 × 150 × 450	5.26	719	1	3	19.5
	3	150 × 150 × 300	5.11	721	1	2	19.6
	4	150 × 150 × 150	5.46	702	1	1	17.5
	5	150 × 150 × 75	5.81	642	1	½	9.8
B	1	70 × 70 × 350	4.84	666	1	5	13.01
	2	70 × 70 × 280	4.90	667	1	4	13.2
	3	70 × 70 × 210	5.36	642	1	3	9.8
	4	70 × 70 × 140	3.91	644	1	2	10.1
	5*)	70 × 70 × 70	3.95	619	1	1	6.4
	5	70 × 70 × 70	4.86	614	1	1	5.6
C	1	150 × 70 × 350	5.25	655	2.14	5	11.6
	2	150 × 70 × 280	5.11	656	2.14	4	11.8
	3	150 × 70 × 210	5.15	672	2.14	3	13.8
	4*)	150 × 70 × 140	4.28	726	2.14	2	20.2
	4	150 × 70 × 140	4.60	679	2.14	2	14.7
	5*)	150 × 70 × 70	3.88	710	2.14	1	18.4
	5	150 × 70 × 70	4.86	652	2.14	1	11.17

*) Firstly performed series, not used for final analyses.

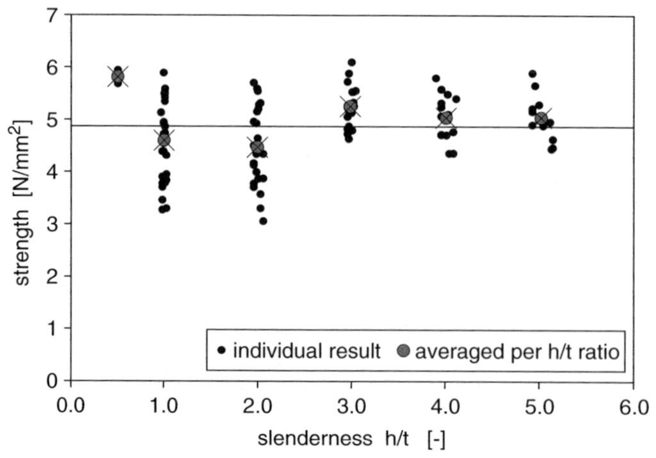

Figure 2. Strength versus slenderness.

performed three series, the average strength equals 5.04 N/mm^2. The strength of each individual specimen is plotted versus its slenderness (height thickness ratio, h/t) in Figure 2.

3.3 *Regression analysis on strength results*

The strength of (stony) specimens depends on several factors. In this project the following aspects were considered:

t: thickness, dimension of the width of the loaded section, 150 mm or 70 mm,
L: Length-thickness ratio (l/t), 1 or 2.14,
S: Slenderness, (h/t ratio), 5, 4, 3, 2, 1, ½,
M: Moisture contents, between 5 and 21% (kg/kg),
P: Position in the mould (rise – effect), layer number 1, 2, 3.

All specimens originated from the same block (one part of a mould filling). The following multiple regression model was employed:

$$f_c = \beta_0 + \beta_1 t + \beta_2 L + \beta_3 S + \beta_4 M + \beta_5 P \tag{2}$$

The analysis was performed on all 108 results. The averaged values are given in Table 2. This resulted in the following equation:

$$f_c = 4.0266 + 0.01358t + 0.3120L + 0.1484S - 0.05443M - 0.2203P \tag{3}$$

with P-values of: $P_t = 0.000$, $P_L = 0.0263$, $P_S = 0.0003$, $P_M = 0.0009$ and $P_P = 0.0002$.

The highest P-value on the independent variables is 0.0263, belonging to the length-thickness ratio (l/t ratio). Since this value is less than 0.05, the term is statistically significant at the 95% confidence level. Consequently, the model does not have to be simplified. However, some series were done in duplo (two times six specimens with the same parameters), so the model may be out of balance. As noted earlier, the three fistly performed series had a relatively high moisture content and deviating boundary conditions (less good parallel load surfaces). Therefore, the firstly performed series, marked in Table 2, were omitted and the resulting 15 series of 6 specimens analysed. This gave the following result:

$$f_c = 4.505 + 0.988t + 0.2617L + 0.05870S - 0.03777M - 0.2008P \tag{4}$$

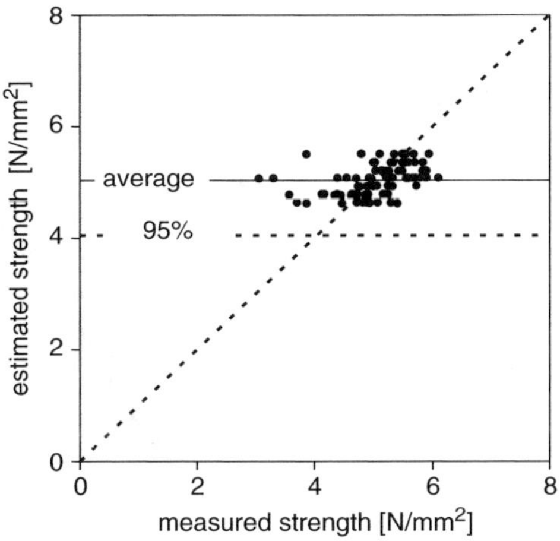

Figure 3. Estimated (Equation 3) versus measured compressive strength (90 results).

P-values: $P_t = 0.0001$, $P_L = 0.0545$, $P_S = 0.1871$, $P_M = 0.0638$ $P_P = 0.0016$.

The highest P-value on the independent variables is $P_S = 0.1871$, belonging to the slenderness (h/t ratio). Since this value is more than 0.05, the term is statistically insignificant at the 95% confidence level. Consequently, the model was simplified. The reduced model, without the slenderness effect was:

$$f_c = 4.6643 + 0.8559t + 0.24081L + -0.02801M - 0.1877P \quad (5)$$

with P-values of: $P_t = 0.0002$, $P_L = 0.0757$, $P_M = 0.1400$ and $P_P = 0.0028$.

In this reduced model, both the P-values for length ratio and moisture percentage are above 0.05, so they also can be removed from the model. This results in the following model:

$$f_c = 4.8516 + 0.5440t - 0.1507P \quad (6)$$

with the following P-values: $P_t = 0.0008$ and $P_P = 0.0092$. This reduced model shows that only size (thickness 70 or 150 mm) and position in the mould are statistically significant. However, their contributions to the estimation are relatively limited, thickness equals 1.5 or 0.7 (*100 mm) which contributes not more than $0.8 \times 0.544 = 0.44$ N/mm^2 (9% of average, 5.04 N/mm^2). Similarly, the position in the mould contributes not more than 0.1507 * (4-1) = 0.45 N/mm^2 (9%). The averaged value of the analyzed 90 test results is 5.04 N/mm^2 with a C.o.V. of 12%. The 95% confidence level lies at 4.84 N/mm^2.

In Figure 3 estimated results (Equation 4, 90 test results) are plotted versus measured results.

The use of the averaged values of each group with similar parameters resulted in the following equation: (the parameter for the position in the mould is left out).

$$fc,a = 3.255 + 0.01480t + 0.4579L + 0.1558S - 0.05681M \quad (7)$$

with P-values of: $P_t = 0.0016$, $P_L = 0.0576$, $P_S = 0.0454$ and $P_M = 0.0576$.

Using averaged results per series, yields in higher P-values because the degree of freedom is much smaller. The P-values on the independent variables are 4.5% or more, except for the P-value of the thickness. This result is similar to that found with the individual results.

3.4 *Young's modulus*

Deformations were measured in axial direction in six tests with four LVDTs per specimen. The averaged results per test are plotted in Figure 4 From this graph a Young's modulus of 2300 N/mm^2 was derived.

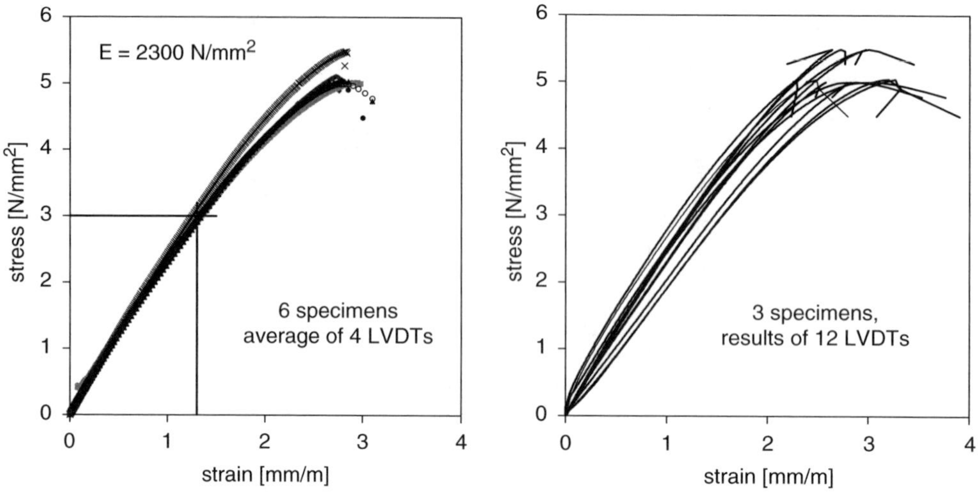

Figure 4. Stress strain diagrams for six, 600 mm high specimens. Strain is the averaged result of four LVDts, gauge length 240 mm. Stress strain diagrams for three specimens, each line is the result of one LVDT, results of twelve LVDTs are plotted.

The graph shows the relatively small variation of the deformations of each of the six specimens. Per specimen the variation of deformation around the circumference of the specimen was of the same order of magnitude.

3.5 *Fracture pattern*

Two types of fracture patterns were found. In the more slender specimens shear cracks in the side surfaces were observed inclined at 20° to 25° to the loading direction, Figure 5. In the more squat specimens an hour glass pattern was observed.

Table 3 shows an overview of the failure patterns, confirming the 'squat' (hour glass pattern) and 'slender' (shear crack pattern) behaviour.

4 DISCUSSION

All specimens were loaded perpendicular to the rise direction.

The modulus of elasticity (E) is probably affected by the loading speed. A higher speed resulted in a higher E value, as observed in one of the six specimens of which deformations were measured.

Moisture content showed to have a negligible effect on compressive strength, however, it was not deliberately varied. Bijen 1995, gives a relationship between moisture content and compressive strength, however this effect was not really significant in the results of the analyzed ninety test results. Some negative effect on strength was observed in the firstly performed three series of six tests, but the boundary conditions were not perfect in these tests. Moisture may have concentrated at the surface of the specimen as a consequence of the cutting process, so the moisture content of the core material may have varied less than averaged over the whole volume.

Aircrete is considered to be non brittle. However, some slender specimens exhibited brittle behaviour. Once shear cracks developed the load decreased rapidly. Squat specimens had a considerable residual load bearing capacity after ultimate load, because the cracked parts are horizontally more confined than in the situation with slender specimens. As known from for instance concrete [Mier], brittleness depends on the shape (slenderness) of the specimen.

Figure 5. Slender specimens after testing showing inclined shear cracks. Deformation measurements with four LVDTs, 240 mm gauge length.

Table 3. Number of specimens per type of fracture, 6 specimens of each series, I = shear crack, c = combination, II hour glass model.

	A:150 × 150 mm^2			B: 70 × 70 mm^2			C: 150 × 70 mm^2		
h/t	I	c	II				I	c	II
5	–	–	–	6	–	–	6	–	–
4	6	–	–	6	–	–	6	–	–
3	5	1	–	6	–	–	6	–	–
2	5	–	1	3	1	2	3	–	3
1	–	2	4	–	1	5	1	2	3
½	–	–	6	–	–	–	–	–	–
			30			30			30

5 CONCLUSION

The average compressive strength of ninety specimens with fifteen different sizes equals 5.04 N/mm^2. Multiple regression analysis yielded the following model:

$$f_c = 4.8516 + 0.5440t - 0.1507P \quad (8)$$

The analysis showed that only size (t = thickness; 70 or 150 mm) and position in the mould (P = layer number) are statistically significant. However, their contributions to the estimation are relatively limited, i.e. a maximum of 9% of the averaged strength which equals 5.04 N/mm^2.

The effect of the moisture content on compressive strength was not significant. However, the moisture distribution may have been unequally over the volume as a consequence of the cutting process.

Brittle behaviour was observed in more slender specimens, more squat specimens behaved more ductile.

ACKNOWLEDGEMENTS

C.H.M. de Louw performed the tests as part of her Master's project.

The specimens were provided by Xella Nederland BV, a producer of calcium silicate and aircrete products under the brand names Silka en Ytong, respectively.

REFERENCES

Bijen, J.M.J.M., 1995, Handboek cellenbeton, Nederlandse Cellenbeton Vereniging NCV, Brouwershaven, Netherlands.

Mier, van J.G.M., 1997, Fracture processes of Concrete, CRC Press. ISBN 0-8493-9123-7.

Eurocodes, European Committee for Standardisation, Brussels.

ENV 2002-1-1:2002, Eurocode No. 6, Common unified rules for masonry structures,

EN 771-4: 2003, Specification for masonry units – Part 4: Autoclaved aerated concrete masonry units,

EN 772-1: 2000, Methods of test for masonry units – Part 1: Determination of compressive strength,

EN 772-10: 2000, Methods of test for masonry units – Part 10: Determination of moisture content of calcium silicate and autoclaved aerated concrete units.

Autoclaved Aerated Concrete – Limbachiya and Roberts (eds)
© 2005 Taylor & Francis Group, London, ISBN 0 415 38356 0

Development of AAC infill blocks in beam and block floors

J. Ward & D.J. Harris
H+H Celcon Ltd, Sevenoaks, Kent, UK

ABSTRACT: The introduction of the European Product Standards for masonry and reinforced AAC components required that manufactures should declare a number of essential requirements so as to meet the Construction Products Directive. In the absence of a recognized procedure to determine the freeze thaw resistance of AAC, the European Co mmittee (CEN TC177) set up to develop the specifications for reinforced AAC products, requested the European AAC Association (EAACA) to develop a procedure to measure the materials durability.

The aim was to develop a simple procedure that could be undertaken without the need for expensive and complex test equipment but still subjected the material to conditions likely to be experienced in service. Using a Standard freezing cabinet and 100 mm cubic samples, various test parameters were examined resulting in a procedure, which determines the volume (mass) loss of a sample with the option to also determine any loss in compressive strength.

1 INTRODUCTION

Suspended floor systems constructed from pre stressed inverted T-beams with concrete block infill first came into widespread use in the UK in the early 1980's. This type of floor was originally developed to overcome the problems of movement causing failure of ground supported slabs. Although only used at first for difficult site conditions, their use has increased in popularity with the majority of ground floors now built in the UK being beam and block. This type of floor construction has a number of advantages in that as a dry construction it reduces the build time and ensures accurate costing. This type of floor also provides an immediate safe working platform for the continuation of the build and remains clear of wet, poor or contaminated ground, in addition to offering a better thermal performance than ground bearing slabs.

The required resistance of floors to imposed loads in buildings in the UK is given in British Standard 6399 which specifies that for domestic applications a floor should resist a point loading of 1.4 kN. In the past it was considered that a factor a safety of 2.5 should then be applied to this load indicating that a satisfactory performance would be achieved if the floor were able to resist an ultimate point load of not less than 3.5 kN.

Initially the only way of assessing the acceptance of the blocks in this situation was to subject them to a simply supported 3-point bending test to determine if they could resist the loads specified. However, a drawback of the 3-point bending test is that it does not predict how different types of blocks behave under point loads or to their mode of failure within a beam and block floor system. Therefore a simulated suspended floor test apparatus was developed by H+H Celcon Ltd to enable a full examination of behaviour of this type of floor system under loading.

2 TEST SET-UP

2.1 *Apparatus*

The test apparatus consisted of a rectangular brick masonry wall five courses high open at one edge to allow access for observation and instrumentation. A damp proof course was laid on top of the

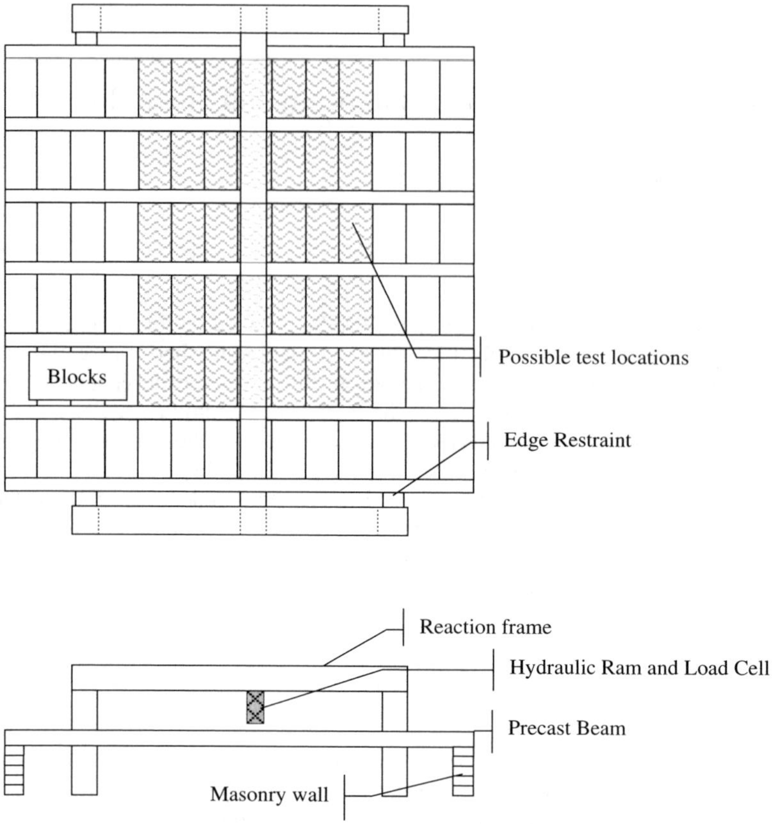

Figure 1a. Schematic of test set up.

supporting brick course onto which the beams were placed, spaced to accept 440 mm by 215 by 100 mm blocks. The size of the test floor was 2.9 m wide, which allowed for six bays of 440 mm wide blocks, and 3m in length, Figure 1. The concrete beams used were pre stressed inverted T-beams designed in accordance with British Standard 8110.

A reaction frame was constructed to allow for test loads to be safely applied as a point load at range of points within the test floor, Figure 1. The reaction frame consisted of four static vertical members bolted to steel channels cast with the laboratory floor, two cross members each spanning a pair of verticals and a third mobile member spanning the floor bed. This third section could be moved along the two cross members allowing for accurate positioning of the test load. Additional edge restraint was supplied in the direction of the beam span to minimise damage to the end wall due to beam and block movement (during test) and to provide a similar degree of restraint as would be provided by a floor in practice i.e. due to greater lateral extent and buttressing from the connecting support walls. The load was applied via a hydraulic pump and cylinder, through a 100 mm square loading plate with a load cell incorporated within the assembly, for load measurement.

For each block tested within the floor the load apparatus was positioned over the centre of the test block and the load applied through a 100 mm square steel loading plate, at a rate of 1.5 $N/mm^2/min$. Failure was judged to have occurred when the rate of the block deformation first exceeded the maximum rate of extension of the loading ram, a typical loading trace is illustrated in Figure 2.

2.2 *Procedure*

A series of tests were carried out using the test procedure outlined above to assess the performance of 440 mm by 215 mm by 100 mm Celcon AAC blocks (table 1) in beam and block flooring with

Figure 1b. Photograph of test set up.

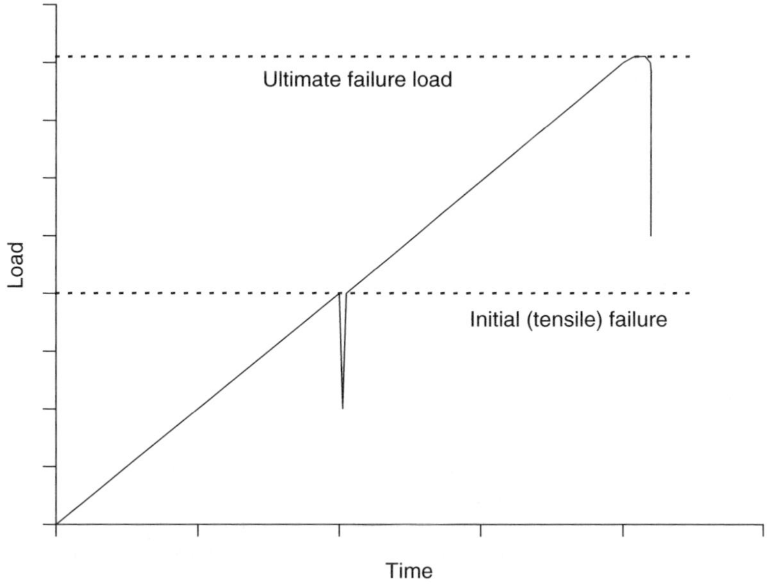

Figure 2. An illustration of a typical loading trace for a block failure on the test apparatus.

respect to various degrees of support and restraint offered at different stages of the construction process, and condition of the block (moisture content). The conditions tested were as follows;

1. Air-dry blocks with dry sand brushed into gaps between block and beams, intended to simulate a floor that has been brush-grouted. No finish.

Table 1. Physical properties of blocks used in tests 1 to 6.

	Mean results for blocks used in tests 1 to 4	Mean results for blocks used in test 5	Mean results for blocks used in tests 6
Compressive strength, N/mm^2	4.55	3.10	5.30
Dry density, kg/m^3	653	438	602
Moisture content after soaking (test 2 only)	49.5	n/d	n/d
Screed compressive strength, N/mm^2	49.9	44.0	n/d

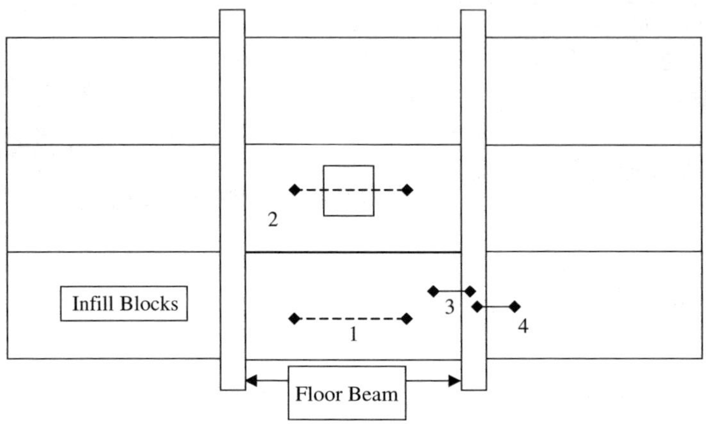

□ 100 mm Square loading plate

Linear transducer measuring floor deflection

1 Gauge line 1
2 Gauge line 2
3 Gauge line 3
4 Gauge line 4

Figure 3. Typical location of strain gauges.

2. Water saturated blocks with dry sand brushed into gaps between block and beams, intended to simulate a floor that has been brush-grouted. No finish.
3. Air dry blocks grouted with a 1:3 sand/cement grout. No finish.
4. Air dry blocks grouted with a 1:3 with sand/cement grout and finished with a 50 mm sand/cement screed.
5. Air dry blocks grouted with a 1:4 sand/cement grout and finished with a 40 mm sand/cement screed.
6. Air dry blocks, with 1:6 sand/cement grout, with 25 mm polystyrene boards and 18 mm tongue and groove chipboard finish.

The physical properties of the blocks used are summarised in Table 1.

In the testing of floor set up number 4, a pre-load was applied up to 35 kN and then removed at a rate 1.5 kN/mm^2/min. Visual inspection of the floor, together with the strain measurements, were

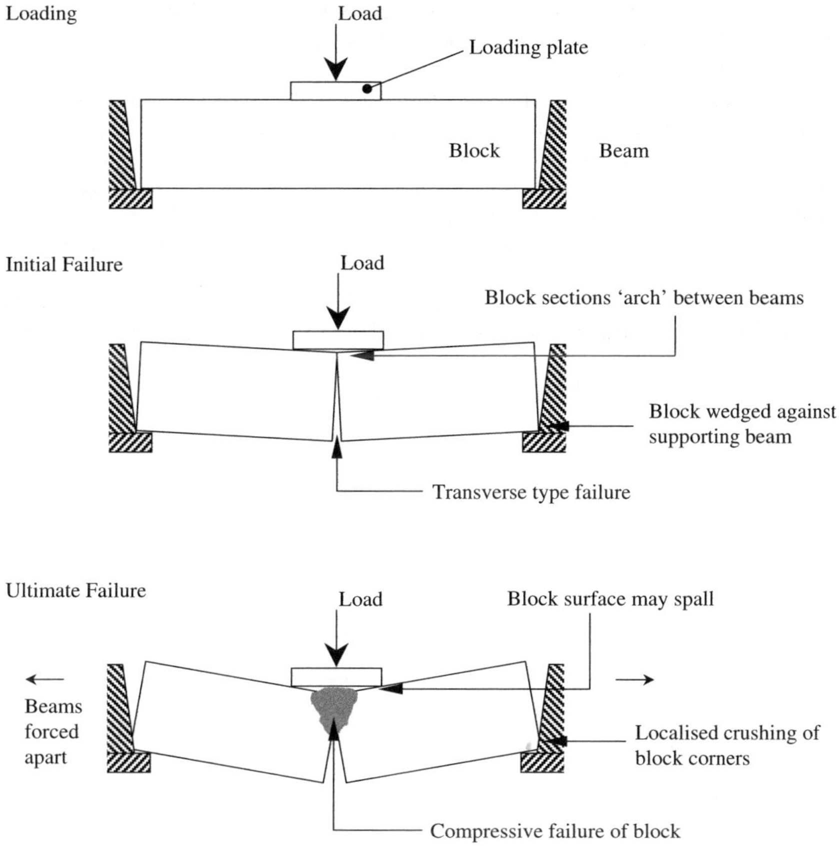

Figure 4. Illustration of the mode of AAC block failure within a beam and block floor.

undertaken to obtain a greater understanding of the mode of failure. The locations of the strain gauges are shown in Figure 3, results in Figure 5.

3 OBSERVATIONS

3.1 *Mode of failure*

The main observations found when applying load to the beam and block floor system were;

(a) With the downward movement of the load beams, the blocks become compressed together providing more frictional restraint between adjacent blocks. This was more noticeable at the centre of the beams where deflection was greatest.
(b) The beams onto which the loaded blocks bear, twisted under loading addition to their vertical deflection causing the top of the beam to rotate towards the load point.
(c) The mode of failure was mainly for a tensile crack to develop followed by "arching" causing local crushing of the compressive face as the block halves rotate down between the beams. This is unlike some of the heavier block types that tend to crack and spall, and fall through between the beams.

The mode of failure is clearly illustrated in Figure 4.

From the initial loading and strain study carried out on the screeded floor (Test 4) the hypothesis that the initial tensile failure of the loaded block, is followed by compressive failure as a result of the blocks "arching" was verified.

This is clearly illustrated by the strain plots of gauges 1 and 2 shown on figure 5. Gauge 1 indicates an increase in tension up to a load of 17 kN at which point initial failure occurs and the tensile strain increase rapidly. The initial failure (crack) line ran across the gauge line, resulting in an increase in strain caused by the fracture opening up due to rotation of the block segment. The strain reading of this gauge after the complete removal of the load was 470 micro strain; this was over a gauge length of 200 mm, corresponding to the fracture width as measured after the load was removed (0.1 mm).

Gauge 2 behaved similarly to gauge 1 up to the point of initial failure after which the block goes into compression. The fracture on this block did not run through the gauge line and thus the

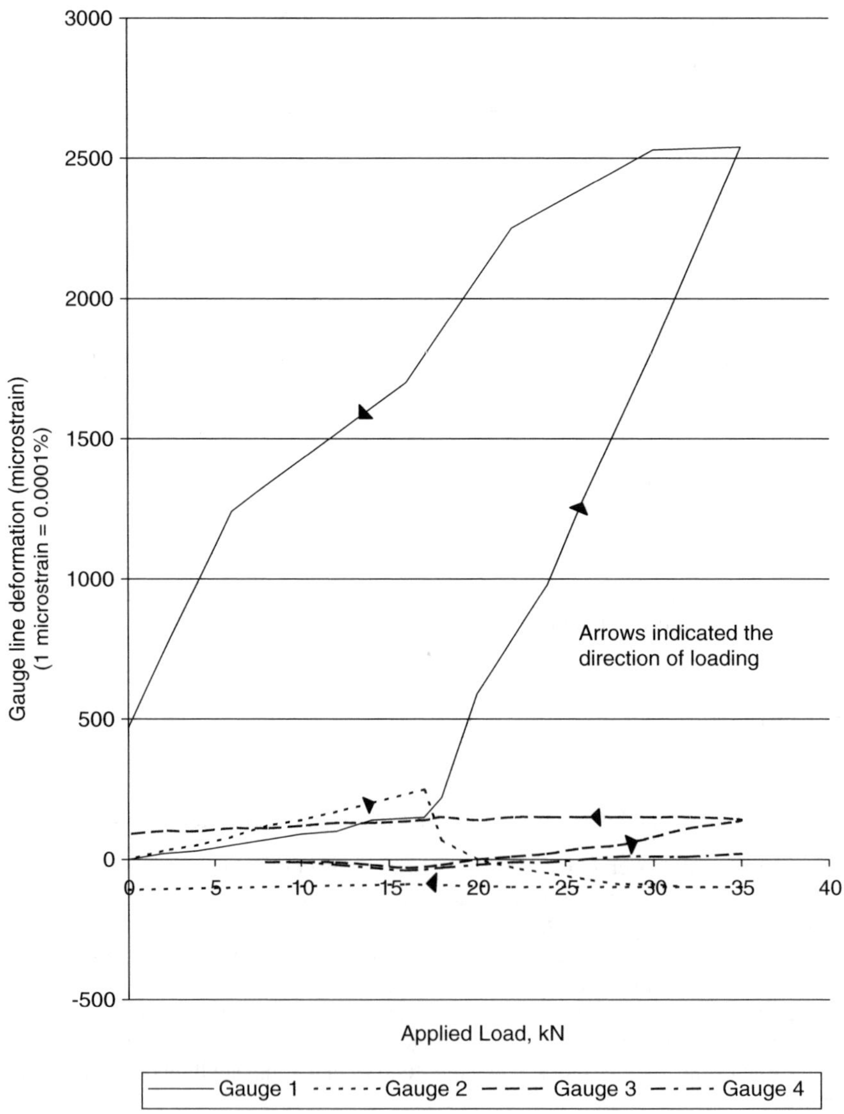

Figure 5. Results from a strain study of test number 4.

gauge output is representative of the behaviour of the material rather than that of the mechanical system.

Gauges 3 and 4 show a small compression prior to the initial failure as the floor bed bowed downwards, but after this, gauge 3 (on the edge of a failed block) showed an expansion as the block segments rotated down requiring the surface screed to accommodate this rotation. Most of the deformation at the peak load on this gauge was retained after the load was relaxed, indicating that the block segments did not rotate back to their original level. Recovery of the blocks would be expected to be near complete if their bonding to the screed was intact; the failure of this block to fully return suggests that debonding of the block from the screed had occurred.

4 RESULTS

4.1 *Main tests*

All the results presented in Table 2 meet the required resistance to point loading, though it can be seen that a sand: cement grout provides the best fixity, with the ultimate failure load, 14.5 kN, exceeding the value specified for domestic loading, by over 10 times (table 2). It is clear that infill blocks are at their most vulnerable after placement between the beams and it is very important to grout in the blocks before applying any loading to the floor.

The application of a floor screed further enhances the load resistance of the floor, with the ultimate failure load of a sand/cement screeded beam and block floor, test 4, being 58 kN, giving a factor of safety of 41 times over the British Standard 6399 requirements.

The sensitivity of load resistance to lower strength materials was evaluated by using a block with strength of 3.1 kN/mm^2 and a 40 mm 1:4 sand/cement screed compared to the original specification of 4.6 N/mm^2 block and a 50 mm 1:3 sand:cement screed, tests 5 and 4 respectively. This resulted in an ultimate failure of 47 kN compared to 58 kN which although is a 19% reduction in load is still well in excess of the load requirement for this type of floor.

To provide information on an alternative floor finish to a sand:cement screed a test was undertaken with a floor finished with 20 mm of polystyrene covered by a 1000 gauge polythene vapour membrane underneath 18 mm chipboard. The ultimate failure load for this type of floor was lower than that measured for a sand:cement screed, Table 2. This is partly attributed to screeded floors

Table 2. Results for tests 1 to 6.

Test No	Description	First crack kN	Ultimate failure load kN	Factor of safety
1	• Air-dry blocks with dry sand brushed into gaps between block and beams, to simulate a floor that has been brush-grouted. No finish	–	4.9	3.5
2	• Saturated blocks with dry sand brushed into gaps between block and beams, to simulate a floor that has been brush-grouted. No finish	–	3.5	2.5
3	• Air dry blocks grouted with a 1:3 sand/cement grout. No finish	5.8	14.5	10.1
4	• Air dry blocks grouted with a 1:3 with sand/cement grout and finished with a 50 mm screed	15.8	58.2	41.6
5	• Air dry blocks grouted with a 1:4 sand/cement grout and finished with a 40 mm screed	15.5	47.5	33.9
6	• Air dry blocks, with 1: 6 sand/cement grout, with 25 mm polystyrene boards and 18 mm tongue and groove chipboard finish.	8.0	18.0	12.9

Table 3. Physical properties of the larger format-flooring units tested.

	Compressive strength N/mm^2	Dry density kg/m^3
H+H Celcon Low density AAC	3.55	470
H+H Celcon Medium density AAC	4.95	610

Table 4. Standard flooring unit test results.

	1st Crack, kN	Ultimate failure load, kN	Factor of safety
H+H Celcon medium density AAC block grouted with 3:1 sand:cement. No Finish	3.80	8.00	5.7
H+H Celcon low density AAC block grouted with 3:1 sand:cement. No Finish	3.55	6.68	4.8

behaving monolithically whereas each layer of a chipboard and polystyrene floor reacts separately to the applied load. As the load was increasingly applied the floor began to deflect with the compressive force between the beam and block resulting in sheering of the underside of the block surface. On dismantling the floor indentations of the edges of the blocks were clearly evident in the polystyrene indicating that the polystyrene absorbed the arching movement of the blocks. The chipboard however, being more rigid broke once the arching became too severe with a hole being finally punched through the chipboard followed by destruction of the floor.

4.2 *Additional testing*

Due to the lightweight nature of AAC, block sizes larger than the traditional UK block size of 440 mm by 215 mm by 100 mm can be safely handled. A block size of 610 mm by 350 mm by 100 mm was evaluated for use as an infill block within a beam and block floor. This larger format has the obvious benefits of reducing the number of beams required for each floor and improving the overall thermal performance.

To assess the performance of this larger format block, a test set up similar to that shown in Figure 1, was constructed, which allowed for the 4 central block to be tested within a floor. Two grades of block were tested, Table 3, in an air-dry condition.

The results from these tests, Table 4, exceed the British Standard 6399 requirement by up to a factor of 5.7.

5 CONCLUSIONS

The ultimate failure load of a well restrained block, within a beam and block floor, is in a compression mode, which brings into doubt the usefulness of the bending strength of the unit as the main guide to performance in this type of construction.

The testing of the standard format AAC blocks within a beam and block floor has shown that even when only grouted they a capable of withstanding up to 10 times the UK required loading for domestic floors as specified in British Standard 6399. This together with AAC lightweight, making it easier for the operators to lay, and the better thermal performance than a concrete ground bearing slab or dense concrete infill blocks, makes AAC the ideal block to use in this application.

As a direct result of this work the UK assessment authorities (British Board of Agrément) has approved the use of H+H Celcon medium density AAC blocks, up to a span of 610 mm, for use within a beam and block floor construction.

ACKNOWLEDGEMENTS

The authors wish to thank, D. R. Packer, J. D. Saunders and G. R. Sargeant for their involvement in the testing presented in this paper.

REFERENCES

British Standards Institute, Code of Practice for dead and imposed loads, BS 6399: Part 1:1996

British Standards Institute, Code of Practice for design and construction of structural concrete, BS8110: Part 1:1997

Autoclaved Aerated Concrete – Limbachiya and Roberts (eds)
© 2005 Taylor & Francis Group, London, ISBN 0 415 38356 0

Tests on loadbearing behaviour of masonry shear walls

U. Schmidt
Scientific employee and head of the working group "Masonry" at the Institute of Building Materials Research, RWTH Aachen University, Germany

B.-G. Kang
Scientific employee at the Institute of Building Materials Research, RWTH Aachen University, Germany

J. Hannawald
Scientific employee and head of the working group "Composite materials and modelling" at the Institute of Building Materials Research, RWTH Aachen University, Germany

W. Brameshuber
Professor, Director of the Institute of Building Materials Research, RWTH Aachen University, Germany

ABSTRACT: Evaluations of test results on the shear load bearing capacity of masonry have shown that in some cases clearly marked deviations exist between experimental test results and calculation formulae which are the basis of the design code in Germany. Within the scope of a research project carried out at the Institute of Building Materials Research tests and numerical simulations were performed on the shear load bearing behaviour of masonry walls. In a large number of small-scale specimen tests the material laws for masonry units and the bonding properties between masonry unit and mortar were determined, in some cases inversely by Finite-Element-Calculation of the small-scale specimen tests. The experimental tests carried out on storey-high masonry walls of autoclaved aerated concrete, in which the inhomogeneity of the masonry was increasingly raised, were simulated with the help of the Finite-Element-Method applying two different load introduction devices.

1 INTRODUCTION, STATE OF THE ART

Masonry building components are subjected to stress from compression, bending or shear forces according to their respective functions. Masonry walls subjected to shear stresses are generally stiffening masonry walls, which are under load resulting from earth pressure, wind, and also earthquakes. Proof of the adequate shear load bearing capacity can be given in Germany according to German Standard DIN 1053-1 (1996) or the Eurocode 6 (1996) with reference to the National Application Document. The basic principles of the shear design equations both in DIN 1053-1 (1996) and also in Eurocode 6 (1996) come from theoretical considerations in the 1970s by Mann and Müller (1973, 1978, 1979), which were verified on masonry test specimens of small-size solid masonry units (model masonry). By means of theoretical considerations on the individual masonry unit four failure criteria were derived, which may be decisive according to the building material properties and stress applied. It was assumed that in the head joints no shear stresses can be transferred and that to maintain the force equilibrium against distortion of the masonry units non-uniform compressive stresses are effective, vertical to the bed joint. A distinction is made between:

- Gaping of the bed joint (failure of the tensile bond strength between the masonry unit and the mortar)
- Failure of the bed joint on shearing off (friction failure)

– Failure of the masonry unit due to exceeding the masonry unit tensile strength as a result of the main stresses
– Exceeding the masonry compressive strength.

The calculation formulae given in Mann and Müller (1973, 1978, 1979) were further developed by various authors. In Mann and Müller (1986) the model from Mann and Müller (1973, 1978, 1979) was extended to such a degree that shear stresses are also transferred in the head joints. Both models and also the Finite-Element-Modellings presented in Schneider et al. (1976) form the basis for the calculation formulae described in Graubner and Simon (2001) and Simon (2002). Essential innovations by Graubner and Simon are the consideration of the overlap and the analytical description of the additional failure criterion "Exceeding the masonry unit tensile strength at the edge of the masonry unit". Experimental tests and numerical calculations were not carried out in Graubner and Simon (2001) and Simon (2002). The above extended model formulae and other have been published in brief summary form in Graubner et al. (2005).

Evaluations in Graubner et al. (2005) have shown that definite differences exist between experimental test results and calculated values from formulae in the current design specifications. The aim of a research project carried out at the Institute of Building Materials Research, RWTH Aachen University, which began approximately the same time as the theoretical investigations of Graubner and Simon, was to determine systematically the states of stress in masonry under shear load with the help of experimental tests and numerical simulations. For this purpose characteristic values of specific geometrical and material-specific importance should be investigated more completely and exactly.

2 TEST PROGRAMME

The experimental tests and numerical calculations were carried out on storey-high, square masonry panels (2.50 m). In these "isotropic" masonry panels, solid units with mortared head joints were tested first. The inhomogeneity of the masonry was gradually increased by incorporating non-mortared head joints and varying the overlap. For all the tests the same masonry unit material – autoclaved aerated concrete units of the compressive strength class 4 – were used without handling device and without a tongue-and-groove system, see Figure 1 left. For the masonry unit material the same material laws were applied in every case and thus the geometrical variables could be deliberately tested. By the use of autoclaved aerated concrete units it was also possible to produce vertically perforated units from the same masonry unit material no longer with isotropic properties and thus increase the inhomogeneity of the masonry further. Such a vertically perforated masonry unit is shown on the right of Figure 1.

Figure 1. Masonry units used (length · width · height = 499 · 300 · 249 mm^3).

For the masonry panels a thin-layer mortar as well as a general purpose mortar were used. The combination of autoclaved aerated concrete units with general purpose mortar is not usual or relevant in practice, however, the relevant failure criteria could be tested both in the case of joint failure and also of masonry unit failure with the selected masonry units of one strength class. An essential part of the research project consisted of tests on small specimens to determine the required material laws of the autoclaved aerated concrete units and the bonding properties of masonry units and mortar for the numerical simulation of the wall tests. For this an independent comparison of the numerical simulation of the wall tests with the experimental results – without adaptation of the material properties for better conformity – is possible.

3 EXPERIMENTAL TESTS ON SMALL TEST SPECIMENS

Tests on small test specimens were carried out to determine the material laws of the masonry unit material and also the bonding properties between the masonry units and masonry mortar. In some cases the material laws could be directly determined from the experimental tests, in some cases an inverse determination of the material laws was made by numerical simulation of the small test specimen results. The tests carried out on the small test specimens and their results are presented below.

3.1 *Compressive strength, load deformation behaviour under compressive stress*

The standardised compressive strength of the high precision autoclaved aerated concrete masonry units was determined in the same way as described in DIN V 4165 (2003) on cubes and on entire masonry units. For the determination of the load deformation behaviour of the autoclaved aerated concrete material and for the derivation of material laws, tests were carried out on slender autoclaved aerated concrete prisms and cylinders, so that influence due to platen restraint was minimized. On the test specimens the axial and lateral deformations were determined with linear variable displacement transducers (LVDT) or a extensometer chain (circumferential extensometer). A test specimen for the determination of the compressive strength is illustrated in Figure 2 left. Figure 2 right shows the stress–strain curves determined. The results of the compressive strength tests are summarized in Table 1.

3.2 *Tensile strength, load deformation behaviour under tensile stress*

For the description of the load deformation behaviour under tensile load, cylinders with a diameter of 100 mm were drilled from the high precision autoclaved aerated concrete units in the direction of unit height and unit length, respectively. Then from these specimens the test specimens shown in Figure 3 left were milled with a tapered cross-section. The load was applied by steel plates attached

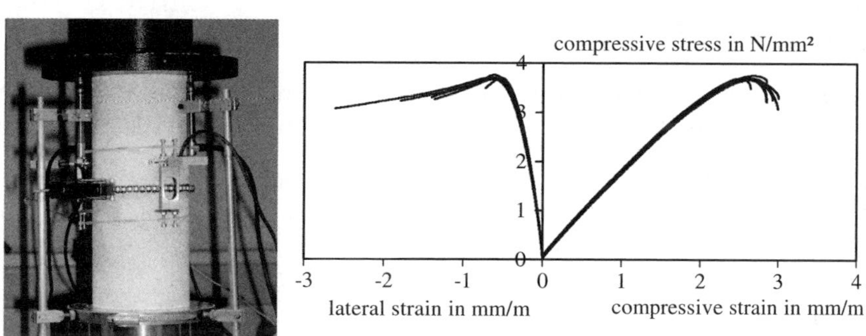

Figure 2. Compression tests on autoclaved aerated concrete cylinders.

Table 1. Properties of the masonry units (compression tests).

Test specimen	Compressive strength (N/mm^2)	Compressive modulus of elasticity (N/mm^2)	Poisson's ratio (–)
Cube*	4.0		
Solid unit*	4.0		
Perforated unit*	3.4		
Prism	3.9	1784	0.11
Cylinder	3.7	1893	0.14

* Without application of the shape factors according to DIN V 4165 (2003).

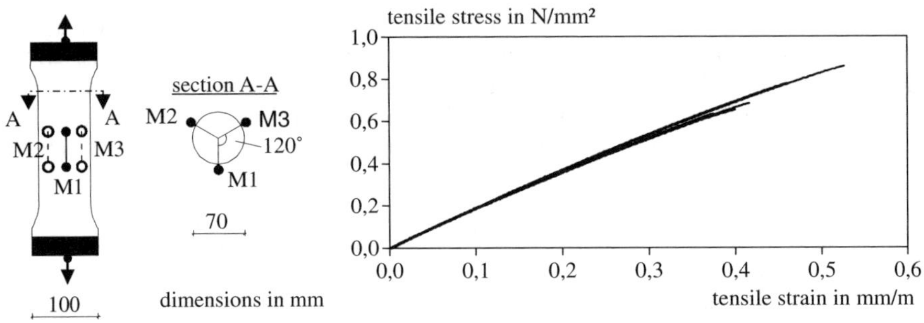

Figure 3. Tensile tests on autoclaved aerated concrete cylinders.

Table 2. Properties of the masonry units (axial tensile tests).

Test specimen/direction of testing	Tensile strength (N/mm^2)	Secant tensile modulus of elasticity (N/mm^2)
Cylinder		
Unit height	0.72	1847
Unit length	0.52	1866
Solid masonry unit		
Unit height	0.45	2182
Unit length	0.35	1877
Vertically perforated masonry unit		
Unit height	0.29	1854
Unit length	0.19	1365

and flexibly jointed to the testing machine and glued onto the test specimens. The load-dependent deformations were determined with linear variable displacement transducers (LVDT). The specific stress–strain curves are illustrated in Figure 3 right.

In addition the tensile strength was determined on the solid and perforated masonry units shown in Figure 1, both in the direction of unit length and also in that of unit height. The test results are given in Table 2.

3.3 *Bending tensile strength, fracture mechanical characteristic values*

In order to determine the post-peak behaviour of the autoclaved aerated concrete under tensile stress, deformation-controlled 3-point bending tests (span 400 mm) were carried out on notched

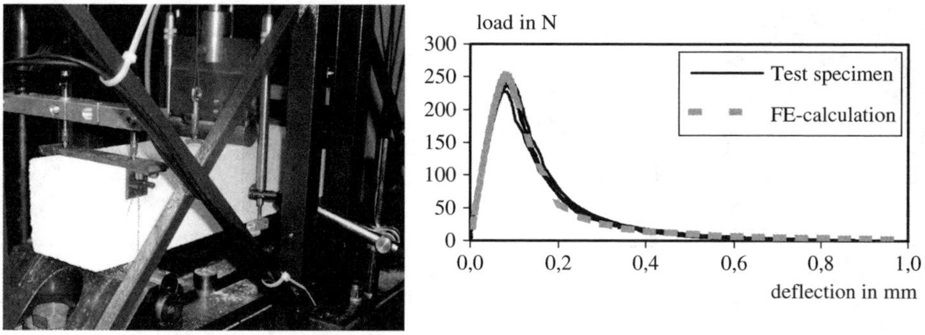

Figure 4. Test equipment for the determination of the bending tensile strength and comparison of the experimental load deflection curves with results from Finite-Element calculations.

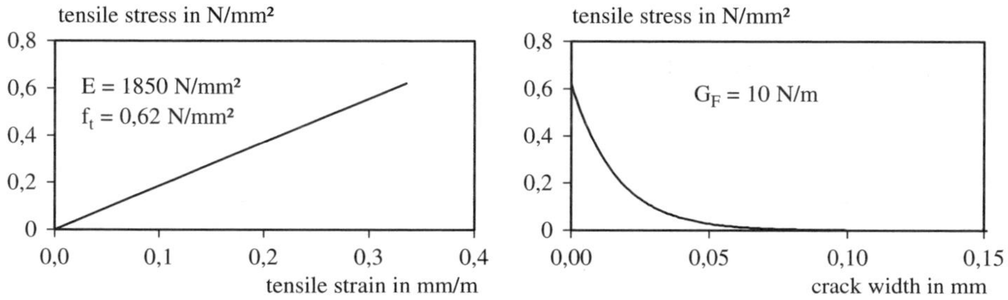

Figure 5. Material model law for the autoclaved aerated concrete.

and unnotched autoclaved aerated concrete prisms ($500 \cdot 100 \cdot 100\,mm^3$) taken in the direction of the masonry unit length. The test equipment, which allows compensation for the specimens dead weight, is presented in Figure 4 left. The load deflection curves determined on prisms with a notch depth of 50 mm are shown in Figure 4 right.

The bending tests were simulated with the help of the Finite-Element-Method. By variation of the stress-crack opening relation, the tensile strength f_t, the calculated tensile modulus of elasticity E_{cal} and the fracture energy G_F in the numerical simulation the load-deflection curves were adapted as well as possible to the experimental test results, see Figure 4 right, and thus inversely determined the material law of the autoclaved aerated concrete material.

The calculated tensile strength $f_t = 0.49\,N/mm^2$ and fracture energy $G_F = 8\,N/m$ are in a good conformity with the test results. The calculated modulus of elasticity $E_{cal} = 2330\,N/mm^2$ is higher than the experimental secant modulus of elasticity E_S determined in the axial tensile tests (see Table 2), because the elastic deformations in the main cross-section in the bending tests are distinctly smaller than the deformations forming the basis for the calculation of the tensile modulus of elasticity E_S. The calculated characteristical length $l_{ch} = 78\,mm$ shows, that autoclaved aerated concrete is less ductile compared to e.g. mortar.

For the calculation of the masonry walls the mean value of the tensile strength of the units in direction of the unit height and unit length $f_{t,mean} = 0.62\,N/mm^2$ were assessed. As the fracture mechanical values were only be determined in the direction of the unit length ($f_{t,mean} = 0.52\,N/mm^2$) the mean values for the material law were extrapolated. The material model law used for the calculation of the masonry shear walls is shown in Figure 5.

3.4 *Adhesive tensile strength*

The adhesive tensile strength between the mortar and the masonry unit was determined in axial tensile tests on autoclaved aerated concrete cylinders (diameter 100 mm) mortared with general

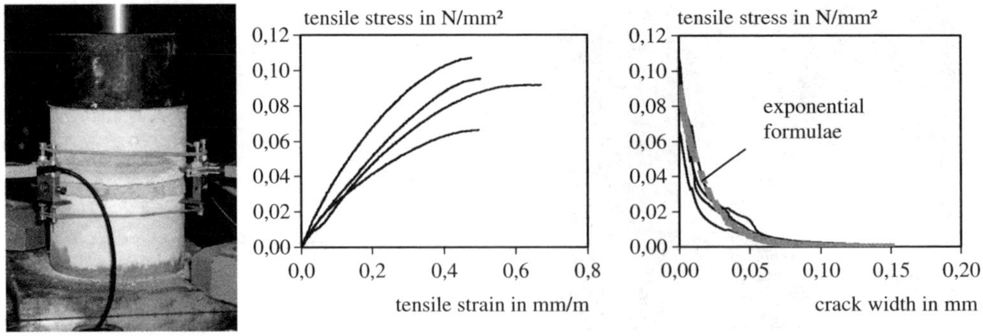

Figure 6. Adhesive tensile strength tests on test specimens of autoclaved aerated concrete with general purpose mortar (head joint).

Table 3. Bonding properties of the axial adhesive tensile strength tests.

		AAC/GPM		AAC/TLM	
Test	Notation	Head joint	Bed joint	Head joint	Bed joint
Adhesive tensile strength	N/mm^2	0.09	0.08	≥0.38	0.36

purpose mortar (GPM) and a thin-layer mortar (TLM), respectively. Tests were made both on the adhesive tensile strength in the bed joints and in the head joints. In the tests with autoclaved aerated concrete and thin-layer mortar failure frequently occurred in the masonry unit. In the tests with general purpose mortar the post-peak behaviour could be determined with the test setup shown in Figure 6 left. In Figure 6 right typical stress-crack width curves of the adhesive tensile strength tests between the head joint surfaces and general purpose mortar are presented. The fracture energies determined from this were taken into account in the Finite-Element-Calculations of the wall tests. The adhesive tensile strength values are summarized in Table 3.

3.5 *Shear strength*

Two methods are used in Germany for the determination of the initial shear strength at the present time – the initial shear strength test according to the German Standard DIN 18555-5 (1986) and the European Standard DIN EN 1052-3/A1 (2005), respectively. The initial shear strengths determined according to the European test method is half of the value determined according to the German test method. Neither the test method according to DIN 18555-5 (1986) nor the test method according to DIN EN 1052-3/A1 (2005) is suitable for the direct determination of the shear stress-displacement-curves due to the shear and axial stress distributions in the specimens, which were determined among other things with Finite-Element-Calculations, Roßbach et al. (2004). In the research project carried out the initial shear strength of the thin-layer mortar is determined exclusively with the method according to DIN 18555-5 (1986). For the general purpose mortar the initial shear strength was determined with both test methods. The difference determined already in earlier investigations between the two test methods has been approximately confirmed. The test according to DIN EN 1052-3/A1 (2005) was made for the general purpose mortar in bond with the bed joint surface both without and with three different loading levels normal to the bed joint, see Figure 7 left. In the tests deformations parallel and perpendicular to the bed joint were determined. The test results were simulated with the Finite-Element-Method from which the material laws were

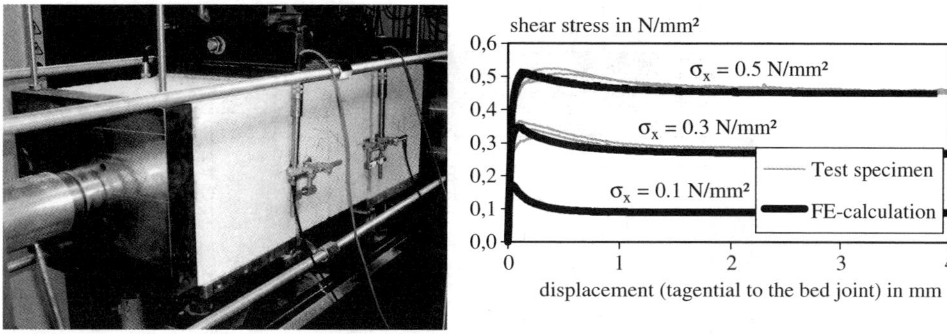

Figure 7. Initial shear stress tests according to DIN EN 1052-3/A1 (2005) and comparison between the experimental shear stress-displacement-curves and results from Finite-Element-Calculations.

Table 4. Bonding properties of the shear tests.

		AAC/GPM		AAC/TLM
Test	Notation	Head joint	Bed joint	Bed joint
German standard				
Initial shear strength	N/mm^2		0.15	0.78
European standard				
Initial shear strength		0.14	0.08	
Shear strength ($\sigma_x = 0.1$)	N/mm^2		0.18	
Shear strength ($\sigma_x = 0.3$)	N/mm^2		0.34	
Shear strength ($\sigma_x = 0.5$)	N/mm^2		0.51	
Coefficient of static friction	–		0.82	
Coefficient of sliding friction	–		0.90	

inversely determined for describing the bonding behaviour under shear stress. The comparison of the shear stress-displacement-curves measured in the middle of the overlap length with the Finite-Element-Calculations is shown in Figure 7 right. The essential test results of the initial shear stress tests are contained in Table 4.

4 THEORETICAL AND EXPERIMENTAL TESTS ON WALL TEST SPECIMENS

In Germany shear tests are carried out on storey-high, square masonry panels (see Fig. 8) within the scope of tests on the general building inspection approval.

The standardized test method appeared for many years to be suitable for determining the shear strength of masonry, even if the test method is disputed to the extent that the load bearing capacity of the masonry is not determined, because the test walls are loaded differently than stiffening walls in buildings. The primarily concern however within the scope of this research project was to test the load bearing behaviour of masonry walls with the most uniform possible shear stress and exactly defined boundary conditions.

With the test structure presented in Figure 8 tests were carried out altogether on 8 different masonry walls.

A large number of deformation measurements were carried out on the test specimens in the masonry unit and joint area, in the direction of the main compressive and main tensile stresses in the masonry and also for determination of the displacements of the corner of the walls (Fig. 8).

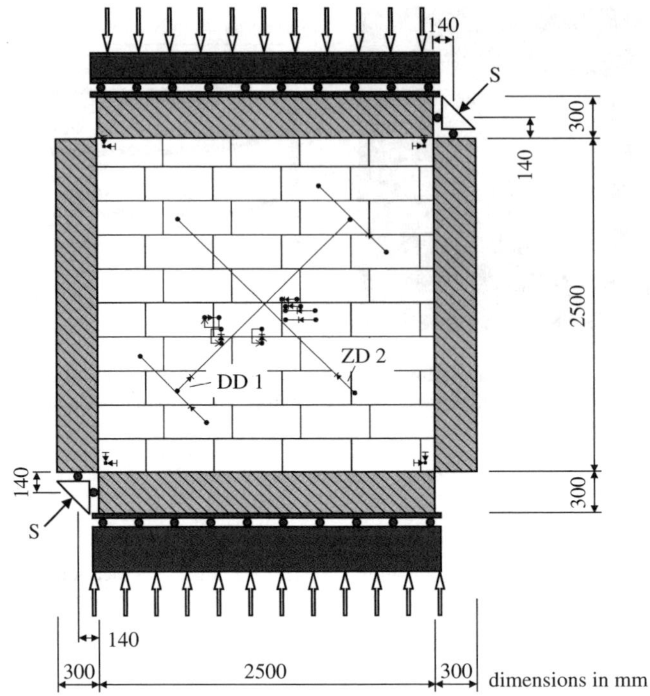

Figure 8. Test method and specimen for the determination of the shear strength with measuring points.

Table 5. Results of the masonry shear tests and FE-calculations.

No.	Unit	Mortar	Head joint	Overlap (mm)	Compressive stress (N/mm^2)	Shear strength (failure criteria*) Observed (N/mm^2)	FE1 (N/mm^2)	FE2 (N/mm^2)
1	Solid	TLM	Filled	250	0	0.12 (u)	0.10	0.60
2	Solid	TLM	Unfilled	250	0	0.14 (u)	0.11	0.17
3	Solid	TLM	Unfilled	100	0	0.09 (u)	0.10	0.14
4	Solid	TLM	Filled	250	1.1	0.23 (u)	0.16	0.81
5	Solid	GPM	Filled	250	0	0.11 (j)	0.10	0.12
6	Solid	GPM	Unfilled	250	0	0.05 (j)	0.05	0.04
7	Solid	GPM	Unfilled	250	0	0.06 (j)	0.04	0.04
8	Perforated	TLM	Unfilled	250	0	0.05 (u)	0.08	0.11

* Observed main failure criteria (u – unit failure, j – joint failure).

The maximum shear strength is calculated from the diagonal force S introduced and the cross-sectional area $A = 2500 \cdot 300\,mm^2$: $\max \tau = S/(1.41 \cdot A)$. The experimentally determined shear strength values are given in Table 5 (observed shear strength).

The shear tests were checked with the help of the Finite-Element-Method employing the material laws previously determined from the small test specimens. In the calculations carried out with the FE-model FE1 the load introduction device was modelled as well so that the Finite-Element-Model of the wall tests selected and the material laws determined from the small tests could be checked. The shear strength values calculated with FE-model FE1 are also given in Table 5 (FE1). Basically the shear strength values determined by calculation are slightly lower than the experimental values,

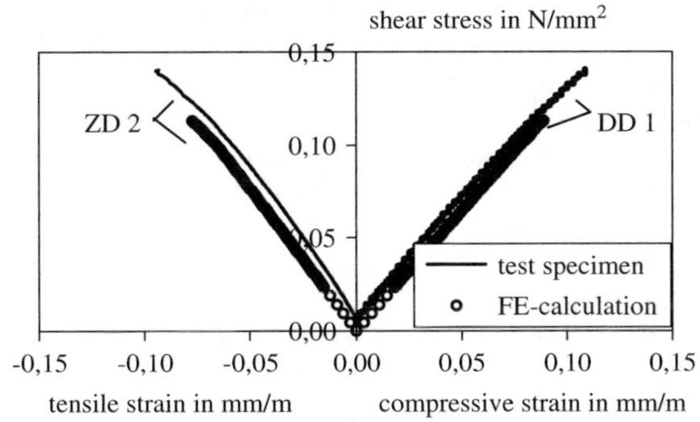

Figure 9. Comparison of the experimental shear stress–strain-curves of wall No. 2 with results obtained by Finite-Element-Calculations (FE-model FE 1).

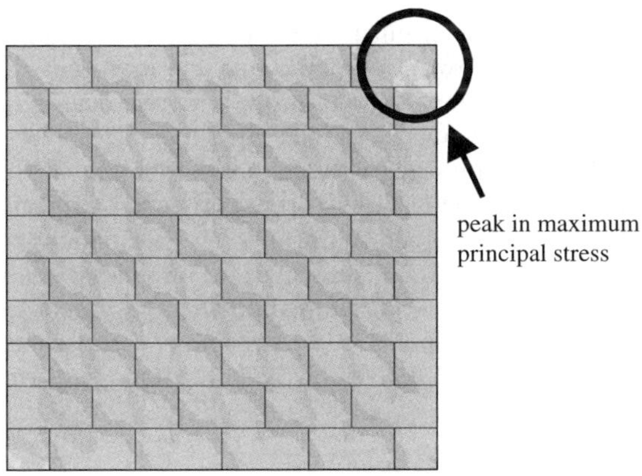

Figure 10. Maximum principal stress in wall No. 2 obtained by FE-calculations with FE-model FE1.

only in the case of Wall No. 4 with a simultaneously acting vertical load and Wall No. 8 with perforated units significant differences between the Finite-Element-calculations and the test values have been observed. But it has to be taken into account that with both variations (vertical load and perforated unit) only one wall was tested. Typical comparisons of the stress–strain-curves determined experimentally and by calculation (diagonal measuring points in the direction of the main stresses ZD 2 and DD 1) of the test specimen No. 2 are presented in Figure 9. The stress–strain-curves determined by calculation are in a good conformity in all cases with the test values.

The analysis of the stress direction in the masonry (see Fig. 10, wall No. 2) showed that the formation of cracks normally starts in the corner units in the area of the load introduction, and after cracking no definite bearing load increase is possible. In particular with high bonding strengths between the masonry unit and the mortar the load bearing capacity of the wall is decisively influenced by the strength of the corner units.

The comparison of the experimental tests and calculations shows that with the Finite-Element-Model selected and in particular the material laws determined in the small-scale tests, the wall tests can be simulated quite well. In further numerical calculations the same masonry walls were modelled, however without modelling of the load introduction device and the load is applied instead by constant displacements to the masonry edges. The simulation of this "optimum" shear test investigates to what extent the load introduction device influences the shear strength of the masonry. The shear strength values calculated with FE-model FE2 are also given in Table 5 (FE2).

The analysis of the stress direction in the masonry showed that the maximum principal stresses are more uniformly distributed without peaks in the corner units. This leads to distinctly larger shear stresses in particular for walls with high bonding strengths between the masonry unit and the mortar. Compared to the results of FE-model FE1 with nearly identical shear stresses of walls No. 1 to 3 (see table 5) caused by the stress peaks in the corner units with FE-model FE2 the influence of filled and unfilled head joints as well as the influence of the overlap can be observed by distinct differences in the maximum shear stresses.

In cases the bond strength is the decisive material property (wall No. 4 to 7) the results of FE-model FE1 and FE2 are almost similar.

In the next step with this "optimum" shear test a parametric study using different variables influencing the shear strengths will be carried out. On the basis of the results, it is intended to derive equations for the determination of the shear strength of masonry, which contain the essential geometrical characteristic values and building material values.

5 SUMMARY AND OUTLOOK

In the research project carried out a determination was first made of the material laws for masonry units and the bonding properties between masonry units and mortar in a large number of small specimen tests. In some cases the material laws or properties were determined inversely by Finite-Element-Calculation of the small specimen tests. The experimental tests carried out on storey-high, shear loaded masonry walls were simulated with the help of the Finite-Element-Method and the application of these material laws. The results of the experimental tests and the calculations with the Finite-Element-Method are in quite good conformity. The calculations show that with the shear testing method as used in Germany, the shear strength of the masonry is considerably influenced by the load introducing process. By means of the simulation of an "optimum" shear test in parametric studies, the essential variables influencing the shear strength of the masonry will be characterized and simple calculation formulae will be derived on the basis of the numerical calculations.

ACKNOWLEDGEMENT

The research project was kindly financially supported by the German Research Foundation.

REFERENCES

DIN 1053-1. 1996. Mauerwerk; Berechnung und Ausführung.

DIN V 4165(Vornorm). 2003. Porenbetonsteine – Plansteine und Planelemente.

DIN 18 555-5. 1986. Prüfung von Mörteln mit mineralischen Bindemitteln; Festmörteln; Bestimmung der Haftscherfestigkeit von Mauermörteln.

DIN EN 1052-3/A1 (Norm-Entwurf). 2005. Prüfverfahren für Mauerwerk – Teil 3: Bestimmung der Anfangsscherfestigkeit (Haftscherfestigkeit); Deutsche Fassung EN 1052-3:2002/prA1.

Eurocode 6. 1996. Bemessung und Konstruktion von Mauerwerksbauten – Regeln für bewehrtes und unbewehrtes Mauerwerk.

Graubner, C.-A. & Simon, E. 2001. Zur Schubtragfähigkeit von Mauerwerk aus großformatigen Steinen. In Mauerwerk-Kalender 26 (2001): 737–752. Berlin: Ernst & Sohn.

Graubner, C.-A. et al. 2005. Festigkeitseigenschaften von Mauerwerk – Teil 3: Schubfestigkeit von Mauerwerksscheiben. In Mauerwerk-Kalender 30 (2005): 7–88. Berlin: Ernst & Sohn.

Mann, W. & Müller, H. 1973. Bruchkriterien für querkraftbeanspruchtes Mauerwerk und ihre Anwendung auf gemauerte Windscheiben. In Bautechnik (1973), Nr. 12: S. 421–425.

Mann, W. & Müller, H. 1978. Schubtragfähigkeit von Mauerwerk. In Mauerwerk-Kalender 3 (1978): 35–65. Berlin: Ernst & Sohn.

Mann, W. & Müller, H. 1979. Cracking Characteristics of Transversely Loaded Brick Masonry in Theory and Practice. In Proceedings of the Fifth International Brick Masonry Conference, Washington, October 5–10, 1979, (Wintz, J.A. ; Yorkdale, A.H. (ed.)): S. 239–246. McLean, Virginia : Brick Institute of America.

Mann, W. & Müller, H. 1986. Nachrechnung der Wandversuche mit einem erweiterten Schubbruchmodell unter Berücksichtigung der Spannungen in den Stoßfugen. In Anlage 2 zum Forschungsbericht "Untersuchungen zum Tragverhalten von Mauerwerksbauten unter Erdbebeneinwirkung", Forschungsbericht IV/1-5-488/86, Darmstadt.

Roßbach, M. et al. 2004. Untersuchungen zur Schubtragfähigkeit von Ziegelmauerwerk. In Mauerwerk 8 (2004), Nr. 2: 72–81.

Schneider, K. H. et al. 1976. Innerer Spannungszustand bei Mauerwerk mit nicht vermörtelten Stoßfugen. In Bauforschungsbericht des Bundesministeriums für Raumordnung, Bauwesen und Städtebau F 1360. Stuttgart: IRB Verlag.

Simon, E. 2002. Schubtragverhalten von Mauerwerk aus großformatigen Steinen. Darmstadt, Technische Universität, Fachbereich Bauingenieurwesen und Geodäsie, Dissertation.

Autoclaved Aerated Concrete – Limbachiya and Roberts (eds)
© 2005 Taylor & Francis Group, London, ISBN 0 415 38356 0

Seismic behaviour of Autoclaved Aerated Concrete masonry constructions

D. Schermer
Technical University Munich, Germany

ABSTRACT: According to the new European earthquake standard Eurocode 8 seismic loads increase significantly in regions of low or moderate seismicity. Combined with the proposed behaviour factor of $q = 1.5$ the earthquake resistance of several common types of unreinforced masonry buildings can't be proven numerically. To examine the seismic behaviour of constructions with loadbearing autoclaved aerated concrete (AAC) masonry walls under realistic conditions, numerical as well as experimental investigations have been carried out. For the numerical investigations a material model based on the theory of plasticity has been developed and for the experimental investigations pseudodynamic tests on full scale masonry walls with dimensions of $1.25\,m \times 0.24\,m \times 2.5\,m$ have been carried out.

1 INTRODUCTION

1.1 *Seismic load-level*

According to the European Earthquake Standard Eurocode 8 in regions of low and moderate seismicity in Central Europe, e.g. Germany, significantly higher seismic loads result. One reason for that is the primary focus of the standard on regions of high seismicity like Southern Europe. On the other hand, e.g. in Germany, in the previous national standard the design seismic action was determined to the maximum in the past observed earthquake and not associated with a reference return period of 475 years.

1.2 *Behaviour factor*

Regarding the behaviour of structures in the case of the design seismic action, generally the non-linear range is reached. To avoid a complex inelastic structural analysis, the capacity of structures to dissipate energy is taken into account by performing a linear-elastic analysis and then reduce the resulting loads with the so called behaviour factor q.

The behaviour factor is an approximation of the ratio of the seismic forces that the structure would experience remaining completely elastic, to the seismic forces that the structure would experience under "real" conditions, i.e. including plastic deformations, energy dissipative effects, a damping different from 5% etc.

For the design of structures classified as non-dissipative, in Eurocode 8 no account is taken of any hysteretic energy dissipation and the behaviour factor is limited to 1.5. Unreinforced masonry is considered to have a low-dissipation capacity due to its low tensile strength and low ductility. Therefore the behaviour factor for unreinforced masonry constructions is determined in Eurocode 8 very conservatively to 1.5.

1.3 *Experiences during past earthquakes*

The experiences in past earthquakes in Germany in 1978 and 1992 – where intensities according the European Macroseismic Scale (EMS 98) near to the design load level appeared – were generally good. At modern masonry constructions been up to standard no relevant damages were found. Only in few structures from the 19th century with weak timber floor systems not acting as horizontal diaphragms problems like decoupling of perpendicular walls were observed. Also on non-structural elements like chimneys damages were observed.

Several calculations of structures with commonly used layouts – e.g. terraced houses – showed, that with the given ground accelerations these masonry structures can't be proven using the design methods given in Eurocode 6 resp. Eurocode 8. This is directly in contradiction to the experiences in past earthquakes.

2 NUMERICAL INVESTIGATIONS

2.1 *Material model*

In a first step numerical investigations have been carried out at the TU Munich. To describe the non-linear behaviour of masonry under cyclic shear loadings a material model according the theory of plasticity has been developed. The basis of the material model was the failure model of Mann/Müller. Regarding a single unit, due to not existing stresses at the face of the units – because of unfilled perpend joints or shrinking of the mortar in combination with no noticeable stresses σ_x in longitudinal direction – a stepwise distribution of the normal stresses in the bed joints $\sigma_x \pm \Delta\sigma_y$ is necessary because of state of equilibrium.

In dependency of the ratio of shear (τ) and normal stress (σ_y) and the geometric properties of the units (height to length-ration $\Delta y/\Delta x$ and the minimum overlapping length) three resp. four failure modes could occur:

- sliding resp. friction/adhesion failure in the bed joints $\quad \tau = \mu \times (\sigma_y - \Delta\sigma_y) + k$
- diagonal tension failure in the brick $\quad \sigma_I = f_{bt}$
- compression failure $\quad \sigma_1 = \sigma_y + \Delta\sigma_y = f_{my}$
- tension failure in the bed joint $\quad \sigma_2 = \sigma_y - \Delta\sigma_y = f_{ty}$

with: μ = friction coefficient
k = adhesion between unit and mortar (initial shear strength)
f_{bt} = effective (diagonal) tension strength of the unit
f_{my} = compression strength of the masonry perpendicular to the bed joints
f_{ty} = tension strength between unit and mortar.

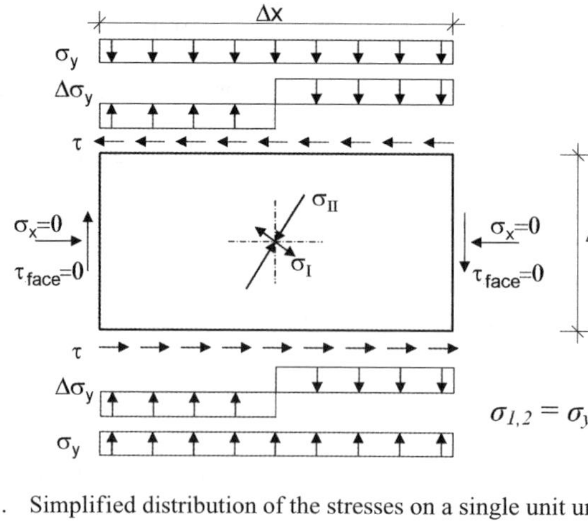

$$\sigma_{1,2} = \sigma_y \pm \Delta\sigma_y = \sigma_y \pm |\tau| \cdot 2 \cdot \Delta y/\Delta x$$

Figure 1. Simplified distribution of the stresses on a single unit under combined shear and normal loadings.

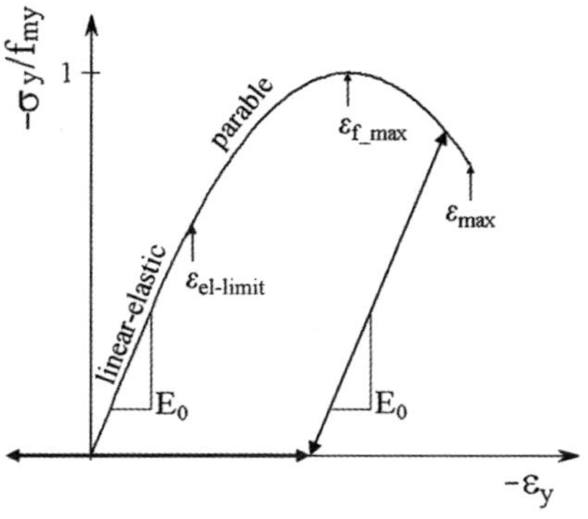

Figure 2. Working hardening under compression perpendicular to the bed joint (loading and unloading).

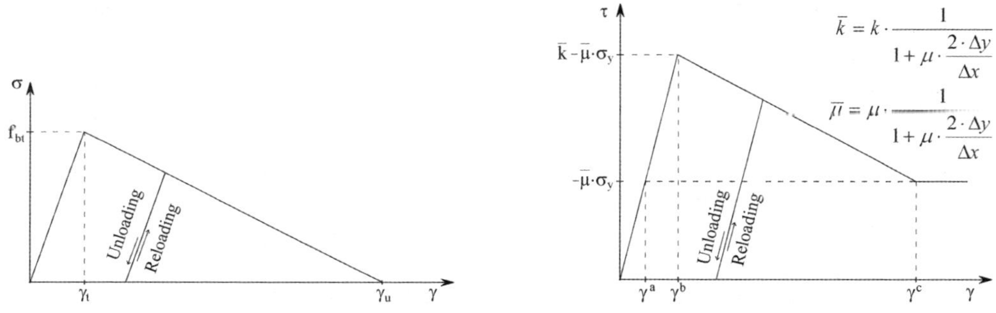

Figure 3. Softening behaviour under loading and unloading (tension failure/sliding failure).

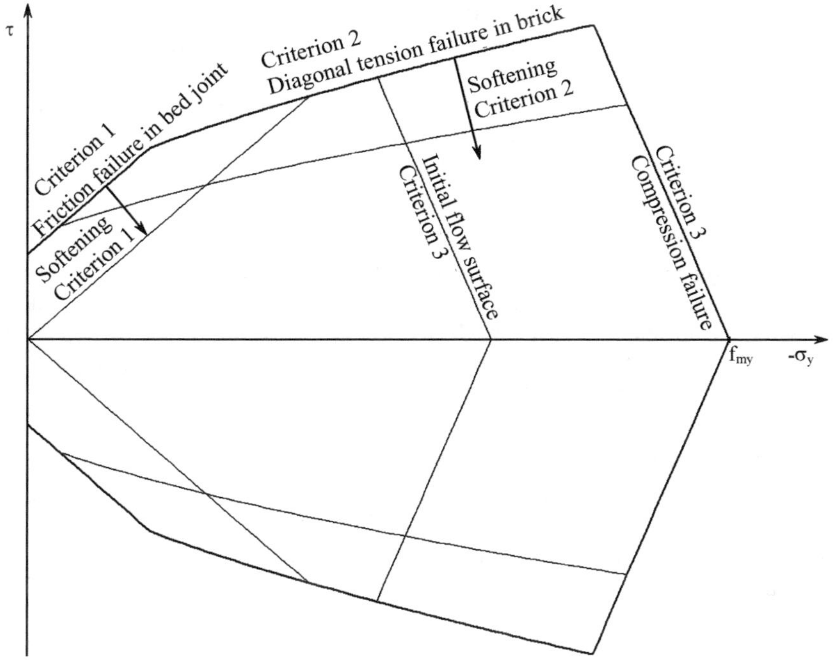

Figure 4. Resulting failure surfaces and flow rules of the material model in the τ–σ_y space.

Table 1. Material properties.

Type of unit, size	Type of mortar	Compression strength of the units	Thickness of the wall	Allowable compression stress of the masonry σ_0
AAC, 500 × 250 [mm]	Thin layer	4 [MN/m^2]	300 [mm]	1,1 [MN/m^2]

Table 2. Detailed input parameters for the material model.

Parameter	Value	Unit
Characteristic strength of the masonry f_{my}	2,9	MN/m^2
Strain $\varepsilon_{el\text{-}limit}$	−0,0015	–
Strain $\varepsilon_{f\ max}$	−0,002	–
E-Modulus perpendicular to the bed joints E_y	1650	MN/m^2
Orthotropic factor n	1,8	–
Shear modulus G_{xy}	320	MN/m^2
Posions ratio ν_{yx}	0,2	–
Friction coefficient μ	0,65	–
Adhesion between unit and mortar (initial shear strength) k	1,2	MN/m^2
Effective diagonal tension strength f_{bt}	0,4	MN/m^2
Fracture energy in the case of tension failure (criterion 2) $G_{f,2}$	0,010	Nmm/mm^2
Fracture energy in the case of sliding (criterion 1) $G_{f,1}$	0,003	Nmm/mm^2
Length of the units (overlapping length = 0.5 × length of the unit)	500	mm
Height of the units	250	mm

The model of Mann/Müller describes only the maximum stress state where the initial failure resp. crack occurs. To describe the post-crack behaviour – important for the behaviour under seismic loadings – further parameters has to be included. For the sliding failure and also for the tension failure of the bricks the corresponding fracture energy is taken to describe the softening effect after exceeding the maximum load. For compression perpendicular to the bed joints a work hardening rule base on uniaxial tests has been implemented. As the adhesion of mortar and AAC-units is very high, the sliding criterion generally remains unattended. The tension failure criterion is generally covered by the friction failure and the tension failure criterion and so can be neglected.

The material model has been included as a makro-model, i.e. with smeared properties, in a finite-element programme.

2.2 *Material parameters*

In the first step the material parameters given in Table 1 have been taken.

The allowable compression stress σ_0 is about 1/3 of the characteristic strength f_{my}. In this value the global safety factor and also reduction factors for the usual wall slenderness $\lambda = 10$ and the eccentricity of the floor load are included. Generally the quasi permanent normal stress in the loadcase earthquake according to Eurocode 0 can be assumed to 75% of σ_0.

2.3 *Representing structure layout*

The calculations have been carried out on a single wall with the dimensions of 2.5 × 2.5 m representing the shear wall in the first storey in a 3-storey structure. The dimensions were also varied to determine the effect of the slenderness of the wall.

2.4 *Simplified system*

At the cap of the wall with a bending spring an elastic restraint representing the floor slab was applied. Damping effects were covered by the viscous damper at the cap of the wall. The point mass

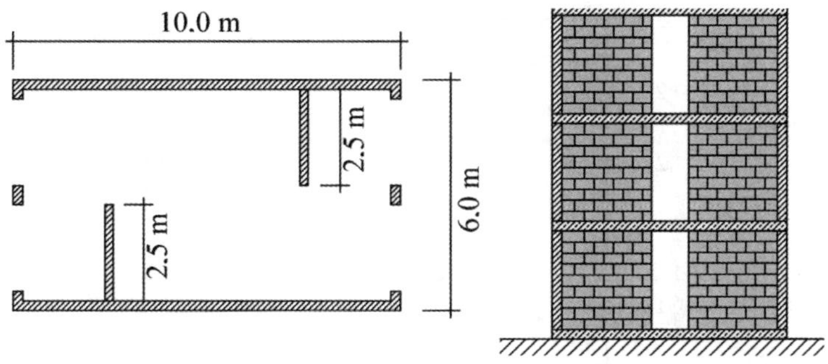

Figure 5. Investigated 3-storey terraced house structure, layout and view (roof not shown).

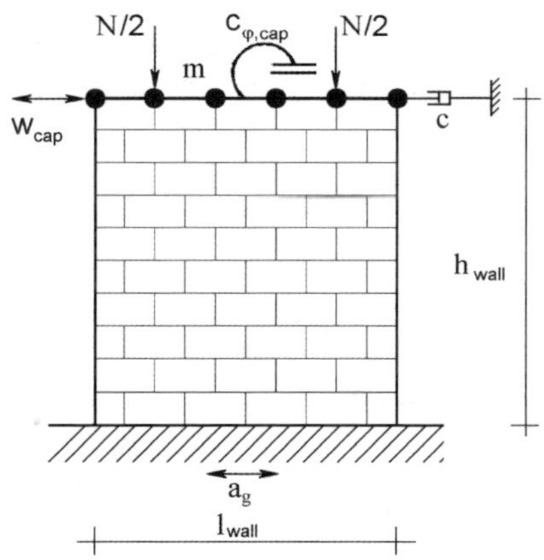

Figure 6. Investigated wall – simplified system.

represents the related mass in the structure. The numerical investigations have been carried out with the permanent existing vertical load level.

At the base of the wall the artificially generated ground acceleration time-history has been applied. The calculations have been performed in time-space with an implicit time-integration method.

2.5 *Results*

The results were analysed to be simplified as a behaviour factor q according to Eurocode 8. Generally also different unit-mortar combinations, i.e. clay bricks and calcium-silicate-units, have been investigated for comparison.

The results showed, that the simple description only by the cap-displacement ductility μ and the derived behaviour factor q_μ is not describing the dynamic behaviour and the load reduction satisfying. For that reason also calculations using the elastic-response spectra with initial stiffness of the wall were carried out. The behaviour factor here $q_{response\text{-}spectra}$ is taken to the ration of maximum horizontal force according the elastic response spectra and the force appeared in the non-linear finite-element calculations. The comparison of the behaviour factor calculated from the displacement ductility and the behaviour factor taken from response spectra calculations is shown in Figure 7.

As an important result it was found, that the comparison of maximum appeared shear load and the maximum load according the design criteria in Eurocode 6 resp. DIN 1053-1 showed significant

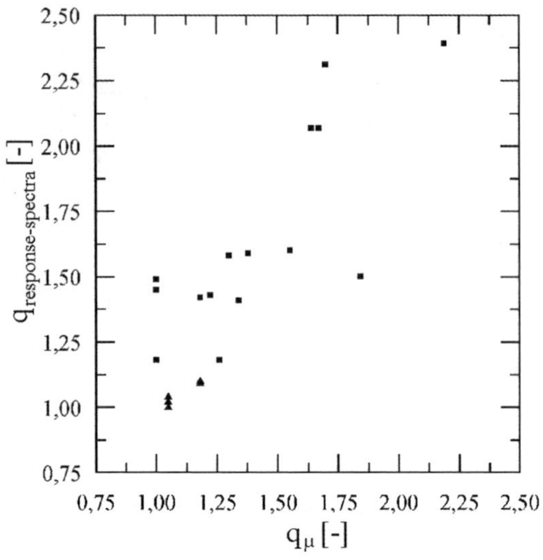

Figure 7. Behaviour factor q as result from the numerical investigations (several unit-mortar-combinations and size ratios).

load bearing reserves. That means that the design method for seismic loads and also for static loads has to be checked and adapted.

Generally it is known, that numerical investigations are restricted to describe the post-crack behaviour exactly. Regarding the dynamic behaviour where the stiffness degradation is essentially, this leads to the perception, that experimental investigations become necessary to validate the numerical result or give the input parameters in a more exact way.

3 EXPERIMENTAL INVESTIGATIONS

3.1 *Test-method*

As experimental investigations became necessary, a test programme to investigate the realistic seismic behaviour of unreinforced AAC constructions was started. As in the numerical investigations it was found that walls with a high ration between height and length of the wall behave more ductile, the test programme focused on slender walls with the dimensions of 2.5 m × 1.25 m × 0.24 m.

For the seismic input several ground acceleration time histories were generated artificially to characterize near-source earthquakes of short duration (9.0 sec), typical for regions of low and moderate seismicity. The test method is basing on the step-wise integration of the equation of motion of the whole structure in discrete time steps (0.01 sec).

3.2 *Investigated structure*

The investigated terraced house structure was reduced to a plane MDOF system, whereas the tests were carried out just on the relevant shear wall section in the first storey. The upper storeys with the walls and the elastic restraint in the floor-slabs (k_{slab}) were considered numerically as a sub-structure.

The stiffness characteristics of the sub-structure – described by nonlinear springs – were determined within preceding static-cyclic tests under adapted normal forces.

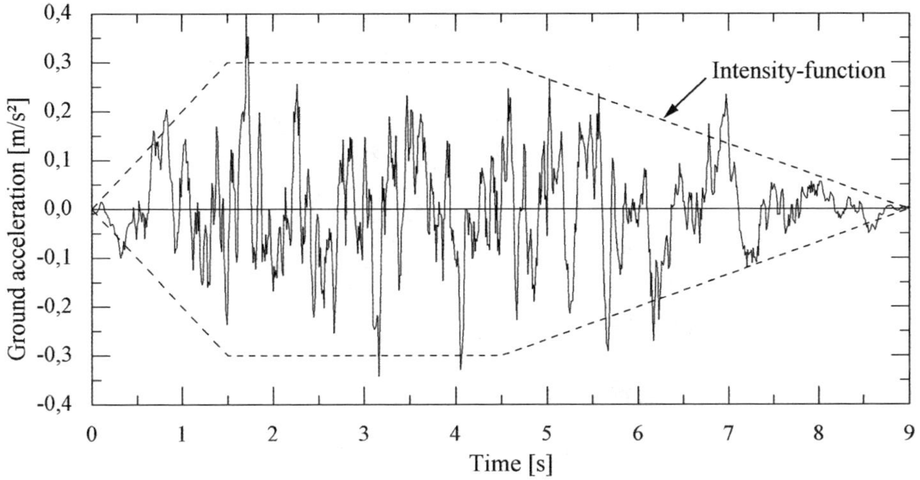

Figure 8. Artificially generated ground acceleration time history with the included intensity-function (time history scaled to several load levels).

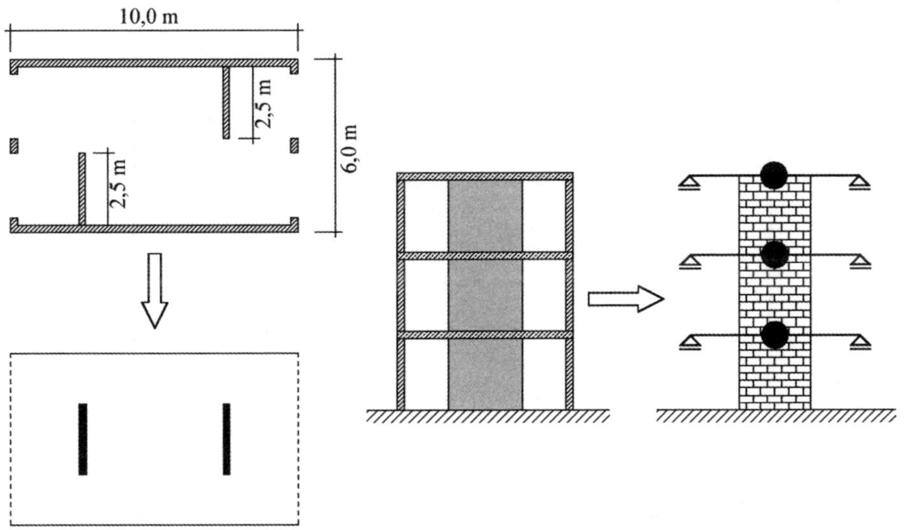

Figure 9. Investigated structure and simplification in plan and elevation.

The damping effect of the total structure was described by three viscous dampers c_1–c_3, leading to a damping ratio of 5% in all three eigenforms.

3.3 *Test set-up*

At the tested wall the normal force, the cap moment and also the horizontal force resp. horizontal cap displacement have to be applied. In the test set-up three computer-controlled hydraulic actuators with a very high accuracy have been applied. With the vertical actuators N_1 and N_2 the normal force and with an unequal distribution also the bending cap moment was applied. The horizontal cap displacement was applied with the horizontal actuator with a very high accuracy of 2/100 mm.

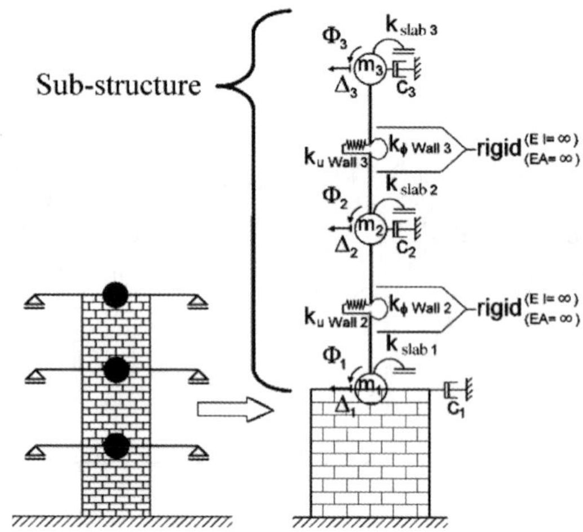

Figure 10. Testes wall with the application of the sub-structure.

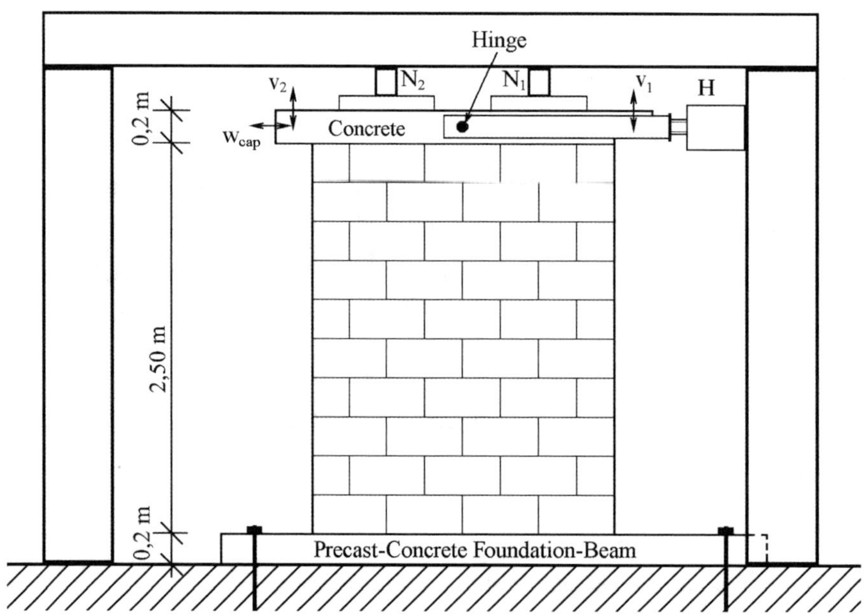

Figure 11. Test set-up.

3.4 *Results*

As the tests started in June 2005, final results can only be presented during the conference. The written results can be downloaded from the web-site www.mb.bv.tum.de/index-engl/forschung/mauerwerk.html.

Autoclaved Aerated Concrete – Limbachiya and Roberts (eds)
© 2005 Taylor & Francis Group, London, ISBN 0 415 38356 0

Mechanical properties of Hungarian AAC elements tested under different conditions

Zs. Józsa
Budapest University of Technology and Economics, Department of Construction Materials and Engineering Geology, Budapest, Hungary

G. Juhász
XELLA Hungary, Budapest, Hungary

ABSTRACT: YTONG (now XELLA) produces several kinds of AAC elements since 1992 is in Hungary. Model tests were carried out to find the influence of moisture content on the compressive strength of small specimens (100 mm cubes, prisms), whole units and wall models wetting them at different height and for different duration of.

Columns with non-rectangular cross sections having concavo-convex corners were tested under eccentric load. In case of 3–5 mm unfilled vertical joint-gaps ultimate loads were found to be 10–15% lower than that of columns prepared in first class quality.

1 INFLUENCE OF WATER CONTENT ON COMPRESSIVE STRENGTH OF AAC

1.1 *Introduction*

In the last decades serious floods attacked Hungary, and lots of buildings were deteriorated or collapsed. Therefore the question arises – what is the behavior of buildings made of AAC elements under such conditions.

All constructions must have stability and enough load-bearing capacity even in case of a flood. For this reason several AAC specimens (100 mm cubes, 100 × 100 × 200 mm prisms, 500 × 300 × 200 mm whole units and wall models) were tested wetting them at different height (¼ height, ½ height and 1 height) and different duration of time (0 hour, ½ hour, 2 hours, 6 hours, 24 hours and 1 week long) (Ring 1999).

1.2 *Test results*

Evolution of water content (m%) is strongly dependent on the shape of specimens, water level and duration of contact with water. Complete water saturation can take place only in the case of 100 mm cubes, all the other specimens contain intact areas. Figure 1 indicates results of evolution of water content for 100 × 100 × 200 mm prisms. The different curves correspond to 3 different water levels (¼ height of specimen, ½ height specimen and equal to the height of specimen) respectively.

Increasing water content resulted decreasing in compressive strength. Table 1 summarizes the average of results for 100 mm cubes. Numerical values are given as percents of the dry strength. Graphical representation of results can be studied in Figure 2.

According to results in Figure 2 slightly different tendencies seems to be found for different water levels. On the contrary, if the decrease of compressive strength is indicated as a function of water content (see Fig. 3), the above-mentioned differences seem to be disappeared. This result is very important for design. It can be demonstrated, that the decrease of compressive strength is independent of the position of saturated areas, however, depending directly on water content. These findings can be demonstrated also for the whole units (Fig. 4).

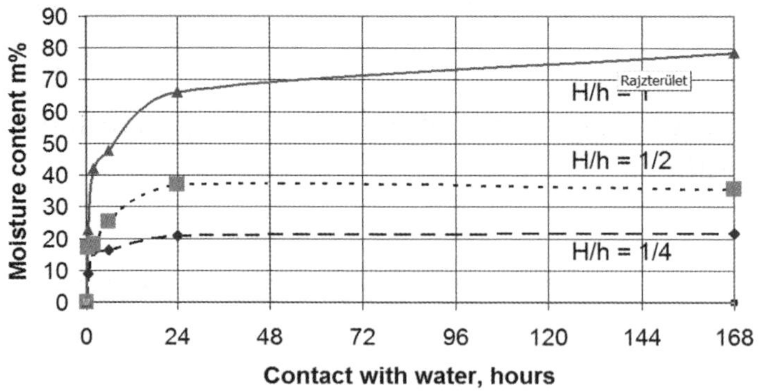

Figure 1. Water suction of 100 × 100 × 200 mm prisms at different water levels.

Table 1. Influence of contact with water on the compressive strength of 100 mm cubes.

Height of water (H)/ height of specimen (h)	Duration of contact with water in hours					
	T = 0 h	T = ½ h	T = 2 h	T = 6 h	T = 24 h	T = 168 h
H = ¼ h	100%	89%	87%	86%	80%	76%
H = ½ h	100%	82%	80%	76%	71%	68%
H = h	100%	75%	74%	75%	67%	64%

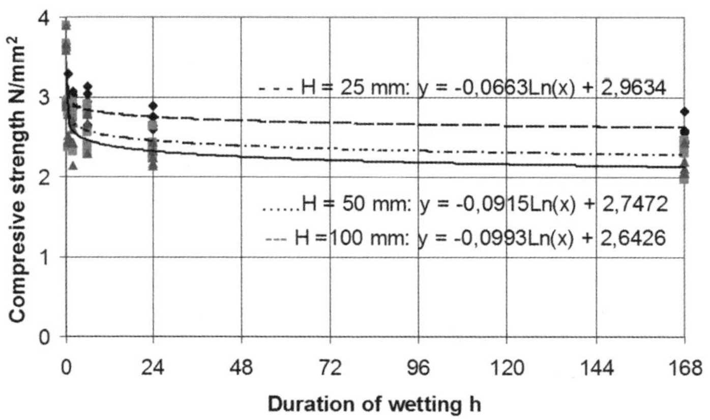

Figure 2. Compressive strength of 100 mm cubes vs. duration of water contact at different levels.

Based on present results a new formula can be developed for the estimation of decrease of strength due to water absorption. Formula is the following:

$$f_{c,w}(t) = f_{c,o} - c \cdot Ln(t) \quad (1)$$

where

$f_{c,w}(t)$ = compressive strength after contact with water after t hours,

$f_{c,o}$ = compressive strength of dry specimen,

c = factor of strength decrease, taking into consideration height of water (H)/height of specimen (h).

Change of factor c is non linear with height of water/height of specimen. As a consequence, decrease of strength is non linear also in time and depth of water front in the specimen (Fig. 5).

Maximal water content of the wall units is much less, than that of cube specimens due to the presence of intact areas in the whole units (Fig. 6).

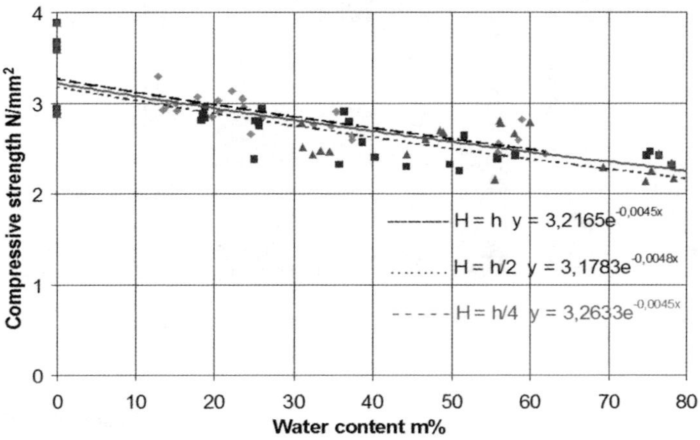

Figure 3. Compressive strength of 100 mm cubes vs. absorbed water content in m% at water contact at different levels.

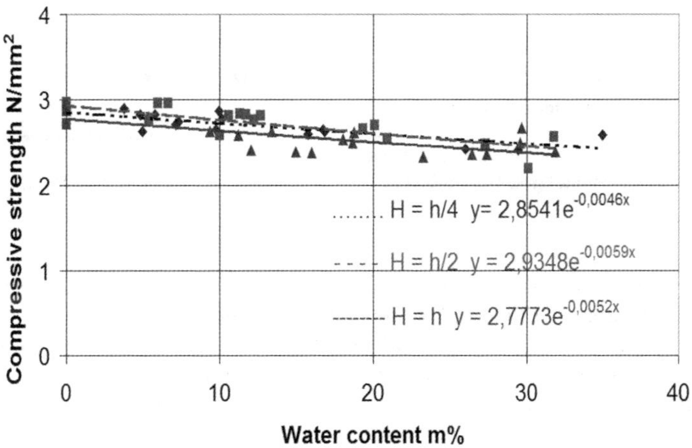

Figure 4. Compressive strength of whole 500 × 300 × 200 mm AAC units vs. absorbed water content in m% at water contact at different levels.

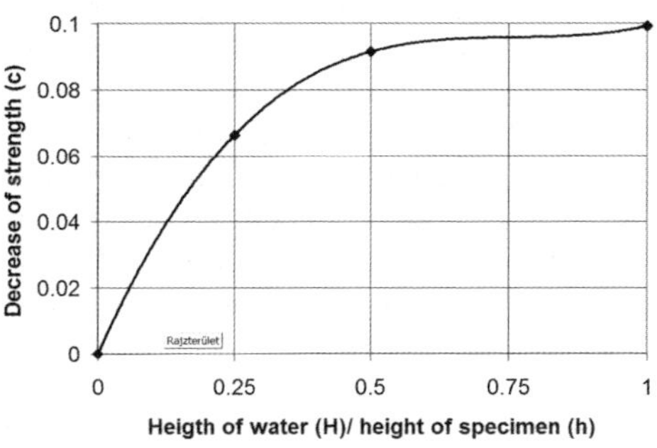

Figure 5. Factor of strength decrease (c) vs. height of water/height of specimen.

Figure 6. Intact areas in the middle of the specimen can be clearly seen even after failure.

Figure 7. Compressive strength test of wet wall models – high water suction in the mortar gaps.

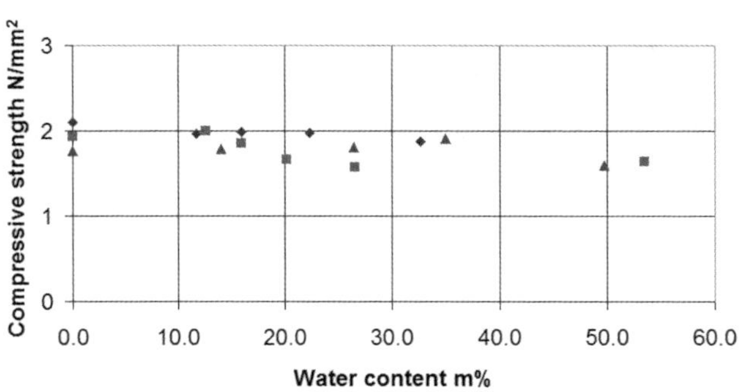

Figure 8. Compressive strength of wall models vs. absorbed water content in m% at water contact at different levels.

Figure 9. Mode of failure.

Compressive strength of wall models was higher than 1.5 N/mm^2 in all cases and failure mode was similar to that of prisms (100 × 100 × 200 mm). Total numbers of tested wall models were 17, Figures 7 shows three of them. It can be realized, that water suction is higher in the mortar gaps than in the AAC elements. Preliminary test results are summarized in Figure 8. Further work is needed in this field.

2 STRENGTH TESTS OF COLUMNS HAVING SPECIAL GEOMETRY

2.1 *Description of models tested*

The research was dealing with columns having 25, 30 and 37.5 cm thickness, height of 2.8 m (with non-rectangular cross section, having concavo-convex corners) tested under eccentric load. Aim of study was to evaluate strength and deformation capacity of these special columns. The specimens aimed to model columns having cross section of 1500 to 2000 cm^2 at corner between two windows. Up to the parapet height 0.3–0.3 m thick additional masonry was applied. Total number of tested specimens was 18, for each wall thickness 3-3, either for eccentric load on inner side or on outer side. The applied eccentricity of the load was according to the dimensions of typical insulated ring-beam used in Hungary recently. The cross sections, dimensions and geometrical characteristics of the tested columns are given in Table 2 and Figure 10.

2.2 *Testing conditions and method*

All the columns were made of P2/0.5 YTONG AAC elements using 50% mixture of specially shaped masonry units with thin layer mortar (TLMB) having 6.0 N/mm^2 compressive strength, and regular

Table 2. Geometrical characteristics of models tested ($l_o = 1 \times m = 2.8$ m).

v mm	$\lambda = l_o/v$	A mm^2	I_s mm^4	y_s mm	m mm	A_{tb} mm^2	y_{stb} mm	A_{tk} mm^2	y_{stk} mm	e_{ob} mm	e_{ok} mm
250	11.2	$150{,}867 \cdot 10^3$	$0.957 \cdot 10^9$	186	50	$116{,}963 \cdot 10^3$	214	$124{,}890 \cdot 10^3$	164	28	22
300	9.3	$157{,}300 \cdot 10^3$	$1.302 \cdot 10^9$	201	100	$96{,}529 \cdot 10^3$	256	$113{,}439 \cdot 10^3$	157	56.3	42.8
375	7.6	$207{,}776 \cdot 10^3$	$2.599 \cdot 10^9$	239	115	$132{,}348 \cdot 13^3$	321	$156{,}533 \cdot 10^3$	191	65.7	47.9

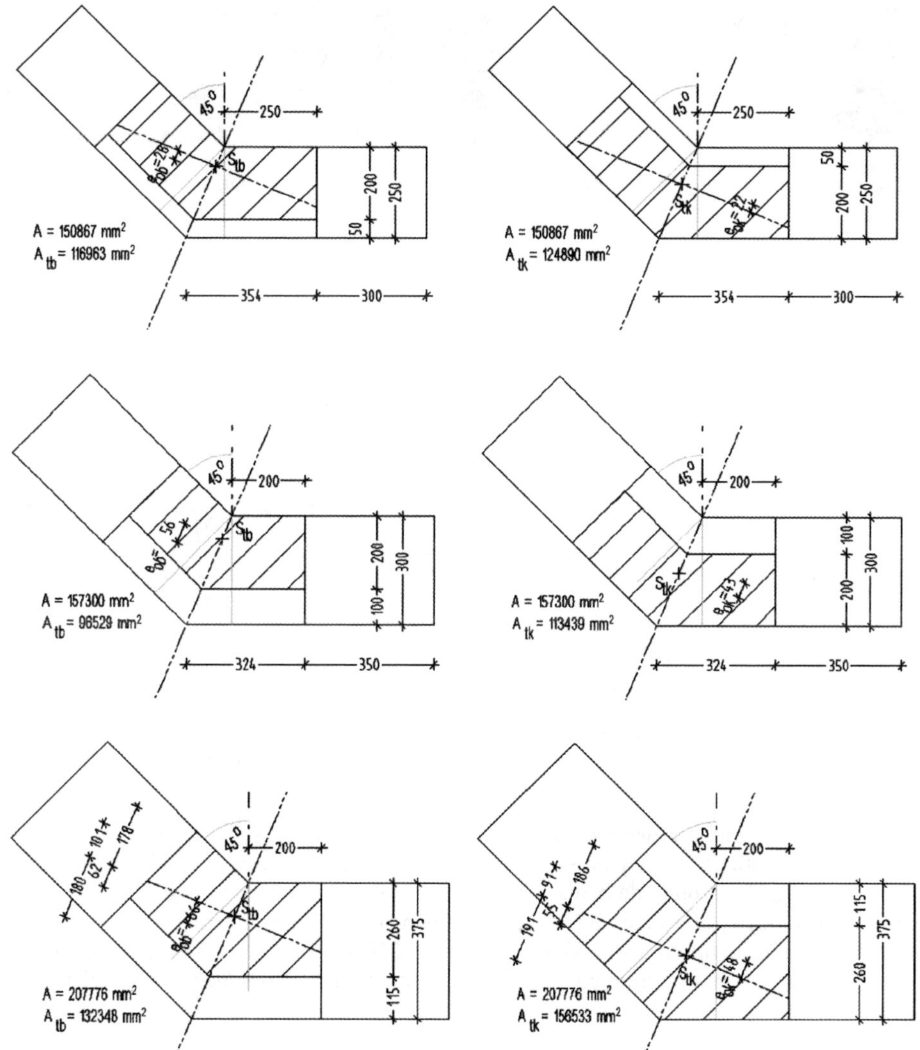

Figure 10. Cross sections, dimensions and geometrical characteristics of the tested columns.

shaped masonry units with lightweight mortar (LM) having 3.0 N/mm^2 compressive strength. To study the influence of the parapet reinforcement on the cracks, zero, one or two reinforcing bars were used in the parapet. Mechanical characteristics of the applied AAC elements and mortars were tested at a laboratory. The compressive strengths, gross dry density and the bending and compressive strength of different mortars were determined. The load carrying capacity of the

Figure 11. Reinforcing bars in the parapet.

Figure 12. Columns before testing.

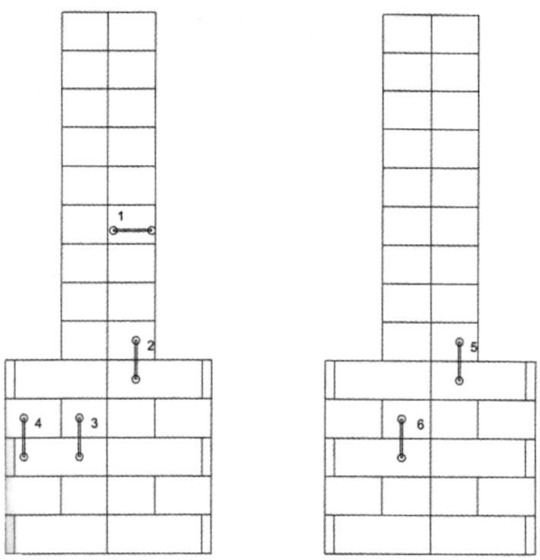

Figure 13. Deformeter arrangement (left outer side, right inner side).

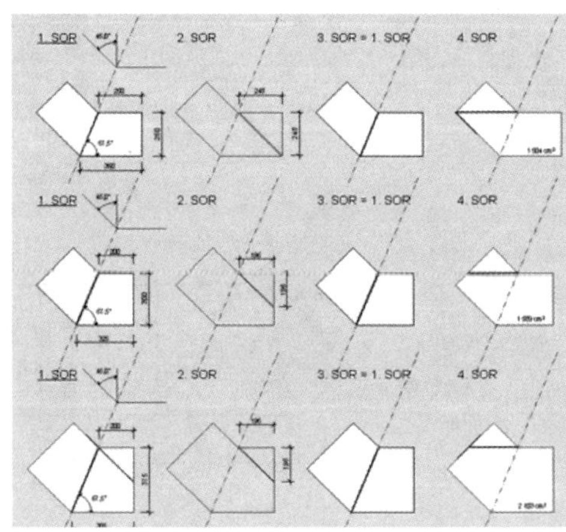

Figure 14. Joints and bonds by layers.

columns were tested with a crushing machine of 5000 kN capacity, applying inner and outer side eccentric loading, with 50 kN steps up to failure load. Five minutes after applying a new loading step the vertical and horizontal deformations were measured at 6 different points of the wall by a 250 mm long Mitutoyo deformeter. Readings were made by 0.01 mm sensitivity.

The reinforcing bar in the parapet is shown in Figure 11. The columns before the test are indicated in Figure 12. The deformeter arrangements and the column joints and bonds in the wall are shown in Figures 13 and 14.

2.3 *Failure modes and cracks patterns*

During the compressive test, in case of majority of columns, few vertical cracks appeared under the last loading step (or at one before the last step) in the upper third of the columns. Soon after the appearance of the first crack the columns suddenly failed by the increasing number of cracks and the dynamic crack opening. These cracks are indicated in Figures 15 and 16.At about 35% of the tests, in addition to the cracks appeared during the final load step, the failure mode was buckling of the columns at around half of the height. Buckling is shown in Figures 17 and 18. Important to note, that cracks were not to be found either on the bottom of the columns – or on the parapet wall).

Figure 15. Progressive cracking of column 30-1 at failure.

Figure 16. Progressive cracking of column 25-1 at failure.

Figure 17. Buckling failure of column 25-2.

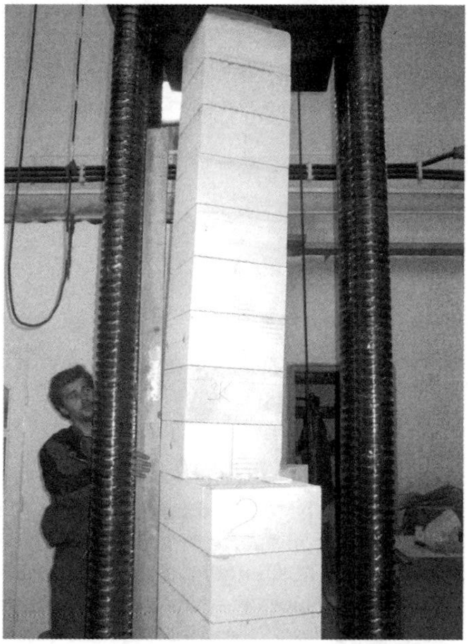

Figure 18. Buckling failure of column 375-1.

Typical load vs. strain diagrams are indicated in Figures 19 to 21.

Table 3 summarizes the failure loads, the bending moment calculated from failure load, mortar type, reinforcing bars, the compressive strengths, the limit forces (according to the relevant Hungarian Standard) and the safety values for different column types. The ratio of the effective local compressive strength, calculated from the failure load, to the limit strength values in the wall was 2.9 to 5.2.

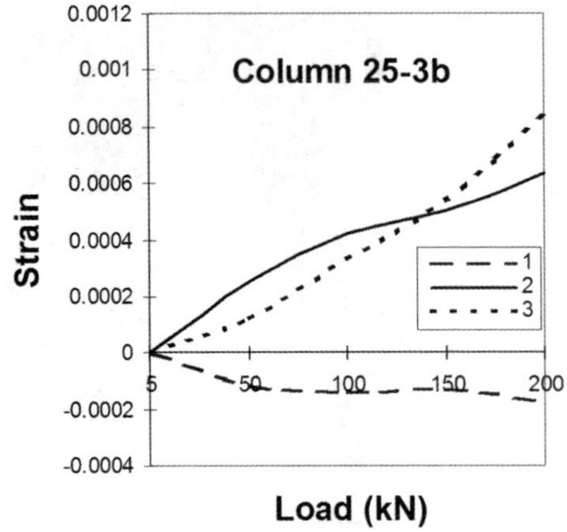

Figure 19. Load–strain diagram of column 25-3b.

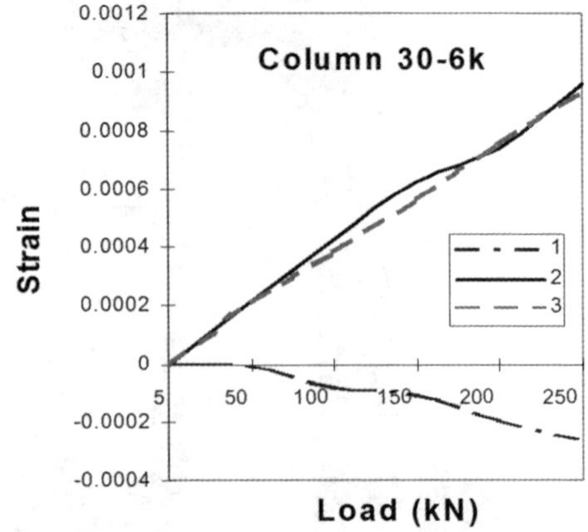

Figure 20. Load–strain diagram of column 30-6k.

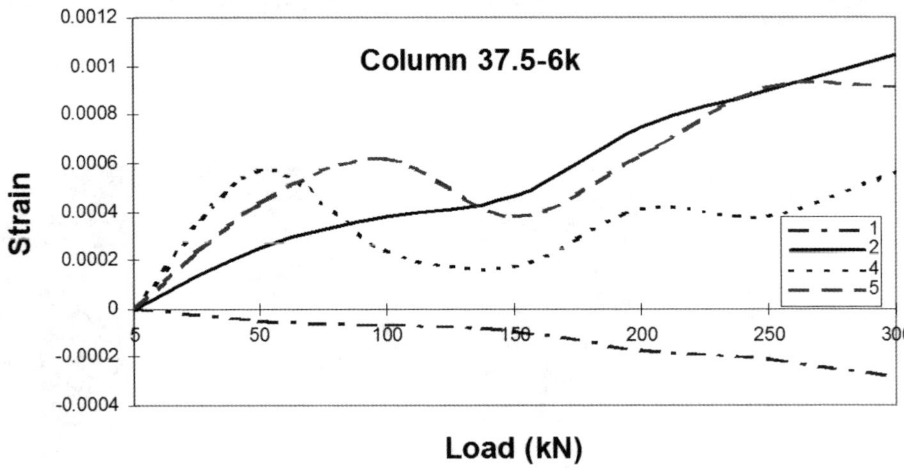

Figure 21. Load–strain diagram of column 37.5-6k.

2.4 *Evaluation*

The mechanical properties of the materials (YTONG category I. masonry units and LM and TLMB mortars) met the requirements of the EN 771-4, EN 998-2 and EC-6. The load bearing cross section of the tested columns were selected as the minimum cross section applicable in Hungary.

Compressive strength characteristic values of the masonry units were greater than 2.0 N/mm^2, and of the LM mortar greater than 3.0 N/mm^2 and of the TLMB mortar were greater then 6.0 N/mm^2. The limit forces of the columns were calculated as if it would be loaded concentrically according to MSZ 15023 (Hungarian Standard).

$$F_H = \varphi \cdot A_t \cdot 0.85 \cdot m_1 \cdot \sigma_f \tag{2}$$

where
φ = value of buckling
A_t = effective loaded surface
0.85 = safety value for columns
m_1 = safety factor for quality of implementation in the range of 0.85–1.15
σ_f = 0.5 N/mm^2 limited stress in AAC P2/0.5 masonry.

Table 3. Test results for columns (Where limit strength $\sigma_H = 0.5$ N/mm^2).

Type of column	Failure load F_t kN	$M_t = F_t e_o$ kNm	Mortar type	Reinforcement in parapet	$\sigma_t = F_t/A_t$ N/mm^2	$n_\sigma = \sigma_t/\sigma_H$	Limit force F_H kN	Safety $n_F = F_t/F_H$
25-1b	225	6.30	LM	2 Ø 8	1.93	3.9		5.3
25-3b	212	5.94	LM	2 Ø 8	1.82	3.6	42.360	5.0
25-5b	170-	4.76	TLMB	2 Ø 8	1.46	2.9		4.0
mean	202	5.66			1.73	3.5		4.8
25-2k	196	4.371	LM	2 Ø 8	1.57	3.1		4.3
25-4k	200	4.460	LM	2 Ø 8	1.60	3.2	45.230	4.4
25-6k	200	4.460	TLMB	2 Ø 8	1.60	3.2		4.4
mean	199	4.430			1.59	3.2		4.4
30-1b	233	13.118	LM	None	2.41	4.8		6.2
30-3b	200	11.260	LM	1 Ø 8	2.07	4.1	37.366	5.4
30-5b	216	12.161	TLMB	1 Ø 8	2.24	4.5		5.8
mean	216	12.180			2.24	4.5		5.8
30-2k*	250	10.700	LM	None	2.21	4.4		5.7
30-4k*	268	11.470	LM	1 Ø 8	2.37	4.7	43.912	6.1
30-6k	294	12.583	TLMB	2 Ø 8	2.60	5.2		6.7
mean	271	11.584			2.40	4.8		6.2
37.5-1b*	250	16.425	LM	None	1.89	3.8		4.8
37.5-3b	300-	19.710	LM	1 Ø 8	2.27	4.5	52.136	5.8
37.5-5b*	250	16.425	TLMB	2 Ø 8	1.89	3.8		4.8
mean	267	17.520			2.02	4.0		5.1
37.5-2k*	300	14.370	LM	None	1.91	3.8		4.9
37.5-4k	330	15.807	LM	1 Ø 8	2.11	4.2	61.663	5.4
37.5-6k	317	15.184	TLMB	2 Ø 8	2.02	4.0		5.1
mean	316	15.120			2.01	4.0		5.1

* vertical gaps unfilled.

Typical load vs. strain diagrams are indicated in Figures 19 to 21.

The longitudinal ε_{break} values were between 0.1% to 0.12%, on the contrary the transverse ε_{expand} values were around 0.02%.

From the results it can be realised, that failure of columns takes place soon after the appearance of the first crack with very small deformation capacity. Values of the safety against failure are in the range of 4.4 to 6.2 as it can be seen in the last column of Table 2. These safety values are acceptable, not too high particularly considering the applied strength of mortars and the fast and brittle failure of the columns.

3 CONCLUSIONS

Majority of the water absorption and decrease of strength take place in the first 24 hours. Decrease of strength in time can be easily estimated by the newly developed formula (Eq. 1). This result is very important, because if the structure survived the first day of contact with water (in case of an accident or flood), it is supposed that stability will not be lost in the next 7 days. All pores of the AAC elements cannot be water saturated.

In case of a small columns having concavo/convex cross sections, the failure load can be calculated by using the effective loaded area and applying concentric loading.

In case of the 37.5 cm thick wall additional insulation of 1 cm is applicable, because it has sufficiently enough load carrying capacity.

If the quality of the masonry work is low – the vertical joints are more then 3 mm thick and do not completely filled – the strength of the columns can be 10–20% smaller than that of the standard quality.

The triangular shape of AAC masonry units applied in the bonds did not influenced the performance of the columns.

REFERENCES

Ring Z. 1999 *Influence of moisture content on the compressive strength of AAC elements* (in Hungarian) MSc Thesis, TU Budapest.

MSZ EN 771-4:2003 Specification for masonry units. Part 4: Autoclaved aerated concrete masonry units.

MSZ EN 998:2003 Specification for mortar for masonry. Part 2: Masonry mortar.

EC-6 – MSZ ENV 1996:2000 Design of masonry structures.

MSZ 15023 – 1987 (Hungarian Standard) Statical design of load carrying masonry constructions of buildings.

Autoclaved Aerated Concrete – Limbachiya and Roberts (eds)
© 2005 Taylor & Francis Group, London, ISBN 0 415 38356 0

Flexural strength of low density blockwork

A. Ahmed, A.N. Fried, M.C. Limbachiya & J.J. Roberts
Concrete and masonry Group, Kingston University, London, KT1 2EE

ABSTRACT: The characteristic flexural strength of low density Aircrete wallettes (2.8 and 2 N/mm^2) incorporating both conventional and thin layer mortar is verified. The wallettes are tested in accordance with BS EN 1052-2. The flexural strength of Aircrete wallettes is derived from the strength of small specimens tested to destruction under four-point loading. The strengths of the wallettes are high with impressive repeatability with the maximum strength being reached for thin layer wallettes within 7 days curing time. In general the strengths of both conventional mortar and thin layer mortar wallettes compare favourably to values reported in BS 5628.

1 INTRODUCTION

Aircrete was developed in Sweden in 1924 and first used in the late 1950's (H+H Celcon Ltd 2003) as an alternative to building with timber. Currently over 30 million m^3 of the material is produced annually. In the United Kingdom it is used extensively by major house builders with block sales of approximately 2.9 million m^3 per annum. It is so extensively used that Aircrete (mainly medium and high density) now accounts for a third of all concrete blocks in the UK.

Aircrete is also known commercially as AAC (Autoclaved Aerated Concrete), Celcon, Durox, Thermalite and Topblock. Aircrete blocks are suitable as vertical load-bearing elements and provide the thermal insulation expected from typical UK wall construction. Aircrete blocks may also be used as non load bearing outer leaves of masonry walls, external walls and walls below ground level, where adequate care is essential to ensure their durability and protection from effects of the environment (Aircrete Bureau 2003a).

The lightweight porous structure and consequent faster build-speed of Aircrete blocks mean even foundations can be constructed quickly, easily and cost-effectively creating an extremely effective moisture barrier with significant thermal insulation properties. Their porous cellular structure and durability (Aircrete Bureau 2003a, Dubral 1992) make them a recognised alternative in most below ground situations. Aircrete is also now more popular than timber or solid concrete floors in new housing at ground level, and is increasingly the preferred solution for upper floors too. Aircrete's superior thermal performance in most cases eliminates the need for expensive cavity insulation.

Aircrete is produced by mixing cementitious materials, cement and/or pulverised fuel ash (PFA), lime, sand, water and aluminium powder. The final process involves autoclaving for approximately 10 hours at high temperature and pressure (Mitsuda and Kiribayashi 1992, Isu and Mitsuda 1992, Pospisil 1992), hence, the material is also known as Autoclaved Aerated Concrete. Aircrete is comprised of 60–85% of air by volume (70–85% for low density Aircrete). The solid material part is a crystalline binder, which is called Tobermorite by mineralogists. Besides the binding phase Tobermorite, grains of Quartz and some other minerals are found in minor amounts. The chemical composition of Tobermorite comprises of silicium dioxide, calcium oxide and water. It is Tobermorite, which provides the high compressive strength of Aircrete in spite of the high proportion of pores in this construction material. This is why low density Aircrete is sufficiently strong for the construction of dwellings in spite of its considerably high air content.

Figure 1. Cellular structure of Aircrete with levels of porosity up to 85%.

Table 1. Physical properties of aircrete blocks.

Aircrete density	Compressive strength (N/mm^2)	Density (Kg/m^3)	Thermal conductivity (W/mK)
Low	2.0–3.5	450	0.09–0.11
Medium	4.0–4.5	620	0.15–0.17
High	7.0–8.5	750	0.19–0.20

1.1 *Physical attributes of Aircrete*

1.2 *Thermal resistance*

Thermal conduction is the phenomenon by which heat is transported from high to low-temperature regions of a substance. The high degree of porosity of Aircrete has a dramatic influence on thermal conductivity; increasing pore volume will, under most circumstances reduce thermal conductivity and increase thermal insulation. Heat transfer across pores is ordinarily slow and inefficient (Callister 2000). Internal pores (Figure 1) normally contain still air, which has extremely low thermal conductivity – approximately 0.02 W/m-k. Furthermore, gaseous convection within the pores is also comparatively ineffective. Hence Low Density Aircrete has outstanding thermal insulation properties (Dubral 1992, Lippe 1992, Daian 1992, Theramalite 2003, Laurant 1992, Millard 1992, Frey 1993, Liu and Wang 1992, Schlegel and Volec 1992) as shown in Table 1.

1.3 *Compressive strength*

The compressive strength of Aircrete is related to its density and increases with increasing density (H+H Celcon Ltd 2003, Wittman 1993a). Commonly produced compressive strengths are 2.8, 3.5, 4.0, 7.0 and 8.4 N/mm^2 (MPa) as indicated on Table 1. In the UK compressive strengths $>4\ N/mm^2$ are commonly used, however, in Europe lower strength Aircrete has been successfully utilised for the construction of dwellings (Wittmann 1993a), implying lower compressive strength will be adequate. The compressive strength of Aircrete is nearly independent of specimen size due to its homogeneity (Wittmann 1993a). Aircrete achieves its final strength during the autoclaving process without further curing being necessary.

1.4 *Lightweight*

The cellular structure (Tada 1992, Jacobs and Mayer 1992, Schober 1992) of the material ensures a lightweight construction. During installation, most Aircrete blocks can be lifted with one hand providing significant productivity advantages. The higher porosity of Low Density Aircrete ensures that the material is extremely lightweight (density $\leq 450\ kg/m^3$), thus, it is even easier to handle in comparison to medium and high density Aircrete. As a result transportation costs would be reduced and furthermore, houses would be built much more quickly.

Table 2. Matrix of specimens tested.

	Mortar designation		Thin layer mortar	
Aircrete block strength (N/mm^2)	iii	iv	A	B
2.0	5B	5B	5B	15B*
2.0	5P	5P	5P	
2.8	×	×	5B	15B*
2.8	×	×	5P	

B wallettes = Flexural strength for a plane of failure parallel to the bed joints.
P wallettes = Flexural strength for a plane of failure perpendicular to the bed joints.
Designation iii = cement:lime:sand ratio of 1:1:6 (volume)
Designation iv = cement:lime:sand ratio of 1:2:8 (volume)
*15 wallettes comprises 5 each for 1, 7 and 28 days curing using Thin Layer Mortar B.

2 EXPERIMENTAL WORK

This paper describes the results achieved for verifying the characteristic flexural strength of low density Aircrete wallettes-built using 2.8 and 2 N/mm^2 compressive strength blocks. Wallettes were prepared using designation iii and iv mortars in accordance with BS 5628 [British Standards a], and using two different makes of thin layer mortar.

Wallettes were tested in accordance to BS EN 1052; Part 2 [British Standards b]. The flexural strength of Aircrete wallettes was derived from the strength of small specimens tested to destruction under four-point loading. The maximum load achieved was recorded. The characteristic value, calculated from the maximum stresses achieved by the samples is considered to be the flexural strength of the blockwork.

A matrix of the different block and mortar combinations built and tested is given in Table 2. The wallette thickness is 150 mm unless otherwise specified.

Given the accelerated setting time for thin layer mortar, it has been suggested that early age flexural testing on thin layer wallettes (up to 7 days curing) could provide some useful results. Professional blocklayers used different thin layer mortars, either type A or B. All specimens were tested at either 1, 7 or 28 days curing as specified in Table 2.

3 RESULTS AND DISCUSSIONS

Results obtained for 2 and 2.8 N/mm^2 Thin Layer wallettes constructed using type A mortar are summarised in Tables 3 and 4. Table 5 shows the results for Thin Layer wallettes constructed using type B mortar. Tables 6 and 7 show the effect of curing time (ageing) on flexural strength for Thin Layer wallettes. Table 8 provides statistical analysis of results obtained for both Thin Layer mortars. Results obtained for 2N Designation iii and iv wallettes are summarised in Tables 9 and 10.

Table 11 provides statistical analysis of results obtained for conventional mortars. Throughout the document, the flexural strength results presented, unless otherwise specified, should be assumed to be after the standard 28 $\pm$ 1 days curing time.

3.1 *Thin layer wallettes*

From the results obtained (Table 8), it can be seen that there is very good repeatability as reflected by low standard deviation and coefficient of variation values.

In BS 5628 [British Standards a] the characteristic flexure strength obtained using 7 N/mm^2 blocks with designation iv mortar is 0.2 and 0.45 N/mm^2 for B and P wallettes respectively. In

Table 3. Flexural strength of thin layer A 2 N/mm^2 Aircrete.

Aircrete block strength (N/mm^2)	Failure stress (N/mm^2) B	Failure stress (N/mm^2) P
2	0.31	0.27
2	0.30	0.29
2	0.28	0.27
2	0.32	0.28
2	0.28	0.30

Table 4. Flexural strength of thin layer A 2.8 N/mm^2 Aircrete.

Aircrete block strength (N/mm^2)	Failure stress (N/mm^2) B	Failure stress (N/mm^2) P
2.8	0.51	0.39
2.8	0.43	0.41
2.8	0.46	0.36
2.8	0.50	0.37
2.8	0.48	0.40

Table 5(a). Flexural strength of thin layer B 2 N/mm^2 Aircrete B-wallettes.

Aircrete block strength (n/mm^2)	Direction	Failure stress (N/mm^2)
2	B	0.35
2	B	0.29
2	B	0.29
2	B	0.33
2	B	0.31

Table 5(b). Flexural strength of thin layer B 2.8 N/mm^2 Aircrete B-wallettes.

Aircrete block strength (N/mm^2)	Direction	Failure stress (N/mm^2)
2.8	B	0.51
2.8	B	0.46
2.8	B	0.47
2.8	B	0.51
2.8	B	0.46

comparison to the data presented in this document, the flexural strength of both the B wallettes (thin layer with 2 and 2.8 N/mm^2 blocks) is substantially higher – 0.30 and 0.48 N/mm^2 respectively. There is remarkable consistency in the results with regards to both thin layer mortars (characteristic strength of 0.27 and 0.26 N/mm^2 for 2 N/mm^2 Aircrete and 0.44 and 0.42 N/mm^2 for 2.8 N/mm^2 Aircrete).

Table 6(a). Flexural strength of thin layer B 2 N/mm^2 Aircrete B-wallettes after 1 day curing.

Aircrete block strength (N/mm^2)	Direction	Failure stress (N/mm^2)
2	B	0.24
2	B	0.24
2	B	0.26
2	B	0.22
2	B	0.26

Table 6(b). Flexural strength of thin layer B 2.8 N/mm^2 Aircrete B-wallettes after 1 day curing.

Aircrete block strength (N/mm^2)	Direction	Failure stress (N/mm^2)
2.8	B	0.23
2.8	B	0.19
2.8	B	0.22
2.8	B	0.21
2.8	B	0.24

Table 7(a). Flexural strength of thin layer B 2 N/mm^2 Aircrete B-wallettes after 7 days curing.

Aircrete block strength (N/mm^2)	Direction	Failure stress (N/mm^2)
2	B	0.31
2	B	0.28
2	B	0.29
2	B	0.30
2	B	0.29

Table 7(b). Flexural strength of thin layer B 2.8 N/mm^2 Aircrete B-wallettes after 7 days curing.

Aircrete block strength (N/mm^2)	Direction	Failure stress (N/mm^2)
2.8	B	0.46
2.8	B	0.42
2.8	B	0.48
2.8	B	0.45
2.8	B	0.48

The normal trend in masonry is that the flexural strength of P wallettes are higher than B wallettes, however, the trend here is to the contrary. This is not altogether surprising as given the very thin and excellent bond strength of the mortar, the B wallettes behave as a large plate. This seemed to be the case as failure was not just confined to the mortar bed but also there was significant material failure, as shown in Figure 2. Further, as the bond in the perpend joints is likely to be inferior to that in the bed joints, it will initiate failure at a lower load in that direction.

Table 8. Statistical analysis of flexural strength of thin layer wallettes.

Aircrete block strength (N/mm^2)	Curing time (days)	Thin layer mortar	Direction	Average failure σ (N/mm^2)	Standard deviation	Coefficient of variation	Characteristic strength (N/mm^2)
2	28	A	B	0.30	0.017	0.060	0.27
2	28	B	B	0.31	0.023	0.075	0.27
2	28	A	P	0.28	0.013	0.046	0.26
2	1	B	B	0.24	0.015	0.063	0.22
2	7	B	B	0.30	0.010	0.034	0.28
2.8	28	A	B	0.48	0.032	0.067	0.42
2.8	28	B	B	0.48	0.023	0.048	0.44
2.8	28	A	P	0.39	0.021	0.053	0.35
2.8	1	B	B	0.22	0.017	0.078	0.19
2.8	7	B	B	0.46	0.022	0.049	0.42

Table 9. Flexural strength of designation iii 2 N/mm^2 Aircrete.

Aircrete block strength (N/mm^2)	Failure stress (N/mm^2) B	Failure stress (N/mm^2) P
2	0.16	0.22
2	0.18	0.20
2	0.16	0.22
2	0.19	0.23
2	0.18	0.24

Table 10. Flexural strength of designation iv 2 N/mm^2 Aircrete.

Aircrete block strength (N/mm^2)	Failure stress (N/mm^2) B	Failure stress (N/mm^2) P
2	0.19	0.17
2	0.17	0.20
2	0.15	0.22
2	0.15	0.23
2	0.18	0.21

Table 11. Statistical analysis of flexural strength of wallettes (conventional mortar).

Aircrete block strength (N/mm^2)	Mortar designation	Direction	Average failure σ (N/mm^2)	Standard deviation	Coefficient of variation	Characteristic strength (N/mm^2)
2	iii	B	0.17	0.013	0.077	0.15
2	iii	P	0.22	0.014	0.067	0.20
2	iv	B	0.17	0.018	0.106	0.14
2	iv	P	0.21	0.023	0.109	0.17

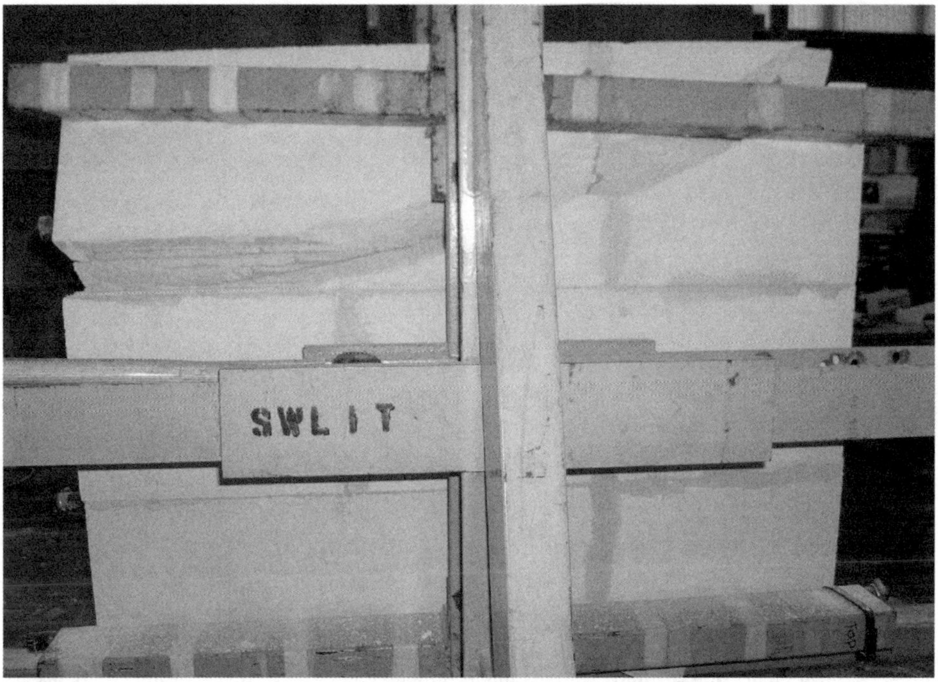

Figure 2. Typical failure of a 2 N/mm^2 thin layer wallette in the B direction showing predominantly material failure.

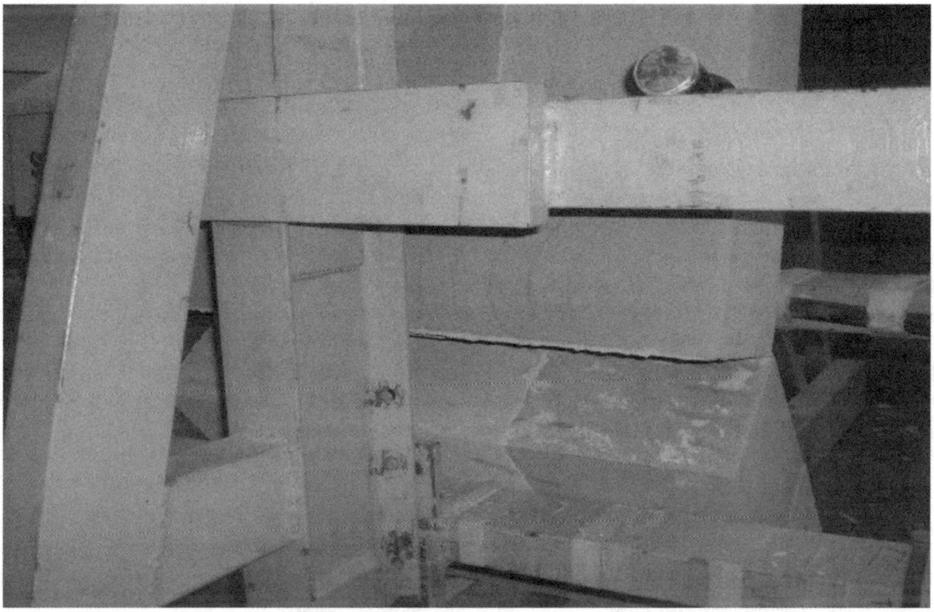

Figure 3. Typical failure of a 2.8 N/mm^2 thin layer wallette (after 1 day curing) in the B direction showing failure occurring along the bed joint.

Figure 4. Typical failure of a 2 N/mm^2 Designation iii wallette in the B direction showing failure occurring along the bed joint.

The results also show that for Thin Layer wallettes, maximum strength is reached within 7 days curing time. The characteristic flexural strengths for both 2 and 2.8 N/mm^2 Aircrete after 7 days is precisely the same as after the standard 28 days curing time. After 1 day curing time 2 N/mm^2 Aircrete wallettes reach 80% of its strength, for 2.8 N/mm^2 Aircrete it is 45%. Figure 3 shows that after 1 day curing time, the failure mode is predominantly confined to the mortar bed as opposed to after 7 and 28 days curing (Figure 2), thus, suggesting bond between mortar and material isn't yet complete.

3.2 *Conventional mortar wallettes*

In BS 5628 [British Standards a] the characteristic strength of B wallettes for (2.8–7) N/mm^2 blocks of 250 mm thickness are 0.15 and 0.1 N/mm^2 respectively for designations iii and iv mortar. The strength of 250 mm thick P wallettes for (2.8–3.5) N/mm^2 blocks are 0.25 and 0.2 N/mm^2 respectively for designations iii and iv. Therefore, the values achieved for 2 N/mm^2 wallettes, 0.15 and 0.20 N/mm^2 in the B and P directions respectively, are respectable.

The strength of these wallettes are in keeping with the normal trend in masonry, i.e. P wallettes are stronger than B ones. Designation iii wallettes are stronger than designation iv, however, the difference is not substantial – especially in the B direction. Further, the failure mode is different to thin joint wallettes: as shown in Figure 4, where cleavage occurs along the mortar with negligible material failure. Hence, the strength is much lower than thin layer wallettes.

4 CONCLUSIONS

a) The characteristic flexural strength of low density Aircrete wallettes (2.8 and 2 N/mm^2 blocks) using thin layer mortar and conventional mortar has been determined.
b) The values of B wallettes in particular are relatively high with very impressive repeatability, especially for the thin layer mortar wallettes.

c) The maximum strength is reached for thin layer mortar wallettes within 7 days curing time.
d) In general conventional mortar wallettes are weaker than thin layer, however, their strengths compare favourably to values reported in BS 5628.
e) The mode of failure for thin layer and conventional mortar wallette specimens are different.

ACKNOWLEDGEMENTS

The authors would like to acknowledge the funding for the project given by the Engineering and Physical Sciences Research Council (EPSRC) and the Industrial Collaborators, Autoclaved Aerated Concrete Products Association, H+H Celcon, Hanson-Thermalite, Quinn-Group, Tarmac Topblock, Building Research Establishment, Catnic-Corus UK, National House-Building Council and Office of The Deputy Prime Minister. Acknowledgement is also given to the significant contributions made by the project Steering Panel, Dennis Coward, Alan Ferguson, Clifford Fudge, Trevor Grounds, Jim Holland, Chris Kirby, Paul Matthews, Gopal Sangarapillai and Richard Shipman.

REFERENCES

Aircrete Bureau, 2003a, Code of Best Practice for the Use of Aircrete Products, www.aircrete.co.uk.

British Standards a, BS 5628: Part 1: 1992 – Structural use of Unreinforced Masonry.

British Standards b, BS EN 1052: Part 2: 1999 – Determination of Flexural Strength.

Callister, W., Jr., 2000, Department of Metallurgical Engineering, The University of Utah, Materials Science and Engineering – An Introduction, Fifth Edition, John Wiley and Sons, Inc, ISBN 0 471 32013 7.

Daian, J-F, 1992, Laboratoire d'etude des Transferts en Hydrologie et Environment (IMG), Grenoble, France, Experimental determination of AAC moisture transport coefficients under temperature gradients, Advances in Autoclaved Aerated Concrete, Wittmann (ed.) © Balkema, Rotterdam. ISBN 90 5410 086 9.

Dubral, W., 1992, YTONG AG, Munich, Germany, Advances in Autoclaved Aerated Concrete, Wittmann (ed.) © Balkema, Rotterdam. ISBN 90 5410 086 9.

Frey, E., Hebel, AG, Emmering, 1993, Germany, Recent results on thermal conductivity and hygroscopic moisture content of AAC.

H+H Celcon Ltd, 2003, Celcon House, Ightham, Sevenoaks, Kent, TN 15 9HZ. www.celcon.co.uk/downloads/RIBA%20CPD.pdf

Isu, N., and Mitsuda, T., 1992, Influence of quartz particle size on the chemical and mechanical properties of autoclaved lightweight concrete, Advances in Autoclaved Aerated Concrete, Wittmann (ed.) © Balkema, Rotterdam. ISBN 90 5410 086 9.

Jacobs, F., and Mayer, G., 1992, Institute for Building Materials, ETH Zurich, Switzerland, Porosity and Permeability of autoclaved aerated concrete, Advances in Autoclaved Aerated Concrete, Wittmann (ed.) © Balkema, Rotterdam. ISBN 90 5410 086 9.

Laurent, J., and Frendo-Rosso, C., 1992, Laboratoire d'etude des Transferts en Hydrologie et Environment (IMG), Grenoble, France, Application of image analysis to the estimation of AAC thermal conductivity, Advances in Autoclaved Aerated Concrete, Wittmann (ed.) © Balkema, Rotterdam. ISBN 90 5410 086 9.

Lippe, K., 1992, YTONG AG, R + D Centre, Schrobenhausen, Germany, The effect of moisture on the thermal conductivity of AAC, Advances in Autoclaved Aerated Concrete, Wittmann (ed.) © Balkema, Rotterdam. ISBN 90 5410 086 9.

Liu, C., and Wang, J., 1992, Suzhou Concrete and Cement Products Research Institute, China, An experimental study on thermal transmission properties of aerated concrete composite panels, Advances in Autoclaved Aerated Concrete, Wittmann (ed.) © Balkema, Rotterdam. ISBN 90 5410 086 9.

Millard, W., 1992, Marley Building Materials Limited, The thermal performance of European autoclaved aerated concrete, Advances in Autoclaved Aerated Concrete, Wittmann (ed.) © Balkema, Rotterdam. ISBN 90 5410 086 9.

Mitsuda, T., and Kiribayashi, T., 1992, Influence of hydrothermal processing on the properties of autoclaved aerated concrete, Advances in Autoclaved Aerated Concrete, Wittmann (ed.) © Balkema, Rotterdam. ISBN 90 5410 086 9.

Pospisil, F., 1992, Unit weight reduction of fly ash aerated concrete, Advances in Autoclaved Aerated Concrete, Wittmann (ed.) © Balkema, Rotterdam. ISBN 90 5410 086 9.

Schlegel, E., and Volec, J., 1992, Bergakademie, Freiberg, Germany, Application of autoclaved aerated concrete for high temperature insulation, Advances in Autoclaved Aerated Concrete, Wittmann (ed.) © Balkema, Rotterdam. ISBN 90 5410 086 9.

Schober, G., Hebel, AG, Emmering, 1992, Germany, Effect of size distribution of air pores in AAC on compressive strength, Advances in Autoclaved Aerated Concrete, Wittmann (ed.) © Balkema, Rotterdam. ISBN 90 5410 086 9.

Tada, S., 1992, Texte, Inc. and Nihon University, Japan. Pore structure and moisture characteristics of porous inorganic binding materials, Advances in Autoclaved Aerated Concrete, Wittmann (ed.) © Balkema, Rotterdam. ISBN 90 5410 086 9.

Thermalite, 2003, Marley Building Materials Ltd, Station Road, Coleshill, Birmingham B46 1HP. www.thermalite.co.uk.

Wittman, F.H., 1993a, Autoclaved Aerated Concrete: Properties, Testing and Design, RILEM Recommended Practice, RILEM Technical Committees 78 – MCA and 51 – ALC.

Autoclaved Aerated Concrete – Limbachiya and Roberts (eds)
© 2005 Taylor & Francis Group, London, ISBN 0 415 38356 0

Strains and constraints in aircrete structures

T. Cornelius
Cornel Consult Aps, Denmark

G. Rasmussen
H+H Celcon A/S, Denmark

ABSTRACT: Worldwide, technical improvements are accelerating and this leads to new advanced methods of structure design. Often various combinations of different materials are used to optimize structures. The interaction between different materials with various stiffness and deformation properties results in unforeseen constraints, causing structure damage.

This paper presents some recent tests carried out within the field of fracture mechanics. The tests show a relatively high amount of strain elongation for aircrete compared to similar brittle materials. The results are used to establish the failure strain of aircrete. Furthermore, the results are used to analyze the behaviour of aircrete in structures constructed with different building materials interacting with each other.

Practical examples are presented and evaluated. These are interactions between various aircrete and wooden connections in modern structures.

This paper intends to agitate for the possibility of further adoption of fracture mechanical theories in aircrete structure design. This is to achieve even higher level of performance, safety and quality in modern aircrete structures.

1 DEFORMATION CONTROLLED COMPRESSION TESTS

To determine the failure strain of autoclaved aerated concrete (in the following denoted aircrete) a large test series was carried out at Technical University of Denmark (Larsen 2002). Different measurement equipment was established to determine the actual deformation of the AAC specimen in compression.

To determine the ultimate strain of aircrete in compression detailed knowledge of the full Load P – deflection u curve is needed. The fact that aircrete is a brittle material represents a problem as to determining the descending branch of the curve. A test rig with sufficient stiffness must be used to obtain a sufficiently stable fracture of the test specimen in compression. The principle of deformation controlled test set up is established by applying the main load upon a very stiff steel column, which displaces vertically together with a test rig applying a load to the tested specimen in compression, in such a way that if the stiffness of the tested specimen is reduced, the load on the test specimen will be reduce similarly. The principle is illustrated in Figure 1. Constant increase in displacement of the loaded test specimen is achieved by using different displacement controlling parameters during the test procedure.

It is rather complicated to measure the actual deformation as part of the tested specimen is cracking which may lead to reverse deformations locally in the tested specimen. Therefore several strain measurements were established as described in the following.

In the test procedure different deformation measurements were used. This was mainly done for two reasons. Firstly, during compression the test specimen will displace differently in certain regions. In the descending part some parts of the test specimen will actually rest more than other parts due to micro cracking near the supports, resulting in negative strain measurements in these

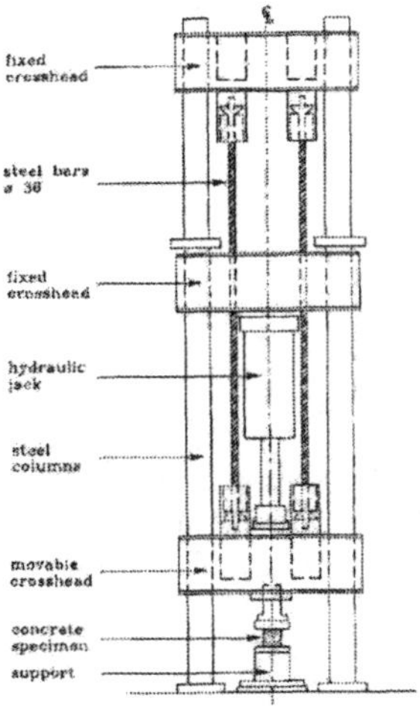

Figure 1. Test setup in principle.

Figure 2. Test specimen mounted with deformation measurement equipment.

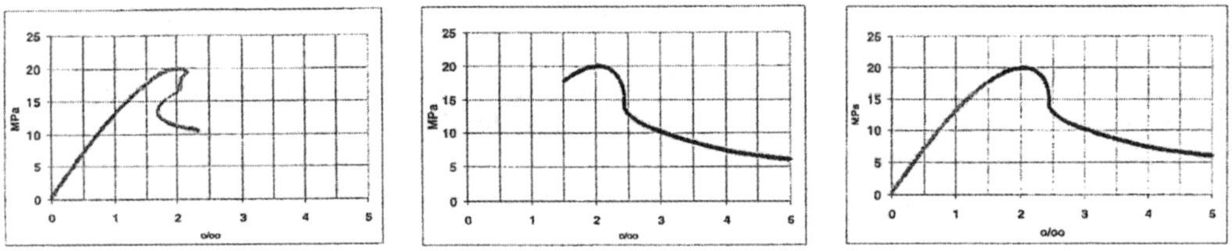

Figure 3. Increasing and descending load deflection curve combined to the full load deflection curve.

regions. Secondly when determining the total deformation of the test specimen it is necessary to eliminate deformation contributions from machinery (machine stiffness) and deformations from the support plate above and below test specimen.

Due to the above, three measurements principles were applied to each test specimen during testing, see Figure 2.

- extensiometer, used to measure deformation in a range of 100 mm over the specimen (one on each side of the specimen)
- displacement measurements between the load plates of the machine (three extensiometers each placed within an internal angle of 120° and with a distance of 200 mm from centreline of testing machine)
- machine measurements, used as reference.

The deformation is controlled by the three extensiometers with a displacement speed of between 0,01–0,15 mm/min. The peak is reached within about 4–5 min. When the load deflection curve increases the displacement is registered using the two extensiometers on the side of the specimens, while displacements between the load plates are used when the load deflection curve descends. The final load deflection curve is then established as a combination of these, see Figure 3.

The specimen tested had a moisture content of between 3,8%–8,4% with an average of 5,5%. The declared compression strength of the specimen is 3,5 MPa, the E-modulus 2000 MPa and the density 650 kg/m^3. The specimen tested is manufactured by H+H Celcon A/S, Denmark.

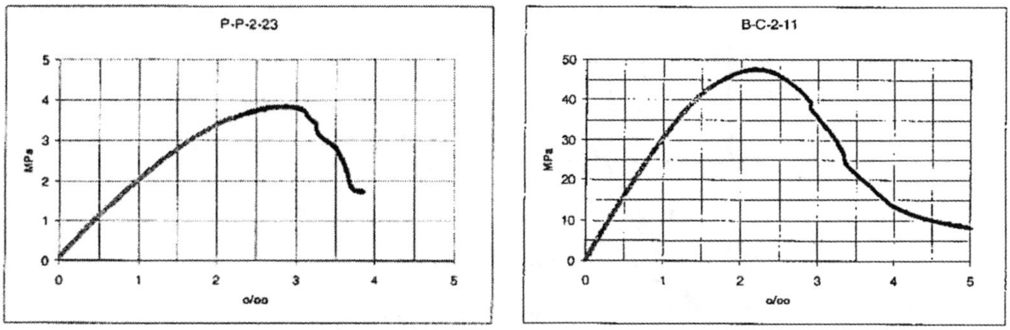

Figure 4. Comparison of load deflection curve for aircrete 3,5 MPa and concrete 25 MPa.

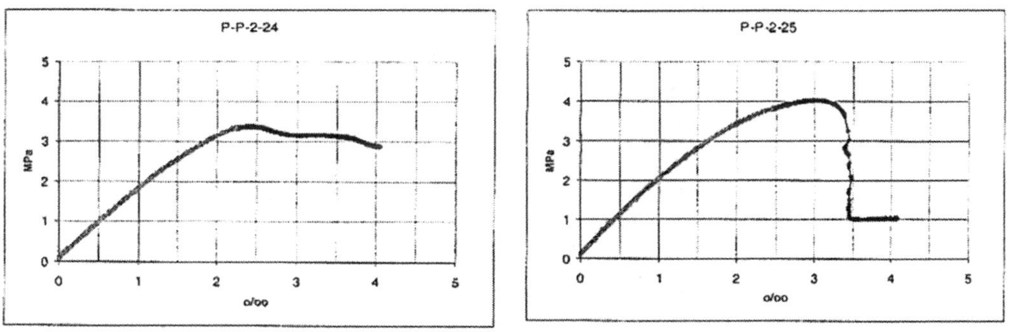

Figure 5. Different load deflection curve for aircrete 3,5 MPa.

The test specimens used were cut from aircrete elements with dimensions w·t·h = 150 · 150 · 300 mm. Furthermore as a reference test, cast concrete cylinders were also, with dimensions: $ø \cdot h = 100 \cdot 200$.

The reference tests using concrete cylinders were carried out with the same deformation control system and deformation measurement principles. Concrete cylinders were made of normal concrete, with average cylinder failure strength of 25 MPa.

In Figure 4 the results are compared with similar tests using an aircrete cube. If we consider the results relative to the ratio of ultimate strength, we observe that the concrete used has a relatively smaller ultimate strain, which may be explained by two facts. Firstly, the relative modulus of elasticity of aircrete is lower than concrete, which leads to a larger peak strain for the aircrete cube tested. Secondly, the peak area of aircrete also contains the plastic crushing effect of aircrete, which increases the strain to some degree.

The effect from crushing was clearer in some tests as indicated in Figure 5. Figure 5 also shows a test illustrating that the test procedure is very unstable and in some cases the descending branch developed very rapidly. Still, the ultimate strain at peak is well determined.

2 DETERMINATION OF COMPRESSION FAILURE STRAIN

It is well-known that the load deflection curves are different for various concrete types. Furthermore, the ultimate failure strain ε_{cu} increases with decreasing ultimate strength as well as shown in Figure 6. See also (Neville 1975).

Nevertheless, this phenomenon depends on the stiffness of the testing machine used to determine the load deflection curve. This means that the actual measured failure strain may vary and in principle be infinite if the stiffness of the machine is infinite.

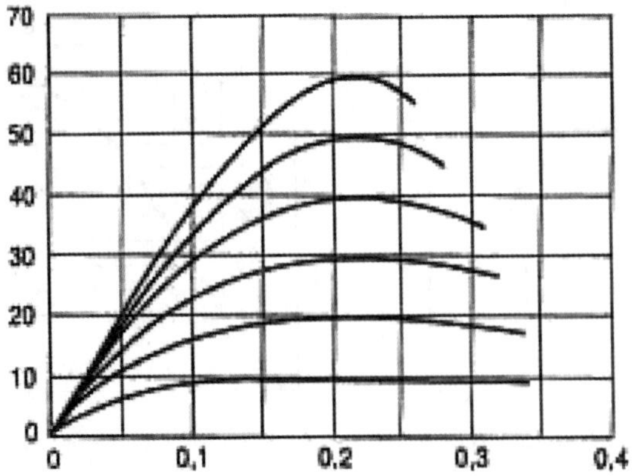

Figure 6. Different load deflection curves of concrete.

Table 1. Ultimate failure strains for aircrete specimen.

Test specimen	Ultimate bending failure strain ε_{cu}[‰]
P-1-12	3.60
P-2-13	3.60
P 2-14	3.45
P-2-15	3.80
P-2-16	3.60
P-2-17	3.70
P-2-21	3.50
P-2-22	3.00
P-2-23	3.20
P-2-24	3.50
P-2-25	3.40

However, the value of the ultimate failure strain may be determined for certain areas of practical use.

The simple case of beams (or slabs) in bending is normally designed limiting the strains to a balanced cross section and assuming yielding in the structural reinforcement, when failure is obtained in the compression zone. This may be established by fulfilling equation 1.

$$\Phi \leq 0{,}8\frac{\varepsilon_{cu}}{\varepsilon_{cu}+\varepsilon_{sy}} \tag{1}$$

where

Φ is the reinforcement degree.

ε_{cu} is the ultimate failure strain.

ε_{sy} is the strain in the structural reinforcement.

If we calculate the bending capacity of a beam using normal plasticity theory, see for instance (CEN 2000), using the measured load deflection curves for the tested aircrete specimen, we may find the optimal bending capacity using the above limitation and thus determine the ultimate failure strain ε_{cu}.

For the test series performed at Technical University of Denmark, (Larsen 2002), the calculated ultimate failure strains are given in Table 1.

The mean value observed for the tested aircrete specimen is 3.50‰. Compared with similar concrete specimen with an ultimate strength of 25 MPa, an average value of the ultimate failure strain in bending of 2,70‰ was observed.

In the same study lightweight aggregate concrete (LAC) was also tested. In this study, average values of the ultimate failure strains between 2.35 and 3.15‰ were observed, depending on density class varying between 1400–2000 kg/m^3.

If we compare the results with limits given in the code for aircrete, we observe a limit $\varepsilon_{cu} = 3,0‰$, (CEN 2000), which seems reasonable based on the above results.

3 CONSTRAINTS OF AIRCRETE STRUCTURES IN PRACTICE

Form stable structures are achieved by controlling the movement between different materials whilst respecting their individual properties. Components in aircrete structures are often connected with organic materials like wood. In particular focus must be centered on the organic materials because movements due to moisture may, for example be over 300% higher in wood than in brittle materials. Furthermore interactions between the building materials may lead to unintended huge increases in tensile stresses, which may have serious consequences.

All building sites risk water damage. Therefore, it is very important to design joints so that every single construction can swell or shrink without damaging other parts of the construction. By correct planning of tolerances, this stress can be avoided without extra cost. It is essential that builders are correctly informed. Therefore, simple structure design methods should be developed and implemented to avoid interaction between materials, thus eliminating unintended stresses. This is even more important when dealing with structures in the building process, as wood is often delivered dry at the building site and can expand substantially if it gets wet.

This problem can be exemplified by the old method used to break rocks when building new road systems. Small dry wooden wedges were placed in drilled holes. When moistened they expanded, extreme stresses developed, and as a result the rocks cracked.

When developing methods to design structures avoiding unintended stresses, it is necessary to establish a form stable structure. It is essential that the force transferring connections are not subjected to unintentional stresses that will result in to small rest capacity. It is necessary to have an overview of the movements in the structure.

4 FRACTURE MECHANICAL EVALUATION OF INTERACTIONS

Below is an example of horizontal interactions caused by moisture reaction in a wooden structure. A wooden top stringer with the dimensions 25 × 100 mm is mounted on the top of the 100 mm inner part of the cavity wall. The same wood type is mounted on top of the connecting partition wall.

In the example a conservative change in strain of 3% is assumed (in the transverse direction) for the wooden top stringer on the cavity wall. The stringer on the partition wall is assumed to have infinite stiffness (this is reasonable given that the modulus of elasticity is considerably higher in the longitudinal direction).

DATA: Characteristic values

AAC = Isotropic

Compression strength = 4,0 MPa

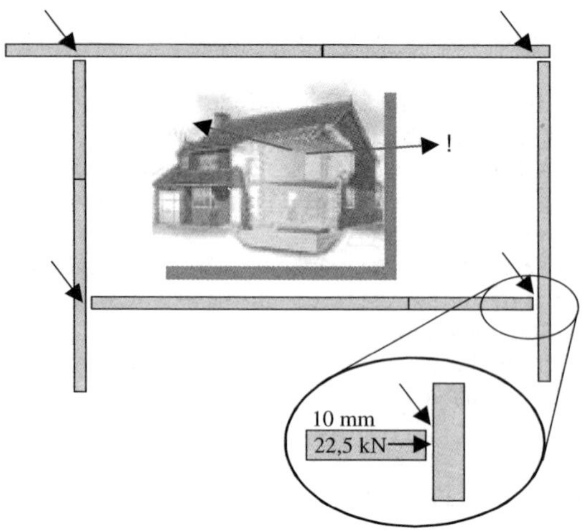

Figure 7. Graphic showing critical connections of the organic parts on the load carrying and the stabilizing walls.

Tensile strength = 1/10 Compression strength = 0,4 MPa

Wood = Anisotropic

E_{90} = 300 MPa at 5% fractile. (Transverse direction).

Wood construction expansion pressure < Tensile strength capacity of the wall.

First we calculate the force in the wooden stringer:

$$\sigma = E \cdot \varepsilon$$

$$\frac{P}{A} = E \cdot \frac{\Delta l}{l}$$

$$P = E \cdot \frac{\Delta l}{l} \cdot A$$

$$P = 300 \cdot \frac{3}{100} \cdot 25 \cdot 100 \cdot 10^{-3} = 22{,}5kN$$

If we now assume that the connection is stabilized by the upper part (250 mm) of the wall joint, avoiding the zipping effect, we may easily calculate the resistance capacity of aircrete:

$$Capacity = t \cdot h \cdot \sigma = 250 \cdot 100 \cdot 0{,}4 \cdot 10^{-3} = 10\,kN$$

Whether the above assumption of 250 mm is reasonable should of course be based upon a fracture mechanical analysis, with reference to an understanding of fracture energy G_F of aircrete (see for instance RILEM 1993) and the geometrical conditions in the actual case given by the stress intensity factor K_I (for further information, see Brook 1986).

An evaluation of the results shows that a wood expansion of 3 mm gives a static pressure of 22,5 kN, and the tensile capacity of the wall is only 10 kN. This construction must be redesigned adding an air gap of at least 10 mm, which will restore the wall to its full capacity.

This is an efficient method to avoid unintended stresses between the building parts, and it is a cost-free building site solution that only requires the information from the engineer to the builder. This is a simple method that will increase building performance, safety and quality.

5 CONCLUSION

This paper presents some recent tests carried out within the field of fracture mechanics. The tests show a relatively high amount of strain elongation for aircrete compared to similar brittle materials.

This paper intends to agitate for the possibility of further adoption of the fracture mechanical theories in aircrete structure design.

In buildings this gives a good stabilizing capacity and usually higher strength than necessary when the surrounding building parts are built correctly (initial stability).

Examples from traditional constructions where different building materials with different characteristics are in tight contact often show dangerously high local contact pressure. These tolerances must be planned correctly to avoid default in the brittle wall construction. High expansion occurs particularly in crossing wooden constructions.

REFERENCES

Broek, D. 1986. *Elementary engineering fracture mechanics*, Martinus Nijhoff, Dordrecht, Netherland.

CEN TC177, 2000. *prEN12602, Prefabricated components of autoclaved aerated concrete*, CEN Bruxelles.

Larsen, H. 2002. *Bestemmelse af arbejdslinier for letklinkerbeton og porebeton*. Technical University of Denmark. Lyngby, Denmark.

Neville, A.M. 1975. *Properties of concrete*. Pitmann London.

Rilem 78-MCA, edited by S. Aroni et al. 1993. *Autoclaved aerated concrete, properties, testing and design*. Chapmann and Hall, London.

Theme Five: Developments in construction techniques

Autoclaved Aerated Concrete – Limbachiya and Roberts (eds)
© 2005 Taylor & Francis Group, London, ISBN 0 415 38356 0

AAC tongue and grooved masonry: performance evaluation

W.A. Ferguson
Building Research Establishment, Watford, Hertfordshire, UK

C.A. Fudge
Autoclaved Aerated Concrete Products Association Limited, Sevenoaks, Kent, UK

ABSTRACT: For the construction of housing and small commercial buildings, solid AAC blockwork with an outer finish that protects against penetration of rain, wind, pests and grime offers an attractive alternative to the complexities of cavity walls. The productivity advantages of thin layer mortar technology and mortar-less interlocking vertical (perpend) joints, when combined with the simplicity of solid wall construction, offers a rapid, economical building system which is less demanding in terms of masons' skills but which can still meet all the requirements of the Building Regulations.

This paper summarises work by the UK Building Research Establishment Limited and some other laboratories to measure the performance of this type of walling in relation to the requirements of the Building Regulations and the background Codes and Standards. Best practice in construction is investigated and construction details, such as those around openings in the masonry, are examined.

1 INTRODUCTION

For the construction of housing and small commercial buildings, solid AAC masonry with an outer finish that protects against penetration of rain, wind, pests and grime offers an attractive alternative to the complexities of cavity walls. The productivity advantages of thin layer mortar technology and mortar-less interlocking vertical (perpend) joints, when combined with the simplicity of solid wall construction, offers a rapid, economical building system which is less demanding in terms of masons' skills but which can still meet all the requirements of building regulations.

This paper sets out the background research database on the key aspects of performance and gives an introductory guide on constructional best practice using this wall system.

2 PERFORMANCE

2.1 *General*

This section summarises work by BRE and some other laboratories to measure the performance of this type of walling in relation to the requirements of the Building Regulations and the background Codes and Standards.

2.2 *Compressive strength of units and mortar*

The mean compressive strength of the blocks used for various tests was determined following the procedure set out in BS 6073: Part 1: 1981 (British Standards Institution, 1981). The value, calculated from ten individual test results, was 3.7 N/mm^2. This comfortably exceeds the minimum requirement contained in the Building Regulations Approved Document AD A (ODPM, 2004) and BS 8103-2 (British Standards Institution, 2005) of 2.8 N/mm^2. The conventional mortar used in the tests was made using a 1: 1: 6 cement: lime: sand mix, referred to here as a general-purpose mortar (GPM). This mortar was adjusted to roughly the same consistency for each mix – around 9 mm – by using the BS 4551: Part 1 (British Standards Institution, 1998) drop-ball test and had a mean strength of 6.03 N/mm^2, when tested in accordance with BS 4551: Part 1 (British Standards Institution, 1998). The thin layer mortar had a mean strength in excess of 5 N/mm^2, when tested, after being air cured.

2.3 *Compressive strength of masonry*

Two sets of test specimens were assessed: one set constructed using the GPM, and the other using thin layer mortar. The thin layer mortar was a proprietary type recommended for use with this product and supplied with the blocks by the manufacturer.

The construction and testing of 255mm thick small walls, termed 'wallettes', followed the recommendations of BS EN 1052-1 (British Standards Institution, 1999), with three replicates being tested for each mortar type. The results of the tests confirmed that the masonry exceeded the design values given in BS 5628-1: 1992 (British Standards Institution, 2002) for both general-purpose mortar and thin layer mortar. This is as expected, since the vertical joints provide little or no effect to the compressive strength of the masonry. This fact is also recognised in the European code of practice for the design of masonry (Eurocode 6), where no differentiation is made between the masonry if the vertical joints are unfilled.

A 255 mm thick solid wall is a very robust wall form and the overall compressive strength of such walling will be comfortably in excess of that required for construction of domestic dwellings of up to two storeys in the Building Regulations AD A1/2 (ODPM, 2004).

2.4 *Lateral strength performance*

The lateral, or flexural, strength of the blockwork in the two orthogonal orientations, perpendicular and parallel to the bed joints, was measured in accordance with the method set out in Appendix A.3 of BS 5628-1: 1992 (British Standards Institution, 2002).

For this test 'wallettes' are constructed and then tested in flexure after 28 days of being stored in a specified temperature/moisture condition. The results of the tests are contained in Table 1.

In Table 1 a comparison is made with the BS 5628-1 design values. However, the British Standard does assume that the vertical joints are filled. The results are not as unexpected since in the parallel direction the effect of vertical joints is negligible. It should also be noted that the BS does not currently recognise thin layer mortar and therefore significantly undervalues this in the design. Future amendments to the BS will have thin layer mortar designs added. For strengths in the perpendicular direction, it could be anticipated the results of unfilled vertical joints would be approximately 50% of those with filled joints, with alternate joints filled; this was found to be roughly true. It this case, it was not anticipated that the tongue and groove would have any mechanical bonding. The values obtained are also of the order which would be predicted when using

Table 1. Results of flexural strength tests on masonry.

Mortar type	Orientation with respect to the bed joints	Mean flexural stress (N/mm^2)	C of V$^{(1)}$ (%)	Characteristic flexural strength: As measured (N/mm^2)	Characteristic flexural strength: BS 5628 value (N/mm^2)
GPM	Perpendicular	0.26	21.8	0.17	0.25
GPM	Parallel	0.17	8.7	0.13	0.15
TLM	Perpendicular	0.19	51.6	0.11	0.25
TLM	Parallel	0.35	10.4	0.23	0.15

[1] *coefficient of variation.*

Eurocode 6, although in the parallel direction, the Eurocode would underestimate the strength of AAC masonry.

In practice, however, these measured strengths would indicate that in a 9 m long by 2.5 m high wall returned at the corners and simply supported at the top and base and required to withstand a wind resistance of 0.89 kN/m^2 would have a factor of safety of over four. This would easily exceed the requirements for the flank wall of a large dwelling with no internal supporting partitions in most inhabited areas of the UK.

2.5 *Shear strength performance*

Specific tests on shear strength of AAC masonry were not carried, since Eurocode 6 guidance exists on the subject. The basic shear strength for in plane design uses a factor of 0.50 where the vertical joints are unfilled. It was thought unlikely that the addition of a tongue and groove would have any significant effect. For shear design, the Eurocode allows for the beneficial effect of additional vertical permanent load on the structure in both cases of filled and unfilled joints.

2.6 *Acoustic performance*

To assess the acoustic performance of the tongue and grooved blocks compared with traditional blocks, comparative testing was carried out on two party separating walls in a laboratory test chamber. One wall was made from the tongue and groove blocks, with their dry perpend joints, while the other used blocks made from exactly the same material but made into traditionally shaped blocks. The wall built using the latter blocks had fully filled perpends. The results of the tests indicated that the two walls provided a comparable acoustic performance.

The use of a wet finish such as plaster on at least one face would, however, be advisable, to prevent passage of fumes between adjacent dwellings. The test walls, though, were built and tested without such a finish.

2.7 *Thermal performance*

Calculations have been made on AAC blocks of size 440 × 215 mm by 255 mm thick. These calculations have been based on units, which are rendered on the outside and plastered on the inside. In these calculations, the thermal conductivity of the aac blocks has been taken as 0.11 W/mK, with a block density of 440 kg/m^3.

The results in the first line of Table 2 represent the U-values for the AAC material as a monolithic slab, without either the tongue and groove or the mortar joints. The second line includes the tongue and groove in the vertical joints between the blocks and the third line takes account of the horizontal thin layer mortar joints. The results are essentially the same, i.e. the tongue and groove has a negligible effect on the thermal performance of the wall and the presence of the mortar causes only a slight reduction.

This assessment assumes that the blocks mate together properly so that they are touching, and that there are render and/or plaster facings to the blocks so that air leakage through the joints is prevented.

Table 2. Effect of joints on U value.

Description	U-value (W/m^{2}K)
AAC assumed to be a monolithic slab (i.e. no joints)	0.401
AAC masonry with tongue and grooved joints	0.401
AAC masonry T&G with (nominally) 3 mm mortar bed joint	0.420

Table 3. Leakage through wall.

Pulse time (minutes)	Pulse differential pressure (Mpa)	Leakage through wall in time pulse (litres)
20	0	0
10	1–150	0
10	1–300	0
10	1–450	0
10	0–600	0

2.8 *Resistance to rain penetration*

Testing was carried out on a wall made with general purpose mortar, incorporating dry tongue and grooved perpend joints. A cement render was made using a 1: 1: 6 cement: lime: sand mix, applied in two coats, was used as the external finish to the wall and testing was carried out in accordance with BS EN 12865: 2001 'Hygrothermal performance of building components and building elements – Determination of the resistance of external wall systems to driving rain under pulsating air pressure' (British Standards Institution, 2001).

The masonry was subjected to a specified sequence of air pressure pulses while being sprayed with water at a rate of approximately 2.0 litres/min/m^2 and the results of the test are shown in Table 3.

Clearly the construction performed extremely well with no leakage occurring. This is perhaps not unexpected, since it is known that AAC as a material has good resistance to the penetration of rain, even in driving rain. The overlap of the tongue and groove also must have a beneficial effect in preventing rain ingress.

The test was repeated using thin-layer mortar and a proprietary type render, but with a window openings contained in the wall. Again under similar test conditions no leakage of water was found in the masonry. The test was used to demonstrate two different methods of installing window frames into solid masonry. In one case the masonry was returned around the face of the front of the external frame and butted against it. This is known as a 'checked reveal' in the UK. There was a mastic seal applied between the frame and masonry. In the second case, the window frame was built in flush against the masonry and an external strip applied at the junction. Both systems of fixing faired equally well and no rain penetration was observed. One comment noted by the testers was that the window seal should be positioned properly to obtain maximum benefit and to prevent rain ingress at the base of the window.

3 CONSTRUCTION BEST PRACTICE

3.1 *Design issues*

Table 4 describes some of the design issues and the solutions to overcome these.

It is recommended that damp proof courses should be installed in accordance with BS 8215:1991 (British Standards Institution, 1991), Code of practice for design and installation of damp proof courses in masonry construction. The weathering (rain resistance) performance requirements are set out in Table 12 of BS 5628-3: 2002 (British Standards Institution, 2002) and in BS DD ENV 1996: Part 2 (British Standards Institution, 2001).

3.2 *Openings*

For detailing large window and door openings should be designed, where possible, to be in modules of whole or half units in width and whole units in height – with boundaries falling at perpend or bed joint lines. However, handed half units will be required to maintain the stretcher bond pattern.

Table 4. Design issues and solutions.

Design issues for solid T&G walling using thin bed mortar	Design solution
A dry vertical joint does is not a reliable barrier against transmission of air, water, fire, insects and fungi	The finishes or auxiliary claddings must provide a seal for the joint
Many of the time-saving and efficiency advantages of the system are lost if significant numbers of units need cutting	Use a modular layout where the boundaries of openings are formed at the end face of units
Course height mismatches at junctions to existing walls formed using thin layer mortar construction for the new wall	Block bonding to existing walls will not be possible and wall-plate connections will be necessary
Design of mortar joint fixings where thin layer mortar construction is used	Any fixings into joints must have a profile that can be accommodated comfortably in the 3 mm thickness Alternatively, a surface fixing system can be used
Movement and cracking control for solid external walls	Solid walls need to be designed to minimise cracking which prejudices structural performance and weather-tightness

Small adjustments of the size can readily be achieved by cutting in-situ. It may be more efficient, in terms of time, to build plain walls and to cut the smaller openings out of them.

For support over lintels, these can be bought 'off the shelf' in standard shapes and dimensions or ordered to bespoke specifications from various manufacturers and outlets. However, it is important that the width of the lintel is at least 95% of the width of the blocks, to ensure adequate bearing. Where part of the lintel is embedded in a joint, thin layer masonry may limit the choice, but conventional mortar could be used in this isolated area. Some guidance on lintels for use with thin layer systems is provided in BRE Good Building Guide (GBG) 58 (BRE, 2003).

3.3 *Using a proprietary wall starter system to connect to an existing wall*

Using a proprietary wall starter system is a less labour intensive and more flexible way to connect to an existing wall than block bonding and gets around any problems of unequal course heights. A specially designed metal plate is fixed on to the existing wall and height-settable tie bars link it to the bed-joints in the new wall. This type of system is suitable for use with either GPM or thin layer mortar, provided low-profile tie bars are used in the thin layer mortar walls.

It may be necessary to use two wall starter systems to provide a suitably strong connection for thick solid walls. This would be equivalent, in terms of the shear strength, to the two connections that would be used for starting a new cavity wall.

Whenever the new wall is exposed externally, the fixing system should be manufactured from austenitic stainless steel to ensure durability against corrosion. Whatever mortar is used, the manufacturer of the wall starter systems' guidance covering such items as closing/sealing the area around the wall starter system should be followed.

3.4 *Metal fixings for thin layer walls*

The standard recommendation in BS EN 845-1:2003 (British Standards Institution, 2003) is that the profile height of metal fixings embedded in mortar joints should not exceed 50% of the joint thickness, i.e. 1.5 mm in a 3 m bed joint. Figure 1 of BS EN 845-1 (British Standards Institution, 2001) illustrates this requirement. Further guidance on suitable fixings, wall ties, straps, joist hangers, lintels etc. is contained in GBG 58 (BRE, 2003).

3.5 *Control of movement and cracking*

In common with all cement-based products, AAC walls are subjected to overall shrinkage movement as the walls dry out. BS 5628-3: 2002 (British Standards Institution, 2002) and BRE Digest 468 Parts 1 and 2 (BRE, 2003) recognise that drying out occurs in all types of concrete masonry and recommends that such movement should be considered at the design stage.

For solid walls, movement should generally be controlled using bed joint reinforcement, which should be placed in every second bed joint in each elevation. Some guidance on this is given in GBG 58 (BRE, 2003). The use of movement joints to control cracking in longer walls is also a feasible option where the perimeter wall abuts an internal masonry partition, which provides vertical edge support at the joint. In these circumstances, ties will be needed to ensure the structural stability of the wall.

The movements joints used should employ a water bar and a suitable sealant system to maintain thermal performance and weather resistance.

4 CONCLUSIONS

The satisfactory use of tongue and groove blocks can be easily achieved both in terms of design and construction. With the appropriate use of finishes, the masonry can be made airtight and weather-tight, as well as providing the builder with a cost effective any simple to use construction. Additional benefits are achieved where the sizes of the masonry units are considered in the design and setting out of the walls to minimise cutting on site and to achieve a fast build sequence.

ACKNOWLEDGEMENTS

This paper was produced as part of a project that was part-funded by the UK Government under the 'Partners In Innovation' programme and is based on BRE Information Paper IP7/05.

BIOGRAPHIES

EN 771-4: Specification for Masonry units: Part 4: Autoclaved Aerated Concrete Units.
EN 772-1:2000: Methods of test for masonry units – Part 1: Determination of compressive strength.
EN 845-2:2001 Specification for Ancillary components for masonry, Part 2 Lintels.
EN 845-3:2001 Specification for Ancillary components for masonry, Part 3 Bed joint reinforcement.
EN 998-2: 2002 British (European) Specification for Mortar for Masonry: Part 2: Masonry mortar.
EN 1052-2: Methods of test for masonry – Part 2: Determination of flexural strength.
EN 1745:2002. Masonry and masonry products: Methods for determining design thermal values.
EN 1996 Parts 1-1, 1-2, 2 and 3.
BRE Digest 337: 1988 Sound Insulation: Basic Principles. BRE.
Code of Best Practice for the Use of Aircrete Products, www.aircrete.co.uk

REFERENCES

BRE, 2003, Good Building Guide 58. Thin layer mortar masonry. Building Research Establishment, Watford.

BRE, 2003, Digest 468. AAC 'aircrete' blocks and masonry. Building Research Establishment, Watford.

BRE, 1994, Good Building Guide 18. Choosing external render. Building Research Establishment, Watford.

BRE, 2005, Information Paper 7/05, *Aircrete tongue-and-groove block masonry Performance Evaluation and guidance*, Watford.

British Standards Institution, 1981, BS 6073: Part 1: 1981. Precast concrete masonry units – Part 1: Specification for precast concrete masonry units. BSI, London.

British Standards Institution, 1991, BS 8215: 1991. Code of practice for design and installation of damp proof courses in masonry construction. BSI, London.

British Standards Institution, 1991, BS 5262: 1991. Code of practice for external renderings. BSI, London.
British Standards Institution, 2002, BS 5628-1: 1992. Code of Practice for Use of Masonry – Part 1: Structural use of unreinforced masonry. BSI, London.
British Standards Institution, 1998, BS 4551: 1998. Methods of testing mortars, screeds and plasters-. Part 1. Physical testing. BSI, London.
British Standards Institution, BS EN 1052-1: 1999. Methods of test for masonry - Part 1: Determination of compressive strength. BSI, London. 1999.
British Standards Institution, 2001, BS 8000-3: 2001. Workmanship on building sites - Part 3: Code of practice for masonry, BSI, London
British Standards Institution, 2001, BS EN 845-1:2001. Specification for Ancillary components for masonry, Part 1 Ties, straps, hangers and brackets. BSI.
British Standards Institution, 2001, BS EN 12865: 2001. 'Hygrothermal performance of building components and building elements – Determination of the resistance of external wall systems to driving rain under pulsating air pressure', BSI, London.
British Standards Institution, BS 5628-3: 2001. Code of Practice for Masonry – Part 3: Materials, and components, design and workmanship. BSI, London.
British Standards Institution, 2001, BS DD ENV 1996-2:2001 Eurocode 6: Design of masonry structures: Part 2: Design, selection of materials and execution of masonry (together with UK National Application Document).
British Standards Institution, 2005, BS 8103: 2005 Structural design of low-rise buildings. Part 2 Code of practice for masonry walls for housing. BSI, London.
Eurocode 6: Design of masonry structures – Part 1-1: Common rules for reinforced and unreinforced masonry structures, published by National Standards Bodies of CEN members.
ODPM. 2004, Approved Document AD A, Office of the Deputy Prime Minister, London.

Autoclaved Aerated Concrete – Limbachiya and Roberts (eds)
© 2005 Taylor & Francis Group, London, ISBN 0 415 38356 0

Inside America

Angelo Coduto
AACPS, Ltd. Florida, USA

ABSTRACT: The world has known about the benefits of AAC as a fire separation wall, but it took some of the biggest tragedies of our time to realize the amazing life saving properties that AAC can provide.

We were hit by surprise but reacted quickly with the strategic use of AAC as an inside product replacing conventional gypsum board fire wall products for stairwells, elevator shaft walls, corridor walls and demising walls. The American design community is beginning to take advantage of this durable, functional, maintenance free and sustainable building product. The masonry industry has begun to see that AAC is the key to their re-entry in the inside walling construction market and they are endorsing its use in the major cities around the country.

The commercial market, multi family housing, student housing, nursing homes and assisted living facilities, hotels and motels have taken to AAC interior walls for all the right reasons including cost competitiveness, which is the driving force of the construction industry worldwide.

It is only fitting that this international forum will take place around the anniversary of the 9/11 Tragedy. My paper will provide some specific insight into the continuing development of the Greatest Fire Wall Product of our time and its growing market share INSIDE AMERICA.

1 BACKGROUND INFORMATION

In order to evaluate the impact to market development for AAC products one must first understand the mind set and industry mentality of why different products, many of them inferior to AAC, are used for FIRE SAFETY and FIRE SEPARATION.

The construction products industry in the United States has been slow to change in accepting new concepts or products for many reasons. The "If it ain't broke don't fix it" mentality has been the status quo for the majority of my thirty five years in the industry.

The industry, that I love, has managed to tweak the basic building products but concrete, steel and wood have been the basic products of choice for commercial and residential construction.

While there are many progressive companies that have created variations on the theme of the basics, very few have ventured into the development of AAC as a viable alternative product. The veil of proprietary design, the absurd idea that a light weight product can be structural, the initial introduction of AAC into the USA, the concept of turning waste products into structural elements seemed to go against the grain of our industry.

The introduction of AAC to the United States provided a very fragmented approach. The experienced European manufacturers made the ultimate business decision to bring their manufacturing technology to the largest potential untapped market in the world. The high cost of building an AAC plant complicated the issue for those companies that have taken the next step and done their due diligence regarding building a plant and entering the United States AAC Market. These brave companies have pioneered the manufacture and sale of AAC products in the United States, fighting the barbs and arrows of the conventional products and their giant lobbying presence in the construction industry. These conventional product suppliers have done their best to question the ability of AAC products as a viable solution to the construction industry. When the AAC industry was able to illustrate our products ability to perform in a similar fashion or better than conventional building products, these suppliers created a myriad of false hoods regarding our products. They decided to scare the construction market with issues from instability and dangerous off gassing, from the use of aluminum powder, to mold and mildew propagation. These serious allegations forced our industry to respond with facts that would counter these issues but at a cost to achieving building code acceptance. This struggle for recognition was carried into the construction code and specification arena where structural ability and sustainability are critical measures of acceptance. Our industry, through the manufacturing and consulting community, was able to perform successfully as a structural and non structural product solution under the highest forms of scrutiny. The AAC industry has now achieved the building code approvals and standard recognition required moving forward into the construction industry with the understanding of the products capabilities.

The conventional construction products suppliers have had the luxury of a strong market and product sales prices have risen exponentially in the last several years, the industry has sat back and let other products take market share. This has happened in structural walling construction for medium and low rise as well as non structural applications for high rise construction. The conventional concrete products industry has given the interior fire separation walling market to the gypsum board and metal studs industry because it was too hard to compete with this apparent low cost method of fire safety separation. This gypsum board side of the industry created a very inexpensive way to address the fire resistance issues and the reality is that the industry is getting what it pays for. The gypsum lobby is very strong and their ability to manipulate the fire standards has moved their products along in fire resistance and separation.

The building design industry has a strategic and specific need to insure fire safety in construction of all types of building projects where the escape from fire and the containment of the source are paramount issues. The commercial construction market multifamily housing, hospitality hotel-motel, student housing dormitories, seniors housing, nursing homes and assisted living facilities are taking to AAC interior walls for all the right reasons including speed of construction, durability, moisture resistance and cost competitiveness, many times the most important reason to the construction industry worldwide.

While Americans are somewhat cynical regarding changes to the construction industry that we know and understand, we are also pragmatic when attempting to resolve important issues and building concerns like fire safety in building construction. The need for fire safety and containment in public and private buildings has manifest itself in the worlds political and social climate and spills over into the way we protect ourselves in constructing safe and secure structures. The American design community has begun to take advantage of AAC's durable, functional, maintenance free and sustainable characteristics for use in solving their building safety and containment requirements. Because of some of the most severe recent dormitory and multi family housing fire losses, the American designers have begun to strategically use AAC as a replacement for conventional gypsum board fire wall products for stairwells, elevator shaft walls, corridor walls and demising walls, in their ongoing attempt to insure safe evacuation of their building structures.

The design community has relied on the fire safety of construction materials through specific measurement of these fire resistive assemblages as specified by American Society for Testing and Methods,(ASTM E 119) and as tested and measured by Underwriters Laboratory, American National Standards Institute, (ANSI/ UL 263). These two standards and inspection bodies have shaped the way construction products are assembled and used for fire resistance and fire safety for all applications including load bearing and non load bearing, fire resistive cladding and mechanical penetrations through these wall assemblies. These assemblies, both gypsum board and AAC masonry products, and their fire resistance ratings are illustrated in Figures 1, 2 & 3 below. These figures will demonstrate the ability of AAC masonry products to provide a superior solution to the fire resistance issues both in load bearing and in non load bearing applications. AAC products utilizing a single width construction process as compared to the metal studs and gypsum board that require up to four layers of gypsum board per side of the wall to meet the same fire resistance requirement. The labor cost and construction effort as well as the multiple inspections required clearly make AAC the better choice for this application.

Design No. U419
Nonbearing Wall Ratings–I, 2, 3 or 4 Hr (See Items 3 & 4)

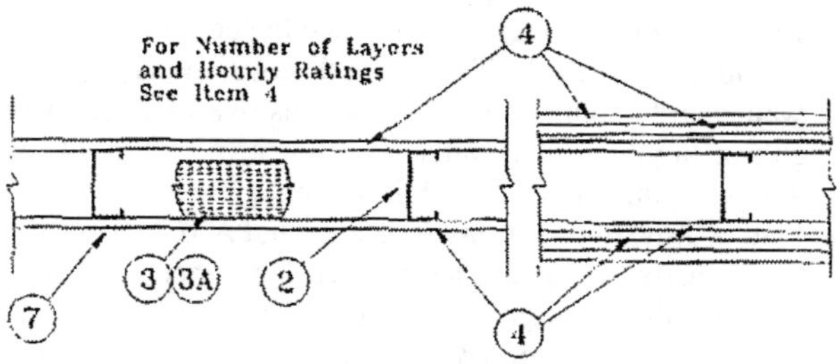

Figure 1. Gypsum Board' – Gypsum panels with beveled, square or tapered edges, applied vertically or horizontally. Vertical joints centered over studs and staggered one stud cavity on opposite sides of studs. Vertical joints in adjacent layers (multilayer systems) staggered one stud cavity Horizontal edge joints and horizontal butt joints on opposite sides of studs need not be staggered. Horizontal edge joints and horizontal butt joints in adjacent layers (multilayer systems) staggered a mm of 12 in. There are a number of thicknesses and number of layers for the I hr. 2 hr. 3 hr and 4 hr ratings as illustrated above. * Underwriters Laboratory Directory

Design No. U919
Bearing Wall Rating – 4 Hr
Nonbearing Wall Rating – 4 Hr

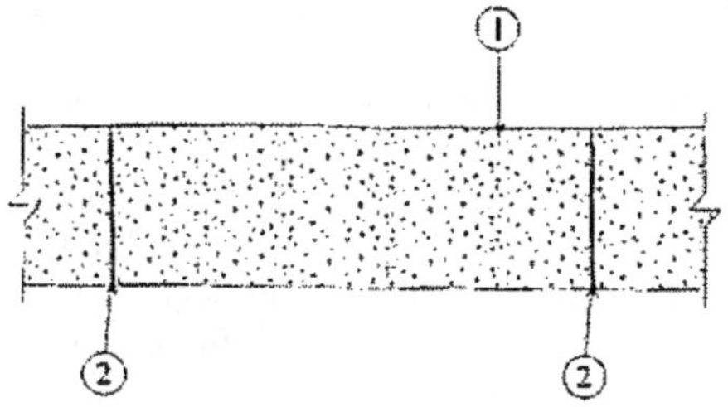

Figure 2. Precast Autoclaved Aerated Concrete Blocks – Min 4 in. thick by 7-7/8 in. high by 23-7/8 in. Blocks for use in nonbearing wall assemblies and 6 in. thick by 7-7/8 in. high by 23-7/8/ in. long blocks for use in bearing wall assemblies. * Underwriters Laboratory Directory

Design No. U917
Bearing Wail Rating – 4 Hr
Nonbearing Wall Rating – 4 Hr

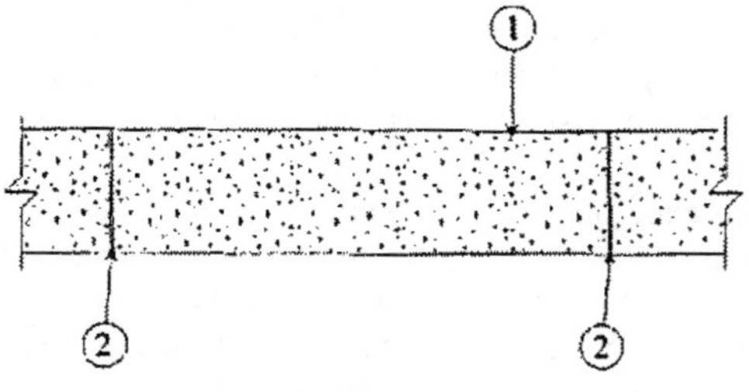

Figure 3. Precast Autoclaved Aerated Concrete Blocks – Mm 6 in. thick By 7-7/8 in. high by 23-7/8 in. long blocks.* Underwriters Laboratory Directory

The use of gypsum wall board, while assembled according to code, is suspect in multiple hour fire resistance ratings and in fact the gypsum wall board disappears during the fire hose stream test. These assemblies have been constructed and tested side by side demonstrating the failure of gypsum wall board after introducing the water hose stream. Our experience in performing these tests illustrate that the AAC masonry wall appears to be unaffected by both the fire and the high pressure of the fire hose stream. This fire resistance performance characteristic as well as the need for dry conditions during their installation has begun to convince the construction design community that there are some better alternatives available. They have discovered that AAC products not only out perform the gypsum wall board products but are not sensitive to weather or most construction conditions, thus taking these highly critical walls out of the critical path of construction.

The illustrations in Figures 1 and figure 2 highlight the simple assembly that AAC provides for load bearing and non load bearing applications. The idea that AAC masonry product walls can be installed during the rough construction phase has positive influence on the cost of this installation to the mason contractor. The light weight and larger product sizes that have developed as the products of choice for this interior fire wall application, allows the mason contractor the ability to increase his production by a considerable percentage, in some cases by over one hundred percent of their budgeted performance. He can schedule his manpower and equipment to maximize their use in performing his scope of work. This ultimately benefits the building contractor and owner providing the construction flexibility of building these walls on their schedule. Thus limiting the risk of damage and rework caused by weather and working conditions and providing insurance against failure in installation or more critically the fire separator walls design performance.

AAC products can be used as load bearing structural walls which clearly separates them from most conventional walling products and specifically gypsum board and metal stud walls. The ability to utilize AAC walls as structural load bearing walls, that require multiple hours of fire resistance, gives the design community the security they need to specify AAC products as the material of choice. The need for life safety in public buildings has created the opportunity for AAC products to be viewed as a solution to an important design concern. The ability to include the sound attenuation characteristic qualities as well as the energy efficient performance of AAC, gives the specifier solutions to many issues with one product application.

These characteristics have separated AAC from the conventional walling product systems for interior wall applications. The hotel and motel industry requires high degrees of sound attenuation in considering demising walls and corridor walls for their projects. AAC can exceed their requirements for sound attenuation with a single application which can be enhanced to even higher standards by insulation and some walling treatments. The student housing, dormitory market has identified sound attenuation and low maintenance as some of the major issues regarding their interior walls beyond the most important fire safety issue. This industry attempts to design and build structures to last for one hundred years. They appreciate the durability of AAC walls with direct applied plaster because of its ability to withstand the years of student progression and turnover as well as the unique behavior of college students in general.

The dormitory designers have begun to specify and draw AAC products into their projects as the first choice of building materials. This significant change from the use of AAC as a possible alternative to the specified construction material of the designer and owner has given our industry a real sense of arriving. The repeated design in the later phase and new student housing projects of these designers confirms that AAC has performed successfully and has met their safety and security needs.

AAC walls by their nature are one width and simply installed creating a clear solution to construction schedule concerns. The design community's increasing development applications and new project specifications has clearly placed AAC products into the mainstream. The design use of AAC as a fire separator wall for elevator shaft walls, stairwell walls, corridor and demising walls has become a part of the mainstream, in interior building design.

AAC has been through the first ten years in the United States as a developing material industry, unlike its European big brother, which has proven to be the material choice for construction in many different applications and many different countries.

AAC continues to gain momentum in developing its presence and because of the superior performance in fire safety and fire separation it will continue to grow in market share. The outlook for market development is very optimistic because the United States has taken AAC INSIDE AMERICA.

Autoclaved Aerated Concrete – Limbachiya and Roberts (eds)
© 2005 Taylor & Francis Group, London, ISBN 0 415 38356 0

Performance of masonry building including AAC products during the 17 August 1999 Marmara Earthquake

Z. Celep
Faculty of Civil Engineering, Istanbul Technical University, Istanbul, Turkey

ABSTRACT: Masonry buildings are widely constructed in Turkey by using local materials. Many of them are non-engineered constructions in the traditional manner without intervention by qualified engineers. In addition to these, numerous buildings are often constructed by using reinforced concrete columns and beams, without proper engineering design. Although these buildings have beams and columns, they can be considered as masonry buildings, because the beams and columns are not properly designed to produce moment resisting frames. In Turkey Autoclaved Aerated Concrete (AAC) blocks are mainly used to produce partition walls within the reinforced concrete frames. Additionally, AAC panels and slab elements are used as prefabricated structural elements and masonry buildings are constructed by using all of these products as well. In the present paper, observations on performance of the buildings made of conventional masonry material and AAC products during the August 1999 Marmara Earthquake are discussed comparatively.

1 INTRODUCTION

In common masonry structures vertical forces due to dead and live loads as well as lateral forces due to earthquake loads are resisted by the structural walls. In masonry structures, besides the structural walls, partition walls are used for the separation of rooms having different functional purposes. Since the contribution of these partition walls to the load bearing capacity of the story is very small in comparison with the structural walls, their capacities are neglected. Structural walls in masonry are very sensitive to tensile stresses; this weakness can be avoided by putting some reinforcement within the masonry units.

In Turkey, masonry buildings are non-engineered constructions which are informally constructed in rural areas as well as in outskirts of the metropolitans in the traditional manner without any or very little intervention by qualified engineers and architects. Such buildings involve field stone, fired brick, concrete blocks, adobe and wood or a combination of these locally available materials. Plain unreinforced masonry is used very widely in rural parts of the country. Although reinforced masonry structures are very common in various parts of the world, they are very rare in Turkey.

In Turkey, buildings are often constructed by using reinforced concrete columns and beams, without proper engineering design, based on the experience of local masonry contractors. Although these buildings have beams and columns, they have to be considered as masonry buildings, because beams and columns are not properly designed to produce moment resisting frames which are of prime importance for resisting carthquake loads. It is difficult to accept these types of buildings as concrete structures; they can be regarded as a type of masonry structures as well (Celep & Kumbasar 2005).

Autoclaved Aerated Concrete (AAC) products are used mainly as AAC blocks within the reinforced concrete frame system as partition walls. AAC panels and slab elements are also used as prefabricated structural elements, such as slabs, exterior or partition walls in the precast structural system as well as cast-in-place structures. In Turkey these elements are used in low-rise masonry residential and industrial buildings. There are also low-rise masonry residential buildings constructed by using only all these AAC blocks as well.

2 EARTHQUAKE BEHAVIOR OF A MASONRY STRUCTURE

Masonry structures can perform well in earthquakes, provided that plain masonry walls are distributed throughout their plan area and acting in both principal plan directions and they have an acceptable level of load-displacement characteristics under cyclic reversed loading. Since they have very little plastic deformation capacities, they should respond primarily elastically and they must be provided with sufficient strength to resist elastic forces. Such masonry buildings are typically low-rise, shear-wall structures. If they are intended to respond inelastically, then the walls should have reinforcement and the walls must be proportioned and detailed to be capable of the reversed cyclic deformations consistent with that inelastic response (Paulay & Priestley 1992, Celep 1998).

3 TURKISH EARTHQUAKE CODE PROVISIONS FOR MASONRY STRUCTURES

Although the new Turkish code for earthquake resistant design essentially focuses on the design of reinforced concrete structures, it contains some general specifications for the plain (unreinforced) masonry structures only (Ministry of Public Works and Settlement, 1998). All masonry buildings have to comply with these minimum requirements given in the code. These requirements deal with the restrictions on the number of stories of the building, dimensions of the door as well as window openings, lengths and thickness of the structural walls, unsupported lengths of the structural walls, dimensions of lintels and tie beams.

The code provides these minimum requirements for the structural design and construction of masonry structures and states that if these minimum provisions are satisfied, an earthquake analysis for masonry building is not required. An analysis can be carried out only for a close examination of the building under earthquake loads. When an earthquake analysis needs to be performed, the code defines the total equivalent earthquake static load (base shear force) as

$$V_t = \frac{W\ A_o\ I\ S}{R_a} \geq 0.10\, W\ A_o\ I \tag{1}$$

where A_o effective ground acceleration coefficient depends on the seismic zone and is in between 0.40 (seismic zone 1) and 0.10 (seismic zone 4), I building importance coefficient (1.0 for buildings of low occupancy), S spectral coefficient (2.5 for masonry buildings), R_a earthquake load reduction coefficient (2.5 for masonry buildings), W total weight of building, which includes G_i gravity loads and Q_i live loads (2.0 kN/m^2 for residential buildings) of each floor. By taking into account n participation factor for live load (0.30 for residential buildings), the total weight of building can be calculated as (Tables 1, 2, and 3)

$$W = \Sigma W_i == \Sigma(G_i + n\, Q_i) \tag{2}$$

The code limits the number of stories of masonry buildings depending on the seismic region. In the first seismic zone where the seismic risk is of highest degree, the maximum number of stories is limited to two, whereas three stories of masonry buildings can be built in the second and third seismic zones. In the fourth seismic zone the maximum number of stories is given as four. However, a masonry building can have additionally a basement floor and a roof floor having maximum 25% of the normal floor area.

The units used in masonry walls can be solid and hollow clay or concrete blocks having a maximum pore ratio of 35% and a minimum compressive strength of 5 MPa obtained by considering the gross cross section of the masonry unit. Natural stones having a minimum compressive strength of 10 MPa can be used in basement walls. Minimum thickness of masonry walls depends on the seismic zones and given in Table 4.

Ratio of total length of the masonry walls in two directions excluding window and door openings to the floor area should not be less than 0.25 I m/m^2. Masonry walls should be laterally supported at

Table 1. Seismic zone coefficient A_o

Seismic zone	1	2	3	4
A_o	0.40	0.30	0.20	0.10

Table 2. Building importance coefficient

Type of building	I
Buildings to be used immediately after an earthquake	1.5
Buildings having high occupancy for a long period of time	1.4
Buildings having high occupancy for a short period of time	1.2
Buildings having low occupancy	1.0

Table 3. Live load reduction coefficient

Type of building	N
Warehouses	0.80
Schools, theaters, garages, commercial buildings	0.60
Private buildings, hospitals, offices	0.30

Table 4. Minimum thickness for masonry walls

Seismic zone	Story	Natural stone (mm)	Concrete (mm)	Others (mm)
1,2,3,4	Basement	500	250	200
	Ground floor	500	–	200
1,2,3,4	Basement	500	250	300
	Ground floor	500	–	200
	First floor	–	–	200
2,3,4	Basement	500	250	300
	Ground floor	500	–	300
	First floor	–	–	200
	Second floor	–	–	200
4	Basement	500	250	300
	Ground floor	500	–	300
	First floor	–	–	300
	Second floor	–	–	200
	Third floor	–	–	200

intervals not exceeding 5.5 m in the first seismic zone and 7.0 m in the other seismic zones. When this condition can not be satisfied, then vertical tie beams should be provided maximum in every 4.0 m including the corners along the masonry wall for which the unsupported length can not be longer than 16.0 m in any case.

Window and door openings should not be closer to the corners (to each other) than 1.5 m (1.0 m) in the first and second seismic zones and 1.0 m (0.8 m) in the third and fourth seismic zones. When vertical tie beams are provided in each side of a masonry wall, these minimum lengths can be reduced by 20%. These window and door openings should not be closer than 0.5 m to any lateral support of the masonry wall. Lintels above window and door openings should have a minimum support length of 0.2 m and 15% of the opening length.

4 AUGUST 17, 1999 MARMARA EARTHQUAKE

Marmara Earthquake of the magnitude 7.4 on the Richter scale occurred on August 17, 1999 with and the epicenter was located at Gölcük. This earthquake broke the western segment of the 1200 km long North Anatolian Fault. The 120 km long break of the August 17, 1999 earthquake occurred on the land between Adapazarı and Gölcük has continued in the Marmara Sea along the shore line between Gölcük and Yalova. This earthquake originated at a shallow depth and generated strong ground motion. Significant lateral offsets reaching up to maximum values of 4–5 m along the surface rupture were observed. In the region, structures were heavily damaged and significant causalities occurred. According to the official estimates more than 16000 people lost their lives and 30000 people were injured. At least 12000 buildings collapsed or suffered heavy damage and 70000 buildings were moderately damaged.

5 PERFORMANCE OF THE CONVENTIONAL MASONRY BUILDINGS

In various small towns in Turkey the masonry structures have been constructed very widely. During recent Marmara Earthquake many of these structures in the rural areas were damaged in various degrees. Some of them collapsed or were damaged heavily and there was no chance to use them any more. These types of structures were torn down in a very short time. The remaining damaged masonry structures were evaluated in view of their structural performance for decision of retrofitting. The primary cause of damages suffered by plain masonry buildings was the tensile stresses in walls which come into being due to seismic forces. Failure due to shear is characterized by diagonal cracks which could be due to diagonal compression and diagonal tension. Such failure may appear either through the pattern of joints or diagonally through masonry units. The cracks usually initiate at the corner of the openings and sometimes at the middle part of the wall segments. These failures can cause partial or complete collapse of the structures. Vertical bending cracks may initiate due to the out-of-plane seismic forces. Since seismic effects occur along both directions of the building simultaneously, the in-plane shear cracks and out-of-plane failure modes are often combined (Arya et al. 1986).

The main weaknesses of materials and masonry constructions are heavy weight and high stiffness which attract large seismic inertia forces, very low tensile strength of masonry units and poor mortars having low shear strength. Brittle behavior of masonry units in tension and compression causes collapse of building without exhibiting significant plastic behavior. Weak connections between walls and roof produce total collapse of the building quickly from a minor damage initiated at one of the weak region of the masonry wall segments. Furthermore, large window and door openings and considerable deviation from symmetry of wall layout cause stress concentration at corners. On the other hand, numerous major damages come into being due to defects in construction such as use of substandard masonry units and mortar, unfilled joints between masonry units, deviation of walls from vertical plane and improper connection between walls, which can be classified as follows:

1. Lack of integrity of the walls,
2. Large openings in structural walls,
3. Irregularities in plane,
4. Discontinuities in the load bearing system between floors,
5. Lack of connection between roof and walls,
6. Absence of rigidity in the roof,
7. Inadequacy of bonding between walls at corners,
8. Use of masonry bricks which do not conform to the corresponding norms,
9. Poor workmanship.

Because of the multi axial nature of the ground shaking under seismic forces, masonry walls are subjected to simultaneous vertical load, out-of-plane and in-plane horizontal loads. While vertical loads are to be a result of the weight of the structure, the inertia yields horizontal forces. The in-plane loads are results of the resistance of the wall rigidity to inertial forces produced from other parts of the masonry structure, such as floor masses, whereas the inertia of the wall itself produces out-of-plane loads. In several cases strength evaluation of masonry building by using sophisticated methods can be of very controversial nature due to assumption done in the analysis. Irregularities in the masonry units not conforming to the corresponding standards and deficiencies in workmanship are very difficult to grasp in modeling (UNDP/UNIDO 1983).

6 PERFORMANCE OF THE BUILDINGS CONSTRUCTED BY USING AAC PRODUCTS

Photograph 1 shows some of typical observations on performance of the buildings made of conventional masonry material and AAC products during the August 17, 1999 Marmara Earthquake. Since the Marmara region is large and has numerous residential and industrial buildings, all types

Photograph 1. Various seismic performance of AAC walls in August 17, 1999 Marmara Earthquake.

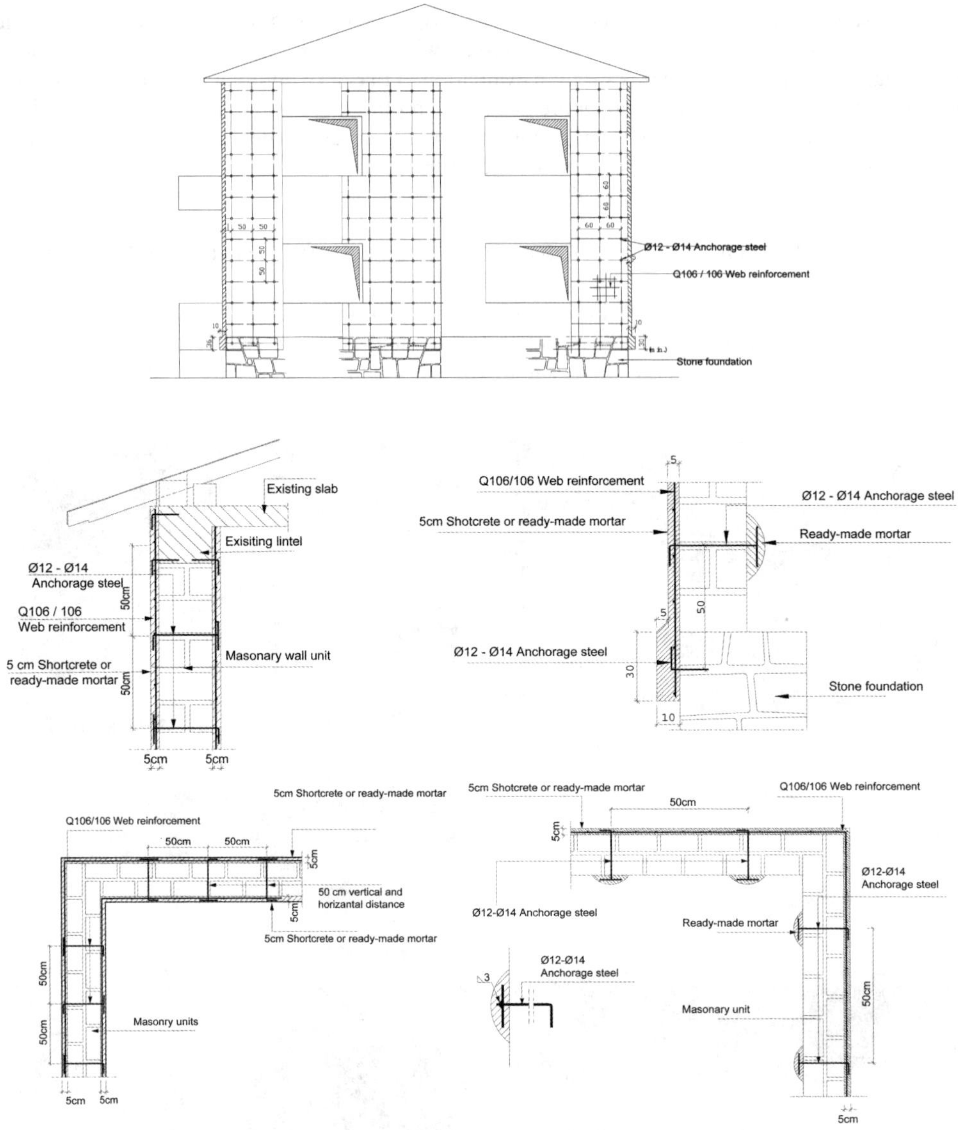

Figure 1. Strengthening of a typical masonry building by using web reinforcement and shotcrete.

of AAC products could be found used in the existing buildings as observation shows. Generally, when a frame of the structural system does not have enough lateral stiffness against lateral forces, in-plane and out-of-the-plane damages may appear in partition walls, because the walls do not have enough deformation capacity to follow the structural system. Furthermore, a minor damage in frame system may lead moderate damage in partition walls. Generally, when the structural system has adequate stiffness, partition walls of AAC products usually do not exhibit any significant damage. AAC panels used in industrial plants exhibited very good behavior, provided that they are connected to the structural system adequately. When a partition wall is within the frame plane, the seismic behavior of the wall was relatively better, compared to the walls which are not in the frame plane. In the later case the inertia forces developed in the wall could not be transferred to the frame and they generate large stresses and consequently damages in the wall. In Turkey, there are relatively less masonry residential houses constructed by AAC blocks. However, their workmanship is relatively better that those of the plain masonry structures. Having better masonry units and construction practice, their seismic performances were very good and generally they did not exhibit almost any sign of damages.

7 REPAIR AND STRENGTHENING OF MASONRY BUILDINGS

Masonry buildings having moderate damages could be repaired and strengthened, provided that they do not display heavy damages and their architectural layout does not display any restriction. Strengthening techniques widely used in Turkey consist of one of the following alternatives:

a. Increasing the shear capacity of the wall by proving additional thickness to the existing wall by applying shotcrete after anchoring steel web reinforcement to the one and/or two faces of the walls. In this case it is important to provide anchorage steel for integration of the existing and added walls to each other in the same floors and between two floors.
b. Increasing the shear capacity of the wall by filling in some door and window openings with masonry units.
c. Improving the shear capacity of the wall by removing damaged part of the walls and repairing it by using repair mortar.

In the strengthening procedure, the shear capacity of partition and masonry walls is increased by applying of shortcrete for repairing existing walls. However, indoor application of shotcrete may cause serious problems. Similar problems may appear when shortcrete is applied in narrow rooms. In these cases ready-made mortar is used instead of shotcrete. Before applying shotcrete, steel web reinforcement is laid on one/or two faces of the wall and it is anchored to the existing wall. Furthermore anchorage bars are provided for the integration of the added concrete layer to each other. These details are presented in Figure 1. The thickness of the added layer is chosen to be minimum 50 mm in one face. However, their extent in horizontal plane and in elevation can be determined by checking the results of strength evaluation (Manzouri et al. 1996, Modena 1994).

8 CONCLUSIONS

Although slight damages in walls made of AAC blocks are observed in the August 17, 1999 Marmara Earthquake, the performance of the masonry buildings produced by using AAC products generally is observed to be excellent. This fact shows that masonry construction has a great potential in the earthquake prone regions of Turkey, provided that they meet the design requirements of the codes and be constructed accordingly.

REFERENCES

Arya, A.S. et al. 1986. *Guidelines for earthquake resistant non-engineered construction*, Tokyo: The International Association for Earthquake Engineering.

Celep, Z. 1998. Post-earthquake rehabilitation of moderately damaged masonry structures in *Repair and strengthening of existing buildings, Second Japan-Turkey Workshop on Earthquake Engineering*, Istanbul: JICA & Istanbul Technical University.

Celep, Z. & Kumbasar, N. 2005. *Introduction to Earthquake Engineering and Earthquake Resistant Design*, Istanbul: Beta Publications (in Turkish).

Paulay, T. & Priestley, M.J.N. 1992. *Seismic Design of Reinforced Concrete and Masonry Buildings*, New York: John Wiley & Sons.

Manzouri, T., Schuller, M.P. & Shing, P.B. 1996. Repair and retrofit of unreinforced masonry structures, Earthquake Spectra, 12 (4): 903–922.

Modena, C. 1994. Repair and upgrading techniques of unreinforced masonry structures utilized after the Friuti and Campania/Basilicata earthquake, Earthquake Spectra, 10 (1): 171–185.
Ministry of Public Works and Settlement 1998. *Turkish Code for Earthquake Resistant Design of Structures*, Ankara.
UNDP/UNIDO 1983. *Repair and Strengthening of Reinforced Concrete, Stone and Brick-Masonry Buildings, Building Construction Under Seismic Conditions in the Balkan Region*, Volume 5, Vienna.

Autoclaved Aerated Concrete – Limbachiya and Roberts (eds)
© 2005 Taylor & Francis Group, London, ISBN 0 415 38356 0

Long-term performance of external renderings on Autoclaved Aerated Concrete walls

H. Kus
Istanbul Technical University, Faculty of Architecture, Istanbul, Turkey

ABSTRACT: Due to its low thermal conductivity and appropriate load bearing capacity, autoclaved aerated concrete (AAC) walls have been used in different climates for both residential and industrial buildings. In-use performance of external walls is greatly dependent on the dose of the agents present within the specific environment to which it is subjected thus different exposure conditions of climatic regions and geography should be taken into account when considering long-term performance, durability and service-life of building materials and components. A research project was carried out in accordance with the European Construction Products Directive and ISO standard 15686 Buildings and Constructed Assets – Service Life Planning in order to investigate the performance of external renderings on AAC through long-term natural and short-term artificial weathering tests. This paper outlines the experimental work and presents general results derived particularly from the microenvironment monitoring carried out at the test cabin built within the DurAAC project.

1 INTRODUCTION

Energy efficient and ecologically sustainable, autoclaved aerated concrete (AAC) walls have been used in different climates for both residential and industrial buildings. Compared to many building materials, AAC offers advantages during the whole service life of the wall component, such as low thermal conductivity, recycleability, high fire resistance, appropriate load bearing capacity etc, and even when environmental aspects are concerned. Due to its high porosity, AAC needs to be protected against the effects of climate, particularly driving rain. In practice, different coatings and external rendering systems are applied as a preventive measure in order to minimise and delay the deteriorating effects of the atmosphere and thus to improve their performance over time, and in turn, to extend their serviceability. Indeed, the properties of these surface coatings determine the long-term performance of the wall component.

Lack of sufficient protection against exposure conditions is one of the reasons for external walls prematurely demonstrating failures, i.e. properties and performance above/below critical levels given in the related standards or regulations. Demands for declared durability and long service life of building materials and components are big and increasing particularly to eliminate ecological and economic impacts of frequent maintenance and repair and thus saving resources, see for example Agenda 21 on Sustainable Construction (CIB 1999). Durability and service life issues are also addressed in European Construction Products Directive (CPD 1989) and the methods for service life planning are standardised in ISO standard 15686 Buildings and Constructed Assets – Service Life Planning. Durability and service life of external walls depend on different factors related to the characteristics of building materials forming the wall component, design and application (workmanship) level, characteristics of exposure environment, and maintenance and in-use conditions (ISO 15686-1 2001). Most factors can be more or less controlled; however the exposure environment remains as an important factor, which should be primarily considered in durability and service life estimations. Data on microenvironment of rendered AAC walls should be systematically registered and gathered in order to be able to estimate the actual in-service performance.

This paper outlines the experimental research work, EUREKA project DurAAC, carried out at the Centre for Built Environment, Gävle, Sweden between 1998–2004 and presents general results particularly based on microenvironment measurements (Kus 2002). Within the framework of the DurAAC project, the effects of different external rendering systems on the durability of AAC façades were investigated. The main objective was to implement and fulfil the requirements set in EU's CPD via ISO 15686. In accordance with the systematic methodology given in ISO 15686, degradation factors and mechanisms were investigated and the long-term performance, durability and service life of AAC walls with different external renderings were assessed.

2 BACKGROUND

All materials experience some changes in appearance, composition or structure with time. The rate of change is dependent on the severity of the agents acting in the exposure environment and the characteristics of the material itself. When designing an external wall for a specified minimum service life, the degradation mechanisms and rates under the prevailing exposure conditions should be known. Different factors are involved on different levels depending on the characteristics of the material and its microenvironment. Facades facing different directions may show differences in condition due to being subjected to degradation agents on different levels.

Many failures of external walls made of porous building materials are caused by excessive moisture content, particularly after driving rain and under long duration of moist conditions. Water ingress into the wall structure is usually regarded to be a critical factor for degradation of rendered AAC systems. The severity of microenvironment conditions together with the absorption and desorption characteristics of the material, i.e. the wetting and drying performance, can be regarded as indicators of the degradation rate. Alternate wetting and drying cycles lead to physical changes as a result of expansion and contraction and chemical changes as a result of dissolution, which may in turn have effects on maintenance of building components in the long run.

In the long-term performance evaluation of external walls, it is important to consider temperature effects as much as the moisture effects. The external surface of the wall material endures large variations in temperature due to ambient air temperature and solar radiation. Besides diurnal and seasonal cycles, rapid temperature and moisture changes in general cause strains in exposed building materials as a result of expansion and contraction, and these cycles may in turn, if the stress level is high enough, result in fatigue damage. The rate, magnitude and frequency of the thermal differentials, the capability of the material to accommodate strain variations and the specific boundary conditions should be considered altogether when assessing degradation by thermal cracking.

3 EXPERIMENTAL

The research work involved continuous monitoring of the microenvironment and periodical field measurements as well as laboratory testing of the initial properties and the performance over time of the material. Investigations were carried out by physical and chemical analysis of fresh samples, samples naturally weathered by long-term field exposure and samples artificially aged by short-term accelerated laboratory tests. Although artificial weathering under carefully controlled conditions gives more useful data and in turn, repeatable and reproducible results, the real life effects of weathering can only be obtained by natural exposure testing. Hence, data from microenvironment monitoring or field measurements are necessary inputs in designing short-term accelerated tests (Martin 2002). On the other hand, the deteriorating effects of air constituents and contaminants are very complex since many variables are involved. Thus, it is too complicated to simulate natural environment conditions through an artificial weathering.

A short summary of the experimental set-up of the research project is given in Figure 1.

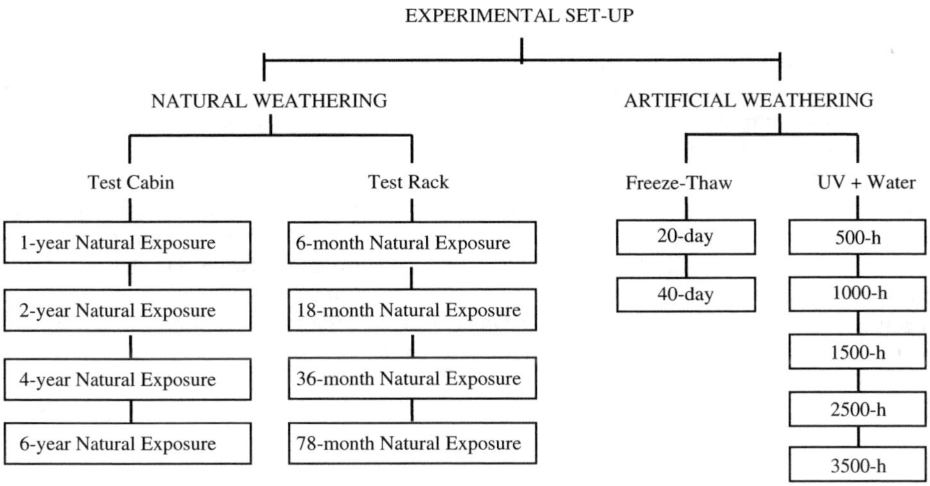

Figure 1. The field exposure and the laboratory testing programme.

3.1 *Natural exposure*

Studies on natural exposure were carried out at two different scales: full-scale experimental building, i.e. panels on the test cabin, and small-scale mock-up testing, i.e. samples on the test rack. The natural exposure set-ups were installed on the roof of the Centre for Built Environment, Gävle, Sweden. The weather conditions at the exposure area were monitored in close proximity to the test rack and the test cabin. According to the microclimate data monitored at the test site for the time period in question, the yearly average air temperature in winter is 0,8°C and in summer 13,6 °C, the min. being −22,5 °C and the max. 34,5 °C.

3.1.1 *Full-scale experimental building*

Within the DurAAC project a test cabin was installed in 1999 in Gävle, Sweden. Each long side of the cabin, north-east respective south-west, consists of twelve unreinforced AAC panels, with a density of 450 kg/m^3 and measuring 600 mm × 1200 mm × 150 mm, with different rendering systems. The systems tested involve inorganic and organic renderings as well as their modifications with water repellants. The products used can be grouped in three categories: cement-lime, hydraulic lime, and cement polymer based renderings; acrylic styrene, acrylic, silicate and silicon paints; and silicon-based water repellants. Characteristics of the test panels and the possible changes over time are investigated through laboratory tests and analysis of samples drilled from each panel at specific intervals.

3.1.2 *Mock-up testing*

Within the mock-up testing, six sample blocks, measuring 400 mm × 300 mm × 150 mm, were exposed on a 45° metal rack facing south and monitored continuously. The systems tested involve inorganic renderings and their modifications with water repellants. The products used were cement-lime, acrylic styrene, and silicon-based water repellants either as surface impregnation or as additive in the rendering. To investigate the actual effect of atmospheric agents and to observe the degradation mechanisms, an additional group of smaller test samples, measuring 130 mm × 130 mm × 150 mm, have also been naturally weathered and subjected to laboratory tests and analysis at specific time intervals.

3.2 *Monitoring*

Microclimate parameters such as air temperature, air relative humidity, driving rain and radiation (UVA + UVB) were monitored continuously. Surface and bulk temperatures and moisture contents of the test panels were also measured continuously. Wetcorr sensors were used to measure the surface moisture and the surface temperature. Copper-constantan thermocouples were used to measure temperatures at different depths in the material. The moisture contents were also measured at the same depths with resistance type nail electrode pairs. The sensors were scanned at 5-minute intervals and the averages of temperature and resistance were stored every hour by means of a data logger/multiplexer device. The resistance measured between the two electrodes together with the temperature in the AAC substrate was converted to moisture content by means of a calibration procedure. The lowest moisture content that can be measured with this experimental set-up, for the AAC material, is approximately 27% (by weight) at 20 °C.

3.3 *Accelerated ageing exposure*

Studies using artificial exposure were carried out by two separate tests; freeze-thaw cycles and UV + water exposure in the Weather-Ometer®. Samples were exposed to 80 freeze-thaw cycles for 40 days in the so called "Malmö Chamber". An other group of samples were exposed to total 3500 hours to UV and water cycles in a filtered xenon-arc lamp type weathering chamber. Both samples were tested and analysed at specific time intervals during artificial weathering.

Figure 2. Test cabin.

Figure 3. Test rack.

Figure 4. Monitoring at the test cabin.

Figure 5. Monitoring at the test rack.

3.4 *Laboratory tests and analysis*

The physical material properties tested in the laboratory and the test methods conducted were capillary water absorption, vapour transmission, evaporation and contact angle. The mechanical material properties tested were adhesion (pull-off test) and compressive strength. The microstructure was studied by (thin section) light optical microscopy (LOM). The surface characteristics were examined by scanning electron microscope (SEM) and the chemical species were identified by energy dispersive X-ray spectroscopy (EDS) and X-ray difractometry (XRD).

4 RESULTS AND DISCUSSIONS

In order to assess the in-service performance of rendered AAC walls, some results particularly based on the long-term temperature and moisture monitoring data obtained from the naturally exposed test panels at the test cabin are reported here.

4.1 *Temperature monitoring*

Measurement data collected through May 1999 to August 2004 indicate clearly that the panels on the south-west (SW) facade experienced higher solar radiation during longer periods than the test panels on the north-east (NE). Temperature variations are thus both faster and greater on the first mentioned panels. The diagram in Figure 6 demonstrates measured surface temperatures for a selected day. Panel surfaces develop much higher temperatures than ambient air temperature when they are exposed to intense direct solar radiation. The temperature changes in AAC panels with surface coatings of different colours were both very fast and could easily be in the order of $\pm 40°C$ many times each year, but the moisture changes were small due to effective surface treatments. Consequently, the panels on the sunny side of the test cabin, namely SW façade, would most likely be susceptible to direct temperature effects such as degradation by thermal cracking.

Rapid variations in surface temperatures occur specifically during which the sky conditions suddenly change between cloudy, rainy and sunny and can be numerous during one day (Figure 7). Such a repeated rapid and extreme fall (or rise) in temperature, which is generally faster than experienced under normal circumstances, might cause fatigue between the surface zone and the innermost parts of the material which is in a relatively steady state condition. After a sufficient number of temperature fluctuations, the cumulative synergistic effects of creep and fatigue stresses and strains would lead to ageing which may in turn result in failure as cracking or fracture.

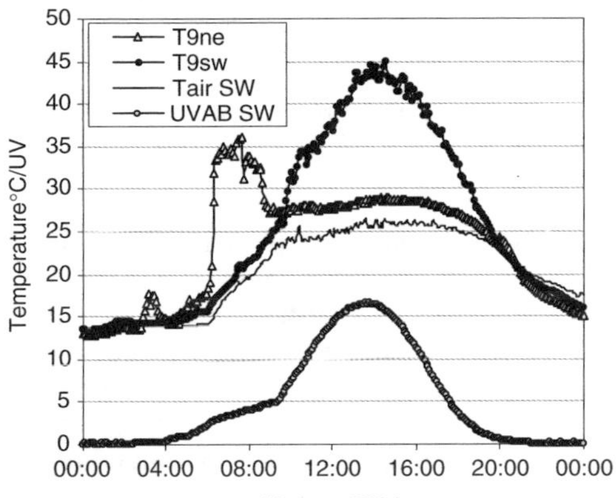

Figure 6. Daily surface temperatures on panel 9 at opposite sides.

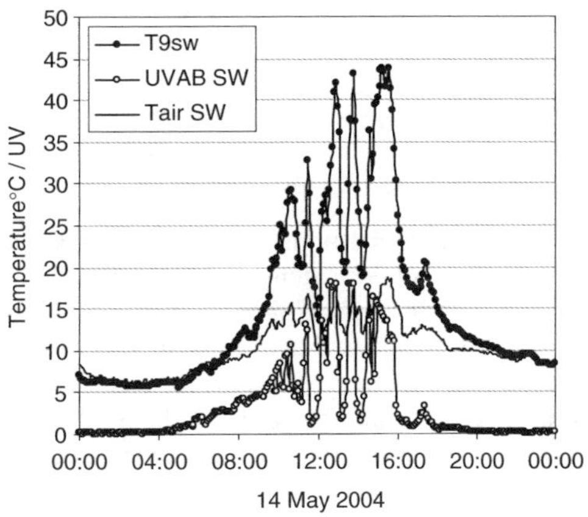

Figure 7. Rapid temperature changes on panel 9 at SW façade.

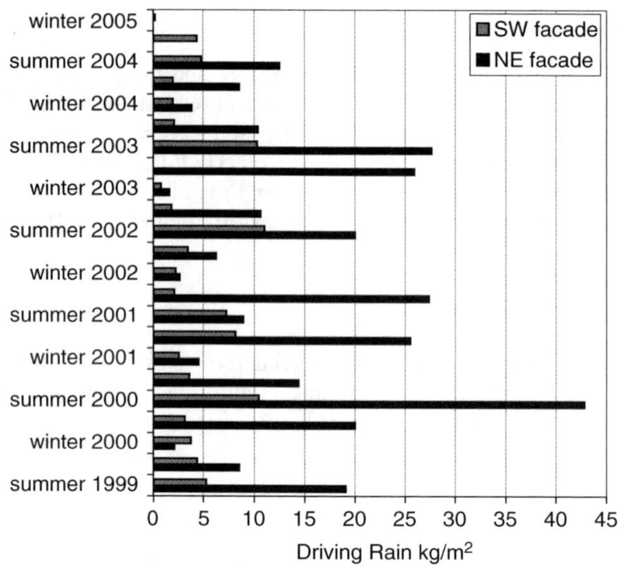

Figure 8. Seasonal total driving rain amounts hit on to the NE and SW facades.

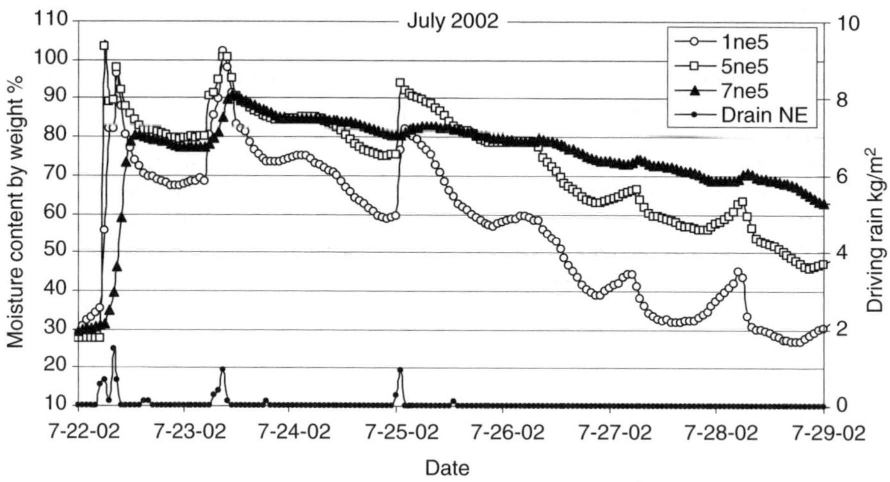

Figure 9. Hourly averages of moisture contents at 5mm depth in the AAC: Inorganic systems.

4.2 *Moisture monitoring*

The seasonal total driving rain loads on both facades of the test cabin are displayed in Figure 8. It is clearly seen that the panels exposed on the NE facade were hardly ever hit by sunlight (Figure 6) and received significantly higher amounts of driving rain (Figure 8). The long-term microenvironment monitoring results indicate also that the SW facade experienced significantly lower amounts of driving rain and thus exhibited relatively low moisture contents. Direct moisture effects are thus of primary concern for the panels at the NE facade, particularly for the inorganic test systems without surface treatments, as these panels experienced significantly higher amounts of driving rain.

4.2.1 *Inorganic rendering systems*

In order to make a better assessment of moisture performances, an example is given below based on measurement data obtained from the NE facade. Figure 9 shows the wetting and drying performance

Table 1. Inorganic rendering sytems.

Panel no.	Substrate	Primer	Undercoat	Final coat	Thickness mm
1	AAC	–	–	–	
5	AAC	Lime-cement 10/90/350	–	Lime-cement 50/50/450	3–4
7	AAC	Lime-cement 10/90/350	Lime-cement 50/50/650	Hydraulic lime	12–15

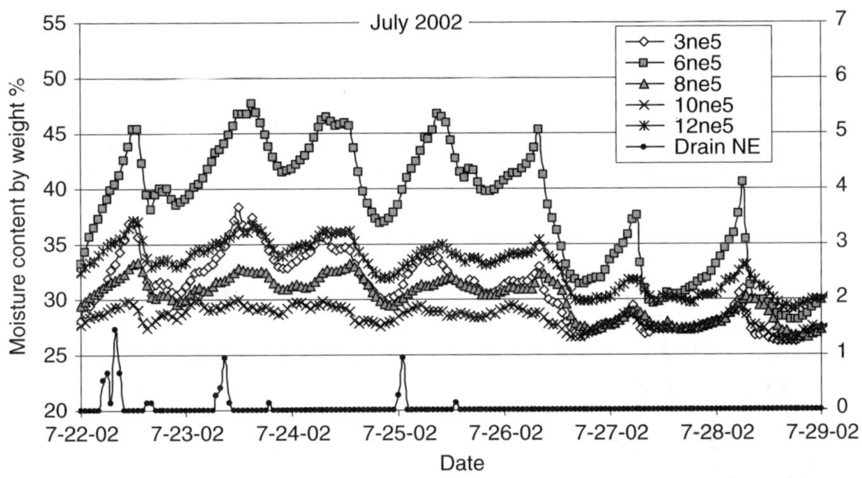

Figure 10. Hourly averages of moisture contents at 5mm depth in the AAC: Organic systems.

of AAC with inorganic rendering systems (Table 1) during a specific time with repeated driving rain. In selection of exposure period, it was also taken into consideration that the systems were in almost dry state just before the rain. In this way, the effects of different doses could be observed and a more reliable moisture performance assessment could be accomplished. However, it should be noted that the given examples reflect only the responses under the specific weather conditions of the selected exposure periods.

The diagram in Figure 9 shows that all three inorganic sysytems (Table 1) absorb high amounts of water. The wetting of AAC substrate is delayed for the rendered systems, the time depending on the thickness and pore structure of the rendering mortar. Once the AAC substrate gets in contact with water, it will rapidly suck up the moisture via its capillary active fine pores. It takes longer time for thick rendering (System 7) to dry out after rain while the unrendered reference AAC panel (System 1) dries out fastest.

4.2.2 *Organic rendering systems*

In Figure 10, the wetting and drying performance of organic rendering systems (Table 2) during the same exposure period given for the inorganic systems. Please note that the scale of y-axis of the diagram in Figure 10 is different from the scale used in the diagram of inorganic systems in Figure 9. These systems absorb significantly lower amounts except for the lime-cement rendering containing silicon additive, namely System 6, which is quite susceptible to all types of driving rain, i.e. heavy, moderate and light driving rain with either long or short duration. System 3, acrylic-styrene paint, and System 12, fiber-reinforced cement-polymer-lime rendering, are the two other systems after System 6, which are likely to be affected (to some extent) when subjected

Table 2. Organic rendering systems.

Panel no.	Substrate	Primer	Undercoat	Final coat	Thickness mm
3	AAC	Acrylic-styrene	–	Acrylic-styrene	<1
6	AAC	Lime-cement 10/90/350 with *Silicone additive*	–	Lime-cement 50/50/450 with *Silicone additive*	3–4
8	AAC	Lime-cement 10/90/350	Lime-cement 50/50/650 with *Silicone additive*	Silicate paint	12–15
10	AAC	–	Lime/cement/fibres with *ethylene/vinyl acetate redispersible pulver*	Silicate paint	7–9
12	AAC	–	Cement/polymer/ limestone-plastic fibres	Cement/polymer/ limestone-plastic fibres	5–7

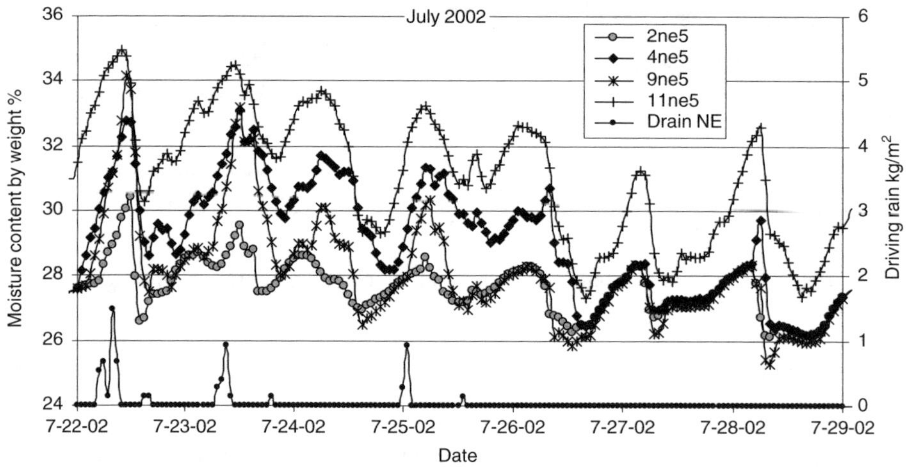

Figure 11. Hourly averages of moisture contents at 5 mm depth in the AAC: Systems on surface treated AAC with water repellants.

to intense and long duration of driving rain. System 10, lime–cement–fibre rendering containing ethylene/vinyl-acetate redispersible powder, seems to have the best moisture performance as it is the most hydrophobic of all the systems tested within the project.

4.2.3 *Organic rendering systems on surface treated AAC with water repellants*

Although they had low levels of absorption, all water repellent treated systems exhibit some changes in moisture behaviour, depending on the frequency, intensity and duration of driving rain conditions. Even under extreme conditions (Figure 11) the moisture contents of these systems increased only upto 34–38% followed by a fast drying. Please note that the scale of y-axis of the diagram in Figure 11 is different from the scale used in the diagrams of Figure 9 and in Figure 10. In this group, in general, surface impregnated AAC with a silicone microemulsion concentrate product (System 2) demonstrated the best performance, i.e. having the lowest absorption and the highest desorption rates.

Table 3. Systems on surface treated AAC with water repellants.

Panel no.	AAC surface impregnation	Primer	Final coat	Thickness mm
2	Silicon microemulsion	–	–	–
4	Silane-siloxane emulsion	Acrylic-styrene	Acrylic-styrene	<1
9	Silane-siloxane emulsion	Acrylic/dolomite-calcite *silikon tillsats*	Pure acrylic copolymer/ dolomite-calcite	<1
11	Silane-siloxane emulsion	–	Silicone resin paint	<1

5 CONCLUSIONS

The research work aimed at investigating the long-term performance of external renderings on AAC walls based on the different testing procedures in accordance with the systematic methodology in ISO 15686. In the experimental study in conformity with the holistic approach, the entire wall construction was taken into account, either by mock-up testing or full-scale experimental building. The overall results indicate that the orientation of buildings plays a major role in the temperature and moisture performance of external walls. Furthermore, there is a significant difference between the moisture performance of the inorganic and organic renderings. Regarding the rendering systems modified with silicon-based water repellants, a long protectability against driving rain effects and in turn a long service life can be expected if properly applied.

It is important to remember that the results regarding the long-term performance of rendered AAC external walls derived from the natural weathering tests give information specific to the geographical and climate conditions of central Sweden which is usually cold and dry. Therefore, the general observations may be helpful to, first of all, those in practice dealing with facade problems encountered particularly in this region. The results in general can be also useful especially for the manufacturers of the tested materials, in the form of feed-back to assess their products, to compare with their own research and development studies, and possibly to improve either the material or the application modes in design or renewal.

REFERENCES

CIB 1999. CIB Report Publication 237. *Agenda 21 on Sustainable Construction.*

CPD 1989. *The CPD Guidance Paper F. Durability and the Construction Products Directive.*

ISO 15686-1 2001. *Buildings and constructed assets-Service life planning - Part 1: General principles.*

Kus, H. 2002. *Long-Term Performance of Water Repellants on Rendered Autoclaved Aerated Concrete*, ISBN: 91-7283-352-1, BMG-KTH, PhD Thesis, Sweden.

Martin, J.W. 2002. *Repeatability and Reproducibility of Field Exposure Results*, Service Life Prediction: Methodology and Metrologies, ed. by Martin, J. W. and Bauer, D. R., ACS Symposium Series 805, Pub. American Chemical Society, Washington DC.

Autoclaved Aerated Concrete – Limbachiya and Roberts (eds)
© 2005 Taylor & Francis Group, London, ISBN 0 415 38356 0

Standard installation methods in Japan for non-load-bearing walls made of AAC panels

K. Suzuki & Y. Hama
Muroran Institute of Technology, Hokkaido, Japan

Y. Aono, S. Shibata, F. Matsushita & H. Taguchi
Sumitomo Metal Mining Siporex Co., Ltd., Research & Development Center, Tokyo, Japan

ABSTRACT: Ever since AAC was introduced in Japan in 1962, original installation methods suitable for the Japanese environment have been developed. Furthermore, the demand for high-resistance walls which can withstand seismic action has been rising, especially after the great Hanshin-Awaji earthquake disaster of 1995. AAC manufacturers in Japan have been promoting the "vertical-wall rocking method," which gives excellent earthquake resistance. The rocking method is now used almost everywhere in Japan, and the conventional "vertical-wall reinforcing-bar insertion method" is no longer a standard method. In the rocking method, the AAC panel is fixed onto an exclusive metal plate with an anchor installed in the longitudinal direction at both ends of the AAC panel; the metal plate is in turn fixed into the building's steel frame with bolts. During an earthquake, each panel becomes slightly rotated due to storey drift. Cracks were observed at 1/300 rad of the storey drift angle with the reinforcing-bar insertion method, but were not observed even at 1/60 rad with the rocking method.

1 INTRODUCTION

The total shipment quantity of AAC panels between 1962, when AAC panels were introduced in Japan, and 2003 was about 84,000,000 m^3 (5,450 ton). During this period, original product systems, applications and installation methods suitable for the Japanese environment were developed.

2 PRODUCT SYSTEM

AAC panels produced in Japan are of two kinds: thick panels over 75 mm in thickness and thin panels 50 mm and 37 mm in thickness. Flat panels and design panels, with artwork printed on the surface, are available for each size.

Thick panels are made to order. The standard width of the thick panels is 600 mm, but it can be varied from 300 mm to 600 mm at a pitch of 10 mm. The panel length can be varied from 610 mm to 6,000 mm at a pitch of 10 mm. The width of the thin panels is fixed at 600 mm, but several standard lengths are available.

At present, the annual output of thick panels is about 7 times that of thin panels.

3 APPLICATION

The proportions of application of AAC are about 70% of the external walls, about 20% of the partition walls, and about 10% of floors and roof, respectively, for both thick and thin panels. Thick panels have been used mostly in the construction of low-rise steel buildings. However, the use of thick panels for external walls in reinforced-concrete and steel-encased reinforced-concrete high-rise buildings has increased in recent years. Meanwhile, thin panels have been used mostly for the external walls of traditional Japanese wooden houses.

The following is an explanation of the installation methods used in Japan for thick AAC panels.

4 INSTALLATION METHODS IN JAPAN

There are two kinds of external walls for which AAC panels are used: "vertical walls" and "horizontal walls". Generally, "vertical walls" are used in multiple dwelling houses, stores and office buildings, while "horizontal walls" are used in factories and warehouses. The "vertical wall" to "horizontal wall" adoption ratio is 7/3.

The standard installation method for AAC panels in Japan was changed from the "vertical-wall reinforcing-bar insertion method" to the "vertical-wall rocking method" in 2002. This is because, after the great Hanshin-Awaji earthquake disaster of 1995, the demand for high-resistance buildings which can withstand seismic action has been rising. Thus, AAC manufacturers in Japan have been promoting the rocking method, which gives excellent earthquake resistance. The following is an introduction of the characteristics of the rocking method in comparison with those of the reinforcing-bar insertion method.

4.1 *History of each method*

The reinforcing-bar insertion method was first used when AAC panels were introduced in the 1960's, and it was a standard installation method for a long time. During the Izu-Ohshima-kinkai earthquake (M 7.0) and the Miyagiken-oki earthquake (M 7.4) of 1978, cracks and fractures

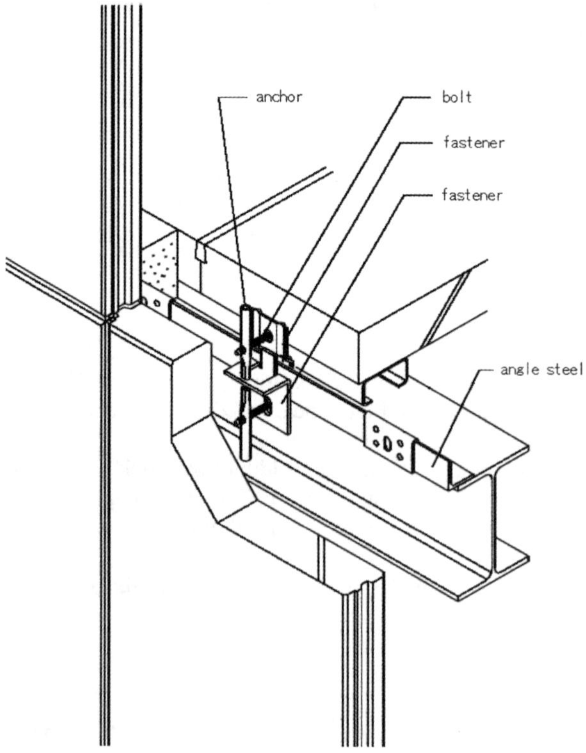

Figure 1. Installation of the rocking method at both ends of AAC panel and building frame.

occurred in the panels of many buildings built with the reinforcing-bar insertion method. This is because the panels could not follow the excessive deformation caused by the earthquakes, due to the insufficient rigidity of the building frames. Therefore, the demand rose for the development of a new method which allows the panels to follow earthquake deformations, and so the AAC manufacturers developed the rocking method in the 1980's.

Still, the high cost of the rocking method prevented it from gaining popularity. In 2001, only 40% vertical walls were built with the rocking method. But the demand for high-resistance buildings which can withstand seismic action has been rising steadily after the great Hanshin-Awaji earthquake disaster of 1995. Also, the demand that builders issue 10-year guarantees against defects in the external walls of houses has been met (the "Housing Quality Assurance Act" was enforced in 2000). Thus, the reinforcing-bar insertion method was definitively judged to be inadequate, and the standard installation method was changed from the reinforcing-bar insertion method to the rocking method.

As a result, the ratio of use of the rocking method for vertical walls has risen from about 40% in 2001 to almost 100% at present.

4.2 *The two methods of attachment and earthquake resistance*

4.2.1 *The rocking method*

In the rocking method, the AAC panels are fixed by simply connecting the anchor installed in longitudinal direction at the both ends of the AAC panel with the metal plate fixed into the building frame with bolts. The dead load of the AAC panel is supported at the center of the panel's lower end. Figure 1 shows the installation of the rocking method at the both ends of the AAC panel and the building frame. Figure 2 shows the installation of the rocking method at the lower end of the panel and the footing. Figure 3 shows the installation of the rocking method at the upper end of the

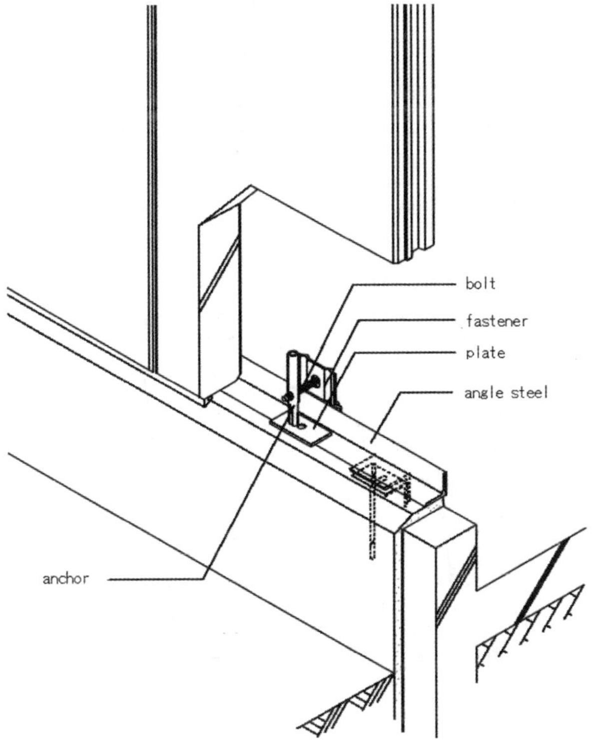

Figure 2. Installation of the rocking method at the lower end of AAC panel and footing.

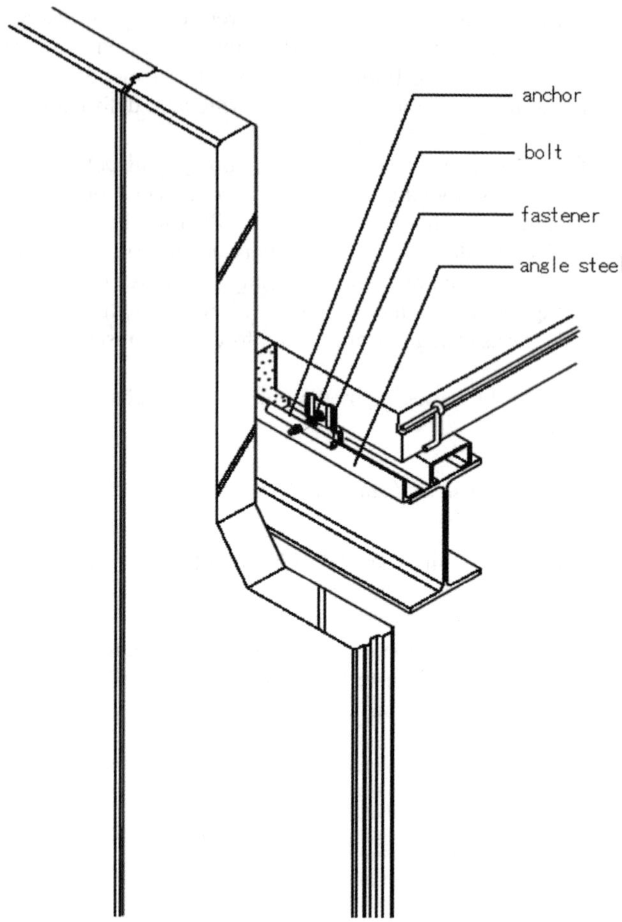

Figure 3. Installation of the rocking method at the upper end of AAC panel and parapet.

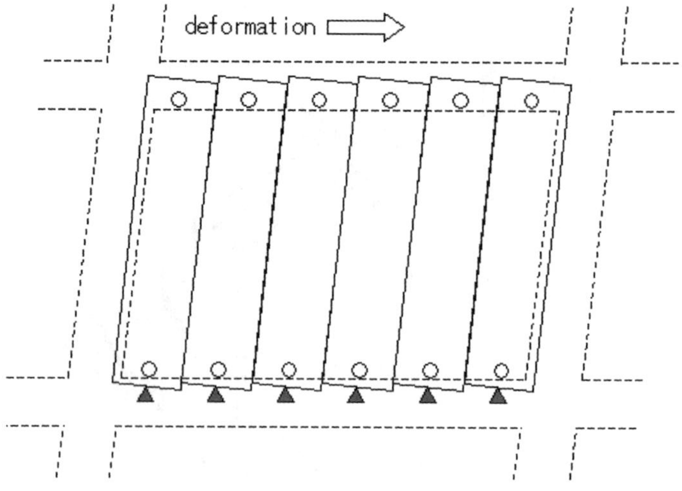

Figure 4. Rocking of AAC panels of the rocking method.

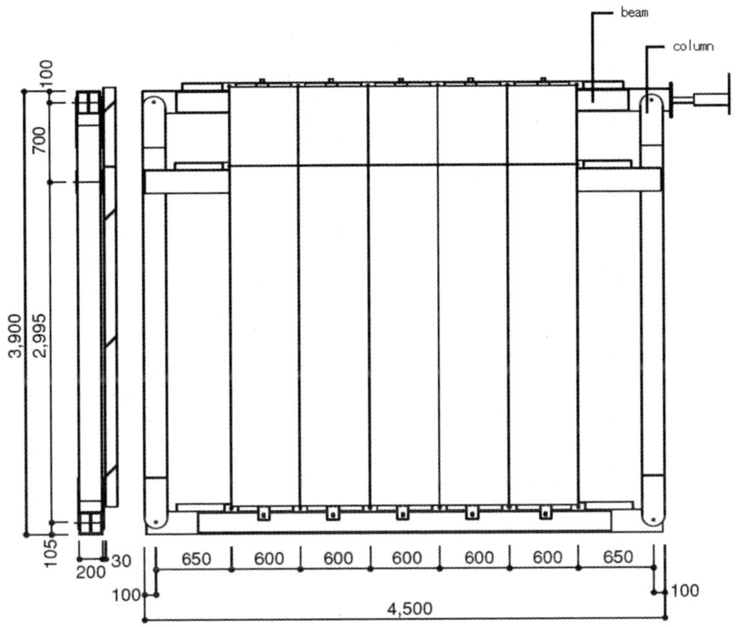

Figure 5. Specimen of in-plane deformation tests for the rocking method of AAC panels.

Table 1. Results of in-plane deformation tests for the rocking method.

Cycle	Drift angle (rad)	Result
1	1/300	No cracks
2	1/200	No cracks
3	1/150	No cracks
4	1/120	No cracks
5	1/100	No cracks
6	1/75	No cracks
7	1/60	No cracks

panel and the parapet. The panel can follow the deformation of the building frame, because each panel "rocks" (slight in-plain rotation) as the earthquake deforms the frame, as shown in Figure 4.

Figure 5 shows the specimen of in-plane deformation tests for the rocking method. Table 1 shows the results of these tests. No cracks in the specimen panel and no damage in the attaching portion were observed at 1/60 rad of the drift angle; thus, excellent earthquake resistance has been demonstrated.

4.2.2 *The reinforcing-bar insertion method*

In the reinforcing-bar insertion method, the AAC panels are fixed together with mortar, which fills the vertical joints between consecutive panels. In addition, the reinforcing bars inserted into the mortar from both ends of the vertical joint are welded to the metal plate fixed onto the building frame. The dead load of the AAC panel is supported at the right and left corners of the panel's lower end. Figure 6 shows the installation at both ends of the AAC panel and the building frame. Figure 7 shows the installation at the lower end of the panel and the footing. Figure 8 shows the installation at the upper end of the panel and the parapet. The frame deformation caused by an earthquake induces a shear deformation in the AAC panel; the whole external wall resists the seismic action instead of following it.

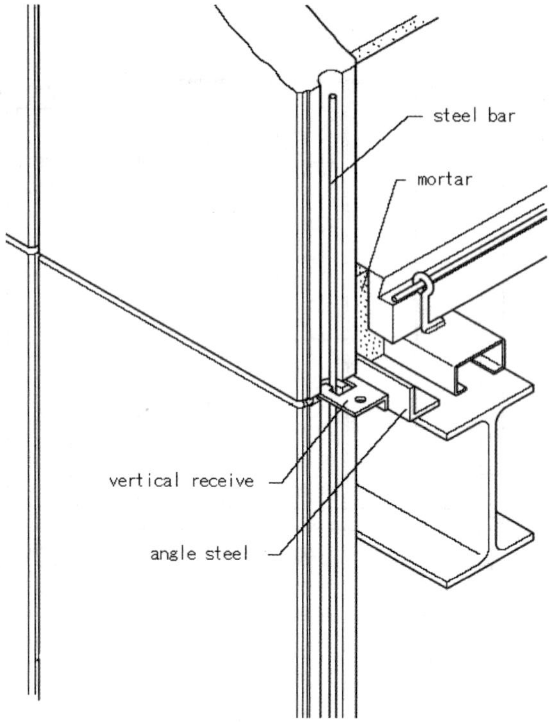

Figure 6. Installation of the reinforcing-bar insertion method at the both ends of AAC panel and building frame.

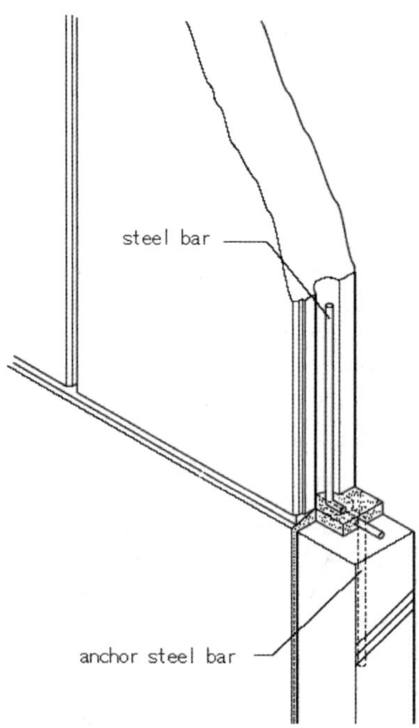

Figure 7. Installation of the reinforcing-bar insertion method at the lower end of AAC panel and footing.

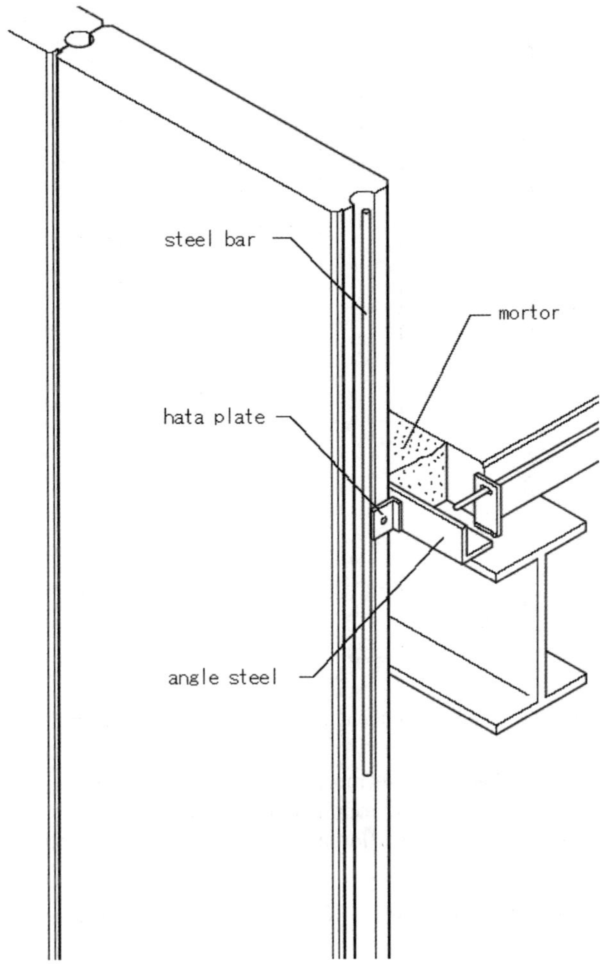

Figure 8. Installation of the reinforcing-bar insertion method at the upper end of panel and parapet.

Table 2. Results of in-plane deformation tests for the reinforcing-bar insertion method.

Cycle	Drift angle (rad)	Result
1	1/300	Slight cracks at the top story.
2	1/200	Increasing cracks at the top story. Slight cracks at the middle and the lowest story.
3	1/150	Increasing cracks slightly at the top and the middle story.
4	1/100	Increasing cracks except the middle story.
5	1/20	Increasing cracks at all stories. No AAC panels falling down.

The results of the in-plane deformation tests are shown in Figure 9 (specimen) and Table 2. Table 2 shows that cracks amenable to repair at the corner of the specimen panel were observed at 1/300 rad of the drift angle.

4.3 *Other advantages of the rocking method*

Not only is the earthquake resistance excellent with the rocking method, but it is also possible to shorten the construction period, since installation is simple. And, since it is a perfectly dry

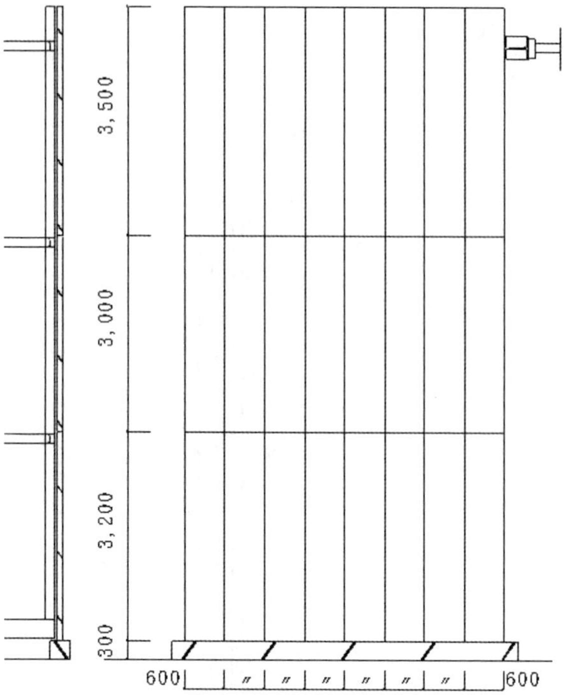

Figure 9. Specimen of the in-plane tests for the reinforcing-bar insertion method.

construction method which does not use mortar, a good work environment can be maintained and there is no debris to damage the environment. The rocking method has many other merits in comparison with the reinforcing-bar insertion method, such as superscription.

5 DAMAGE FROM A RECENT LARGE-SCALE EARTHQUAKES

In Japan, an earthquake occurred in last year in Niigata-Chuetsu (M 6.8) and this year in Fukuoka-seihou-oki (M 7.0). The damage survey results in these two earthquakes show that cracks similar to those obtained in the test were sporadically observed in the AAC panels fixed with the reinforcing-bar insertion method. In contrast, no damage was observed in the external walls built with the rocking method.

6 SUMMARY

The installation methods used in Japan are summarized as follows.

(1) The standard installation method for ACC panels in Japan was changed from the "vertical-wall reinforcing-bar insertion method" to the "vertical-wall rocking method" in 2002.
(2) In the test of in-plane deformation for the rocking method, no cracks and no damage were observed at 1/60 rad of the drift angle; thus, excellent earthquake resistance has been demonstrated.
(3) Also recent great earthquake proved the excellent earthquake resistance of "Vertical wall rocking method".

Autoclaved Aerated Concrete – Limbachiya and Roberts (eds)
© 2005 Taylor & Francis Group, London, ISBN 0 415 38356 0

A fundamental investigation of the composite behaviour of an AAC panel and timber joist floor system

K.J. Graham
CSR Hebel, Sydney, Australia

K.I. Crews
University of Technology Sydney, Sydney, Australia

S.A. Bolden
Hyne & Son P/L, Maryborough, Australia

ABSTRACT: In Australia autoclaved aerated concrete (AAC) panels have been used as external wall cladding and to a lesser extent as the plate element in floors of low-rise structural timber framed construction. In these floors, the timber framing is considered the structural support for the AAC panels.

This paper describes a fundamental investigation of the composite behaviour exhibited by a floor system consisting of AAC panels fixed to a proprietary timber joist framing system subjected to out-of-plane bending. The main objectives of the project were to illustrate the existence of composite action and quantify the magnitude of out-of-plane flexural performance improvement.

The investigation program consisted of a series of flexural tests on full-scale floor specimens at both the serviceability and ultimate strength limit states. This paper presents the results of the investigation in relation to the flexural behaviour under static loadings, dynamic response of floor system, load redistribution under point loadings and slip performance of the connection of panel and joist.

1 INTRODUCTION

In Australia autoclaved aerated concrete (AAC) panels have been used as external wall cladding and to a lesser extent as the plate element in floors of low-rise structural timber framed construction. In Australia, CSR Hebel markets this floor system as SoundFloor™.

This system provides advantages such as longer spans, increased load capacity and flexural stiffness due to composite action, and economic and faster construction. Additionally these decks reduce squeaks, nail popping and costly repairs which are common issues associated with conventional full timber flooring systems. Other advantages are high strength-to-weight ratio, production line quality and efficient use of materials.

In these floors, the timber framing is considered the structural support for the AAC panels. Consequently, no allowance for a composite interaction between the two materials is considered in the design process.

A project was conducted to investigate the flexural behaviour of the CSR Hebel SoundFloor™ flooring system consisting of the CSR Hebel SoundFloor™ panel and Hyne & Son P/L 'I' beam joists (UTS, 2005, Benitez, 2004). The main objectives of this project were to quantify the distinctive characteristic properties of the composite section under serviceability loads and behaviour through to failure, and the dynamic response of the floor system.

Generally in a composite system, the panels act as the compression element and the joists as the tension element. To have a composite section the two elements need to be connected to transfer shear. This was achieved with construction adhesive and metal screws, and monitored during the testing programme. Preliminary research was carried out to investigate that composite action existed between the selected materials by constructing and testing push-out specimens (Daher 2003). The findings of the research are not included in this paper.

This paper describes a fundamental investigation of the composite behaviour exhibited by a floor system consisting of AAC panels fixed to a conventional timber joist framing system subjected to out-of-plane bending, and presents the results in relation to the flexural behaviour under static loadings, dynamic response of floor system, load redistribution under point loadings and slip performance of connection of panel and joist.

2 EXPERIMENTAL PROGRAMME

2.1 *Test specimens*

A total of six (6) full scale simple span floor test specimens were constructed. Each specimen represented a typical CSR Hebel SoundFloor™ configuration. The specimens were constructed from CSR Hebel SoundFloor™ panels bonded together with CSR Hebel Thin Bed Adhesive and fixed to the Hyne & Son 'I' joists (Hyne & Son, 2001), with metal screws and construction adhesive, in accordance with the CSR Hebel literature (CSR Hebel, 2003). Each specimen was 9 panels long and 1.67 – 1800 mm long - panels wide fixed to five (5) 'I' joists spaced evenly at 600 mm centres, resulting in an overall dimension of 5400 mm (length) × 3000 mm (width). The simple span of the test specimens was 5100 mm. The 3000 mm width was achieved with a combination of 1800, 900 and 300 mm long panels. This panel arrangement ensured each joist supported the same width of panel. The arrangement of the panels is shown in Figure 1.

Two joist depths were used to produce different load-deflection, ultimate capacity and dynamic response behaviours. The six (6) test specimens consisted of 2 sets of three (3) specimens constructed with the same depth joist. The two (2) joist depths used were 200 mm and 245 mm. Both joist types had a flange width of 68 mm.

The CSR Hebel SoundFloor™ panel is manufactured from autoclaved aerated concrete (AAC) and strengthened with reinforcement. The Hyne & Son 'I' beam joists are composite high performance joists consisting of solid timber flanges glued to a structural plywood web.

During construction the joists were temporarily braced to maintain squareness. These braces were removed sequentially as the panels were fixed in place. The test specimens were cured for a minimum of 7 days. After curing, the specimens were relocated onto the testing frame. The difference in the support location of the construction bed and testing frame resulted in some joists

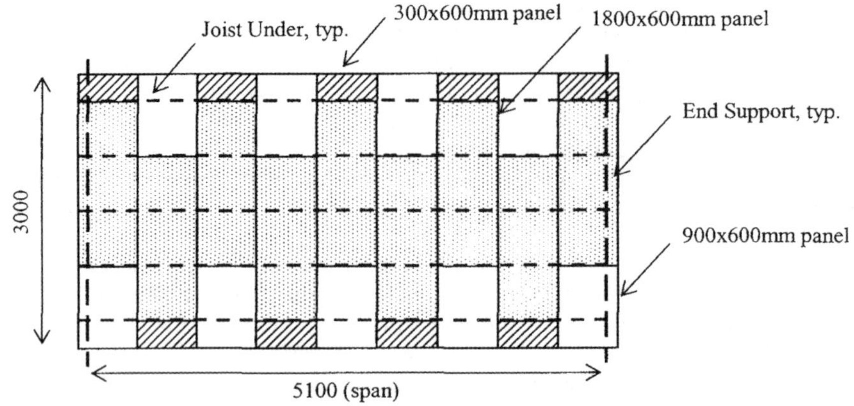

Figure 1. Plan view of the panel arrangement for the test specimens.

not bearing onto the supports. At these locations packers were installed to ensure all joists were in contact with the supports at the start of the test.

2.2 *Flexural tests*

All specimens were tested in the horizontal plane. Four types of tests were performed: three (3) static and one (1) dynamic. Note one static test was used to analysis the dynamic behaviour of the test specimen. The programme of the test specimens is shown in Table 1.

The static test programme consisted of six (6) test specimens subjected to different line or point load configurations to investigate the flexural behaviour under static loadings, dynamic response of floor system, load redistribution under point loadings and slip performance of the panel connection.

The aim of the dynamic tests was to find the first vertical resonance frequency of each of the four test specimens with the existing support set-up used during the Static Load tests. The dynamic tests are referred to as Test Arrangement D in Table 1.

The loading span for all tests was based on AS4063-*Timber-Stress Graded-In-Grade Strength and Stiffness Evaluation* and Designation: D 5055 – 00 *Standard Specification for Establishing and Monitoring Structural Capacities of Prefabricated Wood I-Joists.1*. Both standards described that *"test spans are to be conducted on a span 18xD or span 17 to 21D"*, respectively with D being the depth.

2.3 *Static tests*

The tests load configuration consisted of three different load arrangements (A) *uniformly distributed* across the specimen, (A) *point load* at third span and 3) *point load* at mid span. The three different tests arrangements were by, applying a *Line* and *Point* loads at 3rd span along the longitudinal axis and a *Point* load at the centre of both axes. Figure 2 shows the load arrangement of the three tests.

These test arrangements were chosen due to the following:

- *Line Load* – Provides a uniform bending moment in the middle 3rd span along the longitudinal axis, where traditional serviceability and strength calculation can be applied to the Load-Deflection curves and maximum capacity (Bending Moment) respectively. This load arrangement is referred to as Test Arrangement A, refer to Table 1.
- *Point Load at 3rd span* – Point loads are applied over joists 2 and 4 at 3rd span points along the longitudinal axis. Although similar to Test Arrangement A, bending is induced in both axes. This load arrangement is referred to as Test Arrangement B, refer to Table 1. Note: It was assumed that the pressure was equal in all jacks hence there was no need to used four load cells.
- *Point Load at centre* – A point load was applied at the centre of the specimen over joist 3. This load arrangement induced bending in both directions and imposed localised stresses into the panel. This load arrangement is referred to as Test Arrangement C, refer to Table 1.

Table 1. Summary of test programme.

Test specimen	Joist depth	Test span	Test arrangement A	B	C	D
1	200	5100	✓	✓	✓	✓
2	200	5100	✓	✓	✓	✓
3	200	5100	✓	✓	✓	✓
4	245	5100	✓	✓	✓	✓
5	245	5100	✓	✓	✓	
6	245	5100	✓	✓	✓	

(i) Test Arrangement A

(ii) Load Spreader in Test Arrangement A

(iii) Test Arrangement B

(iv) Test Arrangement C

Figure 2. Applied load arrangement for the static loading test programme.

2.4 *Dynamic tests*

As well as the point load arrangement for investigating the dynamic behaviour of the test specimens. Four of the specimens were assessed using a method that involved inducing excitation using a modally tuned impulse hammer at mid span of the structures. This load arrangement is referred to as Test Arrangement D, refer to Table 1.

2.5 *Deflection, dynamic response and slip measurement*

The vertical deflections were measured with LVDT's and were placed at mid span under each joist, and the bottom of the panel between joists 1 and 2, and under the load at third point at each joist. The horizontal 'slip' was measured by LVDT's at a distance from the supports of 275 mm and 320 mm for joists 3 and 4, respectively.

Strain gauges were attached to the top and soffit of the panel, and to the 'I' joists top and bottom flanges and to the web. Due to the symmetry of the composite flooring and to the loading arrangements only a quarter of the test specimens were instrumented.

Accelerometers are to be mounted at the mid span of each of the joists to measure the structures vertical vibration.

2.6 *Rate of applied loading*

Rate of loading onto each specimen was approximately 4 kN per minute. Specimens were loaded to a maximum of 10 mm of deflection in the linear range for all cases. This gives a *span to depth* ratio limit of L/510 which is on the conservative side considering that AS1684.1 – *Residential timber-frame construction and AS1720 – Timber Structures* code, allows L/400 or 12 mm and L/360 for serviceability calculations respectively. But given the unknown nature and properties of the new composite section it was considered prudent not to subject it to higher deflections.

3 EXPERIMENTAL RESULTS

3.1 *Linear range/serviceability limit state*

In addition, test specimens performance in the linear zone was also analyzed. The Linear limit for each specimen was identified using the following:

- The Linear regression curves were defined by the test data of the service zone.
- A divergence tolerance of 5% was estimated and from this the linear limit was found for each specimen. F_{div}: resistance and δ_{div}: deflection at divergence tolerance.
- The deflection limit δ_e = (Span length)/300 is also included. F_e: resistance at this limit.

A summary of the observed behaviour for the test specimens is as follows:

Test Specimen 1

- linear limit: $F_{div} = 44.3$ kN
- $F_e = 42.2$ kN
- average specimen deflection at F_{div}: $\delta_{div} = 18.3$ mm (bending and shear deflection)
  - 124.1% of theoretical estimating (bending and shear deflection)
  - 99.7% of theoretical estimating (bending deflection only)

Test Specimen 2

- linear limit: $F_{div} = 59.5$ kN
- $F_e = 42.0$ kN
- average specimen deflection at F_{div}: $\delta_{div} = 26.4$ mm (bending and shear deflection)
  - 122.6% of theoretical estimating (bending and shear deflection)
  - 98.5% of theoretical estimating (bending deflection only)

Test Specimen 3

- linear limit: $F_{div} = 48.6$ kN
- $F_e = 39.1$ kN
- average specimen deflection at F_{div}: $\delta_{div} = 22.1$ mm (bending and shear deflection)
  - 119.7% of theoretical estimating (bending and shear deflection)
  - 96.1% of theoretical estimating (bending deflection only)

Test Specimen 4

- linear limit: $F_{div} = 53.23$ kN
- $F_e = 52.3$ kN
- average specimen deflection at F_{div}: $\delta_{div} = 17.9$ mm (bending and shear deflection)
  - 113.9 % of theoretical estimating (bending and shear deflection)
  - 89.3 % of theoretical estimating (bending deflection only)

Test Specimen 5

- linear limit: $F_{div} = 55.8$ kN
- $F_e = 48.6$ kN
- average specimen deflection at F_{div}: $\delta_{div} = 18.5$ mm (bending and shear deflection)
  - 115.3% of theoretical estimating (bending and shear deflection)
  - 90.4% of theoretical estimating (bending deflection only)

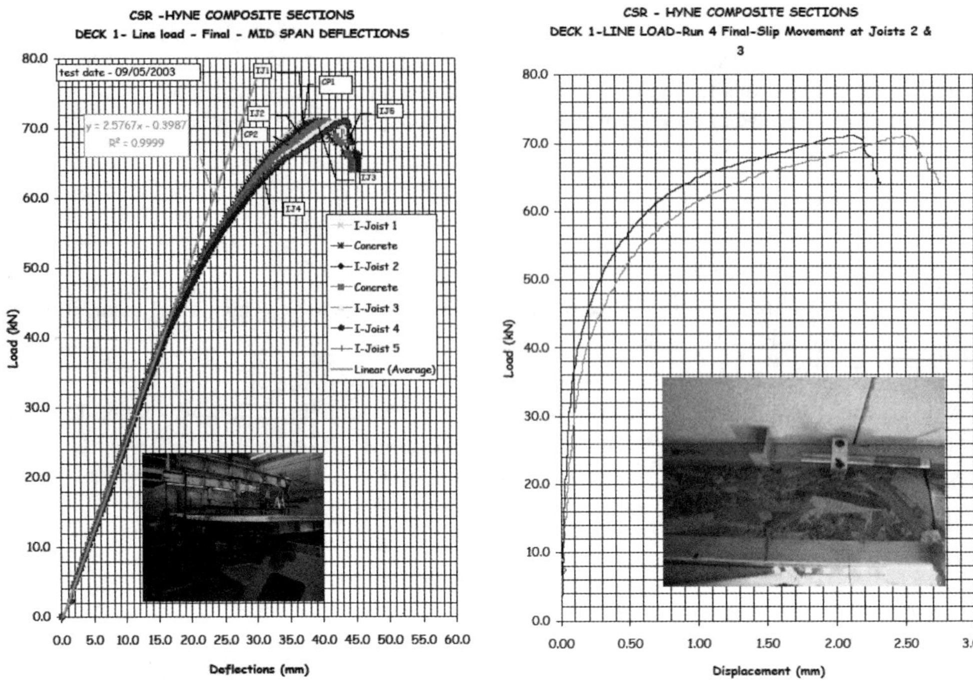

Figure 3. Load-deflection and Load-slip graphs for Test Specimen 1.

Test Specimen 6

- linear limit: $F_{div} = 65.5$ kN
- $F_e = 54.5$ kN
- average specimen deflection at F_{div}: $\delta_{div} = 14.2$ mm (bending and shear deflection)
  - 111.2% of theoretical estimating (bending and shear deflection)
  - 87.2% of theoretical estimating (bending deflection only)

3.2 *Ultimate capacity/ultimate strength limit state*

Four of the specimens experienced tensile failure at the bottom flange within the maximum bending moment zone, and the other two experienced similar failure located in the maximum bending shear zone. The observed ultimate strength behaviour is as follows:

Test Specimen 1

- Tensile failure of the bottom flange of a joist → tensile failure coincided with a knot,
- Failure was located at 1.7 m from support → within the max. bending moment zone,
- Ultimate resistance (ULS): $F_r = 71.1$ kN → 87% of theoretical value,
- Average specimen deflection at ULS: $\delta_r = 41.8$ mm (bending and shear deflection), and
- Joist 5 had the highest deflection at ULS; $\delta_r = 43.7$ mm (bending and shear deflection).

Test Specimen 2

- Tensile failure of the bottom flange of a joist → tensile failure coincided with a knot,
- Failure was located at 1.8 m from support → within the max. bending moment zone,
- Ultimate resistance (ULS): $F_r = 85.9$ kN → 105% of theoretical value,
- Average specimen deflection at ULS: $\delta_r = 59.3$ mm (bending and shear deflection), and
- Joist 1 had the highest deflection at ULS; $\delta_r = 64.2$ mm (bending and shear deflection).

Test Specimen 3

- Tensile failure of the bottom flange of a joist → tensile failure coincided with a knot,
- Failure was located at 2.7 m from support → within the max. bending moment zone,
- Ultimate resistance (ULS): $F_r = 57.8$ kN → 70.9% of theoretical value,
- Average specimen deflection at ULS: $\delta_r = 29.0$ mm (bending and shear deflection), and
- Joist 1 had the highest deflection at ULS; $\delta_r = 34.9$ mm (bending and shear deflection)

Test Specimen 4

- Tensile failure of the bottom flange of a joist → tensile failure coincided with a finger joist,
- Failure was located at 1.23 m from support → within the max. shear zone,
- Ultimate resistance (ULS): $F_r = 89.0$ kN → 89.2% of theoretical value,
- Average specimen deflection at ULS: $\delta_r = 38.3$ mm (bending and shear deflection), and
- Joist 1 had the highest deflection at ULS; $\delta_r = 39.7$ mm (bending and shear deflection).

Test Specimen 5

- Tensile failure of the bottom flange of a joist → tensile failure coincided with a sloping grain pattern,
- Failure was located at 1.23 m from support → within the max. bending moment zone,
- Ultimate resistance (ULS): $F_r = 79.8$ kN → 80.0% of theoretical value,
- Average specimen deflection at ULS: $\delta_r = 38.0$ mm (bending and shear deflection), and
- Joist 5 had the highest deflection at ULS; $\delta_r = 39.3$ mm (bending and shear deflection).

Test Specimen 6

- Tensile failure of the bottom flange of two joists, joist 4 and joist 5 failed simultaneously → tensile failures coincided with knots.
- Failure was located at 1.23 m from support → within the max. shear zone,
- Ultimate resistance (ULS): $F_r = 100.9$ kN → 101.2% of theoretical value,
- Average specimen deflection at ULS: $\delta_r = 48.8$ mm (bending and shear deflection),
- Joist 2 had the highest deflection at ULS; $\delta_r = 49.3$ mm (bending and shear deflection),
- Joist 4 deflection at ULS; $\delta_r = 48.5$ mm (bending and shear deflection), and
- Joist 5 deflection at ULS; $\delta_r = 49.2$ mm (bending and shear deflection).

3.3 *Dynamic tests*

The graph presented below in Figure 4 shows the 1st resonance frequencies for the test specimen. It shows the resonance (frequency) at a range of frequencies and also centrally loaded mass'.

Due to the fact that the fundamental frequency is a function of both stiffness and the mass, Figure 4 has been developed to indicate mass applied, frequency and span length.

The maximum safe span length of 7.3 m for Test Specimen 1 with a 1000 kg static centrally loaded mass is shown by the dotted line. Any span length equal to or less than this will have a first natural frequency of more than 5 Hz, thus safely lying outside the dominant human activity frequencies. With only the SoundFloor™ panel mass present, this span length increases to a maximum span length of 8.5 m.

Additionally, based on typical design rules for *Dynamic Response* where mid span deflection can not exceed 2.0 mm under a 1 kN point load, the composite section performed satisfactory under these conditions, refer to Table 2.

3.4 *Load distribution*

To investigate the load distribution, a concentric concentrated load was applied to the specimen. The load location was at mid span and on centre 'I' beam joist. The load was applied within the service range (linear load versus deflection).

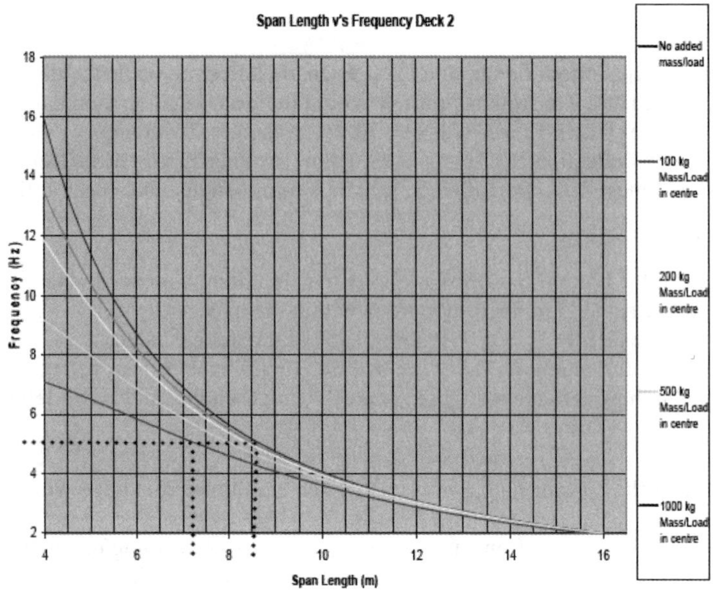

Figure 4. Dynamic properties of flooring systems.

Table 2. Measured joist deflection of the test specimens.

Test specimen	Applied load (kN)	Deflection (mm)
1	1.0	0.75
2	1.0	0.88
3	1.0	0.73
4	1.0	0.58
5	1.0	0.88
6	1.0	0.42

The test results showed that only limited load distribution occurred on all specimens. The load distribution was calculated by using:

Distribution $= \delta_{joist} / \Sigma(\delta_{joists})$ (Crews, 2002)

The graph in Figure 5 shows the extent of load distribution for Test Specimen 2.

The values on the graph in Figure 5 indicates that the joist deflection to the sum of all joist deflections experienced by the joists during the Test Arrangement C with a central joist deflection. The two external joists only received about 25% (12% & 13%) of the total deflection while the centre joist experienced about 30% of the total deflection on its own.

3.5 *Slip behaviour of the connection*

LVDT's were placed on joist 2 and 3 to record the displacement/slip between the joist and the panel. The testing of the specimens showed the connection slip remaining moderate through the service range (linear load verse deflection), showing that the test specimens remained as fully composite. Figure 3 shows typical load-deflection and slip movement results.

Non-linearity in the slip curves indicated that the connections experienced increase movement between the panel and joist. This movement coincided with the end of the linear load-deflection response from the test specimens. This could be because of two reasons; the first being adhesive

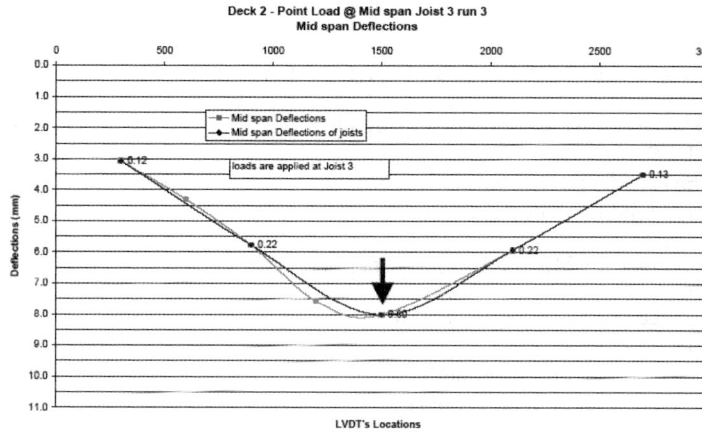

Figure 5. Proportion of load resisted by the 5 joists in Test Specimen 2 for Load Arrangement C.

or cohesion failures and the second being structure member failures such as the concrete and/or wood shear failures.

Adhesive failures are highly un-expected. Bonding failure is more likely to occur with the autoclaved aerated concrete (AAC) due to its lower shear strength. Such a scenario was demonstrated by Daher (Daher, 2003). In his research there was a high ratio of AAC failure with the broken specimen showing a high percentage of AAC material (residues).

Load slip graphs showed that joist 2 and 3 generally had an acceptable variation of slip level within the linear zone. Variation coefficient is inside acceptable range of 30% however the following remarks need to be made:

- Test Specimen 3 connection slip looked unreliable. The load-slip graphs have curves that are untidy and chaotic. The curves of joist 2 appear cleaner than joist 3.
- Test Specimen 2 showed that the slip on joist 3 was about three times that of joist 2, with joist 2 staying within the acceptable value.
- Test Specimen 4 showed slip at joist 2 being four times higher than at joist 3. The slip at joist 3 was 1.3 mm behaving linearly. This may indicate that the bonding was not sufficient. This could have been caused by damage before the completion of the ultimate resistance test.
- Test Specimens 2 and 3 showed that to obtain the slip connection with adhesive present is not a straight forward matter when brittle materials such as light concrete are used. They also show that fabrication of the bond must be controlled; using a sufficient amount of adhesive and also correct pressure applied during curing time, and also care needs to be taken when lifting so not to twist or shock the bond.

4 ANALYTICAL MODELLING OF FLEXURAL BEHAVIOUR

4.1 *Analytical model*

Theoretical estimation of ultimate limit state (ULS) and service limit state (SLS) responses of the specimens was undertaken using the transformed-section method with assumptions that the specimens have constant characteristics over the cross section and linear load-deflection behaviour until ULS.

To determine the characteristics of the composite cross section the tributary width of the panels had to be defined. Two approaches were considered, which are detailed in Eurocode 5 (EC5, 2002) and Australian Standard AS2327 (AS2327, 2003). The result of both calculations was 600 mm. This is the joist spacing of the specimens and was adopted for the calculations.

4.2 *Characteristics of the models*

For the test specimens with 200 mm I-joists (using test arrangement A):

- Ultimate resistance (ULS): $F_r = 81.55$ kN (total load)
- Deflection at ULS:
  - $\delta_r = 44.4$ mm (bending and shear deflection),
  - $\delta_r = 36.6$ mm (bending deflection only)

For the test specimens with 245 mm I-joists (using test arrangement A):

- Ultimate resistance (ULS): $F_r = 99.7$ kN (total load)
- Deflection at ULS:
  - $\delta_r = 38.1$ mm (bending and shear deflection),
  - $\delta_r = 29.9$ mm (bending deflection only)

4.3 *Comparison of analytical model characteristics and test results*

4.3.1 *Linear range/serviceability limit state*

The analytical model gives a satisfactory estimation of the specimen deflections and gives a reasonable estimation of the serviceability of the specimens. It over-estimates the deflection when bending and shear deflections are considered. When only the bending deflection is considered, the model estimates the deflection well. The model accuracy/response vary with test specimens built with 200 mm 'I' beam joists and specimens built with 245 mm 'I' beam joists.

4.3.2 *Ultimate capacity/ultimate limit state*

The location of any joist that experienced rupture was random. There was no trend showing that the failure will occur at a particular joist. All specimens with the 200 mm deep joists experienced failure in the maximum bending moment zone (between the two line loads). These failures occurred at a knot in the bottom flange. Specimens with the 245 mm deep joists showed failure within the maximum shear zone at the bottom flange of a joist.

5 CONCLUSIONS

An experimental and analytical investigation of the flexural behaviour of an AAC panel/timber joist floor system has been described. The significance of the joist system on the deflection and ultimate capacity behaviour of the floor system has been illustrated in the results. The conclusions are based on results of test specimens constructed with Hyne & Son P/L 'I' beam joists. From analysis and testing of the test specimens the following conclusions are made:

- Distance between joists need to be a maximum of 600 mm. This value should be reduced once span lengths are less than 2400 mm.
- A sufficient amount of adhesive should be carefully applied with the correct pressure. Care should be taken as to not break or shock the bond during the curing stage. The adhesive should not change its properties over time.
- Service Limit State can be treated as a fully composite structure.
- Ultimate Limit State should not be considered as fully composite.
- The maximum span length for all specimens that will keep the first vertical resonance frequency above 5 Hz is sufficiently larger than in the static tests. For this test programme, the span length is limited by the static tests, except in the case of very large static loads being permanently on the specimen (more than 1 ton).

ACKNOWLEDGEMENT

CSR Hebel would like to thank Hyne & Son P/L for jointly funding this project and the assistance of the Centre for Built Infrastructure Research of the University of Technology, Sydney for performing the testing and analytical investigation of the project results.

REFERENCES

AS2327. 2003. Composite structures – Part 1: Simply supported beams, Standards Australia.

Benitez, M. 2004. Tests and analytical evaluation of a composite flooring system, *Proceedings of the 8th World Conference on Timber Engineering, Lahti, Finland, 14–17, 2004.*

Crews, K.I. 2002. Behaviour and Critical Limit States of Transversely Laminated Timber Cellular Bridge Decks, PhD Thesis, University of Technology, Sydney – Faculty of Engineering, Sydney.

CSR Hebel. 2003. Design Guide SoundFloor™. Sydney: CSR Hebel.

Daher, G. 2003. Composite Floor System Using Hebel Panels & Timber 'I' Beams, Undergraduate Project, University of Technology, Sydney – Faculty of Engineering, Sydney.

EC5, 2002. Eurocode 5 Design of timber structures – EN1995-1.1 Common rules and rules for buildings, 2002–08.

Hyne & Son. 2001. Hyne I Beam Engineered Timber Products-Span Table, Maryborough: Hyne & Son, P/L.

University of Technology, Sydney (UTS). 2005. Investigation of Composite Action Behaviour in a Lightweight Autoclave Aerated Concrete-Timber Composite Floor System, University of Technology, Sydney – Centre for Built Infrastructure Research of the University of Technology, Sydney, Sydney.

Autoclaved Aerated Concrete – Limbachiya and Roberts (eds)
© 2005 Taylor & Francis Group, London, ISBN 0 415 38356 0

The use of AAC blocks for foundations in UK construction

T. Grounds & A.H. Riza
Tarmac Topblock Limited, Wolverhampton, UK

ABSTRACT: Over a number of years the combination of testing (laboratory and field) and trouble-free in-service performance, has demonstrated the benefits of the use of AAC blocks for solid foundation walls below damp proof course (DPC). This has led to the widespread adoption of this form of construction in the UK particularly in housing.

The benefits include:

- Improved construction productivity due to the ability to manually handle larger blocks and ease of working with AAC.
- Improved thermal insulation performance for ground floors, which has been recognised within the UK Building Regulations.
- Excellent durability performance as demonstrated by resistance to freeze thaw attack and the actions of sulphate-bearing groundwater, both of which have been recognised in UK construction guidance and certification.

1 INTRODUCTION

During the last 20 years, the use of solid AAC blocks for the foundation walls to external and internal walls in UK construction has grown significantly, particularly for housing which is the major market for AAC or 'Aircrete' as it is known in the UK. This paper discusses the technical background work to support this development. The terms 'foundation' or 'trench' blocks are often used to describe blocks used below DPC.

Figure 1 shows a typical application detail for foundation blocks used to support an external cavity wall, the predominant form of external wall construction used in the UK.

The depth below ground level to which the foundation walls are taken depends on site conditions and the depth at which a good bearing strata is reached, usually less than 1.5 m. For economic reasons, the use of relatively shallow, strip concrete foundations are preferred to mass filling the trench with concrete. Masonry foundation walls are then used to raise the loadbearing elements to ground level. Typically, foundation blocks are produced in a range of densities and strengths of 450–750 kg/m^3 and 3.5–7.0 N/mm^2, when measured according to the current British Standard, BS 6073. The bed heights are usually 215 mm or 140 mm, which correspond to traditional sizes of UK brick coursing heights (65 mm). Block lengths are between 440 and 620 mm.

During the last 10 years the market for this type of product is estimated to have doubled and, although there are large regional variations, has reached up to 25% of production in some areas. This growth rate has been achieved through research and technical demonstration of fitness for purpose supported by the Autoclaved Aerated Concrete Products Association (AACPA) and through recognition by the end user of the potential benefits of AAC in this application. For example:

- Increased productivity,
- Reduced material cost,
- Ease of handling and working,
- Improved thermal performance.

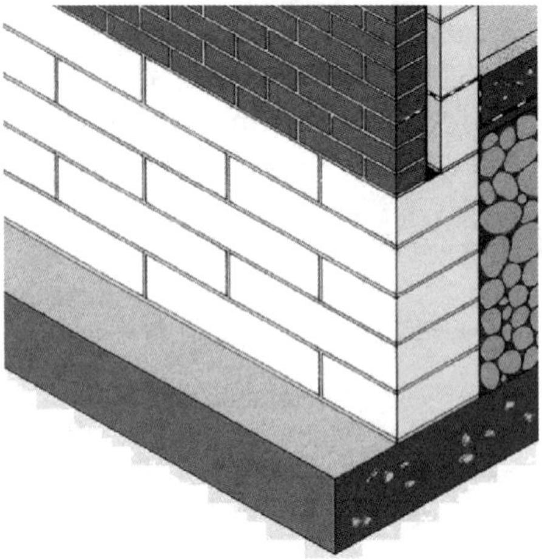

Figure 1. Typical application of solid AAC foundation blocks used to support a cavity wall.

2 COMMERCIAL DRIVERS

In common with construction industries worldwide, one of the major objectives of the builder is to reduce costs, maximise efficiency and hopefully improve profits whilst maintaining risk free construction methods.

Due to their improved productivity and simplicity in use, solid AAC foundation walls have provided an extremely cost effective construction when compared to the typical alternative constructions.

2.1 *Reduced material cost*

Traditionally, foundation walls would have been constructed in solid brickwork to at least the same thickness as the wall being supported. As the advent of cavity external walls resulted in wider wall zones, it became common for the load bearing leaves of the cavity construction to be carried down to the concrete foundation. These leaves should be tied with cavity wall ties and may be of brick or block construction, however, the cavity between the two leaves is required to be filled with a concrete mix for the section of wall below ground to prevent inward collapse as the trench is backfilled with soil. The use of solid foundation walls simplifies the construction process and removes the need for additional materials, such as infill concrete and cavity wall ties. However, the use of dense masonry units tends to be less cost effective given that they would need to be small enough to be easily handled.

2.2 *Manual handling requirements*

The UK Health and Safety Executive launched, some 12 years ago, major new guidance aimed at improving the health, safety and welfare record of the construction industry. This included advice on the repetitive lifting of masonry units intended to reduce the incidence of back injuries in brick and block layers (Construction Sheet 37). The advice given recommended that, for a unit weight above 20 kg, a manual handling assessment should be carried out to ensure safe working practice. With inherent lightweight properties, most AAC foundation blocks fall below 20 kg, giving a significant incentive to use these products instead of heavier types of unit.

2.3 *Productivity gains*

Independent productivity trials (Percy Howes, unpublished) have shown that a significant time saving could be achieved using solid AAC walls depending on the depth of foundation walls and

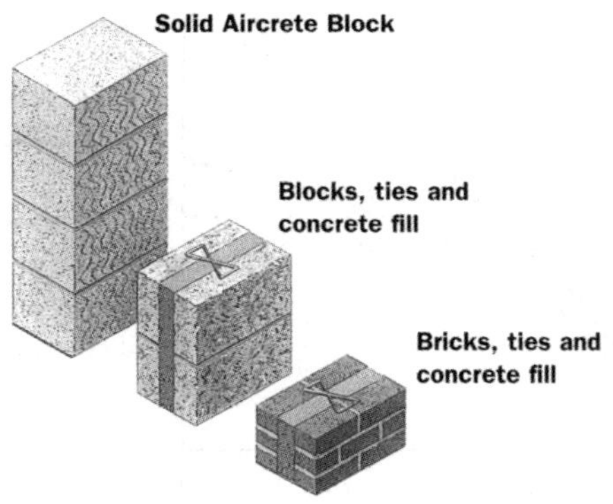

Figure 2. Comparative laying rates for foundation walls per unit of time.

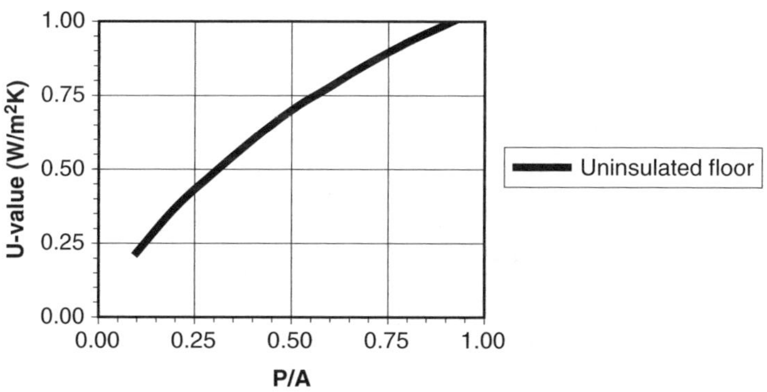

Figure 3. Relationship of ground floor U-value and perimeter/area ratio.

that the average saving was approximately 70% compared to cavity brickwork and 45% compared to aggregate concrete blocks, Figure 2. This means that up to twice as much solid AAC foundation walling can be built in the same time as cavity block walls and, more significantly, up to four times the area of brick cavity constructions.

Coupled with ease of working the AAC, and a reduction in block layer fatigue due to the lighter weight of the product, the potential to reduce time on ground works is easily apparent, thus reducing construction costs. This situation is likely to improve further with the increased use of tongue and grooved blocks, which would not require vertical joints to be filled with mortar, as is common in mainland Europe. The second major commercial and technical benefit is a potential improvement in the thermal performance of the building.

2.4 *Thermal insulation benefits*

2.4.1 *Thermal transmittance (U-value)*

The transmission of heat through ground floors is related to the size and shape of the floor (BS EN ISO 13370). This relationship is often expressed as the ratio of the floors perimeter length, P, to its surface area, A. Figure 3 plots the U-value against P/A.

It can be seen that larger P/A ratios (i.e. greater perimeter lengths relative to area) will have greater (and therefore worse) U-values. Consequently, insulation positioned around its perimeter can significantly reduce the overall heat loss through the floor, particularly where a suspended ground floor is used to incorporate a beam and AAC infill block floor (Fig. 4).

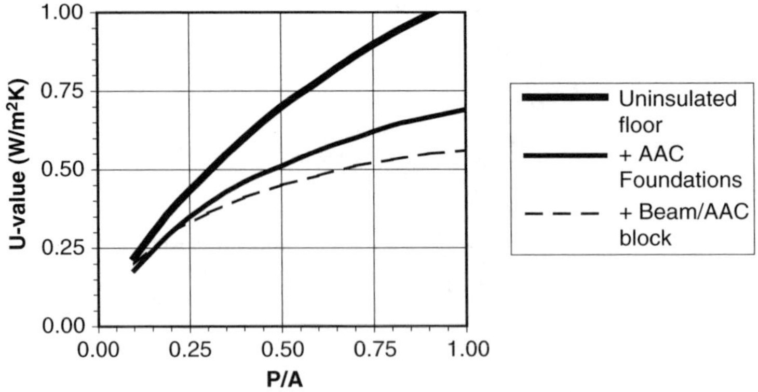

Figure 4. Effect of AAC foundation blocks of ground floor U-value.

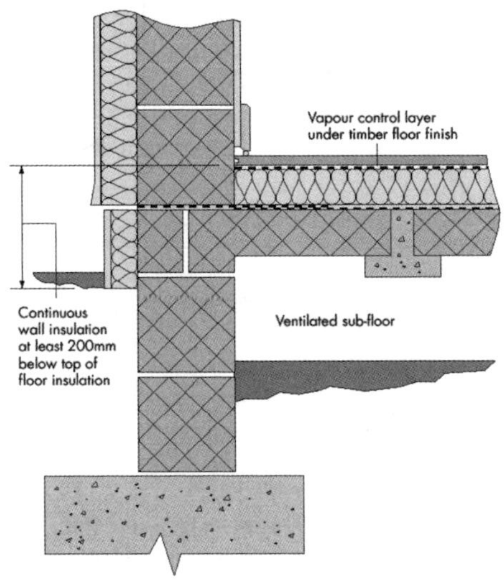

Figure 5. Typical Robust detail using foundation blocks and beam and floor.

In this situation it is possible to minimise the need for expensive additional insulation by utilizing the combination of good structural performance and thermal insulation properties of AAC.

2.4.2 *Limiting thermal bridging*

One of the important aspects of thermal efficiency of buildings is to limit thermal bridging and air leakage both of which can undermine the overall performance of otherwise well insulated structures. The effects of cold bridging can be calculated in accordance with BS EN 10211 but the UK Regulations include a simplified approach based on constructions acknowledged to perform well in reducing both thermal bridging and air leakage. These are known as the Standard Robust Details and include reference to AAC foundation walls which allows the builder to adopt this form of construction without the need for further proving calculations (Figure 5).

The key factor is that the AAC foundations provide continuity by giving a similar insulation performance to the wall above DPC. This can remove the need for additional layers of insulation needed to avoid thermal bridging for denser foundation materials. This simplifies the construction detail, saves cost and improves in service performance.

3 TECHNICAL SUPPORT

The advantages outlined above can only be used if the product can meet other technical performance requirements for the application i.e. fitness for purpose. The main requirements are for structural performance and durability.

3.1 *Structural performance*

The majority of housing in the UK is not individually designed for structural performance. Instead the fabric of the house is specified and built to comply with Building Regulations using simple design rules that have been proven over the years of experience in service to provide structurally sound houses. Approved Document A to the England and Wales Building Regulations gives minimum block strengths which are suitable for most domestic situations, provided that certain limiting dimension stipulated within the Document are met.

In brief, loadbearing walls of two storey houses must be constructed of blocks with a minimum compressive strength of 2.8 N/mm^2 when tested in accordance with BS6073, for three storeys the lowest storey must consist of 7 N/mm^2 blocks (these strengths relate to 2.9 N/mm^2 and 7.3 N/mm^2 respectively when tested to EN 771-4). The main load bearing wall element in UK housing is the inner leaf of the external cavity wall, which is usually of 100 mm thickness. AAC blocks can meet these structural requirements at the narrow thicknesses of the inner leaf and therefore can easily achieve the required capacity in thicker foundation walls. This effectively allows foundation blocks to be used at 2.8 N/mm^2 for structural purposes however, currently, for durability reasons the practical strength of AAC foundations is currently no lower than 3.5 N/mm^2. This is in part due to the conservative nature of the assessment methods for durability and, as the assessment methods are refined, could provide further opportunities for optimisation of the benefits of AAC in this application.

3.2 *Durability*

It is obviously essential that building materials should maintain the required performance in service for a long period of time. This period will depend to some extent on the ease of maintenance and/or replacement of the particular element. For blocks below ground, it is essential that their performance match, at the very least, the intended lifetime of the building.

The key aspects of durability in this case are resistance to degradation by freezing and thawing cycles and resistance to potential aggressive substances, which may be present in the ground, in particular sulfates.

3.2.1 *Frost resistance*

Although the UK is not prone to particularly severe winter weather conditions, foundation blocks are often used in the worst positions within the building that are subject to freezing and thawing cycles while potentially being at high moisture content.

BS5628 Part 3 states that most types of AAC blocks would be suitable, reference is also made to consult the manufacturer.

As BS 6073 does not contain, or refer to, a test method for assessing frost resistance for blocks, and as there is currently no agreed European test method, the AACPA consulted the British Board of Agrement (BBA), a member of EOTA, for a suitable method.

The method chosen is detailed in BBA MOAT 12, 1977. In this test a random sample of 20 blocks is taken and divided into four batches of five blocks. The first batch is tested for compressive strength in accordance with BS 6073 while the remaining blocks are immersed in water for 48 hours. They are then placed in a tray containing 25 mm water (oriented as they would be in a wall) with one bed face immersed in the water. In this condition the blocks are then cycled for 60 times through a regime of $-5°C$ for 16 hours, $+15°C$ for 6 hours and then re-immersed fully in water for 2 hours. They are assessed for damage and strength loss at 20, 40 and 60 cycles. This is, therefore, a relatively severe test and blocks are deemed to show satisfactory performance if the volume loss is less than 5% and the strength loss less than 10%.

By subjecting products to this assessment with results certified by the BBA, the industry has been able to demonstrate that blocks at a density of at least 450 kg/m^3 can be suitable for foundation walls subjected to the normal freezing and thawing cycles which can be expected over the lifespan of a building.

3.2.2 *Resistance to sulfate bearing ground waters*

For many years the Building Research Establishment (BRE) has published guidance for the UK construction industry regarding the requirements for concrete in the ground and subject to a sulfate-bearing environment. This stems from concerns about the potential for attack of the concrete predominantly through the reaction of sulfate with the calcium hydroxide and tri-calcium aluminate components in hydrated cement. This potentially leads to the formation of expansive ettringite (calcium tri-sulfoaluminate hydrate) and disruption of the concrete (BRE 2001). In addition, as planning policy in the UK has encouraged the development of infill sites and 'brownfield' sites (sites which have been previously been built on), concerns over the greater potential for site conditions to include higher levels of sulfate and other possible contaminants have risen and led to more stringent recommendations for concrete in the ground.

In addition, approximately 5–6 years ago, the mechanism known as the thaumasite form of sulfate attack was identified in deterioration of bridge abutments in a particular part of the UK. This is a potentially more serious form of sulfate attack in that it can lead to a progressive deterioration of the hydrated cement in concrete (DETR 1999). As this mechanism can occur in conditions where the temperature tends to be below 15°C and where there is a source of sulfate and carbonate ions, it needs to be considered for all concrete used in the ground.

It is well documented (Tada 1992) that the autoclaving stage of the AAC process alters the form of the reaction products usually associated with the hydration of cement. The calcium hydroxide and calcium aluminates present in the mixture of cement, lime and silica are converted into crystalline forms of complex calcium silicate hydrates, calcium aluminate hydrates and calcium sulphoaluminate hydrates. This conversion imparts a significantly improved resistance to the types of sulfate attack outlined above. The latest BRE document on this subject, Special Digest 1 2001, classifies sulfate conditions from DS1 to DS5 based on the concentration of sulfates within the groundwater, DS1 being the lowest level at <0.4 g/l and DS5 the highest at >6.0 g/l of sulfate. This is further subdivided depending on the acidity and other site conditions.

To assist in the development of their guidance, in 1970 the BRE began a research project to assess the long term resistance of various types of concrete buried in sulfate bearing ground conditions at Northwick Park Hospital, north of London (Harrison et al, 1974). The highest level of sulfate here equated to DS4 sulfate conditions but the general level was classed as DS3. This project assessed the deterioration of concrete both in these conditions and in laboratory simulated acceleration tests.

AAC blocks were included in this study, which confirmed that AAC at density greater than 600 kg/m^3 could be used without question in up to Class 3 sulfate conditions. Because the AAC had actually performed well in more severe sulfate conditions, the manufacturers have again turned to BBA certification to demonstrate a higher level of capability.

Through a combination of examination of the results from Northwick Park and further accelerated laboratory testing, the BBA has assessed individual products from the range produced in the UK. The assessment method used by the BBA was developed at the BRE (Osbourne, 1989) and included two stages, accelerated testing over 12 weeks and long term testing over 3 years.

The accelerated testing comprised the immersion of different shaped samples in a solution of 2% magnesium sulfate, a very aggressive environment that was maintained by replacement of the solution every two weeks. Acceleration was enhanced by the use of small specimens.

10 mm cubes were visually assessed for deterioration. Flat prisms, $10 \times 40 \times 160$ mm were measured for expansion and square prisms $10 \times 10 \times 60$ mm were tested for bending strength, in both cases results are compared to water stored control samples. The criteria for showing adequate

resistance to the conditions were, no visual deterioration, expansion of <0.05% and bending strength of at least 80% of the control samples.

The long term tests comprised the immersion in the solution of 100 mm cubes for three years tested for compression strength at 3, 6, 9, 12, 18, 24 and 36 months with comparison to water stored control cubes. The solution was renewed at each of these test dates. The criterion for adequate resistance was a strength for the sulfate stored cubes of at least 80% of the control.

After using this method, the BBA were able to provide certification that AAC blocks can be used in conditions beyond the levels given in the original BRE guidance. The result is that certain AAC products are considered suitable for up to Class 4 sulfates or allowed for up to Class 3 but at 450 kg/m^3 density, thus optimising the benefits of using AAC in the conditions. The latest version of SD1 recognises this additional guidance by reference to BBA certification. To date we are not aware of any site failures through sulfate attack of AAC foundations.

4 SUMMARY

The success of AAC used in foundation walls demonstrates its flexibility as a construction material. Tradition and conservatism, particularly in the UK construction industry, suggested that a material as light as AAC would not be accepted for use in such a situation. However, the combination of potential benefits for both builder and building owner and the demonstration of the ability of AAC to perform beyond its perceived limits, has given users the confidence to adopt this form of construction.

In conclusion, it is hoped that such good experience of performance in use will encourage a wider use of AAC for solid external walls above ground in the UK as is common practice elsewhere in Europe and Worldwide.

REFERENCES

BBA M.O.A.T. 12 1977, British Board of Agrement, 1977

BRE Special Digest 1 Concrete in aggressive ground, CRC Ltd, 2001

BS 5628-3 2001: Code of Practice for use of masonry – Part 3: Materials and components, design and workmanship, BSI 2001

BS 6073 1981 Precast Concrete Masonry Units, BSI 2003

BS EN 10211-2:2001 Thermal bridges in building construction – Calculation of heat flows and surface temperatures – Part 2 Linear thermal bridges

Construction Sheet No 37 Handling Building Blocks, HSE 1993

DETR, 1999. The thaumasite form of sulfate attack: Risks, diagnosis, remedial works and guidance on new construction. Department of Environment, Transport and the Regions, 1999

EN ISO 13370: 1998 Thermal performance of buildings – Heat transfer via the ground – Calculation methods, BSI 1998

European Standard EN 771-4 2003: Autoclaved Aerated Concrete masonry units, CEN 2003

Harrison W H, Teychenne DC, Forrester J A and Spratt B H, 1974 Long term investigation of sulphate resistance of concrete at Northwick Park. Field installation and initial testing. CIRIA Technical Note 57. Construction Industry Research and Information Association, 1974

Limiting thermal bridging and air leakage: Robust construction details for dwellings and similar buildings, The Stationery Office, 2001

Osbourne GJ, 1989. Determination of the sulphate resistance of blastfurnace slag cements using small-scale accelerated methods of test. Advances in Cement Research, January 1989, 2 (1) 19–25

Percy Howes, Chartered Surveyors, www.thermalite.co.uk

Tada S 1992 Bibliography on autoclaved aerated concrete 1982–1992, in Proceedings of the 3rd RILEM international symposium on autoclaved aerated concrete, Zurich, 1992. Balkema, Rotterdam 1992

The Building Regulations 2000, Approved Document A Structure, The Stationery Office, 2004

Autoclaved Aerated Concrete – Limbachiya and Roberts (eds)
 ISBN 0 415 38356 0

The use of polymer mesh reinforcement in AAC walls constructed with thin layer mortar

I.J. Gray
Tarmac Topblock Limited, Wolverhampton, United Kingdom

ABSTRACT: The control of shrinkage movement in AAC blockwork panels constructed with conventional sand/cement mortars is normally accommodated by a combination of vertical movement joints and the use of lower strength mortars (1:1:6).

The use of thin layer mortar techniques in the UK, whilst gaining momentum, has been slow. This has mainly due to the reluctance of the trades to adopt new site practices. Experience has shown that the use of stainless steel bed joint reinforcement (Murfor) to control shrinkage movement has proved difficult with blocklayers, despite its wide spread use in other parts of Europe. This resistance has been two fold, firstly the material cost and secondly the practicality in use.

In recent years polymer mesh reinforcement has been developed which is similar to the mesh-cloth used in projection renders. It overcomes both of the perceived problems with traditional reinforcement in that it cost a fraction of stainless steel and is very simple to use on site.

A series of testing, monitored by John Moores University School of the Built Environment, was conducted to assess the performance of the polymer mesh compared to conventional stainless steel bed joint reinforcement in thin layer mortar walls.

1 INTRODUCTION

During the last 30 years the use of AAC blockwork laid in a thin layer mortar has been used for a variety of construction projects ranging from housing, to leisure projects, to refurbishment, to hospitals and schools. Throughout this period the adoption of this form of construction has been sporadic an often dictated by the trends within the construction industry and certainly slowed by the reluctance of the industry to embrace change. A typical example of this was during the early 1990's when a number of the larger house builders conducted site trials.

The housing market was buoyant and the builders where looking to increase their productivity whilst retaining a "masonry" construction method. Traditionally, the external cavity walls were largely faced with brickwork with aircrete as the inner leaf, bedded in conventional sand and cement mortar, which met the thermal regulations of the day without the need for additional insulation. Replacing the conventional mortar with thin layer mortar offered the opportunity to construct the inner leaf in advance of the outer leaf which in turn meant the internal floors and roof could be installed and the building may weather proof. This allowed the other trades, carpenters, plumbers, electricians and decorators to start and finish their work whilst the external brickwork was being completed thereby significantly reducing the build time.

When the housing market went into decline the house builders no longer need a quicker build form and reverted back to more conventional building methods. Since that time all of the UK Aircrete manufactures have introduced their version of this type of product, and with government initiatives focusing on Modern Methods of Construction (MMC) renewed interest is focused on this form of construction.

So whilst this form has been available for some time it is still seen as being new to the UK. Clearly, other parts of Europe are far more advanced and much of the technology is copied from this experience. However, the UK building industry is largely very conservative and reluctant to accept change. This is certainly the case with the brick and blocklaying trades and as a consequence we have consciously tried to adapt the accepted European practice so that it doesn't detract too far from "traditional" blockwork. It could be argued that this was a mistake but set against the back-drop of a "skills shortage" it appeared to be a reasonable strategy.

2 CONTROLLING MOVEMENT IN BLOCKWORK

When using conventional blockwork, regardless of the type of masonry, the routine use of any form of reinforcement in the mortar beds is not common practice in the UK. In general terms drying shrinkage or thermal movement is accommodated by introducing on or more of the following measures:

a. Vertical movement (contraction) joints at a nominal spacing of around 6 m in plain wall panels and at half this distance from corners and returns.
b. Where panels are punctured by openings for windows and doors it is recommended that the spandrel panels are protected with 2 courses of suitable bed joint reinforcement extending some 600 mm past the reveals.
c. The use of the correct designation of mortar. For AAC this is accepted as being a 1:1:6 cement/lime/sand designation iii mortar or more commonly a 1:6 cement/sand mix with the addition of a proprietary plasticiser. The blockwork is able absorb a certain amount of movement by the formation of micro cracks with in the mortar joints.

In practice vertical movement joints and bed joint reinforcement are used in larger works but are not commonly used in housing. The control of site batched mortars can be variable and so the consequence is the acceptance of some hairline, non-structural, cracking below openings particularly in housing.

Pre-mixed, dry packed, thin layer mortars formulations are by nature stronger than conventional mortars, typically with mix proportions in the order of 1:2 cement/sand with the addition of water-retaining and workability agents. This means that the resulting blockwork is likely to be less able to accommodate the normal movement associated with drying shrinkage as the building dries out. As the mortar is stronger than the blocks any excessive movement may result in cracking within the blockwork. For this reason the use of a bed joint reinforcement, together with movement joints, becomes an integral part of the design.

3 REINFORCEMENT OPTIONS

In most parts of Europe, where thin layer blockwork is a conventional technique, stainless steel reinforcement is introduced into the bed joints as a matter of practice. This may typically be in every joint or placed in a number of joints producing a ring beam effect and often in association with movement joints. Corner pieces are supplied to provide continuity around corners.

In the UK where bed joint reinforcement is not common practice there has been reluctance in some case to insert the steel. This is simply due to the fact that brick/blocklayers don't carry tools to cut the reinforcement to length and also they do not appreciate why it is being used. Consequently, on many early sites the reinforcement was either left out completely or put in less frequently than required. Similarly, the detail at corners was not always completed correctly.

Commercially the cost of the various accessories that are needed for this type of blockwork became a high proportion of the overall cost. This started to become a barrier to the take up of this type of construction albeit that the potential benefits of improved productivity would compensate for the apparent additional initial cost.

It was clear that whilst there was a good technical solution to the need to reinforce the blockwork it was not ideally suited to the UK construction industry from a practical and commercial point of view.

4 POLYMER CRACK CONTROL MESH

The use of fabrics to provide a degree of "reinforcement" in building components is fairly wide spread, ranging from simple scrims and mats in plastering and renders to complex high strength continuous fibres of carbon, aramidic and glass in polymer matrices. Typically, these fibre reinforced plastics are used the restoration of historic buildings improving their behaviour under seismic conditions.

A manufacturer of thin layer mortar, working with the UK distributor of their product, looked into the use of a polymer mesh that would have sufficient strength to resist, and limit, the effects of drying shrinkage in AAC blockwork panels.

Various types of mesh were investigated, although the purpose there is not to look the constituency of the mesh but rather to examine its performance in use. Practically, it was found that they were easy to handle on a construction site, which lead to them being consistently inserted. In general terms the mesh was placed in at least 2 courses above and below opening and more frequently in plain gable walls. Vertical movement joints where located at the nominal centres in line with normal UK site practice. The mesh was not used to increase the distance between movement joints, as can be the case with traditional bed joint reinforcement.

The results from the use on sites were encouraging in so much as there was little evidence of cracking, and where it had occurred it was generally hairline. This was often associated with unfilled perpendicular (perpends) mortar joints. The view was taken that the absence of a full perp joint had created a stress point, by reducing the wall section, thereby inducing a crack.

Earlier work with steel reinforcement indicated that the blockwork was prone to larger movement when it become saturated due to prevailing weather conditions during the construction of the blockwork. The wetted condition had not been experienced during the polymer site trials but needed further investigation.

5 LABORATORY TESTING

In order to substantiate the performance of the polymer mesh a series tests where commissioned, with the monitoring by John Moores University, to replicate the worst case scenario. Nine test walls where constructed, in a small industrial unit, using conventional sand and cement mortar as a control panel and thin layer mortar which was either plain or reinforced with steel or the polymer mesh, see Table 1 below. The walls where 8.0 m in length, six courses (215 mm) high, restrained at both ends and isolated from the floor by a polymer dpc bedded in conventional mortar (see Figure 1).

Table 1. Details of mortar and reinforcement used.

Wall reference	Type of mortar	Reinforcement to bed joints	Open perpends
A1	Sand & cement	None	None
A2	Thin layer mortar	None	None
A3	Thin layer mortar	Steel every course	None
A4	Thin layer mortar	Polymer every course	None
A4C	Thin layer mortar	Polymer every course*	None
B1	Sand & cement	None	Three
B2	Thin layer mortar	None	Three
B3	Thin layer mortar	Steel every course	Three
B4	Thin layer mortar	Polymer every course	Three

* Polymer mesh placed on the bedding face before the mortar was laid.

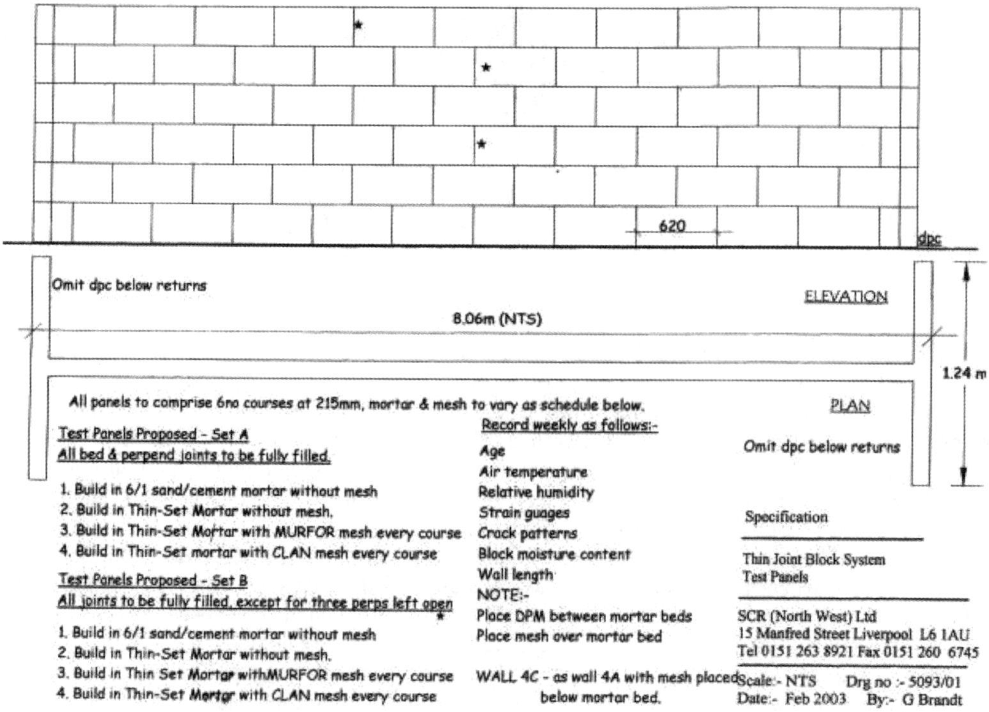

Figure 1. Layout of test walls.

Selected perp joints were not filled with mortar to investigate the theory that inconsistent workmanship could increase the risk of cracks developing. Where reinforcement was inserted it was placed into the mortar once it had been spread, with the exception of wall A4C* in which it was placed under the mortar. Prior to the start of the measurement process the walls were saturated and allowed to dry naturally during the test period.

5.1 *Terms of reference*

Liverpool John Moores University School of the Built Environment was commissioned to instrument and monitor the series of test walls during the period May to November 2003. The following were required to be recorded:

Age
Air temperature
Relative humidity
Strain gauges (dial indicators or similar)
Displacement across the open perpends
Crack patterns
Block moisture content
Wall length

The frequency of measurements were set at the following intervals:

Month 1	weekly
Month 2	fortnightly
Month 3-4-5	monthly

5.2 *Instrumentation*

Instrumentation points were set up along the walls after the panels had been saturated and monitoring commenced on day 6. The following measurements were taken, as the frequency indicated above, to monitor the drying of the walls and any subsequent movement and cracking.

5.3 *Environmental conditions*

Air temperature and relative humidity were measured using electronic sensors at three positions along each wall and the average value was recorded.

The moisture content of the blocks was measured to determine the rate of drying of the blockwork. Surface readings (electronic) were taken along the length of the walls and drilled samples taken from the return ends. These were analysed by using a "Speedy" moisture meter and also by weight and oven drying. As the electronic meter is non-destructive it enabled more reading to be taken directly from the test walls rather than the return ends.

5.4 *Effectiveness of the end restraint*

In order to ensure that the panels were effectively restrained at the ends they were bonded into 215 mm solid blockwork cross walls. Unlike the panels these were bonded directly to the floor base with sand/cement mortar. The overall length of the walls was measured using a tension steel tape with corrections being made for the calibration constants of the tape for temperature and tension.

5.5 *Slippage along the base/dpc*

At four positions along the wall digital calliper gauges were use to ensure that the dpc was acting as an effective slip plane.

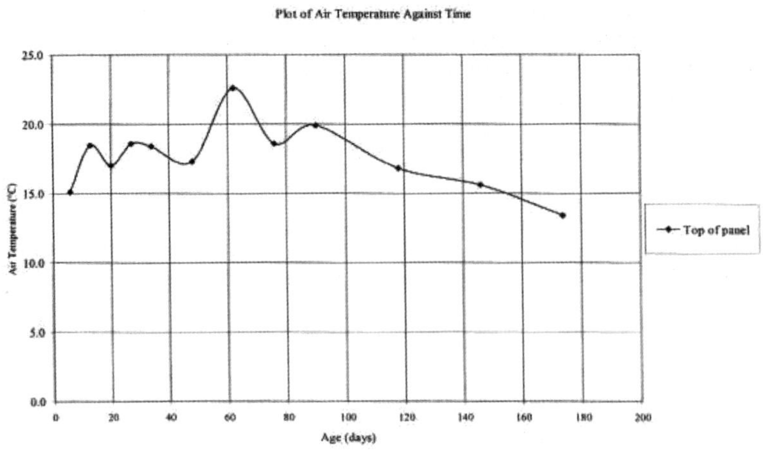

Figure 2. Variation of air temperature during the test period.

5.6 *Displacement across the perpends*

At eight locations along the walls pair of small studs were glued in place either side of the perpend joint. Where perpends were left open two pairs of studs were used at the top and bottom of the joint.

5.7 *Cracking*

Close visual inspection was used to monitor any cracking that occurred. The crack width was measured using an optical crack microscope and the extent of the crack was marked on the wall after each visit.

6 RESULTS OF OBSERVATIONS

6.1 *Relative humidity and temperature*

After a brief cold period, the air temperature in June/July remained at 18 ± 1°C with the exception of a warmer period in mid-July. Temperature then gradually declined after the summer to a level of approx. 13°C in November (see Figure 2). No significant variation was observed in temperature in different parts of the building.

After initially high humidity readings of 80 to 90% RH following wetting of the walls, the relative humidity remained mostly between 70 and 80%, with peaks and troughs reflecting ambient weather conditions (Figure 3).

6.2 *Moisture content of blockwork*

The moisture content of all of the walls reduced steadily from a value above 28% (the upper limit of the instrumentation) over the first 50 to 60 days, levelling off thereafter as expected as the blocks came to equilibrium with the surrounding air. Fluctuations in moisture content during this later stage followed changes in air humidity, particularly where the latter was brought on by persistent rain. Results from samples drilled from the returns confirmed the trend monitored using the non-destructive electrical resistance moisture meter.

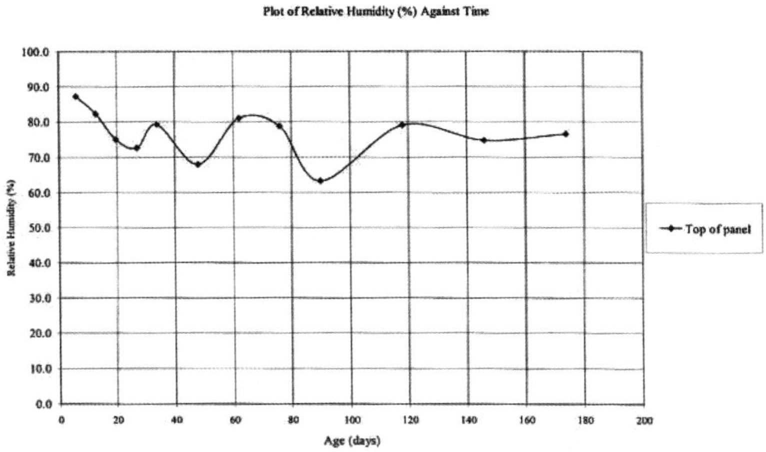

Figure 3. Variation of air temperature during the test period.

6.3 *Effectiveness of returns in providing restraint*

The overall length of the walls as measured near their bases remained essentially constant, within the reading error of the tape (±0.5 mm), indicating that the bases of the returns had not de-bonded from the floor. At the top of the wall, an initial overall shrinkage of up to 2 mm was noted over the first month, thereafter fluctuating with changes in air humidity. This is to be expected, as the top of each return will be less well fixed than the bottom, which is cemented to the floor. It was considered that the returns were providing adequate restraint, similar to that produced in the construction of a domestic or small industrial building enclosure.

No evidence of significant rotation of the returns was noted when comparing the length changes with the movement of the ends of the returns. Movement indicators at the ends of the returns picked up a general inward movement at this position of up to 1 mm, with fluctuations corresponding to changes in relative humidity in a similar manner to length measurements.

Movement measurements across the dpc demonstrated the tendency of the walls to move under the influence of (tensile) shrinkage stresses towards the restraining end walls. The central monitoring points showed little other than fluctuating movement, whereas the outer points showed consistent movement of the base of the wall towards the ends of between 0.1 and 0.6 mm with most results within the range 0.2 to 0.4 mm. This confirmed the effectiveness of the returns in providing restraint to the walls.

6.4 *Crack formation*

Visible through-block (structural) cracks were observed in three of the test walls: B2 (thin layer mortar without reinforcement), B4 (thin layer mortar with polymer reinforcement every course) and A4C (thin layer mortar polymer reinforcement every course*).

In both B2 (no mesh) and B4 (polymer mesh) cracking grew downwards from the open perpend in the top course, being first observed at the 20-day visit. An apparently independent crack also opened up lower down the wall, vertically below the first crack, later joining together to form one continuous crack running the entire height of the wall. This may have been the result of increasing stresses in the wall below the first crack, as cross-sectional area was reduced, or may indicate that only one crack was forming but that parts of it were too fine at the surface for observation, even when using a microscope.

* Polymer mesh laid into thin layer mortar.

A comparison of the crack widths exhibited in wall with open perpends B2 (no mesh) and B4 (polymer mesh) shows that, although the mesh did not prevent cracking from the open perpend in the top course, it did significantly reduce the width of cracking from about 1 mm to 0.1 mm. This indicates a significant bridging of the crack by the polymer mesh.

In all three cases, the crack had extended essentially the full height of the test wall by the time the walls were 48 days old. Once they had formed, the cracks tended to close up as the blocks re-expanded in response to increases in relative humidity. No further cracks appeared in these walls once the primary crack had developed within the first three weeks.

Although wall A4C* did not include open perpends, cracking nevertheless proceeded from one of the joints in the top course of blocks. This is not unusual, as movement would be expected to be concentrated at joints and the top course inevitably suffers from the lack of an upper bed joint to carry stresses across its upper end.

In all cases, no cracking was detected around the open perpends in enclosed courses of blockwork.

6.5 *Displacement across perpends*

Displacement across the perpend joints in the two walls constructed using sand/cement mortar generally followed the expected pattern whereby, many perpends throughout the wall open up slightly (up to 0.2 mm) due to micro-cracking between mortar and block as tensile shrinkage stress are applied. Thus resulting in the spreading of shrinkage cracks throughout the wall. This is accompanied by minute amounts of shear (slip) along the bed joints, allowing this traditional construction to accommodate relatively large amounts of movement without visible damage.

This movement was less noticeable in the walls built using thin layer mortar. Here, opening displacements were smaller (generally less than 0.1 mm) and a greater proportion of the joints monitored showed a tendency either not to move at all or even to close up very slightly. This indicates that greater adhesion exists between the blocks and the thin layer mortar, reducing bed joint slip and perpend joint cracking.

7 DISCUSSION OF RESULTS

7.1 *Possible reasons for shrinkage behaviour*

In a composite structure such as blockwork, the behaviour of both components (block and mortar) will contribute towards shrinkage behaviour as a whole. Shrinkage behaviour of cement-based mortars is complex and depends upon the cement content, water content, air content and also the presence of admixtures. Contributions to shrinkage include:

- Suction: capillary absorption of water from the fresh mortar into the (dry) blocks.
- Initial drying shrinkage: irreversible shrinkage as the set mortar first loses water to the atmosphere.
- Moisture movement: reversible shrinkage/expansion. As the set cement in the mortar releases/absorbs water from the atmosphere in response to changes in humidity.
- Carbonation: reaction of the lime in the mortar (produced as a by-product of cement hydration, as well as that added as a plasticiser) with atmospheric Carbon Dioxide to produce Calcium Carbonate.

Of these, the first and last can be discounted for the present purpose. Aircrete blocks absorb relatively little through capillary suction and this takes place whilst the mortar is still plastic, making cracking or de-bonding unlikely unless suction is severe (as is the case with, for instance, Fletton bricks), and carbonation takes effect over a period of years rather than months.

Typical values for initial drying shrinkage (irreversible) and moisture movement (reversible) of sand/cement mortar quoted in the literature are 0.06% and 0.04% respectively. These figures are not additive, and taking the highest value of 0.06% for first drying of the mortar would result in the shrinkage of a 12 mm mortar joint by 0.007 mm.

The same sources give moisture movement values for aircrete blocks of typically 0.06%, similar to that for the mortar. Over a block of length 620 mm this would result in shrinkage of 0.37 mm, which is of the same order as the micro cracking observed in the mortar joints of walls A1 and B1 (both walls were sand/cement without reinforcement).

It must therefore be concluded that the dominant factor in producing shrinkage stresses is the shrinkage of the blocks themselves, rather than the material in the joints.

At the time of writing, no data was available on the shrinkage behaviour of Thinset mortar alone, though it can be postulated that its effect will be even smaller than that of sand/cement mortar because of the greatly reduced joint width.

8 PERFORMANCE OF BLOCKWORK WALLS SUBJECTED TO SHRINKAGE STRESSES

Observation of the performance of conventional block/mortar or brick/mortar walls suggests that higher bond strength between the blocks and the mortar (as produced by increasing the cement content of the mortar) results in improvement of the structural performance of the wall. This is only if shrinkage cracking can be controlled and prevented from spreading into the blocks themselves as structural cracks. This is the basis of the conventional wisdom in masonry construction that the mortar should be "no stronger than necessary".

In the case of the thin layer mortared walls, where reinforcement of the bed joints is successful or where uniform adhesion is obtained along the bed joints, tensile stresses are spread evenly throughout the wall, with micro-cracking more likely to take place within the blocks themselves. Because of their complex cellular structure, AAC blocks are capable of sustaining relatively high amounts of internal damage without either deteriorating in performance or exhibiting visible signs of cracking.

Any variation in the strength of the bed joints in this case will lead to the concentration of stresses and an increased probability of structural cracking. This has occurred in walls B2 and B4, where the presence of the open perpends in the top course has precipitated cracking. The greater stiffness of the stainless steel fabric has allowed the bridging of this potential weakness, since the modulus of elasticity of steel (typically 200 GPa) is considerably higher than that of the blocks (typically 2 to 10 GPa depending upon density). This means that the steel can carry stresses across the base of the open perpend without significant extension of the reinforcement.

Polymers, on the other hand, generally have much lower elastic modulus (less than 1 GPa) and in order to carry significant tensile loads they must therefore extend relatively far. If the bonding in the bed joint around the area of weakness is good, this therefore means that stresses will be transferred directly to the block in the course below, resulting in a higher probability of structural cracking. Nevertheless, bridging of the shrinkage crack by the polymer mesh in wall B4 has greatly reduced the crack width by one order of magnitude.

Wall A4C was constructed of the same materials as wall A4, but with the polymer fabric laid on the blocks before application of the mortar. This would appear to produce a bed joint with greater variability in strength, since the mortar is less likely to penetrate the mesh consistently than is the case if the mesh is laid on top of the toothed surface of the mortar bed. It is not simply a matter of weakness of the mortar bed; were the bed joint to be very weak, there would be no stress transfer between courses and hence less probability of structural cracking.

In replacing the conventional poor bonding of sand/cement mortar with a joint of far superior strength, it would appear that the thin layer mortar technology also brings with it the requirement for a higher level of consistency in the strength of the bed joints. Conventional block walls avoid shrinkage cracking by having all the joints consistently weak, allowing micro cracking to occur in the mortar throughout the wall. Thin layer mortar joints must, in equal measure, be consistently strong since tensile stresses will be higher and the wall thus more susceptible to the presence of weaknesses which act as stress concentrators.

It must be emphasised that the effect of such stress concentrations is a matter of probability. In many cases, where the walls do not become saturated after construction or where weather conditions allow for more gradual shrinkage, then damage may not occur even in the presence of open perpends. The design of this series of experiments deliberately attempted to produce the worst case scenario, where movement is constrained and the walls become saturated early in their life. Undoubtedly, such conditions do not always prevail on site, even under British weather conditions.

9 CONCLUSIONS

1. Use of sand/cement mortar in aircrete blockwork resulted in the formation of multiple hairline shrinkage cracks at perpend joints throughout the wall.
2. Use of thin layer mortar prevented the appearance of hairline shrinkage cracks at perpend joints.
3. Incorporation of steel mesh in bed joints was effective in preventing the formation of shrinkage cracks from isolated open perpend joints.
4. Incorporation of polymer mesh in bed joints significantly reduced the width of shrinkage cracks where these occurred.
5. Polymer mesh bed reinforcement was more effective when laid on top of the thin layer mortar rather than underneath it.

10 RECOMMENDATIONS

From the results of this research it would seem appropriate to recommend that:

1. Particularly where there is a risk of rain saturation, polymer mesh should be incorporated in the bed joints to reduce the occurrence of shrinkage cracks and minimise crack width and the mesh should be laid on top of the mortar and not below it.
2. Isolated open perpends should be avoided as they act as stress concentrators and increase the risk of the formation of structural shrinkage cracks.
3. A more radical step might be to develop a mortar formulation for non-structural blockwork walls, which is weaker that the current product allowing movement accommodation to occur within the joints themselves.

REFERENCES

Shrinkage behaviour of Aircrete blockwork wall, Final report. John Moores University School of the Built Environment for Clan Products (North West) Ltd.
Everett, A "Materials", Batsford, 1991.
Illston, JM "Construction Materials", E&FN Spon, 1994.

Autoclaved Aerated Concrete – Limbachiya and Roberts (eds)
© 2005 Taylor & Francis Group, London, ISBN 0 415 38356 0

Impact response of Autoclave Aerated Concrete/FRP sandwich structures

N. Uddin, K.V. Shelar & F.H. Fouad
Department of Civil & Environmental Engineering, University of Alabama at Birmingham, Birmingham

ABSTRACT: A novel concept of hybrid fiber-reinforced polymer (FRP) – Autoclaved Aerated Concrete (AAC) sandwich panels have proven to be a structurally efficient combination for lightweight structural components. The FRP composite material was made of carbon reinforcing fabrics embedded in an epoxy resin matrix. The carbon fiber reinforced polymer (CFRP) reinforcement was applied on the top and bottom faces of the AAC panel and several innovative processing techniques were used including hand lay up as well as vacuum assisted resin transfer molding (VARTM). Beside the need to have adequate flexural properties, the AAC-FRP sandwich structures need to be evaluated in their ability to withstand localized damage. During service, the structural members in the building structures are subjected to impact loading that varies from object-caused impact, blast due to explosions, to high velocity impact of debris during tornados, hurricanes, or storms. The objective of this paper is to evaluate the response of AAC-FRP sandwich structures to low velocity impact and to assess the damage performance of the panels. Low velocity impact (LVI) testing serves as a mean to quantify the allowable impact energy that the structure is able to withstand and to assess the typical failure modes encountered during this type of loading. Impact testing was conducted using an Instron drop-tower testing machine.

1 INTRODUCTION

The possible uses of composite materials in infrastructure related application is an area receiving considerable attention, and *Autoclave Aerated Concrete* (AAC) is a proven building material that will become much more widely used in the U.S. for both residential and commercial construction. With the growing threat of international and domestic terrorism, construction materials that help mitigate the effects of impact and blast are immediately needed. It is also important that the building material be cost effective, energy efficient and available throughout the world. A sandwich panel with AAC core and *Fiber Reinforced Plastics* (FRP) composites as a face sheet has the potential to be an excellent energy absorbing construction material plus it is relatively inexpensive and available over much of the globe. While the resin infiltrates into the concrete micro-pores of the AAC panel, additional strengthening /hardening is provided by using non-crimped glass and carbon fiber. FRP/AAC sandwich can be fabricated in form of panels for roof and floor decks, panels for exterior walls and interior partitions, and lintels. It also has some properties, in addition to those mentioned above, that makes it ideal for tornado shelter and safe houses and may be of interest to the military and others responsible for protecting live.

AAC is an ultra-lightweight concrete with a distinct cellular structure. The raw materials used in production are simply cement, lime, aluminum paste/powder, and water, plus sand or fly ash as a silica source. The dry bulk density of the material ranges from 400 to 800 kg/m^3, and the compressive strength ranges up to 6.9 MPa. Entrained air bubbles are the main reason behind the enhanced physical properties of the material. The macroscopic air cells (0.1 to 1.0 mm) are the result of a chemical reaction between the cement hydration products and the aluminum paste/powder in which hydrogen gas is liberated. Hydrogen gas causes the fresh material to rise in the molds and expand to about twice the original volume, thus creating cellular structure. Due to the cellular structure and reduced weight, the material is highly fire resistant and very durable compared with conventional construction material, and has unique thermal insulation properties. It requires significantly less energy for heating and cooling. The potential advantages of panelized construction using AAC core as building material are lightweight, energy efficient, easy to use, fire resistant, environmental friendly, weather resistant, pest resistant, durable, acoustically efficient, and versatile. It can be formed into blocks and reinforced panels of practically any usable shape and size. Ideal for the protection against harsh environments including corrosion, freeze-thaw, deicing salt and marine environment. Also ideal for tornado shelter and safe houses against natural hazards. As a versatile and readily available building material it has the potential to serve construction against terrorism uses such as in construction of hardened buildings and bunkers, retrofit hardening, munitions storage facilities, etc, at low cost.

FRP composites on the other hand are anisotropic and are characterized by high tensile strength in the direction of the fibers. They are corrosion resistant and offer high flexural resistance, and therefore should perform better than other construction materials in terms of weathering behavior. The use of fiber reinforced composites, as internal as well as surface reinforcement is effective in controlling deflection and cracking, well within the permissible limits, without affecting load carrying capacity for fatigue loading as well. FRPs offer a strong potential in meeting needs in terms of high strength, low weight, high resistance to corrosion, and high impact as well as fatigue strength. FRPs are also attractive from a standpoint of modularity as carbon or glass reinforcement results in material that are five times lighter than steel, and hence easy to install. When FRP/AAC panels are used in construction, the thin-bed mortar provides strength between blocks that is greater than the tensile and shear strength of the material that the blocks are made of. Thus, unlike walls of standard concrete blocks, a wall made of FRP/AAC panels results in a homogeneous wall or floor panel with no weakness at joints. Furthermore, FRP/AAC panels of any practical size and shape can be manufactured. The reinforcement for such panels is placed during manufacturing and can be deigned to be effective against blast. Thus, again, an entirely homogeneous panel is formed, similar to that of a reinforced concrete tilt-up panel.

This paper is concerned with the application of FRP/AAC panels for emergency shelter safe houses and related issues. In order for such materials to gain acceptance in such applications, improvements are required in their impact properties. The latter is the subject of this study. In the major structural applications of FRP reinforced AAC to civil structures, the laminates will be used most often as the skins of sandwich panels. These sandwich panels may be either used for cladding or load bearing purposes. Under these circumstances, impacts from small objects, such as hand tools at low velocities, or blast fragments at high velocities, will involve local indention and damage without global deformation of the sandwich panels.

1.1 *The 'VARTM' processing*

Most infrastructures related applications of FRPs use unidirectional fiber reinforcement using expensive carbon fibers, with traditional hand lay-up technology. The hand lay-up is tedious, expensive, labor-intensive and prone to personnel skill level to obtain a uniform quality product. Any anomalies developed during inefficient hand lay-up of concrete panels, roofs, walls and floors can lead to corrosion, water seepage and degradation due to fatigue cracking. From a material standpoint, several advances have been made in vinyl ester and epoxy resins, as well as different types of textile fibers that are suited for building construction applications. However, there is complete lack of any systematic studies that investigate the utilization of newly developed fibers/fabrics and resins for building structures. There is significant room to improve cost and quality over traditional hand lay-up of FRP for infrastructure applications, while incorporating advances in materials development.

In this study, a newly emerging liquid molding VARTM technique was used to manufacture FRP/AAC panels. It is a promising low cost method of composites manufacturing and producing integrated structures such as FRP/AAC sandwich panels. The VARTM is a cost-effective/

affordable alternative to hand lay-up or fiber spray processing technique for panel construction. This new approach can be implemented at a semi-automated level in the field or factory, and not influenced by personnel skills. It uses single-sided tooling (which results into significant savings in tooling costs) and vacuum-bag technology. Resin is infused into dry fabric preform that is "closed" with an inexpensive vacuum bag film (Fig. 1). This process has very attractive cost-advantages in comparison to spray-up or hand lay-up impregnation methods, and it is far less expensive than conventional manufacturing methods. It is particularly advantageous to large infrastructure applications. Other advantages of VARTM are low process volatile emissions (and therefore environmentally friendly/safer), high fiber-to-resin ratios, good quality and process repeatability. As a result of savings in time, material, and labor 60% cost-reduction is achievable as compared to traditional hand lay-up and/or autoclave molding.

1.2 *FRP/AAC sandwich panels*

Unidirectional carbon fiber Sikawrap HEX 103 (Manufacturer-Hexcel) was used as the facing material with an epoxy resin Sikadur 300 (Manufacturer – Sika) matrix. The face sheets were bonded to the core using hand lay-up. Resultant properties for the composite laminate used in the faces and approximate values for the AAC provided by the manufacturer and compressive testing are presented in Table 1.

In this project, FRP/AAC sandwich panels were investigated for its damage tolerance using new developments in processing and constituent materials. The unique contribution of the effort is the application of cost-effective VARTM (Vacuum Assisted Resin Infusion Molding process) (Uddin & Vaidya 2002) processing using innovative FRP hardening schemes with recently developed carbon fiber architectures as well as epoxy resin types. The effort also has the potential to eliminate traditional hand lay-up processing techniques which prove to be expensive and time-wise inefficient, and simultaneously advanced the state of the art of usage of advanced composites in civil infrastructure.

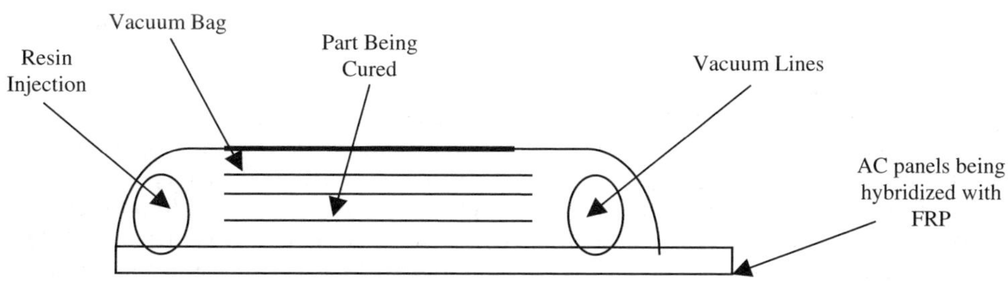

Figure 1. A simplified illustration of the VARTM process.

Table 1. Physical and mechanical properties of carbon/epoxy face and Autoclaved Aerated Concrete core.

Face sheets-Sikawrap hex 103C/Sikadur 300			Core- AAC 640 kg/m^3		
Tensile strength	1.41E+9	Pa	Tensile strength	8.62E+05	Pa
Tensile modulus	1.13E+11	Pa	Tensile modulus	1.90E+09	Pa
Shear strength	7.00E+09	Pa	Shear strength	1.00E+06	Pa
Shear modulus	43.6E+09	Pa	Shear modulus	9.48E+08	Pa
Poisson's ratio	2.14E−01		Compressive strength	3.45E+06	Pa
			Poisson's ratio	0.2	

2 LOW VELOCITY IMPACT TESTING

The focal point of this paper is to investigate the effectiveness of confining AAC panels with uniaxial carbon fiber wrap against impact loading, and to compare the energy absorption characteristics. It is to be noted that the damage tolerance of AAC structures to localized loading is poor and should be evaluated in each application prior to its implementation. Low velocity impact (LVI) testing serves as a mean to quantify the allowable impact energy that the structure is able to withstand and to assess the typical failure modes encountered during this type of loading.

The testing was conducted using a Dynatup 8250 impact-testing machine equipped with a load cell of capacity 25.13 kg. The following data illustrates the results obtained with the hemispherical impactor at 40 cm. A flat striker was used for this test and had an impact area of 0.076 m × 0.102 m.

2.1 *Specimen and equipment*

Impact testing was conducted using a total of 9 panels: 3 AAC specimens, 3 FRP/AAC specimens processed with hand lay up, and 3 specimens processed with VARTM. One layer of CFRP was used for all the face sheets. Table 2 describes the specimens. In an effort to illustrate a similar loading situation as would be seen in reality, the specimens were placed horizontally inside a testing jig and subjected to load.

The hammer (tup) contained an internal load cell, which was used to record the contact load between the falling assembly and the column during the impact event. The load cell was rated for a maximum load of just over 10 kips (44 kN). A drop height of 40 cm was used for all tests, since the combination of this height and the weight of the striker assembly produced loading close to that of the maximum allowed by the specimens before failure. In previous studies using this machine, load-time plots reported peak-loads several times the expected value. Conclusions from similar tests, drawn by Suaris and Shah (1983), illustrated that this loading was not indicative of the material properties but instead was a result of inertial effects of the samples. Though these effects have been accepted and calculated before for testing metals, AAC creates a more complex problem due to the relatively small fracture strain and increased size of test specimen (Server et al. 1977). As in previous testing, a rubber pad was added to the striker to eliminate the inertial effects or "ringing" being generated from the impact of the steel hammer and AAC specimens. A summary of test data is given in Table 2.

The testing machine used DynaTup software to generate the data produced during impact loading. Load versus time and impact velocity are measured directly from the software. For a forced object traveling in a straight line, standard equations of motion are used to express velocity, deflection, and energy absorption. Variables were $p(t) =$ load value measured by the 830-I at time t, $m =$ hammer mass, $g =$ acceleration due to gravity for vertical motion (i.e., drop towers) 0 for horizontal motion (i.e., pendulum machines), $f(t) =$ total force acting on the hammer at time t, $a(t) =$ resultant acceleration on the hammer at time t, $v(t) =$ velocity of the hammer at time t,

Table 2. Summary of impact test results.

Specimen ID	Peak load (kN)	Maximum deflection at mid span (mm)	Total energy (joule)
P-1	7.18	1.61	6.33
P-2	10.30	1.72	10.26
P-3	21.07	1.84	18.36
H-1	7.02	1.85	12.86
H-2	14.15	2.24	19.03
H-3	19.04	3.02	78.28
V-1	13.65	1.75	15.11
V-2	16.93	2.05	21.41
V-3	23.39	2.27	66.56

x(t) = position (deflection) of the hammer at time t, K(t) = kinetic energy of the hammer at time t, $E_a(t)$ = energy absorbed by the specimen at time t, and E(t) = total energy of the hammer/specimen system at time t.

$$f(t) = mg - p(t) \tag{1}$$

$$a(t) = \frac{f(t)}{m} = \frac{g - p(t)}{m} \tag{2}$$

$$v(t) = \int_0^t a(t)dt = gt - (1/m)\int_0^t p(t)dt \tag{3}$$

$$x(t) = \int_0^t v(t)dt = v_i t + 0.5gt^2 - 1/m \int_0^t \int_0^t p(t)dt \tag{4}$$

$$E(t) = K(t) + v(t) + E_a(t) = constant \tag{5}$$

$$E_a(t) = (m/2)(v_i^2 - v(t)^2) + mgx(t) \tag{6}$$

The first letters are used to denote the type of specimen, "P" for plain AAC, "H" for FRP/AAC processed with hand lay up, and "V" for FRP/AAC processed with VARTM. The next letter are used to denote thickness of specimens, "1" for 1 in. (2.54 cm), "2" for 2 in. (5.08 cm), and "3" for 3 in. (7.62 cm). All specimens are 24 in. (0.61 m) long and 8 in. (0.203 m) wide.

2.2 *Results and discussions*

As shown in the table higher energy is absorbed for FRP/AAC compare to plain AAC. In addition VARTM process consistently demonstrates a superior performance compare to hand lay up. Over all increase in energy absorption is on the average 2.5 times the plain AAC and 1.25 times the hand lay up specimens. It is also interesting to observe that deflection of hand lay at failure is higher than both VARTM and plain AAC. This can be explained by the fact that VARTM processed specimen produced more integrated and sound structure where FRP/AAC acted as a single unit. On the other hand, in the hand lay up separation between AAC and FRP occurs during the failure. As expected the dynamic bending load increase was highest for the less stiff specimens and gradually declined in value with increased stiffness. This trend was also observed by testing conducted by Jerome and Ross (1996).

Load versus time graph is presented in Fig. 2a. The inertial effects have been reduced and are not visible in this plot due to the addition of the rubber pad to the striker. This phenomenon is evident in tests conducted by Erki and Meier (1999). Several things can be observed from the figure. First, the initial peak is the actual peak load. The subsequent peaks of smaller amplitude are simply rebounds of the tup. Since it is the stiffest, VARTM processed specimen has the largest peak load of all the specimens with a value of 23 kN. As expected, plain AAC have lowest peak loads of 11 kN since these specimens are less stiff than FRP/AAC. Energy versus time data is presented in Figure 2b. It is obvious for this figure that there is a significant change in the slope of the energy plot by the time CFRP face sheet breaks. This seems to indicate that the core is absorbing a considerable amount of energy while taking the shear loads and while crushing. Also from the the energy versus time curve (Fig. 2b), it appears that energy absorption of the VARTM is comparable to that of the hand lay up. The displacement results indicate that the hand lay up produces higher deflection than the VARTM. This effect is due to the ability of thc material to compress and separated from FRP skins during failure.

Table 3 shows a comparison of LVI data for AAC and FRP/AAC specimen with a 1 in thickness and Figure 3 plots the peak load and energy data for the same specimens. FRP/AAC specimens show a considerable increase in energy absorption (almost 6 times for VARTM specimen compare to plain AAC), velocity slow down, maximum load and load to failure. On the other hand VARTM demonstrates a superior performance over the hand lay up in all categories. Maximum load value recorded show that VARTM achieved a value of 23.4 kN compare to hand lay up which is 19 kN. Similarly velocity slow down for VARTM is about 8.6% compare to hand lay up which is 7.5%.

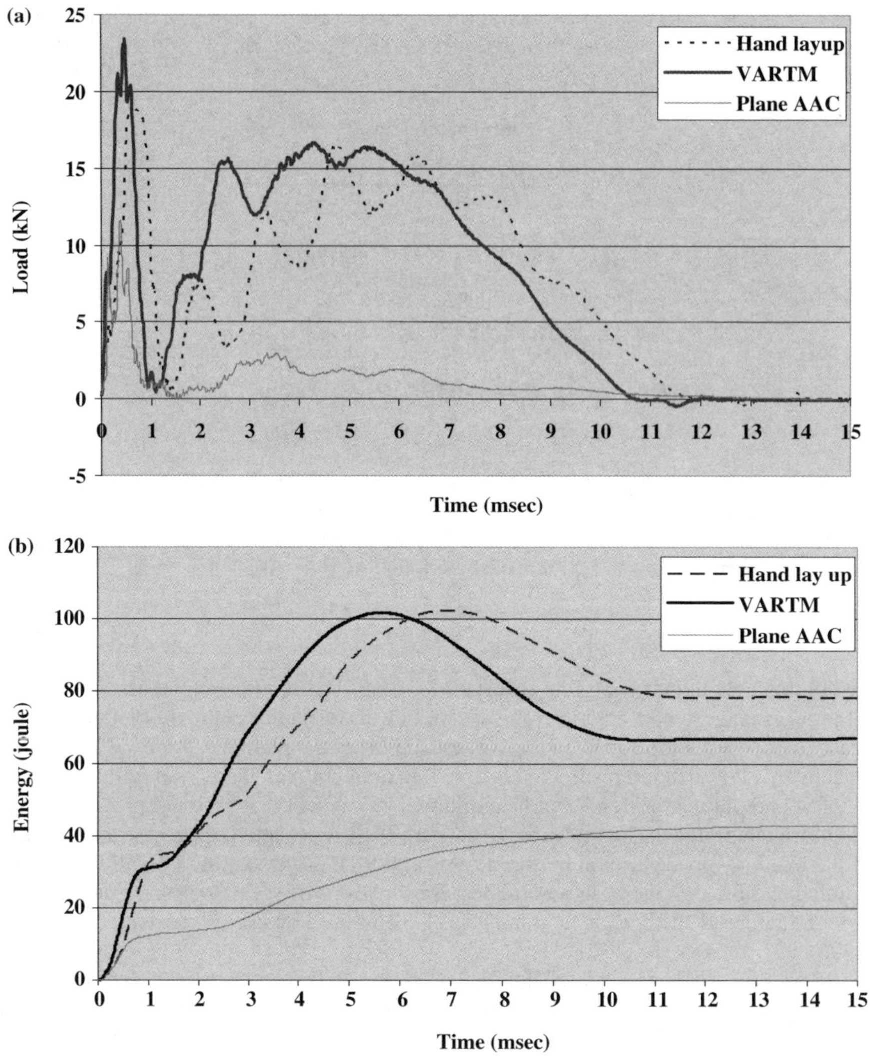

Figure 2. (a) Load vs. time and (b) Energy vs. time obtained from impact test results with flat impactor at 40 cm.

2.3 *Failure modes of specimens*

Plain AAC: The sample breaks in very abrupt and sudden failure. The AAC failed completely in the impact area by crushing and in the surrounding area by shear (Figure 4a).

Hand Lay Up Sample: This sample showed the same trend as observed in plain AAC but the damage was more visible and stable (Figure 4b). The top facesheet presented matrix rupture in the form of two major cracks (due to the nature of the unidirectional laminate the plane of fracture propagates through the fiber tows) and some localized fiber breakage. The core failed completely, in the impact area by crushing and in the surrounding area by shear. The interface between the bottom facesheet and the core failed and separated, and the bottom laminate failed showing a continuous matrix crack almost in the middle of the panel. By visual inspection of the post-mortem sample it is visible that the AAC ruptured in a progressive nature and that the damage area in the

Table 3. LVI data for plain AAC, hand lay up and VARTM processed specimens with 1 in. thickness.

Specimen ID	Unit	Plane AAC	Hand lay up	VARTM
Impact energy (1)	joule	99.5505	99.7398	99.6249
Energy to max load (1)	joule	5.6483	14.7911	16.6678
Total energy (1)	joule	13.1111	78.287	66.5676
Energy to yield (1)	joule	0.5896	0	0
Energy to failure (1)	joule	10.9824	33.8693	30.2643
Impact velocity (1)	m/sec	2.8803	2.883	2.8813
Velocity slow down (1)	%	2.7428	7.493	8.5757
Maximum load (1)	kN	11.646	19.0459	23.3916
Load at yield (1)	kN	7.2199	−0.0941	0
Load at failure (1)	kN	2.2444	3.4829	4.4869
Time to max load (1)	msec	0.379	0.598	0.463

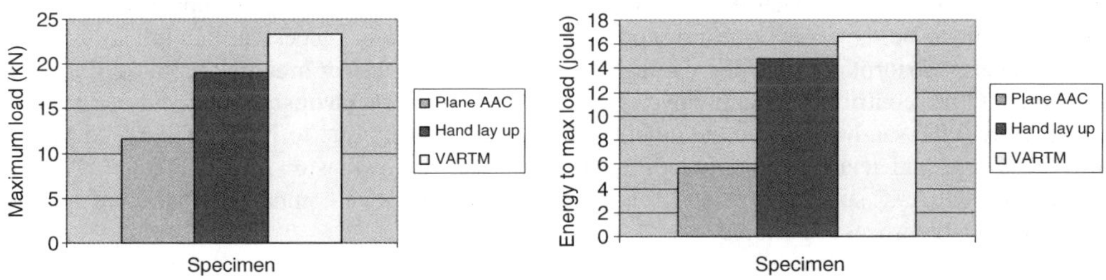

Figure 3. Plot of LVI data for plain AAC and FRP/AAC specimens Plain Sample.

Figure 4. (a) Failure modes of P-1, (b) Failure modes of H-1, and (c) VARTM processing and failure of V-1.

bottom laminate is significantly larger than the damage area in the top. This may be the mechanism why the AAC is absorbing energy as shown in Figure 2.

VARTM Sample: This sample showed no visible sign of failure of the sandwich structure. The core did not show any disintegrated area around the impact area and the failure could not be observed except from the load deflection graph. The top facesheet did not split into parts as observed in the hand lay up specimens. The core did not crush visibly and perhaps failed by bending and shear, and in a progressive failure mode. The interface between the bottom facesheet and the core remained intact. This seems to indicate that the core is absorbing a considerable amount of energy while taking the loads and while deflecting without crushing. In the same time interval there is a significant increase in load capacity.

3 SUMMARY

Low velocity impact testing provides meaningful information about the energy absorbing characteristics of a sandwich structure. It illustrates how the sandwich structures will respond to the explosion, drop or impact of an object falling from a certain height. This loading scenario is of significant importance since AAC/FRP panels are being considered as possible candidate for damage tolerant structures.

Typical failure of AAC subjected to low velocity impact includes core localized crushing, however it was observed that when FRP/AAC core is subjected to low velocity impact, progressive failure seems to be the typical failure mode. This failure mode is successful in applications that require energy absorption since the damage is spread around a larger area on the backside of the sandwich. Other contributions to the overall energy absorption mechanisms observed during testing of AAC/FRP sandwich structures include face sheet rupture due to matrix failure, localized fiber breakage, and delamination of facesheet/core interface in the tension face. Moreover VARTM processed FRP/AAC specimen demonstrated superior performance compares to hand lay up, and has proven to be process of choice.

REFERENCES

Abrate, S. (1998). *Impact on Composite Structures*. Cambridge University Press, 135–160.

Allen H. G. (1969). *Analysis and design of structural sandwich panels*. Pergamon press. Oxford.

Erki, M. A. and Meier, U. (1999). "Impact loading of concrete beams externally strengthened with CFRP laminates." *J. Comp. Constr.*, ASCE, 3(3), 117–124.

The Babb companies. (2004). http://www.babb.com/aac/, last accessed 08.2004.

Pan Pacific engineering PTY Ltd. Technical information on aerated concrete. www.litebuild.com, last accessed 08.2004.

Sika, construction products catalog, c-160, 2002.

Hazinan Md. A and Cantwell, W.J. (2002). The low velocity impact response of foam based sandwich structures. *Composites part B: Engineering*. 33. 493–204.

Hazinan Md. A and Cantwell, W.J. (2003). The low velocity impact response of aluminum honeycomb sandwich structures. *Composites part B: Engineering*. 34. 679–687.

Jerome, D. M. and Ross, C. A. (1996). "Dynamic response of concrete beams externally reinforced with carbon fiber reinforced plastic (CFRP) subjected to impulsive loads." *Structures Under Extreme Loading Conditions*, ASME, PVP 325, 83–94.

Server, W. L., Wullaert, R. A., and Sheckhard, J. W. (1977). "Evaluation of current procedures for dynamic fracture toughness testing." *Flaw Growth and Fracture, American Standards for Testing and Materials*, STP 631, 448–461.

SikaWrap (2002). "Composite fabrics for structural and seismic strengthening." Products, http://www.sikausa.com (June 5, 1998).

Suaris, W. and Shah, S. P. (1983). "Properties of concrete subjected to impact." *J. Struct. Eng.*, ASCE, 109(7), 1727–1741.

Silva, M. A. G, Cismasiu, S, and Chiorean, C.G. (2000). Low velocity impact on laminates reinforced with polyethylene and aramidic fibers. 1–10.

Uddin, N. and Vaidya, U. K. (2002). "UTCA Project Proposal", *Unpublished Report*.

Yang, S. H. and Sun, C. T. (1982). Indentation law for composite laminates. *Composite Materials: Testing and Design (sixth conference), ASTM STP 787.1* American Society for Testing and Materials. 425–449.

Zenkert, D. (1997). *An introduction to sandwich construction*. Emas publishing. London. 7–37.

Autoclaved Aerated Concrete – Limbachiya and Roberts (eds)
© 2005 Taylor & Francis Group, London, ISBN 0 415 38356 0

Mortar for low density aircrete applications

A.N. Fried, A. Ahmed, M.C. Limbachiya & J.J. Roberts
Concrete and masonry Group, Kingston University, London

ABSTRACT: Mortar for masonry is important because it provides the linkage between masonry units so enabling the composite to behave as a single material. The type of mortar used determines the flexural and compressive strength of the masonry so in this paper, a range of mortars is examined. These include traditional designation (iii) (1 cement : 1 lime : 6 sand), designation (iv) (1 cement : 1 lime : 9 sand) mortars as defined in BS 5628; Part 1, and two thin layer mortars denoted A and B. The conventional mortars were formed using either 42.5 N Portland cement (PC) to BS EN 197; Part 1- CEM I or 32.5 N PC to BS EN 197; Part 1- CEM I in order to ascertain the difference these two cements have on the properties of mortar.

1 INTRODUCTION

Experimental work was undertaken to establish the mechanical properties of three mortars types (designations iii, iv and thin layer). A series of tests were carried out to evaluate the cube compressive, tensile and flexural strengths of conventional mortar prepared using 32.5 N and 42.5 N Portland cement and two types of thin layer mortar designated type A and B. The two cements were selected as there is no guidance on cement choice in the British code although there is some data on mortar strengths in Table 1 of BS5628: Part 1[1], the relevant parts of which are reproduced as Table 1 of this paper. Flow properties of these mortars were also established. Sample preparation and testing were carried out in accordance with appropriate Standards.

The paper is divided into two main parts. The first part gives details of test materials and mortar properties and this is followed by the test results.

1.1 *Test materials*

1.1.1 *Cement*

The first set of tests were carried out using 42.5 N PC to BS EN 197; Part 1- CEMI[2], these being repeated using 32.5 N PC to BS EN 197; Part 1- CEM I[2].

1.1.2 *Lime*

Bulk Hydrated Lime was used in the mortar production. It contains between 95.0–97.0% Calcium Hydroxide.

Table 1. Designations iii and iv mortar composition and respective mean compressive strength at 28 days, as per BS 5628[1].

Mortar designation	Cement:lime:sand (Volume)	28-day compressive strength, N/mm^2
iii	1:1:6	3.6
iv	1:2:8	1.5

Table 2. Sand grading test results.

Sieve aperture size	Mass of sand passing sieve (g)	Mass of sand retained by sieve (g)	Cumulative sand passing sieve (%)
6.30 mm	1160.5	0.4	99.97
5.00 mm	1160.5	0.0	99.97
2.36 mm	1158.7	1.8	99.81
1.18 mm	1151.3	7.4	99.17
600 μm	980.2	171.1	84.43
300 μm	199.4	780.8	17.18
150 μm	34.2	165.2	2.95
75 μm	8.2	26.0	0.71

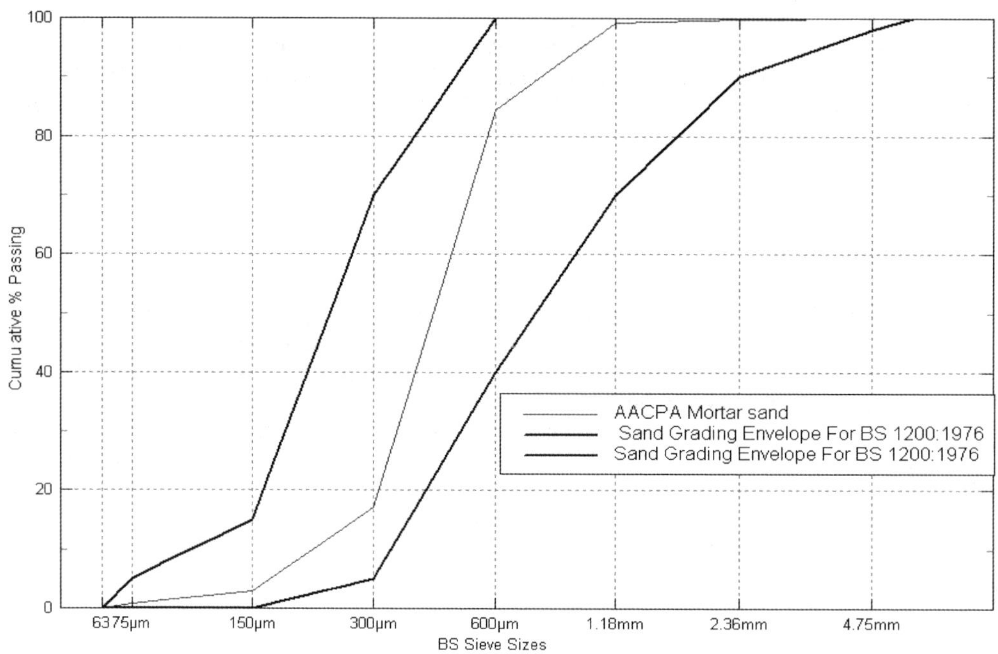

Figure 1. Particle size distribution of sand.

1.1.3 *Sand*

Soft building sand was used. The particle size distribution of the sand is given in Table 2 and schematically shown plotted in Figure 1. Tests were carried out in accordance with BS 1200[3] and the results indicate that the sand used complies with the requirements of the BS 1200:1976 [3].

1.2 *Specimen preparation*

1.2.1 *Conventional mortar*

Conventional mortar samples of designation (iii) and (iv) were produced to establish fresh and mechanical properties. Water was added so that the workability was consistent and corresponded to a 10 mm penetration of the dropping ball test as suggested in BS 5628. The flow properties were determined in accordance with EN 1015; Part 3[4]. The flow values obtained for all mixes fell within a range of between 186–188 mm.

Table 3. Mortar mixing preparation.

	Mortar A	Mortar B
Mortar weight (kg)	25	25
Water content (litres)	4.4	5/5.5

Table 4. Mortar properties and testing regimes.

Mortar property	Specimen	Test ages
Compressive cube strength	100 × 100 × 100 mm	1–28 days
Tensile strength	Dog bone	28 days
Flexural strength	40 × 40 × 160 mm	28 days

Table 5. Compressive strength results.

	Compressive cube strength, N/mm^2			
	42.5 N PC		32.5 N PC	
Curing age, Days	Designation iii	Designation iv	Designation iii	Designation iv
1	3.7	1.9	1.9	0.6
2	5.2	3.9	2.7	0.9
5	7.6	6.6	5.0	2.0
7	9.2	8.0	6.0	2.5
14	13.9	11.8	8.2	3.4
28	15.2	14.2	9.1	3.8

1.2.2 *Thin layer mortar*

Again, mortar samples were produced to establish fresh and mechanical properties. Manufacturer's mixing guidelines, given in Table 3 were strictly followed – the mixture was stirred for approximately 10 minutes until a lump free paste was obtained and the workability was consistent and corresponded to a 9.5 mm penetration of the dropping ball test. The flow values obtained for all mixes fell within a range of between 154–156 mm. Both thin layer mortars were manufactured in accordance with prEN 998-2:1997[5].

1.2.3 *Properties examined*

A range of properties were examined during experimental work as shown in Table 4. In all testing, three specimens were broken at each test age (Table 5). Tests were carried out in accordance with EN 1015; Part 11[6].

Test specimens were demoulded after 24-hours of casting and then transferred into an Environmental Chamber where a constant temperature of 20°C and relative humidity of 95% was maintained throughout.

2 RESULTS AND DISCUSSIONS

2.1 *Compressive strength development of mortar*

The compressive strength development of both conventional mortars (designations iii and iv) for both cement types are given in Table 5 and Figure 2 while Table 6 summarises the 28-day cube compressive, flexural and tensile strength test results of both designations iii and iv mortars.

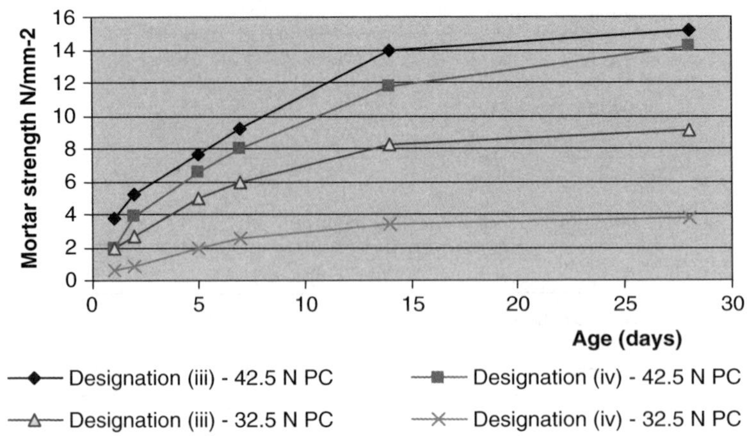

Figure 2. Strength development of conventional mortars.

Table 6. 28-day strength results of designations iii and iv mortars.

Cement type	Mortar designation	Compressive strength (N/mm^2)	Flexural strength (N/mm^2)	Tensile strength (N/mm^2)
42.5 N PC	iii	15.2	5.1	4.8
	iv	14.2	4.8	4.5
32.5 N PC	iii	9.1	2.9	2.5
	iv	3.7	1.5	2.0
	iv*	2.4	–	2.6

iv* = work carried out outside Kingston University by Corus Catnic UK.

Table 7. Compressive strength results (thin layer mortar).

Curing age, Days	Compressive cube strength, N/mm^2	
	A	**B**
1	7.5 (2.08)	2.9 (0.81)
3	11.9 (3.31)	5.8 (1.61)
7	14.9 (4.14)	8.6 (2.39)
10	16.0 (4.44)	10 (2.78)
14	17.0 (4.72)	11.5 (3.19)
21	17.4 (4.83)	11.8 (3.27)
28	17.6 (4.88)	12 (3.33)

2.2 *Discussion*

With 42.5 N PC, the compressive strengths for designations (iii) and (iv) mortars are approximately 5 and 10 times greater than that specified in BS 5628. With the 32.5 N PC cement mortar, the 28 day strength of Kingston mortar is 2.5 times that specified in the code whilst the testing undertaken outside the university still gave strengths 1.6 times that specified in the code. Variations are probably due to different mortar consistencies.

2.2.1 *Thin layer mortar*

The compressive strength results of thin layer mortars up to 28-day curing are given in Table 7 and plotted on Figure 3 while Table 8 summarises 28-day compressive cube, flexural and tensile strength test results of these mortars.

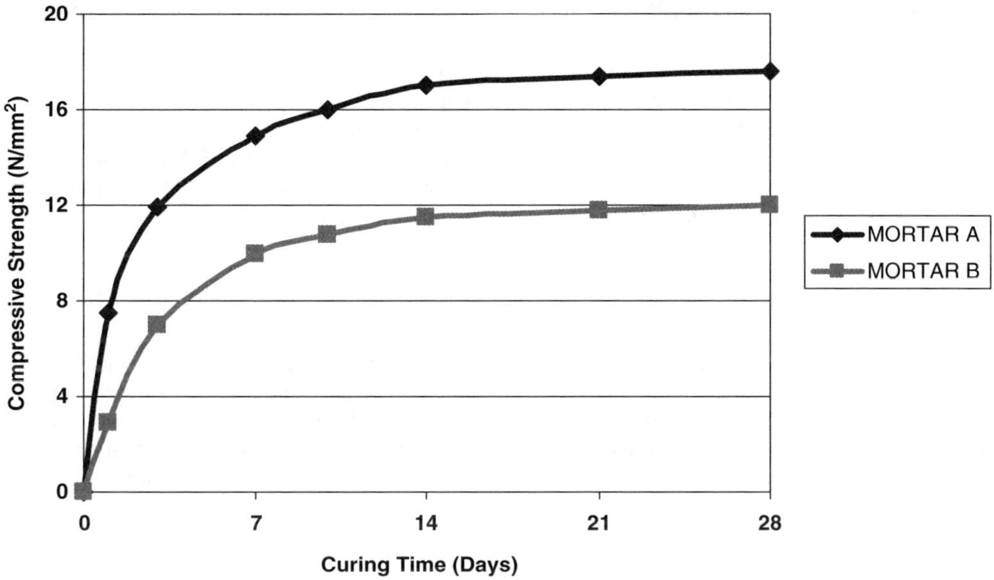

Figure 3. Compressive strength development for thin layer mortars.

Table 8. 28-day strength results of thin layer mortars.

Thin joint mortar	Compressive strength (N/mm^2)	Flexural strength (N/mm^2)	Tensile strength (N/mm^2)
A	17.6	4.6	1.9
B	12.0	3.6	1.7

Bracketed numbers indicate how many times the thin joint mortars are stronger than the BS5628: Part 1[1] specification for a designation (iii) mortar.

Thin Layer Mortar **A** has a compressive strength nearly 50% greater than Mortar **B** but both mortars exceed the strength requirement of designation (iii) mortar as specified in BS 5628: Part 1[1] by significant amounts. The bracketed numbers in Table 7 indicate how many times stronger these mortars are than the 28 day strength of designation (iii) mortar. Mortar A at 1.0 day, is twice the 28 day BS specified strength, whilst mortar B at 1.0 day is 0.85 the BS 28 day strength and at 3 days is 1.6 times the 28 day strength. Both mortars give remarkably consistent flexural strength results using wallettes despite their discrepancy in strength. For the conventional mortars reported in this paper, the strength development is approximately 60% after 7 days, however, for the thin layer mortars, nearly 75% of the final strength is reached after 7 days curing.

3 CONCLUSIONS

(a) Key strength properties of three mortar types (designations iii, iv produced using 32.5 N and 42.5 N PC) and thin joint have been established.
(b) Strengths of mortars produced using 32.5 N PC are considerably lower than those obtained for 42.5 N PC mortars.

(c) Mortar strengths from both 32.5 PC and 42.5 PC mortars exceed the values given in BS 5628; Part 1.
(d) 28 day compressive strength results from an independent laboratory were $\frac{2}{3}$ Kingston values but still achieved the values given in BS5628 : Part 1.
(e) The Thin Layer Mortar designation **A** was stronger than Thin Layer Mortar type B at all ages.
(f) With both thin layer mortars 70% of the total strength was reached after 7 days curing.
(g) With both thin layer mortars, the strength at 3 days was at least 1.6 times (3.3 for mortar A) the 28 day strength for designation (iii) mortar as required in Table 1 of BS 5628:Part 1.

ACKNOWLEDGEMENTS

The authors would like to acknowledge the funding for the project given by the Engineering and Physical Sciences Research Council (EPSRC) and the Industrial Collaborators, Autoclaved Aerated Concrete Products Association, H+H Celcon, Hanson-Thermalite, Quinn-Group, Tarmac Topblock, Building Research Establishment, Catnic-Corus UK, National House-Building Council and Office of The Deputy Prime Minister. Acknowledgement is also given to the significant contributions made by the project Steering Panel, Dennis Coward, Alan Ferguson, Clifford Fudge, Trevor Grounds, Jim Holland, Chris Kirby, Paul Matthews, Gopal Sangarapillai and Richard Shipman.

REFERENCES

[1] BS 5628-1:1992 – Code of practice for use of masonry – Part 1: Structural Use of Un-reinforced Masonry.
[2] BS EN 197; Part 1- Cement Composition, Specifications and Conformity Criteria for Common Cements.
[3] BS 1200:1976 – Specifications for Building Sands from Natural Resources.
[4] BS EN 1015-3:1999 – Methods of Test for Mortar for Masonry – Part 3: Determination of consistence of fresh mortar (by flow table).
[5] BS EN 998-2:2003 - Specification for mortar for masonry. Masonry mortar.
[6] BS EN 1015-11:1999 - Methods of Test for Mortar for Masonry. Determination of Flexural and Compressive Strength of Hardened Mortar.

Autoclaved Aerated Concrete – Limbachiya and Roberts (eds)
© 2005 Taylor & Francis Group, London, ISBN 0 415 38356 0

Bed joint reinforced AAC masonry applications and developments

P. Timperman
Bekaert – Belgium

ABSTRACT: The reinforced masonry technology has obtained a wide experience and acceptance in Europe. The system allows to realize two objectives for which often people are reluctant to apply it.

First of all, bed joint reinforcement is the most adequate mean to prevent masonry from cracking. Cracking which is due mostly from shrinkage but also from a number of other causes as seismic events. Murfor bed joint reinforcement will control the crack width and bring this down to acceptable levels.

Second to this technology are the structural applications which offer the designers, architects and contractors alternatives to the classical reinforced concrete constructions. The benefit of this technology can be economical, technical or esthetical.

1 INTRODUCTION: REINFORCED MASONRY: THE PAST AND PRESENT

The use of reinforcement in masonry structures is already more than 75 years old. The first recognised use of reinforced masonry was in 1813 when a British engineer Marc Brunel built a chimney for an industrial production plant near London. In 1825 that same technique was used by Brunel to construct a hydraulic tunnel under the Thames.

After that time, the use of reinforced masonry was put aside until the end of the First World War, when it was reintroduced in Canada, UK, USA and Australia. The more recent architect, Eladio Dieste – Argentina, built a lot of important structures in reinforced masonry (even churches) and as such motivated the architects to use the full benefits of this construction technology.

The use of reinforced AAC masonry is from a more recent period and has almost developed in Belgium. when the former AAC producers where participating at a global programme investigating the different pathologies in masonry.

Nowadays the use of AAC block work has passed the level of purely crack control and offers opportunities as well for structural applications. Unfortunately at present the development and applications is mostly limited to Europe. The important vector which will increase the potential is the new codes for masonry as foreseen in EC-6 and EC-8. As such the European CEN Standards will regulate the use of reinforced masonry for all major application, even for the construction in seismic areas.

At this time greater attention is being given to the potential of masonry structures, not only because of the growing knowledge of the material and its composition, but also because of the economic benefits of the technique.

This paper deals with the use of bed joint reinforcement for crack control and structural applications.

In total, 3 research projects are presented in which bed joint reinforcement has been used to increase the overall performance of AAC block work:

- The use of bed joint reinforcement in laterally loaded aac block work. And documented by a practical case
- Comparison between steel bed joint reinforcement and synthetic reinforcement
- The analysis of the different reinforcements in aac block work submitted to seismic actions.

2 THE USE OF BED JOINT REINFORCEMENT

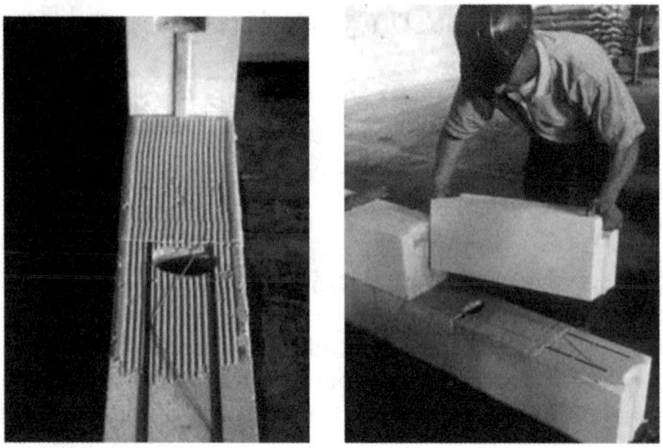

2.1 *Crack control*

Masonry cracks may be classified by:

- Structure type;
- Masonry type;
- Location;
- Pattern;
- Width and Cause.

For a given type of structure and material, the location, pattern, and width often provide clues to the cause. Cracks result from strain, which induces stress in excess of strength in compression, tension, or shear. Strain may be induced by the imposition of loads or by restraint of volume changes in the masonry materials. Volume changes include those induced by change in temperature, moisture, water or salt crystallization, or corrosion. Movements of foundations, structural frames, shelf angles may impose loads, roof slabs, spreading of pitched roofs, wood expansion, or retaining wall deflection. Vibration, blasting, and fire may also cause cracks. Seismic events are even the best example of the creation of cracks.

The type of cracked masonry units include all masonry materials,

Cracks are located in the masonry units, in the mortar joints, between units and mortar, and continuous through units and mortar.

Crack smaller in width than 0.1 mm are insignificant to water presence because that is the least width through which wind-driven rain will enter.

The application of horizontal bed joint reinforcement will largely contribute to the reduction of crack widths and bring them down to acceptable levels. Typical application where bed joint reinforcement have proven evidence are in the

- Shrinkage
- Around door and window openings
- Areas of differential settlements
- Excessive deflection in the support
- In long wall construction
- Seismic constructions

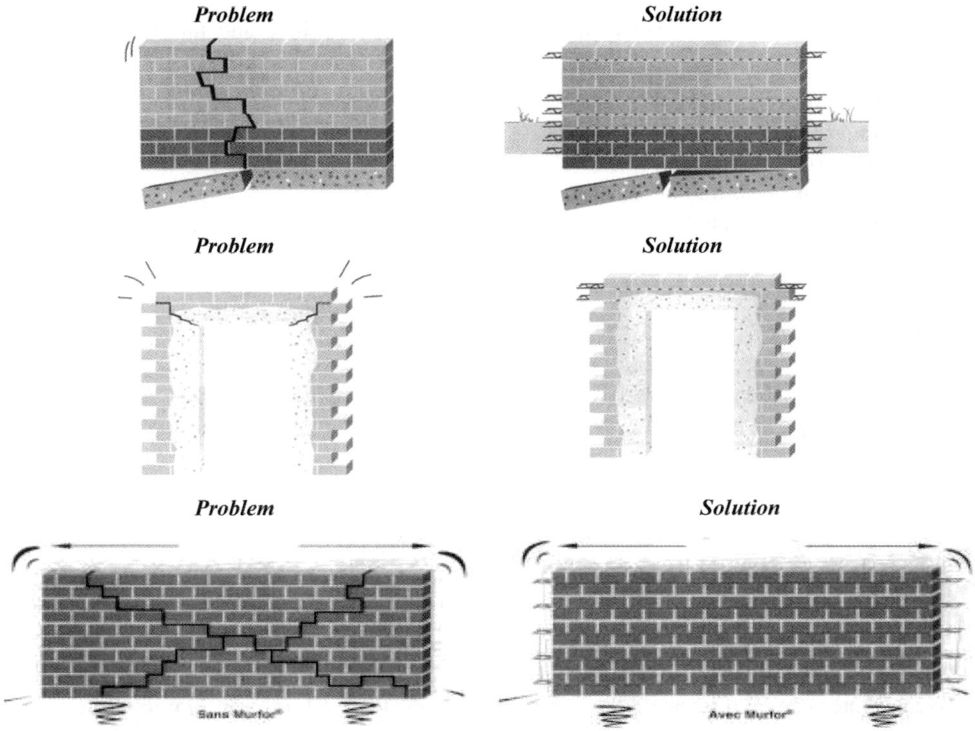

The amount of reinforcement required, largely depend on the materials used and type of construction type.

3 THE STRUCTURAL APPLICATIONS

Reinforced masonry = reinforced concrete

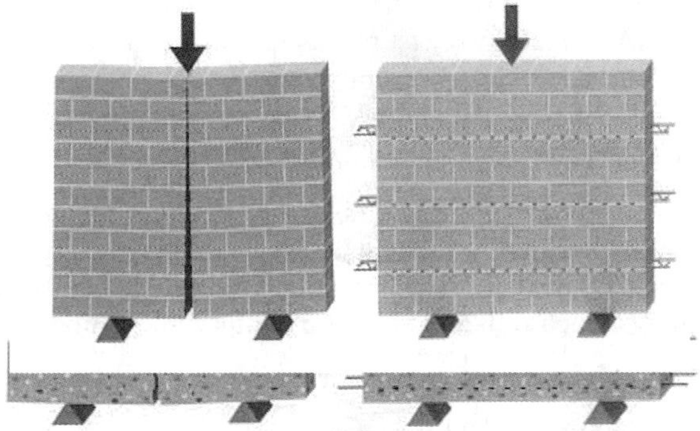

More and more acceptance is obtained for the application of the structural use of reinforced masonry. In recent time more analysis, research and development is carried out at different universities in Europe. The main argument is the need to prove these constructions have a similar or equivalent behaviour to standard reinforced concrete. The results are positive and this for a number of reasons:

- Increase the ductile behaviour of the masonry structure
- Prevention of thermal bridges
- Homogenous reinforced structures
- Architectural benefits and increased creative freedom
- Optimal cost/performance

4 STEEL WIRE BED JOINT REINFORCEMENT

Murfor is a prefabricated steel wire reinforcement to be embedded in the horizontal joints of the masonry. Its unique shape and quality increase the tensile strength of the masonry.

Murfor masonry reinforcement was first introduced into the European Market more than 35 years ago. Over this period, Murfor was developed in Belgium at a time of prolific building and changing construction methods.

5 THE RESEARCH INTO THE POSSIBILITIES OF THE BED JOINT REINFORCED AAC MASONRY WALLS

This paper describes 4 research cases of AAC block work reinforced with murfor:

- RUG – University Gent Belgium: the performance of AAC reinforced walls subjected to lateral load (+test case)
- IBAC – Germany: comparison between Murfor and Synthetic masonry reinforcement
- Seismic behaviour of AAC blockwork: test results of Uni Pavia - Italy and NTUA Athens: the use of Murfor in filled masonry wall of seismic constructions

5.1 *Murfor reinforced walls subjected to lateral load*

At the Belgian State University in Gent, a research was carried out to verify the performance of bed joint reinforced masonry subjected to lateral loads.

The objective was to verify to what extend partition walls could be used absorbing high wind loads in industrial buildings. Such wall also serve for compartimentation or fire walls.

Test set up :

- 5 walls: dimensions of 2400 × 1250 mm
- AAC – Blocks: 600 × 150 × 250 mm: average compressive strength (by 6% humidity) = 4.20 N/mm^2
- Thin joint mortar: – Flexural strength = 3.63 N/mm^2
– Compressive strength = 15.1 N/mm^2
- Bed joint reinforcement Murfor efs 100 mm in every second joint.

5 walls unreinforced compared to 5 walls murfor reinforced, as given in Table 1 below.

	Unreinforced walls		Reinforced walls	
Wall	Max load (kN)	Flex strength (N/mm^2)	Max load (kN)	Flex strength (N/mm^2)
1	11.8	0.67	34.7	1.96
2	12.2	0.69	31.7	1.73
3	12.6	0.71	30.7	1.73
4	13.1	0.74	30.4	1.72
5	10.7	0.60	30.3	1.71

Average load at 1st crack for the reinforced walls = 13.6 kN
Average stress at first crack: 0.77 N/mm^2.

Conclusion:

- Average increase in max load: 3 X
- Average increase in flexural strength: 2.5 X
- Crack width: regularly spread over the total surface : <0.10 mm.

The results have allowed that AAC block work can be designed (according EC-6) in walls submitted to wind loads and as such they are at presently a standard solution in projects such as warehouses.

Market Case: The Value Retail Outlet Shopping Centre In Maasmechelen – Belgium

Considering the test results and in alignment with the EC-6 design rules, the verification of wind loads on the external aac walls was made for this project. Second to this wind loads, the project was realized on an old mining area which could provoke differential settlements.

For the design of aac walls submitted to wind loads a verification of the moments and shear stresses is required.

As such a calculation is needed of the compression stresses parallel to the bed joints, the tensile strength of the steel wire t reinforcement, the flexural strength perpendicular to the bed joints and the shear stresses.

Based on the outcome and obtained data, the definition of the required reinforcement can be calculated according to:

$$As \geq \frac{\gamma_m . \gamma_f . M_m}{0.8.d.F_{yk}}$$

The wall dimensions could be increased to 5 m long and 8 m high, exposed to a wind load of 0.7 kN/mm^2, the required cross sectional area is 96 mm^2/m wall height or 1 reinforcement every 250 mm.

5.2 *Comparison between the use of steel wire reinforcement and synthetic grid reinforcement*

The research was carried at the IBAC institute of the RWTH of Aachen Germany. The objective of the research was verification to what extend synthetic grid could be used as alternative to the steel wire reinforcement.

5 walls were submitted to a horizontal bending test on walls of 2500 mm × 800 mm.

- Block dimensions: 625 × 300 × 300 mm
- Mortar: – Compressive strength: 17.1 N/mm^2
  – Flexural strength: 4.18 N/mm^2

- Bed joint reinforcement: type Murfor EFS/Z – 100 mm and 200 mm - Rm: 916 N/m^2 every second layer
- Synthetic mesh: tensile strength: 2800 N/mm^2 in each layer

The results

Wall	Level arm	Shear load at 1st crack	Shear load at failure	Moment at 1st crack	Failure moment
		kN		kNm/m	
Unreinforced	496 mm	4.08	4.08	2.53	2.53
Murfor 100	498	6.63	13.66	4.11	8.47
Murfor 200	500	5.92	14.42	3.68	8.95
Murfor connected	500	7.26	7.26	4.51	4.51
Synthetic	496	2.56	9.97	1.58	6.13

Crack width at max load

- Unreinforced: 2 mm
- Murfor reinforced: 0.56 mm
- Synthetic reinforcement: 1.58 mm

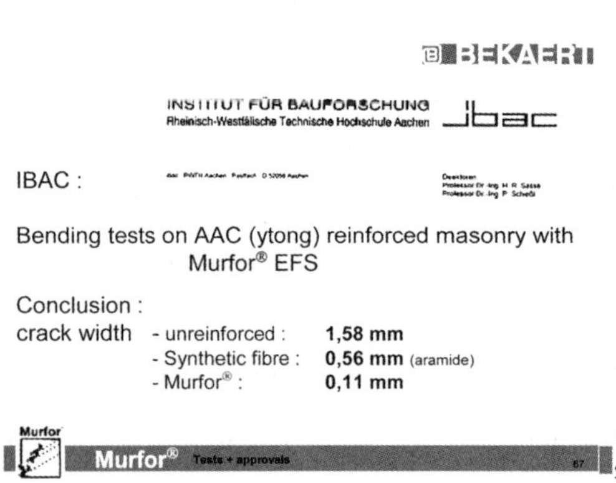

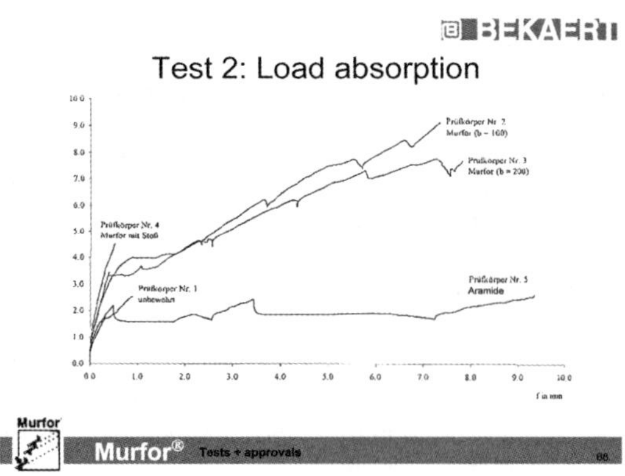

5.3 *Murfor reinforced infill walls for seismic constructions*

5.3.1 *Test results of the RDB research at the University of Pavia (2005)*

A Recent research ordered by the Italian AAC block producer RDB at the University of Pavia (Italy) has compared ao Murfor reinforced masonry infill walls with rebar, RC-beam reinforcement and non-reinforced masonry walls.

The objective of the research can be summarised in following points:

- To compare the in plane response of unreinforced and murfor reinforced masonry infills, for different intensity level earthquakes, in order to assess the potentially different damage level attained;
- To asses the potential for out of plane expulsion of unreinforced and murfor reinforced masonry infill panels, at different level of damage induced by in plane action;
- To evaluate the effects of different properties of the infill on the response of buildings with different geometrical configuration and different distribution of the infill panels, in terms of PGA required to induce given level of damage.

In plane test results:

- Unreinforced infill

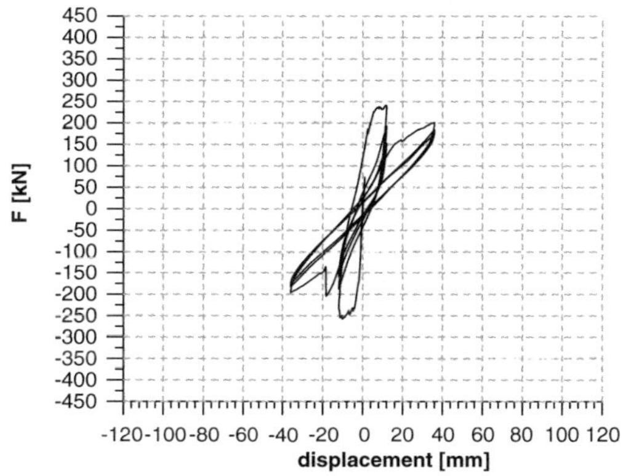

Murfor reinforced infill

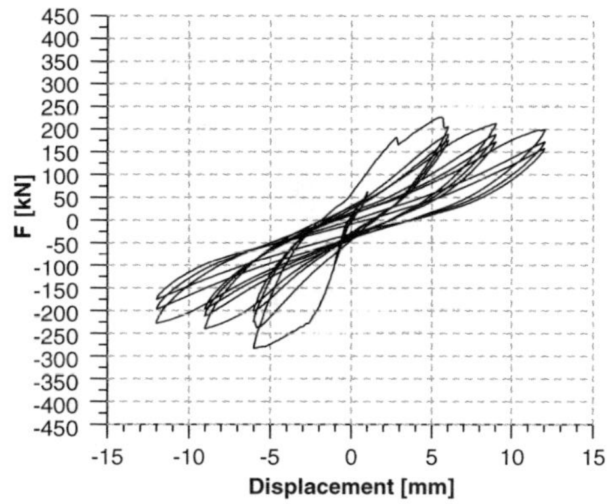

- Out of plane results (results after in plane tests)

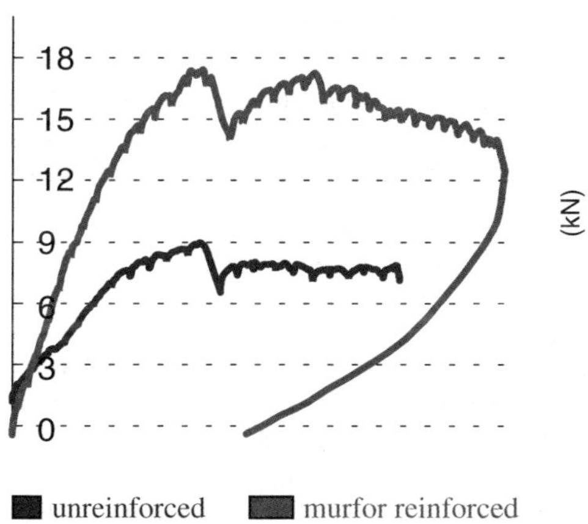

The use of murfor reinforcement, significantly improves the response of a single infilled frame, particularly for what concern the damage limit state.

The crack pattern showed clearly the benefits of the bed joint reinforcement, where more and smaller cracks appear and which as such does not influence thc serviceability of the wall. After the seismic event, renovation of the wall is still possible. For the unreinforced walls, rebuilding the wall is the only solution as the cracks are too big (>2 mm)

The use of murfor, with a geometrical percentage lower than 1%, results in an increase of the peak ground acceleration level to 0.35–0.70 g depending the damage level.

5.3.2 *Reinforced concrete frames with infill walls from AAC block work*

The results of the NTUA research programme ordered from Ytong Greece - 2003.

1. *Experimental program*

The experimental work was undertaken at the National Technical University of Athens (NTUA). It involved the testing of two RC frame models under dynamic loading on the shaking table of the Laboratory of Earthquake Engineering (LEE) at NTUA. The two RC frames consisted of one level with the four infill walls each made from 'YTONG' type of bricks. The compressive strength of

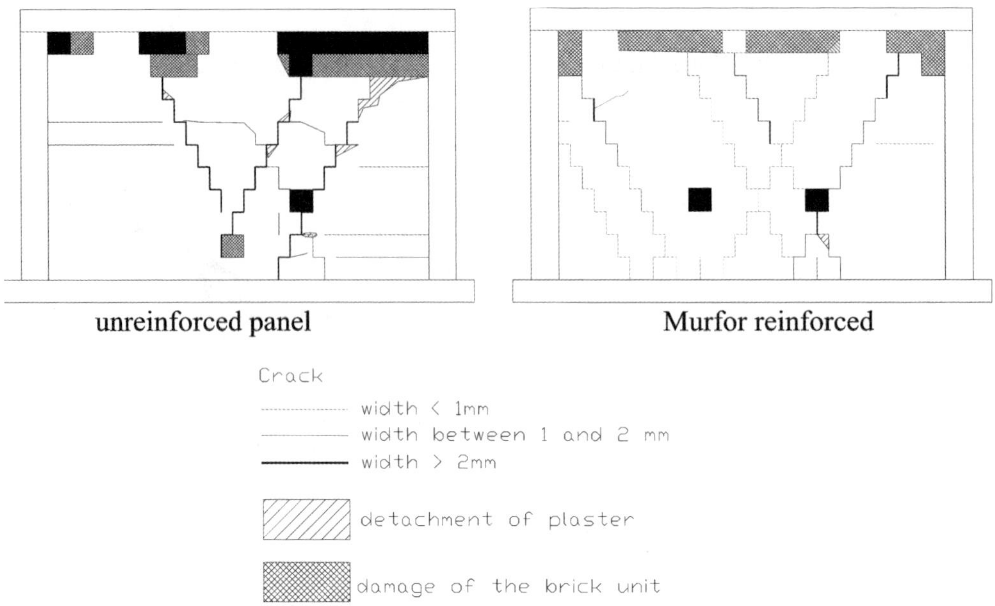

Damage representation

these bricks is 2.7 MPa and weight 420 kg/m^3 (in dry conditions). The dimensions of each model were $2.85 \times 2.30\,m^2$ in plan and had a height of 3.16 m. The thickness of the walls was 0.20 m.

The total mass of each model, including its base which was made of steel, was 9.45 Mgr.

Model I: Walls 1,2 and 3 had no openings whereas wall no. 4 had a window situated at the upper half of the frame. In wall no. 1, a layer of aac blocks with a channel (half-pipe type of opening along the length of the brick on the top side) filled in with concrete was placed at the mid-height of the wall. In wall no. 2, a concrete beam was placed between two layers of bricks at mid-height of the wall. In wall no. 3, Murfor reinforcement was placed at every two layers of bricks. Finally, wall no 4. consisted entirely of a non reinforced block work.

Model II: Walls 5, 6 and 8 had no openings whereas wall no. 7 had a window situated at the mid height of the frame. In wall no. 5, a thin glass fibre mesh was placed on every bed joint. Wall no. 6 consisted from only from bricks without any type of reinforcement. In wall no. 7, an aac prefab lintel was placed above the opening. Finally, in wall no. 8, the bricks were locked to the columns with mortar.

2. *Testing procedure*

The models were tested under a dynamic loading on three axes (X, Y & Z) with a different time history applied along each of these axes. The time histories were artificial and had a time length of 2048 secs and a time step of 0.01 sec. The maximum ground acceleration was 0.16 g.

3. *Experimental results*

In model no. II, cracks appeared at an earlier stage compared to model no. I. In both models, at the end of the testing the infill walls had partly separated from the RC frame. Also, there were horizontal and diagonal cracks and the bricks were fractured in many cases. More specifically, the damages for walls 2, 3 and 5 were as follows:

In wall no. 2, the RC beam at the mid height of the frame was separated from the frame columns and they were also diagonal cracks at the joints of the columns and frame beam.

In wall no. 3, there were some horizontal and diagonal cracks on the bed joints and the bricks but, no damage on the frame.

In wall no. 5, apart from the damage described above, there was also out of plane displacement of the wall and partial collapse of the upper half.

Therefore, the outcome of this experimental procedure is that the wall reinforced with Murfor (wall no. 3) had the less damage. Most of the damage was due to cracking of the 'YTONG' bricks (due to low compressive strength).

6 CONCLUSIONS

Reinforced AAC block work can be more widely used in application which is expected out of the framework for this kind of block work.

At present it is largely proven that steel wire reinforcement largely fulfils the requirements for structural applications and as such contributes to the expected quality levels of the AAC block work.

Any other material proposed as an alternative does not meet the expected outcome.

BIBLIOGRAPHY

G. Vandeloock: Test on reinforced brick masonry in Belgium – 1979

P. Timperman: Murfor – Reinforcement for Masonry – 1996

P. Timperman, T. Rice: Bed joint reinforcement in Masonry – 1999

P. Timperman: Murfor: La maçonnerie armee – 1999

St. Schaerlaekens, La maçonnerie armee – CSTC – 2001

Biegeversuche an bewehrten porenbeton Mauerwerk – Ibac – 1998

RC with infill walls from AAC block work – NTUA – LLE lab – 2002

Buigproeven op gelijmd cellenbetonmetselwerk – RUG 93/0236

O. Pfeffermann, B. Haseltine, P. Timperman: The use of Murfor reinforced masonry to prevent cracking problems: 25 Years of experience

G.M. Calvi, D. Bolognini; RDB : Seismic response of reinforced concrete frames infilled with AAC block work – Pavia 2005

The use of bed joint reinforcement in structures submitted to differential settlements – Jan Kubica – TU Gliwice Poland 2003

G.M. Calvi: Reinforcement in seismic design of masonry structures – Pavia 1999

EN 845-3: ancillary components

Pr EN 1998-1-3: Eurocode 8 – Design provisions for earthquake resistance of structures – part 1-3: General rules – Specific rules for various materials and elements

Pr EN 1996-1-1: Eurcode 6: Design of Masonry structures

Part 1-1: Common rules for reinforced and unreinforced masonry structures.

Theme Six: Environment and building performance aspects

Autoclaved Aerated Concrete – Limbachiya and Roberts (eds)
© 2005 Taylor & Francis Group, London, ISBN 0 415 38356 0

Building energy consumption and thermal performance of the Autoclaved Aerated Concrete (AAC) exterior walls

R. Lindberg & V. Leivo
Tampere University of Technology, Structural Engineering, Tampere, Finland

ABSTRACT: This paper presents results of field survey concerning the energy consumption and the thermal performance of the AAC exterior walls. According to the measurements the actual heat loss through AAC exterior wall was smaller than calculated. The massive thermal mass of the AAC wall absorbs energy surpluses both from the solar gains and from the heat produced by internal energy sources. Based on the measurements even at the cold Nordic winter days the solar radiation raises the temperature of the outer wall surface. Also the values of the thermal conductivities and wall area for calculations are chosen safely which factors causing the difference. The internal heat energy sources such as lightning and electric devises are producing nowadays even more energy surpluses than in the past. Thus the heat capacity of the exterior walls is as important than the thermal conductivity. The heat storage capacity can cut the peak heating and cooling energy demand.

1 INTRODUCTION

The building sector consumes about 40% of the total energy in European Union. Heating, cooling, warming of water and lightning consumes most of it. The EU directive on energy performance of buildings has been created to reduce this consumption and related CO_2 produce. The national regulations based on that directive should be valid by the beginning 2006. In practise this will change building regulations controlling energy use and the energy consumption of buildings will be more emphasized.

Energy consumption of building consists of following parts: energy used for heating and cooling and energy used in air conditioning, service water, heating, cooling, lightning and other electric systems. Transmission and ventilation losses from the internal to the external environment and useful internal heat gains and solar gains form the energy used for heating.

2 FIELD TESTS

2.1 *Test buildings*

The research project, started 1996, was established in order to determine the thermal behaviour of different type of outer wall construction and compare how accurately the current calculation methods for heat losses can predict the actual heat loss. The measurements lasted 5 years. 6 test buildings are constructed in a moderately exposed parking area in Tampere University of Technology. The floor area of each test building was 2.4 m × 2.4 m and the free floor to ceiling height was 2.6 m. The ceilings and floors of the all buildings are composed of two layers of foamed polyurethane elements, total thickness 200 mm. The external walls of the test buildings were constructed of following building materials: polyurethane insulated wood framed wall (T-B 1), insulated cavity

Table 1. The thermal transmittances of the structural components of the test buildings.

Building component	U-value () (W/m^2 K)
Ceilings (200 mm thick)	0.12
Floors (200 mm thick)	0.12
External door (two consecutive doors)	0.35
Polyurethane insulated wood framed wall (T-B 1)	0.17
Insulated cavity brick wall (T-B 2)	0.27
Insulated log wall (T-B 3)	0.29
Plastered massive brick wall (1.5 bricks)(T-B 4)	0.86
Autoclaved aerated concrete (ACC) block wall (T-B 5)	0.35
Massive log wall (T-B 6)	0.67

brick wall (T-B 2), insulated log wall (T-B 3), plastered massive brick wall (T-B 4), autoclaved aerated concrete (ACC) block wall (T-B 5) and massive log wall (T-B 6). The ACC wall consist of 375 mm width ACC blocks with 1 to 5 mm smoothing outer surface and 2 to 4 mm fibre mortar and 2 mm sprayed mortar finish inner surface. The coefficients of thermal transmittances (U-values) of various walls and structures are presented in Table 1. The U-values have been calculated using λ_n-values of structural components.

All test buildings were heated with radiators and the indoor temperature during heating season was constantly maintained at +20°C. Test buildings 1, 3 and 5 were ventilated by balanced mechanical ventilation system with heat recovery and buildings 2, 4 and 6 by exhaust mechanical ventilation system. The air change rate in all the test building was 0.5 1/h. For simulating occupancy, moisture supply of 0.2 g/m^3 was added to the indoor air.

2.2 *Measurements*

A weather observation station constructed at the test buildings' site measured the outdoor temperatures, wind speed and, direction and relative humidity of the air. The intensity of solar radiation was also measured on-site by a pyranometer (solar meter). Inner and outer surface temperatures and temperatures inside the wall at different depths were monitored at three heights. Airflow rates and temperatures of supply and exhaust air were also continuously monitored. The monitoring and data collection system was automatic.

3 RESULTS

Total energy consumption (kWh) of the test building was measured with electricity meter. The actual transmission heat losses were determined reducing the ventilation losses from the meter reading. The calculated transmission heat losses are determined using calculation method presented on Finnish building regulation. The calculation method is mostly similar with EN 832. The calculation method proposed that the wall areas determined using outer wall line should be used in calculations. The measured indoor and outdoor temperatures was used in calculation. The area of corners compared with the total wall area is larger in the test buildings than actual buildings.

Difference method was used to calculate heat losses through corners (Lindberg, R. et al., 1998). Using values of heat losses through corners and walls it can be determined the proportion of the intersecting wall depth, which should be used in calculation heat losses. Two different calculation values was determined, one without (calculated) and one taking into consideration the heat losses

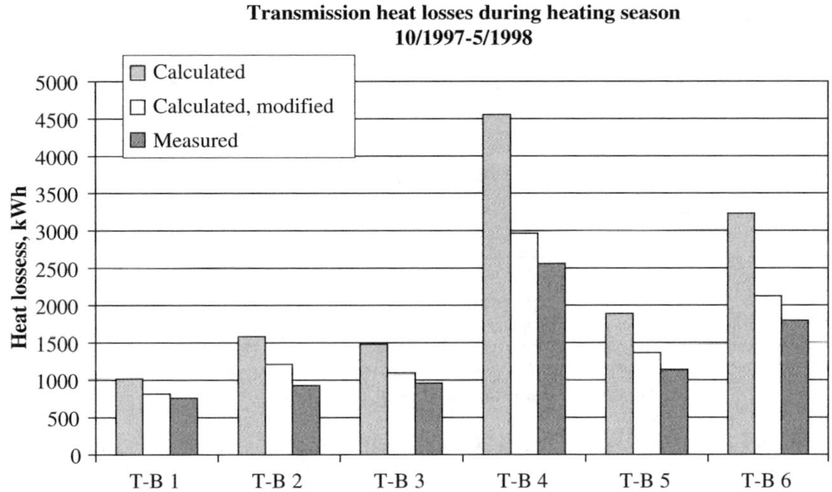

Figure 1. Transmission heat losses during heating season 10/1997-5/1998.

of corners (calculated, modified). The thermal transmission heat loss of the test buildings during one heating season has been presented on Figure 1.

It can been seen at the figure that the actual (measured) transmission heat losses are smaller than calculated. The difference without corner modification on about 65% and even with corner modification about 21%. There are 3 reasons for the difference. The wall areas calculated using outer wall lines conduct to too large heat losses. Secondly the thermal transmittance values (U-values) used in calculations are determined using normal values for thermal conductivity (λ_n), which is determined having safe marginal compared to the actual values. The third reason is that the massive thermal mass of the AAC wall absorbs energy surpluses both from the solar gains and from the heat produced by internal energy sources.

The effect of solar gains can seen on Figure 2. The figure presents temperatures distribution curves of ACC wall located at the southern façade during one winter day. Average outdoor temperature was −11.6°C. Temperature of the outer surface was at the morning (4:00 o'clock) same as the outdoor temperature (−18°C).

When the sun raises the surface temperature raises and at the afternoon (16:00 o'clock) the surface temperature was about +27°C while the outdoor temperature was −2°C. Still few hours after sunset (20:00 o'clock) the surface temperature was −7°C while the outdoor temperature was −20°C. This means that heat flux is from outside to inside and solar gains produces energy surpluses even at the cold Nordic winter days. At the field measurements only the massive wall structures with high heat storage capacity was able to absorb energy surpluses from the solar gains.

4 DISCUSSION

Trend in energy efficiency in building construction has been that the energy used for heating has been tried to minimise by reducing heat losses improving thermal conductivity and air tightness of structures and making ventilation and heat recovery systems more efficient, Figure 3. At the same time different internal heat energy sources such as lightning and electric devises are producing more and more energy surpluses than in the past.

Consequence of all this is that heating season shortens and need for cooling increase. Structural solutions, such wall structures with good heat storage capacity are usable for cutting the peak cooling and heating energy demand.

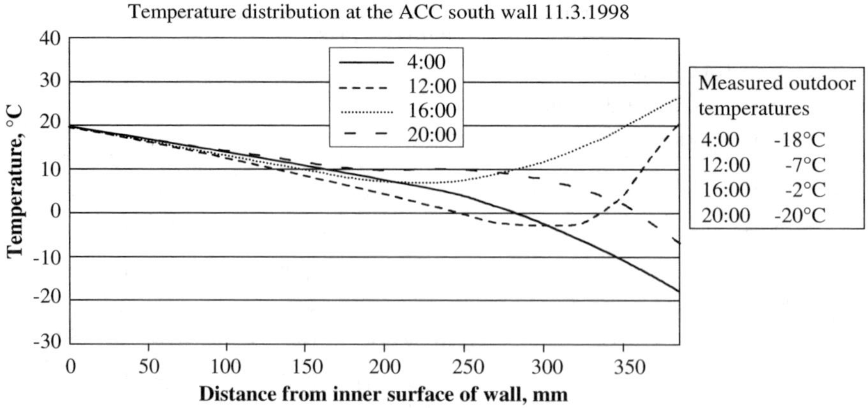

Figure 2. Temperature distribution at the ACC south wall 11.3.1998. Average outdoor temperature was −11.6°C (−20.5 … −1.8°C).

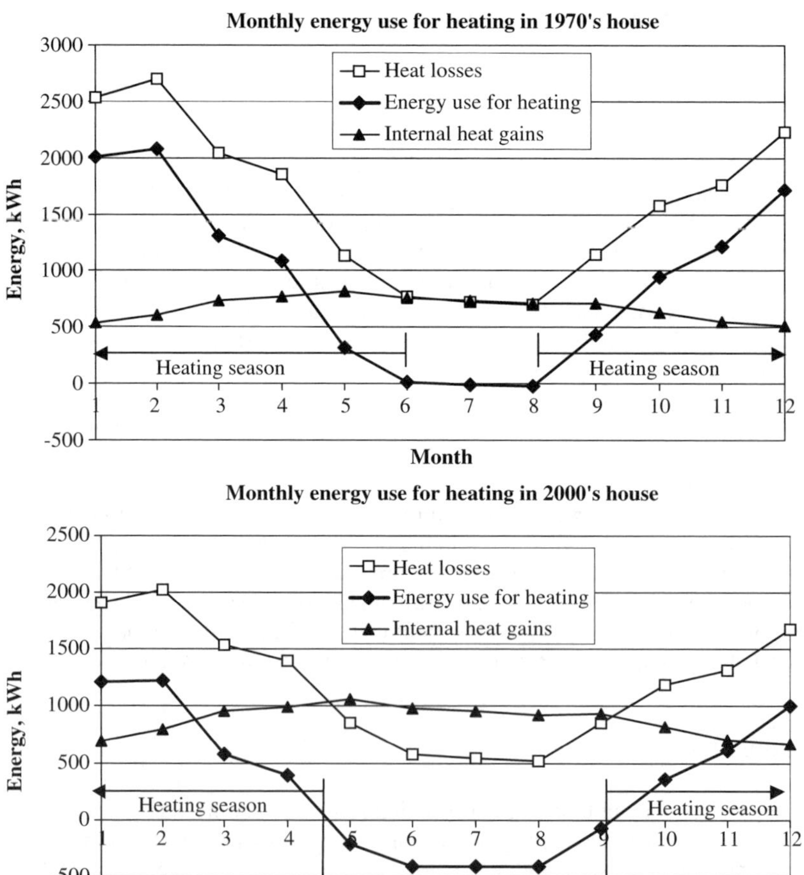

Figure 3. Schematic presentation of the annual energy use for heating in different decades.

5 CONCLUSIONS

According to the 5-year field measurements the actual heat loss through AAC exterior wall is smaller than calculated. The massive thermal mass of the AAC wall absorbs energy surpluses both from the solar gains and from the heat produced by internal energy sources. Based on the measurements even at the cold Nordic winter days the solar radiation raises the temperature of the outer wall surface. There are also two other factors causing the difference. Values of the thermal conductivities and wall area are chosen safely while the real energy consumption should be calculated using average values. The internal heat energy sources such as lightning and electric devises are producing nowadays even more energy surpluses than in the past. Thus the heat capacity of the exterior walls is as important than the thermal conductivity. The heat storage capacity can cut the peak heating and cooling energy demand.

REFERENCES

EN 832. 1998. Thermal performance of buildings – Calculation of energy use for heating – Residential buildings. European Committee for Standardization.

Lindberg, R. et al. 2004. five-year data of measured weather, energy consumption, and time-dependent temperature variations within different exterior wall structures. Energy and Buildings 36 (2004) 495–501. Elsevier.

Autoclaved Aerated Concrete – Limbachiya and Roberts (eds)
© 2005 Taylor & Francis Group, London, ISBN 0 415 38356 0

Environmental product declaration of Autoclaved Aerated Concrete

H.B. Walther
Xella Technologie- und Forschungsgesellschaft mbH, Kloster Lehnin, Germany

ABSTRACT: Type III Environmental Product Declarations (EPD) are a new instrument to supply quantified environmental data for a product using predetermined parameters according to ISO CD 14025.3. These parameters are based on the ISO 14040 series of standards and may be accompanied by other relevant quantitative and qualitative environmental data. EPD can cover the full life cycle of a product or only selected stages of it. This results in different levels of comparability and completeness. It is difficult to acquire reliable data of a building material because of the overwhelming influence of the utilization phase, the influence of the construction unit and the long life time of a building.

The AUB develops EPD in Germany. These EPD are based on product specific patterns i.e. Product Category Rules (PCR). The first PCR that has been developed is the PCR of AAC which started in late 2001. This was done in a AUB product forum. Producers and experts from different institutions as DIBt (Deutsches Institut für Bautechnik) and UBA (Umweltbundesamt) as well as representatives of federations discussed the points of interest. Relevant data should be given in a clear and brief way. In 2004 a third party advisory board verified the PCR of AAC.

1 ENVIRONMENTAL INFORMATION IN THE PAST

1.1 *General development*

Fifty years ago, nearly nobody asked about the environmental impact of production processes and the influence to the environment during the utilization of the goods produced. In the late sixties and the early seventies of the last century environmental product information started to become interesting but were normally not given according to distinct rules and schemes. They followed mainly the intentions of the producers to display only positive aspects of their products and it was not wanted to allow a comparison between different products on the same scale. Naturally, each product has individual advantages and disadvantages. Consequently, the information provided were not complete and "apples were compared with oranges".

1.2 *LCA – A new quality in environmental information*

A new quality of environmental information was reached about 20 years ago, when Life Cycle Assessments (LCA) were introduced in environmental information. This new instrument included relevant product information about energy consumption from cradle to grave. The main aspect was to look not only into the production process of the product but also into the fabrication process of raw materials and preliminary products used. Products containing materials with high energy consumption for their production display a huge environmental impact – a small influence of the production itself but a "heavy" ecological backpack. A typical example for that are metals. This aspect is generally accepted by specialists but is difficult to communicate. "Why do I have radioactive waste in my LCA? I do not have any radioactive waste, never had!"

However, any product fabricated using electricity contributes to the waste from power stations including radioactive waste from nuclear power stations.

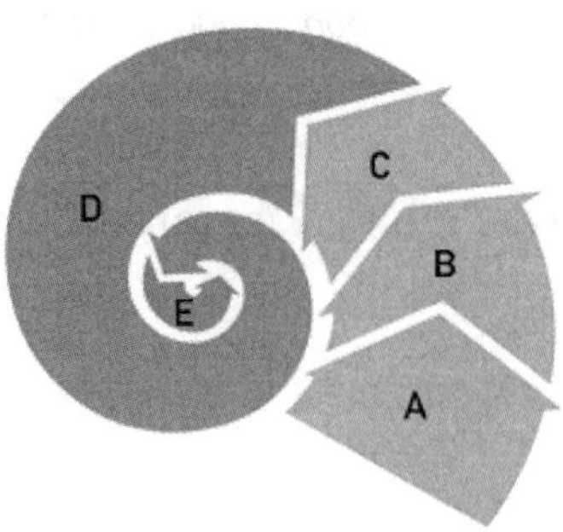

A Mining and processing of raw materials
B Production of building material
C Erection of the building
D Use of the building
E Demolition and recycling

Figure 1. Content of an LCA for building products according to ISO EN 14040.

Over the years LCA were developed for an increasing number of products and preliminary products. Hence, the quality of input data increases consequently the quality of LCA themselves. Life cycle assessment today is an appropriate tool for analyzing and assessing the environmental impacts that are caused by production, utilization and disposal of products or product systems for specific applications, see Figure 1. However, the results of LCA normally are very complex and appropriate for specialists but not suitable for customers because they are not clear-cut assertions.

2 DEVELOPMENT OF RULES FOR ENVIRONMENTAL PRODUCT DECLARATION

2.1 *The relevant standards*

There are two main families of standards being relevant in this context, i.e. the ISO 14040–14043 concerning LCA and ISO 14020 concerning Environmental declaration and followed by ISO 14025 specialized to building products and ISO 21930 provided for the development of Product category rules (PCR).

The standard ISO 14040 defines an LCA as follows: "LCA is the compilation and evaluation of inputs, outputs and potential environmental impacts of a product system throughout its life cycle."

The definition from ISO CD 14025.3 covers "quantified environmental data for a product using predetermined parameters" and specifies:

- The predetermined parameters are based on the ISO 14040 series of standards.
- The quantified environmental data may be supplemented by other relevant quantitative and qualitative environmental information.

2.2 *LCA of building material*

The LCA normally claims to cover the whole life cycle of a product. This can easily be calculated for technical products with a clearly defined utilization as for example a fridge. The situation is much more complicated if products are concerned having different potential applications as building products. In this case an LCA frequently covers only distinct parts of the life cycle – not cradle to grave but cradle to gate.

The environmental impact of building materials can be only calculated correctly for the first steps in their life cycle. The influence of raw material acquisition, processing and production including the transports can be exactly defined. At the gate of the plant the products were spread out into different construction sites and different applications. Hence, LCA of building materials do often end at the gate of the producer.

In the 1990's the University of Stuttgart carried out an extensive study to LCA of building materials, together with 14 associations of relevant producers of building materials. The rules of balancing the environmental impact according to ISO and especially for building products were defined in the first phase of the project (Kreißig et al 1997), detailed calculations were carried out in the second one (Kreißig et al 2000). Consistent LCA of the building materials as well as building products like windows or heating systems were the result of this. In order to exclude the huge influence of user behaviour the LCA ended with the production.

2.3 *LCA of AAC*

The utilization phase normally is the most important step in the life cycle of a building product being of overwhelming influence (see fig. 1). The advantages of AAC, mainly its excellent thermal insulation, reduce the environmental impact during use and this is the main impact of a building product. This situation is not very satisfying for the producer of AAC and consequently they looked for ways to solve this problem.

In 1997 YTONG published an LCA containing information about the complete life cycle of AAC related to the functional unit of 1 m^2 outside wall. In order to show differences and to highlight the thermal insulation properties impact analyses of wall thicknesses of 24 cm as well as 36.5 cm were compared with a standard wall fulfilling the relevant legal criteria. It could be proven that a slightly higher environmental burden during raw material acquisition and production is resulting in a drastically reduction of impact during use and the whole life cycle.

2.4 *The need of environmental product declaration of building materials*

Building materials are something special in context to LCA because the product that has to be analyzed is the building erected at site and being unique. There is no alternative to evaluate this building itself because the number of varieties in utilization even the intended one of building materials is not restricted at all as well as the possible scenarios of impacts to the environment.

Consequently, it is of great importance to supply consistent data sets of the individual building materials because the quality of the building calculation only is as good as the quality of worst data set. A consistent quality of the data is even more important, if construction alternatives were calculated for evaluation of environmental impacts.

3 ECO LABELLING

Marketing activities require striking information, labels or single numbers are preferred – consequently, a wide variety of labels exist. Even if marketing needs these simple "good" and "not good" conclusions, life is not that easy, because a building material can be used in different places and functions and under different circumstances. Any building material has its advantages and there are applications to which it is nearly unique. Therefore, specialists need substantial information about the relevant product.

Table 1. Eco labels according ISO 14025.

Type I	Type II	Type III
Information to customers	Information to customers	Information to planners, authorities and customers
Display one or two important environmental aspects	Do often concentrate on one single, environmental aspect	Give substantial, quantitative information
Can be used in public procurement function	Singular marketing information	Based on an LCA
High credibility		Display environmental impacts without evaluation
Awarded from independent associations	Accounted from producer	Independently checked information accounted from producer

4 ENVIRONMENTAL DECLARATION IN EUROPA

4.1 *Environmental declaration in the Netherlands – MRPI system*

The Netherlands have implemented an advanced system of environmental information, sometimes known as MRPI system. This institute developed an information sheet containing the most relevant ecological LCA-information in a condensed form. The functional unit is described in detail (production, transport, deposition) concerning

- intended use
- product description
- utilization of energy and raw materials (organic and inorganic)
- Greenhouse effect
- Ozone depletion potential
- Acidification potential
- Eutrophication potential
- Human toxicity
- Eco toxicity
- Photochemical ozone creation potential.

4.2 *Environmental declaration in Germany – role of AUB*

Numerous organisations and associations provide environmental information in Germany. One of the most active ones is the AUB, a consortium of producers has been dealing with environmental problems in a highly professional way for decades. The AUB in Germany was one of the pioneers discussing advantages and disadvantages of building materials. They started to bring objectivity into "green discussions" and finished the division between "good" product and "bad" product. AUB provides substantial information about building materials for more than 20 years. Previously, the facts given mainly concerned

- Emission of gases and dangerous particles
- Emission of radiation
- Pollution of water and soil
- Waste treatment
- Production process
- Raw material and other ingredients
- Durability.

The assessment includes the whole life time of a building product including production of raw materials and preliminary products, erection of buildings and reuse or recycling after demolition.

4.2.1 *Developing the first EPD*

These criteria were broadened and changed in recent years, LCA started to play a major role and the requirements were harmonized with the international standardization. This new procedure was developed and tested on the declaration of AAC. In order to develop the EPD of AAC product specific patterns had to become developed i.e. Product Category Rules (PCR). This was done in the first AUB "product group forum".

Producers and experts from different institutions such as the DIBt (Deutsches Institut für Bautechnik) as well as representatives of government as UBA (Umweltbundesamt) discussed the points of interest as from 2001. Relevant data should be given in a clear and brief way. The discussion of PCR and EPD of AAC ceased in 2002. In 2003 a third party advisory board was established and this board verified the PCR of AAC together with that one of other wall materials in 2004.

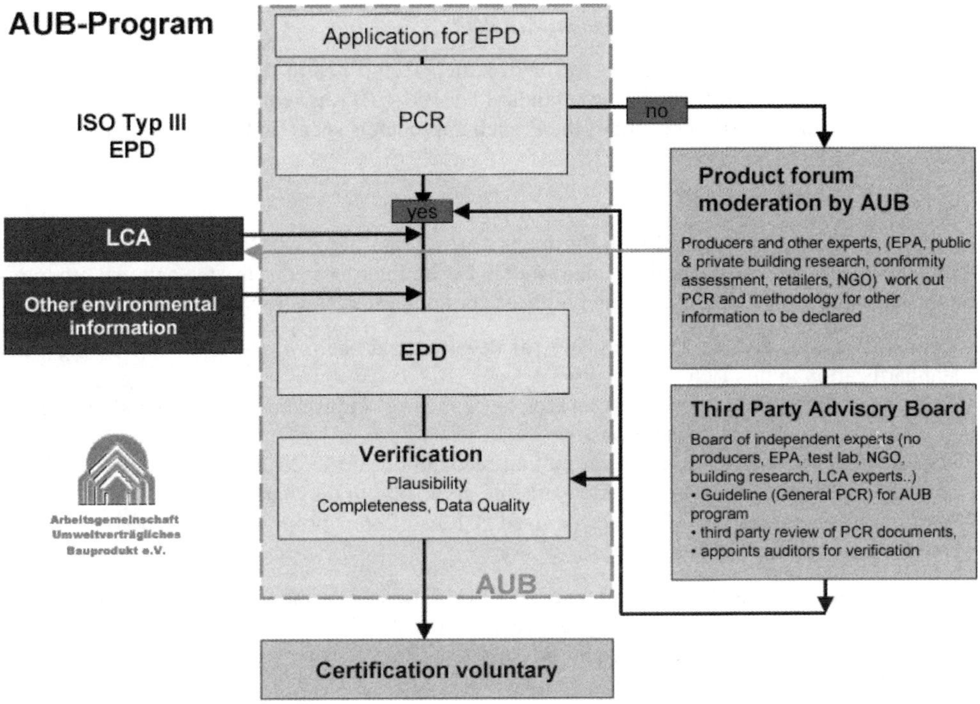

Figure 2. Steps in developing an EPD in Germany.

4.2.2 *Routine steps in developing an EPD*

The procedure of getting the label of type III according to the AUB scheme is as follows (see Figure 2).

The producer compiles the relevant product information. If the relevant PCR already exist, an independent third part advisory board checks the data with respect to completeness and plausibility and asks for additional information, verification or test certificates, if needed. If there are no PCR available, a "product group forum" is established with the aim to develop the specific requirement being relevant to this group of products. The third party advisory board finally validates the PCR-draft from the product group forum and the public has the possibility to comment on it. PCR exist now and the process of checking data, verifications and test certificates runs as described above.

Only the AUB currently develops a type III label in Germany. Up to now PCR exist for

- AAC
- Bricks
- Metals
- Mineral wool
- Sand-lime bricks
- Wood-materials.

4.2.3 *Officials about the AUB*

German governmental institutions do intensively support the AUB by man power. Officials from the building ministry, the environmental department and other contribute to the process by working in the independent third part advisory board of the AUB.

Dr. Manfred Stolpe, German minister of traffic, building and housing recently wrote: "We cannot adequately appreciate the AUB work in the frame of our concerted endeavour to encourage sustainable, ecological and health building activities."

4.3 *Environmental declaration in French – AFNOR*

The development of French program for environmental and health declaration of construction products is already pending. The French standard NF P01-010 was released according to ISO CD 14025.3 in December 2004. The aim of the French approach is scenario based cradle to grave.

4.4 *European harmonization of EPD*

The harmonization process of EPD in Europe is starting right now. At March 24th 2005 delegates from the Netherlands, France and Germany met in Berlin in order to discuss the national programs and to start cooperation. The following points of common interest were highlighted:

- Cooperation was seen as a prerequisite for developing a useful input to the upcoming CEN standardization in this field.
- Exchange of practical experience was seen as chance for improvement of managing programs as well as the quality of declarations.
- A goal for cooperation is to arrive at mutual recognition of the different programs in order to reduce barriers to trade, avoid double work and confusion in the market.
- Arriving at comparable EPD from harmonized PCR in different programs was seen as a necessary step towards mutual recognition.

The different programs and their influence on EPD of the same product are currently investigated in two pilot studies, one of them being carried out on AAC. First results of this will be available late this summer and will be presented at the conference.

REFERENCES

Anonymus 2005. Bauen und Umwelt. ed. by Arbeitsgemeinschaft Umweltverträgliches Bauprodukt e.V.. Königswinter

Anonymus 2005. PCR Porenbeton. Product Category Rules. ed. by Arbeitsgemeinschaft Umweltverträg-liches Bauprodukt e.V.. Königswinter

Kreißig, J.; Baitz, M.; Beetz, M.; Eyerer, P.; Kümmel, J.; Reinhard, H.-W. 1997. Baustoff-Ökobilanzen. Leitfaden zur Erstellung von Sachbilanzen in Betrieben der Steine-Erden-Industrie. Frankfurt

Kreißig, J.; Eyerer, P.; Kümmel, J.; Reinhard, H.-W. 2000. LCA of Building Materials. Impact Assessment and Interpretation in the Building Materials Industry. Frankfurt

Autoclaved Aerated Concrete – Limbachiya and Roberts (eds)
© 2005 Taylor & Francis Group, London, ISBN 0 415 38356 0

Sustainability in Autoclaved Aerated Concrete (AAC) construction

E. Nunez
Southern Company Generation, Birmingham, Alabama, USA

S.A. Nunez & F.H. Fouad
The University of Alabama at Birmingham, Birmingham, Alabama, USA

ABSTRACT: Sustainability is a new trend in the construction industry intended to minimize environmental impact by reducing resource consumption, emissions, and waste production. Sustainability also implies recycling of materials after their use.

Autoclaved aerated concrete AAC, originally introduced in Sweden in the late 1920s, is a lightweight material with no coarse aggregate, produced by mixing cement, lime, sand or fly ash, aluminum powder and water. In the United States, the use of AAC has been somewhat limited due to marketing conditions, lack of applicable building codes, lack of familiarity with design methodologies, and limited understanding of the advantages of AAC by the design and construction industry. This paper provides a review of the issues associated with the use of AAC in comparison to other construction materials from a sustainability perspective, with the intent of providing further information to highlight the advantages of AAC as an environmentally friendly material.

1 INTRODUCTION

Sustainability is a new trend in the construction industry intended to minimize environmental impact by reducing resource consumption, emissions, and waste production. Sustainability also implies recycling of materials after their use.

Sustainability is becoming an important factor when deciding about construction materials. Hence a discussion about sustainability issues related to autoclaved aerated concrete AAC is relevant, in order to illustrate the potential of this material from an environmental perspective.

AAC was originally introduced in Sweden in the late 1920s. AAC is a lightweight material with no coarse aggregate, produced by mixing cement, lime, sand or fly ash, aluminum powder and water. The use of AAC is widespread in Europe and it is relatively common in some Asian countries while marginally used in Australia. In the United States, the use of AAC has been somewhat limited due to marketing conditions, lack of applicable building codes, lack of familiarity with design methodologics, and limited understanding of the advantages of AAC by the design and construction industry.

This paper provides a review of the existing literature associated with sustainability of AAC construction in comparison to other construction materials, including comparisons of embodied energy with the intent to show the advantages of AAC as an environmentally friendly material.

2 AAC AS A SUSTAINABLE MATERIAL

One of the basic steps in sustainable construction is the proper choice of materials according to the characteristics of the project. An analysis of sustainability of a given material generally includes considerations such as energy consumed in production, energy consumed in transportation to the job site, indoor air quality during and after construction, life span, disposal and/or recycling after use. All of these considerations are usually included in the Life Cycle Assessment LCA of a given material.

Several authors (Glover 2001, Kesik 2002, Lawson 1996) have conducted comparisons of LCA for different materials used for residential construction. Most of the comparisons include wood, steel and concrete, with some discussion of AAC.

In terms of sustainability, AAC compares well with other construction materials since its manufacturing requires common constituents such as cement, lime, sand, water, and aluminum powder to facility the bubble formation that leads to the characteristic porous structure. Because of its lightweight, AAC has a clear advantage in transportation energy in comparison to denser materials like concrete or masonry. AAC waste during construction is minimal since pieces can be cut to exact dimensions using standard wood cutting tools. AAC is a highly durable material and can be easily recycled after disposal. Because it is made with inert materials, AAC does not affect adversely indoor air quality.

In order to compare several construction materials to AAC from a sustainable perspective, it is necessary to discuss their energy demands during manufacturing.

3 EMBODIED ENERGY

Embodied energy is the energy consumed during all the processes involved in the production of a given material or building (Milne & Reardon 2004). In general, highly processed materials have higher embodied energy than less processed materials.

The embodied energy accounts for a substantial amount of the total energy requirements of a low energy building (Thormark 2002). Therefore, the embodied energy can be used as an indicator of the environmental impact of building materials, assemblies, or systems (Kesik 2002).

The embodied energy can only be fully quantified through a life cycle assessment LCA including resource depletion, inherent pollution and overall environmental degradation. A complete measurement of the embodied energy also known as Gross Energy Requirement GER, typically includes the following (Milne & Reardon 2004)

- Energy consumed in transportation of the material and construction workers to the jobsite
- Energy required to produce the material
- Energy necessary to maintain the machines that make the material
- Energy required to install and maintain the urban infrastructure.

In general, GER is very difficult to quantify, therefore, a simpler measurement of the embodied energy is the Process Energy Requirement PER. PER is a measurement of the energy required to produce a given material. PER usually can account for as little as 50 percent and as much as 80 percent of the GER of a material (Milne & Reardon 2004). The embodied energy measurements are affected by factors such as geographic location, project type, and calculation methodology.

Several authors (Lawson 1996, Kesik 2002, Glover 2001) have published comparisons of embodied energy for different materials. Because of the variability in values of embodied energy, these comparisons are best analyzed in relative rather than absolute terms.

Figure 1 shows the PER embodied energy of some construction materials by weight. The highest value shown in Figure 1 corresponds to galvanized steel, because of the high energy requirements of the manufacturing process. AAC exhibits a relatively low PER embodied energy similar to that of kiln dried softwood and only slightly higher than that of clay bricks. On a by weight comparison, AAC exhibits a higher PER embodied energy then concrete.

Figure 2 shows the embodied energy of some construction materials by volume as shown on several references (Kesik 2002, Milne & Reardon 2004). The value of embodied energy by volume for AAC is based on a PER embodied energy by weight value of 3.6 MJ/kg (Milne & Reardon 2004) multiplied by an average density of 440 kg/m^3.

On a by volume comparison, the embodied energy of AAC is considerable smaller than that of concrete block and brick, while only slightly higher than the embodied energy of lumber. This

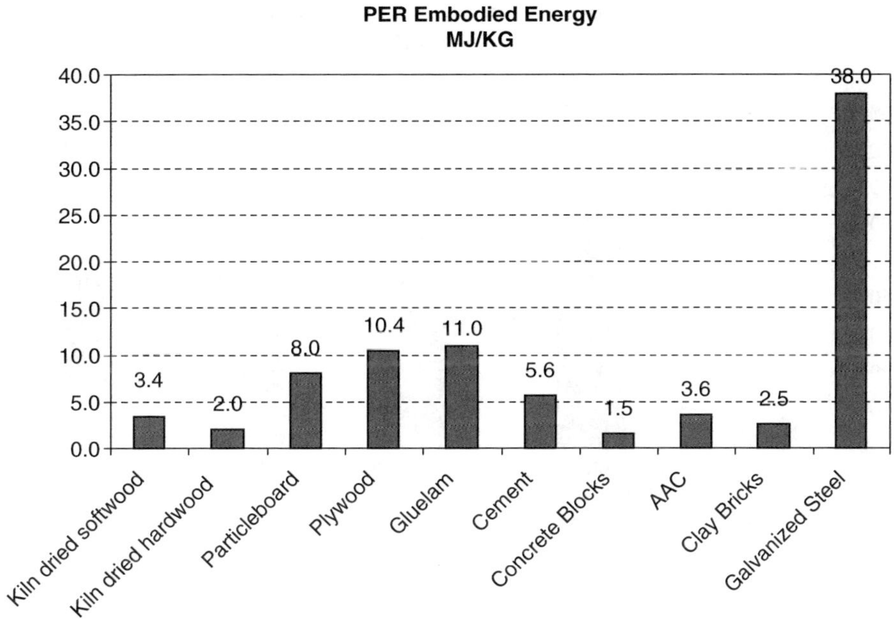

Figure 1. PER Embodied energy of different construction materials (data from Milne & Reardon, 2004).

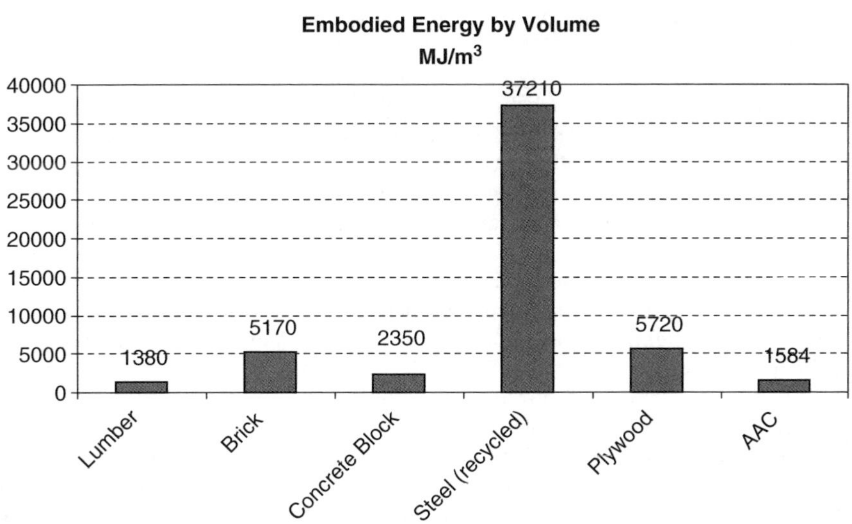

Figure 2. Embodied energy by volume (with data from kesik, 2002 and Milne et al., 2004).

comparison shows a clear indication of the relative advantage of using AAC with respect to other similar materials from a sustainable perspective.

Figure 3 shows the embodied energy of typical residential assemblies, based on the values published by Lawson (1996). The highest embodied energy value shown corresponds to unpainted double brick assemblies. Painted AAC assemblies exhibit comparable values to those of painted fibre cement siding on steel frame and slightly smaller than the value for unpainted brick siding on wood frame. This observation is noteworthy since unpainted brick siding on wood frame is a very common type of residential construction in the United States.

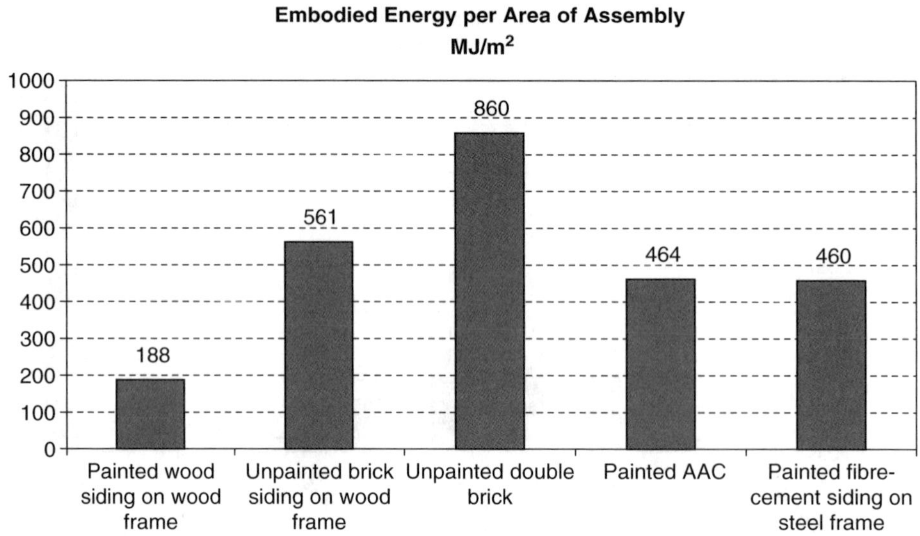

Figure 3. Embodied Energy of typical residential assembly (Lawson, 1996).

In order to conduct a more meaningful comparison of entire assemblies, Lawson (1996) proposed the use of a Building Material Sustainability Index BES. The BES is based on a LCA approach but includes a set of parameters defining environmental impact based on a five point scale of perceived importance. The computation of the BES index for embodied energy presented by Lawson (1996) includes three parameters: PER energy requirement, transport energy, and construction energy with a rating of importance of 5, 3, and 1 respectively.

Figure 4 shows PER embodied energy BES index values for AAC and other typical building materials. Both AAC panels and blocks show a comparable embodied energy BES index to that of wood studs, and higher embodied energy BES index than clay and cement products. As expected, steel products exhibit the highest embodied energy BES index because of the intense energy requirements for their production.

Figure 5 shows the embodied energy BES index for entire ground floor assemblies of three typical materials. On this Figure, it is apparent that AAC panels on AAC blocks with concrete footings exhibit only a slightly higher embodied energy BES index than a concrete raft slab.

Figure 6 shows the embodied energy BES index for upper floor assemblies of three typical construction materials. From this figure it is apparent that AAC slabs exhibit considerable lower embodied energy index than comparable concrete slabs.

Figure 7 shows that AAC block with render exhibits an embodied energy BES index slightly smaller than that of reinforced concrete and considerably smaller than that of clay brick with render.

Based on the embodied energy data found in the literature, it is apparent that AAC is a viable alternative to similar construction materials such as concrete or brick in order to reduce the embodied energy of residential or commercial. This assessment is based on comparisons including AAC made without by-products such as fly ash or recycled steel reinforcement. The use of these alternative components tends to reduce the total embodied energy of AAC. Finally, the embodied energy of AAC can be further reduced by replacing the autoclaved process by chemical curing (Kohlhaas 2003).

4 AAC WITH FLY ASH

Fly ash from coal burning generation plants can be added to AAC production as a substitute for powdered sand. Unlike concrete however, fly ash cannot replace cement on AAC production

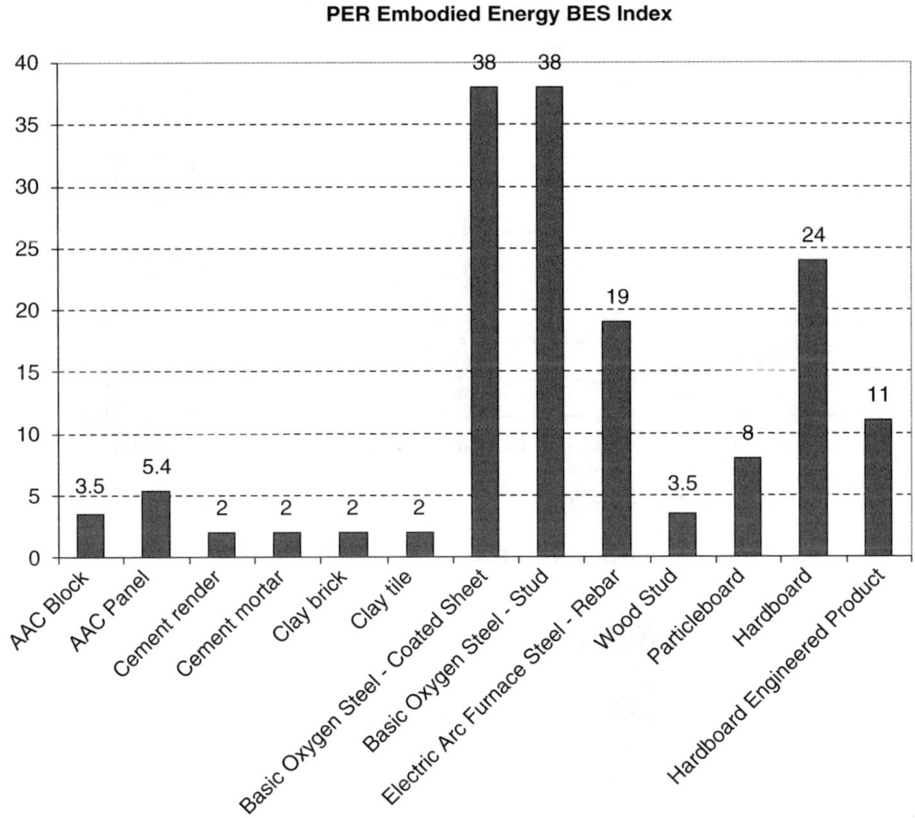

Figure 4. PER Embodied Energy BES index for typical building materials (Lawson, 1996).

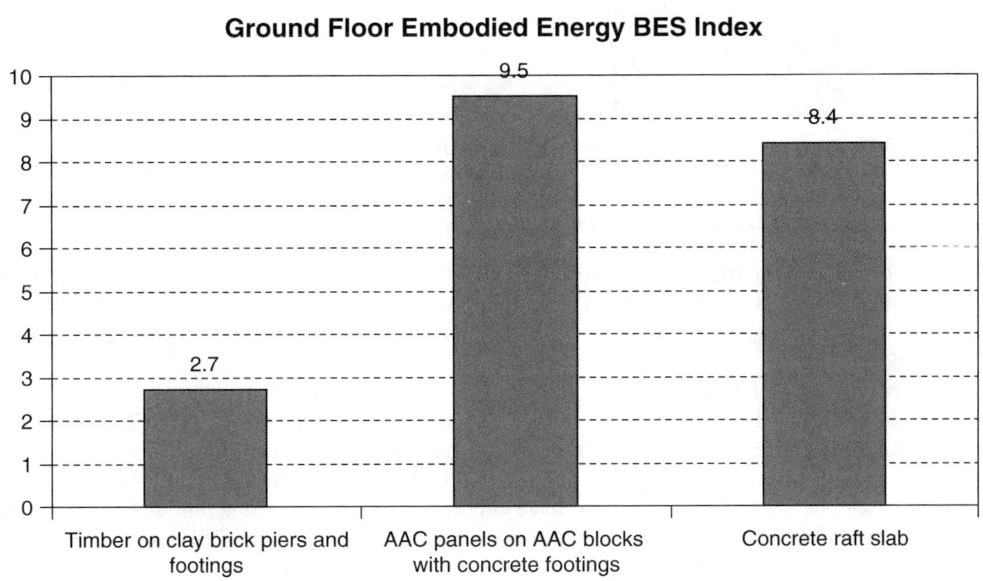

Figure 5. Embodied Energy BES index for ground floor assemblies (Lawson, 1996).

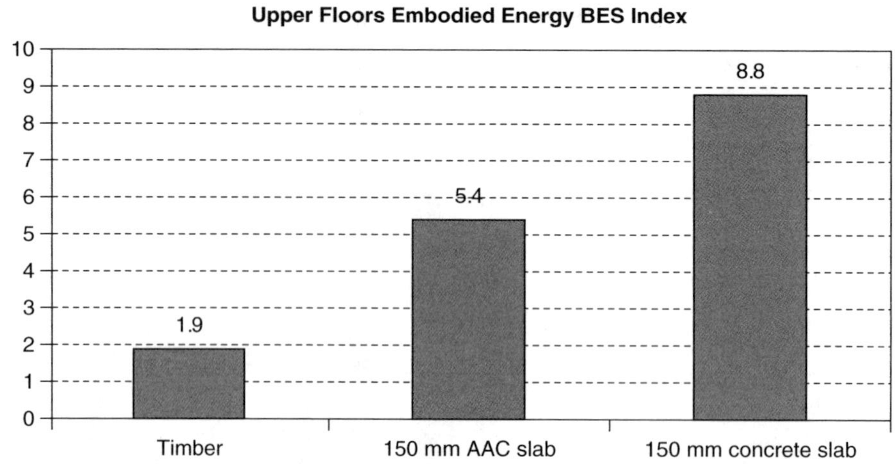

Figure 6. Embodied Energy BES index for upper floor assemblies (Lawson, 1996).

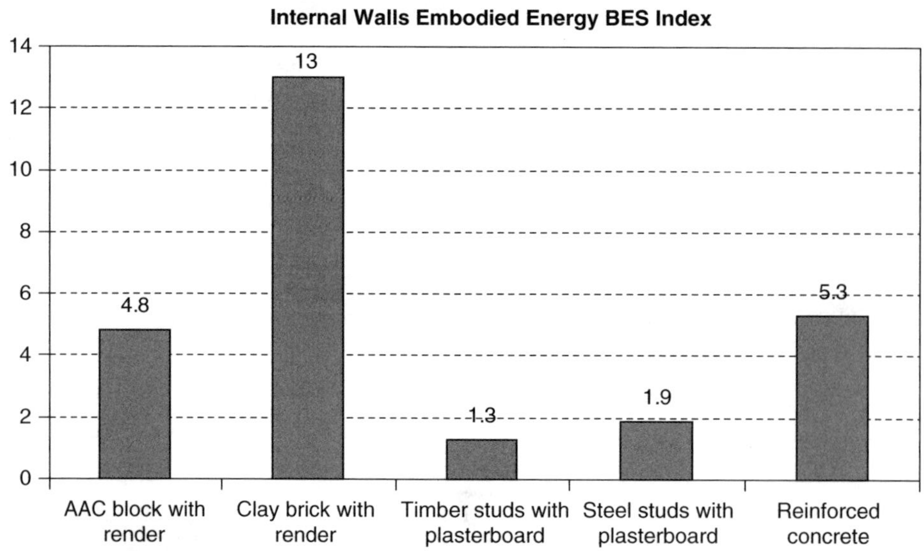

Figure 7. Embodied Energy BES index for internal wall assemblies (Lawson, 1996).

(EBN 1996). One of the major issues preventing the widespread use of fly ash on AAC production in the United States, is the difficulty in obtaining consistent physical characteristics of the final product, and the change in color from white on sand based AAC to gray on fly ash based AAC (EBN 1996).

Another concern is the presence of ammonia in fly ash from coal fired power plants fitted with NOx control equipment such as low NOx burners and selective catalytic reactors or SCRs. However, a recent study (Dilmore & Neufeld 2001) has shown that it is possible to produce acceptable quality AAC with fly ash from coal fired power plants fitted with low NOx burners or SCRs.

Other researchers (Hauser et al. 1999) have conducted studies on production of AAC with fly ash from the cellulose industry, with mixed results depending on the type of fly ash used.

From an environmental perspective the use of fly ash on AAC production makes sense since it helps reducing the embodied energy of AAC, however it appears the very little information has being published in the literature regarding quantification of the embodied energy of fly ash based AAC.

5 CONCLUSIONS

This paper provides a review of the issues associated with the use of AAC in comparison to other construction materials from a sustainability perspective. An important parameter to quantify the sustainability of a construction material is the embodied energy. A comparison of values of embodied energy for different materials was made based on information published in the literature. Based on the information found, it is apparent that AAC compares favorably against most materials except wood that has the lower embodied energy in most of the categories reviewed. When compared to similar construction materials such as brick or concrete, AAC appears to be a viable alternative with lower embodied energy. Very little information was found on sustainability issues associated with AAC in the United States. Also very limited information was found on life cycle assessment LCA and embodied energy quantifications for fly ash based AAC. Overall, it may be concluded that based on embodied energy comparisons, AAC can be considered as an environmentally friendly material.

REFERENCES

ASCE Committee on sustainability of technical activities committee. 2004. *Sustainable Engineering Practice*. Reston, VA, USA: American Society of Civil Engineers

ASTM E-917-95. 1994. *Standard practice for measuring life cycle costs of buildings and building systems*. West Conshohoken, PA, USA: ASTM International

Dilmore, R.M. & Neufeld, R.D. 2001. Autoclaved aerated concrete produced with low NOx burner/selective catalytic reduction fly ash. *American Society of Civil Engineers Journal of Energy Engineering* 127 (2): 37–50

Downton, P. 2004. Construction systems. In *Your Home – Technical Manual – Design for Lifestyle and the Future*. Australia Greenhouse Office

EBN. 1996. Autoclaved aerated concrete: is North America finally ready? *Environmental Building News* 5 (2)

Glover, J. 2001. *Which is better? steel, concrete, or wood: a comparison of assessments on three building materials in the housing sector*. Department of Chemical Engineering. University of Sidney.

Hauser, A.; Eggenberger, U. & Mumenthaler, T. 1999. Fly ash from cellulose industry as secondary raw material in autoclaved aerated concrete. *Cement and Concrete Research* 29: 297–302

Kesik, T. 2002. Perspectives on sustainability. *Canadian Architect*. January

Kohlhaas, M. 2003. Autoclaved aerated concrete, *Sustainable Building Coalition* 10 (3): 1–2

Lawson, W. 1996. *Timber in building construction – ecological implications*. Final report project PN017b.96. Forest and Wood Products Research and Development Corporation. Victoria. Australia

Milne, G. & Reardon, C. 2004. Embodied energy. In *Your Home – Technical Manual – Design for Lifestyle and the Future*. Australia Greenhouse Office

Thormark, C. 2002. A low energy building in a life cycle – its embodied energy, energy needed for operation and recycling potential. *Building and Environment* 37 (4): 429–435

Author index